Evolutionary Relationships Among Protozoa

The Systematics Association Special Volume Series

Series Editor

Alan Warren

Department of Zoology, The Natural History Museum, Cromwell Road, London, SW7 5BD, UK.

The Systematics Association provides a forum for discussing systematic problems and integrating new information from cytogenetics, ecology and other specific fields into taxonomic concepts and activities. It has achieved great success since the Association was founded in 1937 by promoting major meetings covering all areas of biology and palaeontology, supporting systematic research and training courses through the award of grants, production of a membership newsletter and publication of review volumes by its publishers Chapman and Hall. Its membership is open to both amateurs and professional scientists in all branches of biology who are entitled to purchase its volumes at a discounted price.

The first of the Systematics Association's publications, *The New Systematics*, edited by its then president Sir Julian Huxley, was a classic work. Over 50 volumes have now been published in the Association's 'Special Volume' series often in rapidly expanding areas of science where a modern synthesis is required. Its *modus operandi* is to encourage leading exponents to organise a symposium with a view to publishing a multi-authored volume in its series based upon the meeting. The Association also publishes volumes that are not linked to meetings in its 'Volume' series.

Anyone wishing to know more about the Systematics Association and its volume series are invited to contact the series editor.

Forthcoming titles in the series:

Molecular Systematics and Plant Evolution
Edited by P. Hollingsworth, R. Bateman and R. Gornall

Homology in Systematics
Edited by R. W. Scotland

Other Systematics Association publications are listed after the index for this volume.

The Systematics Association Special Volume Series 56

Evolutionary Relationships Among Protozoa

Edited by

G. H. Coombs
K. Vickerman
Institute of Biomedical and Life Sciences,
University of Glasgow, Glasgow, UK

M. A. Sleigh
School of Biological Sciences,
University of Southampton, UK

and

A. Warren
Department of Zoology,
The Natural History Museum, London, UK

KLUWER ACADEMIC PUBLISHERS
DORDRECHT / BOSTON / LONDON

Library of Congress Cataloging in Publication Card Number: 98-70277

ISBN 0 412 79800 X

Published by Kluwer Academic Publishers,
P.O. Box 17, 3300 AA Dordrecht, The Netherlands.

Sold and distributed in North, Central and South America
by Kluwer Academic Publishers,
101 Philip Drive, Norwell, MA 02061, U.S.A..

In all other countries, sold and distributed
by Kluwer Academic Publishers Group,
P.O. Box 322, 3300 AH Dordrecht, The Netherlands.

Printed in Great Britain

Contents

Contributors

C. Adje
Research Unit for Tropical Diseases, International Institute of Cellular and Molecular Pathology and Laboratory of Biochemistry, University of Louvain, B-1200 Brussels, Belgium.

A. Adoutte
Laboratoire de Biologie Cellulaire (URA 2227 du CNRS), Bâtiment 444, Université Paris-Sud, 91405 Orsay Cedex, France.

C.L. Anderson
Department of Pure and Applied Biology, Imperial College at Silwood Park, Ascot, Berkshire SL5 7PY, UK.

A.C. Barbrook
Department of Biochemistry, University of Cambridge, Tennis Court Road, Cambridge CB2 1QW, UK.

G.W. Beakes
Department of Biological and Nutritional Sciences, The Agriculture Building, The University, Newcastle upon Tyne NE1 7RU, UK.

E.U. Canning
Department of Biology, Imperial College at Silwood Park, Ascot, Berkshire SL5 7PY, UK.

T. Cavalier-Smith
Department of Botany, University of British Columbia, Vancouver, BC, Canada, V6T 1Z4.

C.G. Clark
Department of Medical Parasitology, London School of Hygiene and Tropical Medicine, Keppel Street, London WC1E 7HT, UK.

G.H. Coombs
Institute of Biomedical and Life Sciences, Joseph Black Building, University of Glasgow, Glasgow G12 8QQ, UK

J.O. Corliss
P.O. Box 2729, Bala Cynwyd, Pennsylvania 19004, USA.

L.S., Diamond
Laboratory of Parasitic Diseases, National Institutes of Health, Bethesda, MD, USA.

T.D. Edlind
Department of Microbiology and Immunology, MCP-Hahnemann School of Medicine, Allegheny University of the Health Sciences, 2900 Queen Lane, Philadelphia, PA 19129, USA.

J.T. Ellis
Molecular Parasitology Unit, Department of Cell and Molecular Biology, Faculty of Science, University of Technology, Sydney, Westbourne Street, Gore Hill, NSW 2065, Australia.

T.M. Embley
Department of Zoology, The Natural History Museum, Cromwell Road, London SW7 5BD, UK.

W. Foissner
Universität Salzburg, Institut für Zoologie, Hellbrunnerstrasse 34, A-5020 Salzburg, Austria.

M. Gouy
Laboratoire de Biométrie, Génétique et Biologie des Populations, Université Claude Bernard, 43 Boulevard du 11 Novembre 1918, 69622 Villeurbanne Cedex, France.

J.H.P. Hackstein
Department of Microbiology, Faculty of Science, University of Nijmegen, Toernooiveld, NL-6525 ED Nijmegen, The Netherlands.

V. Hannaert
Research Unit for Tropical Diseases, International Institute of Cellular and Molecular Pathology and Laboratory of Biochemistry, University of Louvain, B-1200 Brussels, Belgium.

G. Hide
Wellcome Unit of Molecular Parasitology, Glasgow University, Anderson College, 56 Dumbarton Road, Glasgow G11 6NU, UK.

R.P. Hirt
Department of Zoology, The Natural History Museum, Cromwell Road, London SW7 5BD, UK.

D.S. Horner
Department of Zoology, The Natural History Museum, Cromwell Road, London SW7 5BD, UK.

C.J. Howe
Department of Biochemistry, University of Cambridge, Tennis Court Road, Cambridge CB2 1QW, UK.

A.C. Jeffries
Molecular Parasitology Unit, Department of Cell and Molecular Biology, Faculty of Science, University of Technology, Sydney, Westbourne Street, Gore Hill, NSW 2065, Australia.

J.F. de Jonckheere
Protozoology Laboratory, Biosafety and Biotechnology, Department of Microbiology, Institute of Hygiene and Epidemiology, B-1050 Brussels, Belgium.

P.J. Lockhart
Molecular Genetics Unit, Massey University, Palmerston North, New Zealand.

U. Mackenstedt
Institut für Zoologie, Fachbereich Parasitologie, Universität Hohenheim, Emil-Wolf-Str., D-70593 Stuttgart, Germany.

P.A.M. Michels
Research Unit for Tropical Diseases, International Institute of Cellular and Molecular Pathology and Laboratory of Biochemistry, University of Louvain, B-1200 Brussels, Belgium.

D.A. Morrison
Molecular Parasitology Unit, Department of Cell and Molecular Biology, Faculty of Science, University of Technology, Sydney, Westbourne Street, Gore Hill, NSW 2065, Australia.

M. Müller
The Rockefeller University, 1230 York Avenue, New York, NY 10021, USA.

F.R. Opperdoes
Research Unit for Tropical Diseases, International Institute of Cellular and Molecular Pathology and Laboratory of Biochemistry, University of Louvain, B-1200 Brussels, Belgium.

H. Philippe
Laboratoire de Biologie Cellulaire (URA 2227 du CNRS), Bâtiment 444, Université Paris-Sud, 91405 Orsay Cedex, France.

J. Rosenberg
Lehrstuhl für Tierphysiologie, Ruhr-Universität-Bochum, D-44780 Bochum, Germany.

J.D. Silberman
Marine Biological Laboratory, Woods Hole, MA 02543, USA.

M.A. Sleigh
School of Biological Sciences, University of Southampton, Bassett Crescent East, Southampton SO9 3TU, UK

M.L. Sogin
Marine Biological Laboratory, Woods Hole, MA 02543, USA.

N.J. Tourasse
Laboratoire de Biométrie, Génétique et Biologie des Populations, Université Claude Bernard, 43 Boulevard du 11 Novembre 1918, 69622 Villeurbanne Cedex, France.

K. Vickerman FRS
Division of Environmental and Evolutionary Biology, University of Glasgow, Glasgow G12 8QQ, UK.

G.D. Vogels
Department of Microbiology, Faculty of Science, University of Nijmegen, Toernooiveld, NL-6525 ED Nijmegen, The Netherlands.

F.G.J. Voncken
Department of Microbiology, Faculty of Science, University of Nijmegen, Toernooiveld, NL-6525 ED Nijmegen, The Netherlands.

A. Warren
Department of Zoology, The Natural History Museum, Cromwell Road, London SW7 5BD, UK

M. Wilkinson
School of Biological Sciences, University of Bristol, Bristol BS8 1UG, UK.

R.J.M. Wilson
National Institute for Medical Research, Mill Hill, London, NW7 1AA, UK.

Preface

Improvements in methods for assessing genetic similarity have led to breathtaking changes in recent years in our understanding of protozoan phylogenetics. These developments have shattered many previously accepted concepts. This book attempts to bring together the current views of the major players in the science and to provide a compendium that will be essential reading for all interested in the phylogenetics of lower eukaryotes.

The book arises from a joint meeting of The British Section of the Society of Protozoologists, The Systematics Association and The Linnean Society that was entitled 'Evolutionary Relationships Among Protozoa'. The meeting itself was most stimulating and whereas its delivered papers provided the framework for the reviews in this book, the majority have evolved since then to include even more recent ideas. Thus we hope that they provide as up-to-date a view as it is possible to achieve in a book of this type. Clearly, however, there are a few very recent findings or interpretations that are not covered fully, but we feel that these omissions will not detract from the collection as its main purpose is to detail the marked changes that have revolutionized ideas over the last few years and provide a reference text to act as a foundation for later developments.

It was anticipated at the outset that the symposium would be stimulating and interesting, but the first speaker, André Adoutte, promptly went one stage further and promised that we would be shocked by some of the new data to be presented. Many of the audience were indeed astounded to learn that the 18S rRNA phylogenetic tree of the protozoa, which has become widely accepted in recent years, is probably unreliable. Mutational saturation of sequences and inequalities in evolutionary rates were cited as being primarily responsible to the generation of falsely resolved trees. Fortunately, for taxonomists, a number of solid monophyletic groups can still be identified but their order of emergence remains uncertain.

Among the most compelling pieces of evidence which reduce our confidence in the 18S rRNA tree are the deeply incongruent trees that have been obtained using sequences from proteins and the genes which encode them. Many of the pertinent data are included in the reviews dealing with relationships among protozoa based on tubulin and on enzymes of central metabolism. There is now widespread acceptance that the evolution of the eukaryotic cell can be visualized as a series of endosymbiotic events involving the acquisition of new genetic material and new cytoplasmic compartments. The origin and evolution of organelles such as mitochondria, plastids and hydrogenosomes are therefore of central importance when considering protozoan phylogenetics. These matters are explored in several chapters. Many of the reviews concern individual groups which are phylogenetically interesting or otherwise important, and discuss interrelationships both within and between the groups. The relationships between protozoa, fungi and animals are discussed primarily in two reviews. The authors give an overview of the exhaustive analysis of both morphological and molecular data and conclude that animals and fungi each arose by radical transformation of uniflagellate choanozoan-like ancestors.

The final chapter surveys historical and recently-proposed classification schemes and highlights the dramatic changes that have taken place at all levels in the taxonomic hierarchy. The sanguine conclusion is that protozoan taxonomy is in a state of flux, rather than in a state of chaos as some have suggested. We are left to ponder whether, by the beginning of the 21st century, we shall arrive at a classification scheme for the protozoa that is compact, clear, uncomplicated and understandable to users while at the same time accurately reflecting known phylogenetic relationships.

Clearly, many key problems remain to be resolved, not least of which is a satisfactory taxonomic circumscription of 'Protozoa' and whether the term has any phylogenetic meaning. At the conclusion of the meeting Keith Vickerman addressed this point and in an unexpected display of knowledge of popular culture drew an analogy between protozoa and the pop star Prince who is now known by a sign (which cannot be conveyed verbally) and so is universally referred to as 'the artist formerly known as Prince'. Unless John Corliss's goal of a universally acceptable classification scheme for the Protozoa is realized, we may find ourselves having to refer to 'organisms formerly known as Protozoa'. We hope, however, that the production of this book will play some part in preventing such taxonomic anarchy.

As expected for such a vibrant field, it has been difficult to get consensus on all issues. The reader needs to bear this in mind when trying to put together the data provided in the different chapters. We have attempted to obtain uniformity of nomenclature in the various reviews, but in a few cases that has not been possible. Some helpful guidance is provided in Keith Vickerman's introductory chapter and we strongly advise that this overview is read before the others.

We are grateful for the financial support provided for the meeting and the book by the Wellcome Trust and the Systematics Association.

Graham H. Coombs
Keith Vickerman
Michael A. Sleigh
Alan Warren

1

Revolution among the Protozoa

Keith Vickerman

Division of Environmental and Evolutionary Biology, University of Glasgow, Glasgow G12 8QQ, UK. E-mail: k.vickerman@bio.gla.ac.uk

1.1 INTRODUCTION

Human consciousness, philosophers tell us, is inseparable from the use of metaphor. We simply cannot discuss certain concepts without recourse to it. And the depiction of family descent in the form of a tree must be one of the oldest metaphors around. It is not surprising, then, that attempts to reconstruct the evolutionary history of life on earth in the form of a branching tree appeared soon after the publication of Darwin's *Origin of Species* in 1859. In 1866, Ernst Haeckel, in addition to coining the word phylogeny, also presented phylogenetic trees for all the then known groups of living organisms in his book *Generelle Morphologie der Organismen* (Figure 1.1). These trees were largely based on evidence from comparative anatomy, the discipline of the correlation of parts fathered by Georges Cuvier and fostered by Richard Owen who introduced the valuable concept of homology.

Haeckel started a mania for phylodendrography – the drawing of evolutionary trees. In league with the palaeontologists, the demonstration of evolutionary descent became a major preoccupation of zoologists in particular and remained so until after the Second World War (Bowler, 1996).[1]

Students of the single-celled animals – or Protozoa – were not able to participate in this intellectual pursuit, however. Protozoa had little in the way of anatomy to compare. Homologies were uncertain and with

[1] Few biologists would agree with the harsh conclusion of the philosopher Michael Ruse (1997) that this vast endeavour was a complete waste of time on the part of second-rate scientists who would have been more properly employed in devoting their energies to quantitative study of the principles of evolutionary biology!

Evolutionary Relationships Among Protozoa. Edited by G.H. Coombs, K. Vickerman, M.A. Sleigh and A. Warren. Published in 1998 by Chapman & Hall, London. ISBN 0 412 79800 X

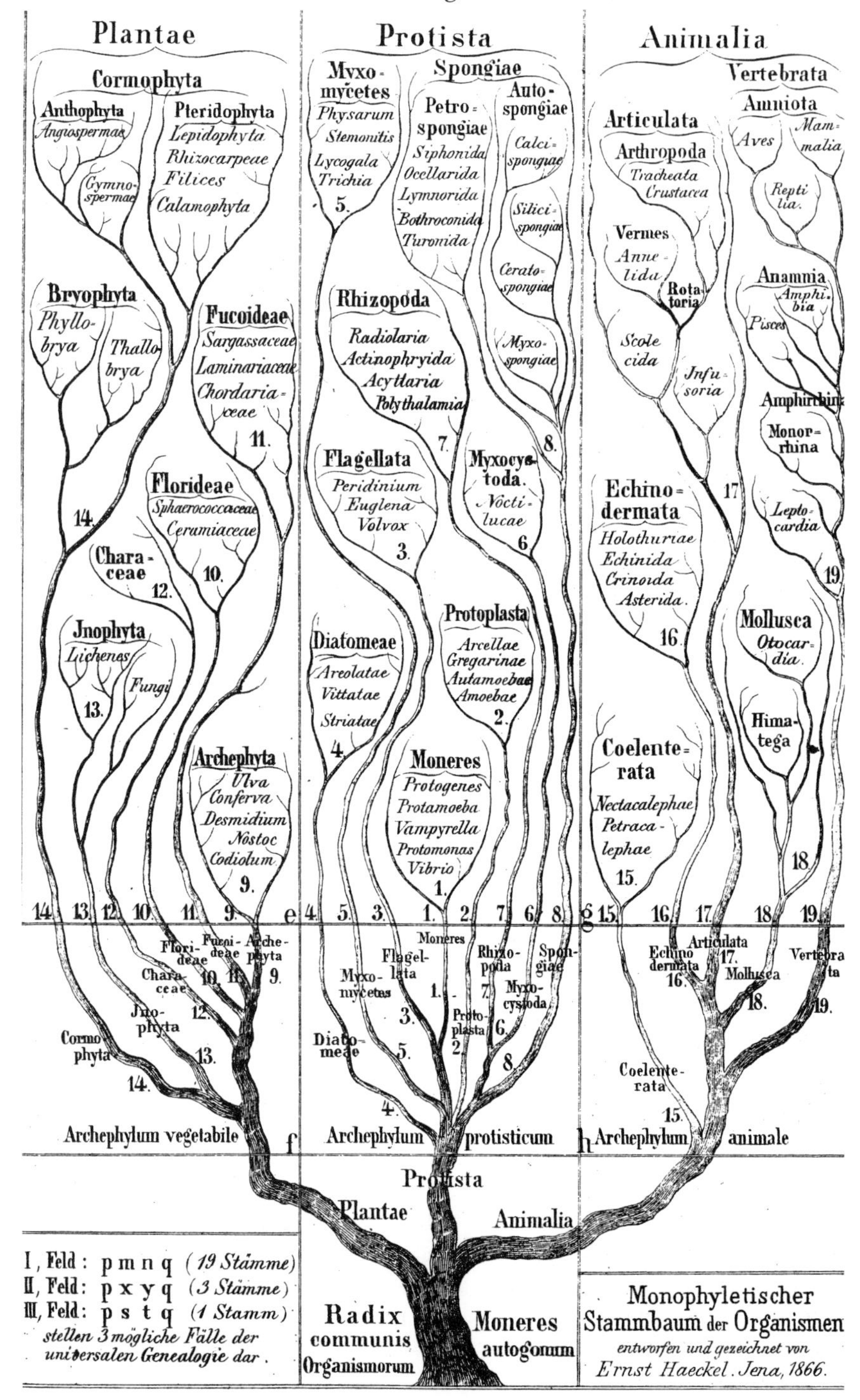
Plantae
Protista
Animalia
Cormophyta
Anthophyta
Angiospermae
Gymnospermae
Pteridophyta
Lepidophyta
Rhizocarpeae
Filices
Calamophyta
Bryophyta
Phyllobrya
Thallobrya
Fucoideae
Sargassaceae
Laminariaceae
Chordariaceae
11.
Florideae
Sphaerococcaceae
Ceramiaceae
10.
14.
Characeae
12.
Jnophyta
Lichenes
Fungi
13.
Archephyta
Ulva
Conferva
Desmidium
Nostoc
Codiolum
9.
Myxomycetes
Physarum
Stemonitis
Lycogala
Trichia
5.
Spongiae
Petrospongiae
Siphonida
Ocellarida
Lymnorida
Bothroconida
Turonida
Autospongiae
Calcispongiae
Silicispongiae
Ceratospongiae
Myxospongiae
Rhizopoda
Radiolaria
Actinophryida
Acyttaria
Polythalamia
7.
8.
Flagellata
Peridinium
Euglena
Volvox
3.
Myxocystoda
Noctilucae
6
Protoplasta
Arcellae
Gregarinae
Autamoebae
Amoebae
2.
Diatomeae
Areolatae
Vittatae
Striatae
4.
Moneres
Protogenes
Protamoeba
Vampyrella
Protomonas
Vibrio
1.
Vertebrata
Amniota
Aves
Mammalia
Reptilia
Articulata
Arthropoda
Tracheata
Crustacea
Vermes
Annelida
Rotatoria
Scolecida
Infusoria
Anamnia
Amphibia
Pisces
Amphirhina
Monorrhina
17
Leptocardia
19
Echinodermata
Holothuriae
Echinida
Crinoida
Asterida
16
Mollusca
Otocardia
Himatega
Coelenterata
Nectacalephae
Petracalephae
15.
18
14 13 12 10 11 9 e 4 5 3 1 2 7 6 8 g 15 16 17 18 19
Floridae
Fucoideae
Archephyta
Characeae
Jnophyta
Cormophyta
Moneres
Flagellata
Myxomycetes
Rhizopoda
Spongiae
Myxocystoda
Protoplasta
Diatomeae
Echinodermata
Articulata
Mollusca
Vertebrata
Coelenterata
Archephylum vegetabile
f
Archephylum protisticum
h
Archephylum animale
Protista
Plantae
Animalia
I, Feld: p m n q (19 Stämme)
II, Feld: p x y q (3 Stämme)
III, Feld: p s t q (1 Stamm)
stellen 3 mögliche Fälle der universalen Genealogie dar.
Radix communis Organismorum
Moneres autogonum
Monophyletischer Stammbaum der Organismen
entworfen und gezeichnet von
Ernst Haeckel. Jena, 1866.

the notable exception of one or two groups (e.g. foraminiferans, radiolarians, silicoflagellates) the protozoan fossil record was abysmal. This period of abstinence came to an end when electron microscopes with their vastly improved resolving power became widely available in the 1950s and 1960s. A whole new world of comparative microanatomy was uncovered and protozoologists could generate as much phylodendrographia as their vertebrate zoologist colleagues.

Today there is probably more interest in phylogenetic tree construction than at any time in the past and this book on evolutionary relationships among protozoa reflects it. But tree construction is not the arcane guessing game that it was. Biologists have long felt that the supreme arbiter of evolutionary relationships would be a direct measurement of similarity of the genetic material of organisms and now such measurement is possible. In the past phylogenies have been constructed using a battery of anatomical, behavioural and biochemical characters on the assumption that these reflect genetic affinity. The most widely accepted method of constructing trees of relationships (or phylogenies) is derived from the cladistic approach developed in the 1950s by the German zoologist Hennig (Hennig, 1966) who distinguished between 'ancestral' and 'derived' characters. The former set of characters, shared by a relatively large set of taxa, reflect common ancestry of all these taxa, while the latter, shared by fewer taxa, reflect a more recent common ancestry of this particular subset.

Over the last few decades the concept of homology has been extended to the molecular level. Nucleotide sequences in homologous DNA regions or amino acid sequences in homologous proteins can be compared and used in tree construction. The rapidly burgeoning molecular data would be meaningless, however, without parallel development of objective methods of phylogenetic inference and the construction of computer hardware and software which is up to the job of applying the new methods to the new data. Even so, the reconstruction of phylogenies is not straightforward. Sequences need careful interpretation because evolution may occur at different rates along different arms of a branching phylogeny as well as in different regions of the

---

Figure 1.1 Haeckel's most famous tree depicts the three Kingdoms (or Archephyla, as he called them) – animal, plant and protist – the third being a creation of his own to embrace the simplest organisms. His Protista included most of the organisms known later as Protozoa – flagellates, amoebae, gregarines, slime moulds – but also diatoms, sponges and heterotrophic bacteria (Monera). The ciliates, or 'infusorians', strangely were not included but assigned to the animal kingdom in a phylum Articulata that also embraced worms and arthropods! (From Haeckel, 1866.)

genome; certain base substitutions may be more likely to occur than others and silent substitutions are more likely than those that change the amino acid sequence of a protein. In addition reverse mutations are probably quite common. Various statistical models have been developed to handle these problems (Hillis, 1997). The critical approach to inference is an all important part of phylogenetic studies, as indicated in this book by the chapters of Philippe and Adoutte (Chapter 2), Tourasse and Guoy (Chapter 3), and Horner, Hint and Embley (Chapter 7), in particular. A particularly important problem is congruence between trees derived from different data sets, for example nucleic acid versus protein sequences, as discussed by Edlind (Chapter 5). The question that enters the reader's mind on finishing each chapter in this book is 'can we believe it?'.

I shall leave these nagging doubts, however, for others to assess and be concerned here simply with the impact of all this evolutionary tree building on the study of protozoa in the broader context. And the impact is potentially enormous. Two aspects strike the general protozoologist as exceptionally important. The first, is the revolutionary effect of phylogenetic studies on the classification of these organisms, which has never been satisfactory. The second is the contribution of studies on protozoa to general basic biology. The latter has undoubtedly been underplayed in the past. Yet protozoa are the great 'rebels', they can often provide the exceptions that test the rules proclaimed with confidence by workers with higher eukaryotes. These two aspects are not unrelated, as the spread of interest in protozoa has been hindered by unsatisfactory systematics. A further aspect, but one not covered in any detail in this book, is the early history of unicell life on earth and the debate beween proponents of the molecular clock and palaeontologists (Martin, 1996).

Darwin (1859) contended that a sound classification must be based on two criteria: common descent (genealogy) and degree of similarity or difference, i.e. amount of evolutionary change. If Darwinism has a pictorial symbol, it is the phylogenetic tree. It carries with it powerful implications for systematics but also connotations of competition and survival of the fittest. Whereas the former are currently welcomed with open arms, the latter may not be so well-received in some quarters where evolutionary biology is perceived as entering a new phase, dominated by ideas of cooperation and control and of conflicts between units of selection on different levels of organization. According to adherents of this creed, the focus of evolutionary biology has begun to shift from explaining the origin of species to the modelling of processes through which autonomous entities cooperate to form systems of greater complexity (Maynard Smith and Szathmáry, 1995; Weiser, 1997). Many of the studies presented in this book illustrate just how much protozoa can contribute in this respect and especially to our understanding of the evolution of the eukaryotic cell through cooperation of one-time autonomous microorganisms. They also shed light on such seemingly distant fields as the nature of parasitism and the epidemiology of parasitic diseases caused by protozoa. But first, let it be said that protozoa are not what they were, and we should discuss what is meant by that.

1.2 REVISING KINGDOMS: WHAT ARE THE PROTOZOA THESE DAYS?

> ...the Protozoa of Von Siebold ... has undergone much modification at the hands of biologists since its first institution in the year 1845. Great diversity of opinion exists, even at the present day, with respect to the deliminations both of its own borders and those of the minor sections and orders into which it may be most conveniently and naturally subdivided.
>
> *Manual of the Infusoria (1880–1881)*

These sentiments expressed by W. Saville-Kent in the second chapter of his mighty *Manual of the Infusoria* (1880–1881) could well be re-expressed today. Kent followed his master, T. H. Huxley, in regarding the Protozoa as a subkingdom of the Kingdom Animalia, but the difficulty of distinguishing unicellular animals from unicellular plants had long been evident and Hogg's Protoctista (1861) or Haeckel's Protista (1866) acknowledged the fact that both should be united in a kingdom of their own. Since Whittaker's (1969) hallowed Five Kingdoms proposal, biology textbooks have recognized the Monera (Bacteria), Protista, Plantae, Fungi and Animalia as the kingdoms of Life, but of late phylogenetics has demanded a dramatic redrawing of kingdom boundaries. Our foremost living taxonomist, Ernst Mayr, has recently (1997) decreed that whereas 'it may still be convenient to speak of unicellular eukaryotes as protists, a formal taxon Protista is no longer defensible'. His view echoes that of most of those who would call themselves protistologists today (see Corliss, Chapter 22). He recommends recognition within the Empire Prokaryota of two kingdoms – the Eubacteria and the Archaebacteria (or Archaea) and within the Empire Eukaryota six kingdoms – the Archezoa, Protozoa, Chromista, Metaphyta, Fungi and Metazoa. To redefine the Protozoa as a kingdom calls for some redefinition of other kingdoms too, and this has largely come about, anyway, through application of the concept of monophyly (all members descended from a single ancestor) to the kingdom concept.

Most phylogenetic studies on eukaryotes have been conducted comparing small subunit (18S) ribosomal RNA gene (SSUrDNA) sequences.

There is now good evidence from such sequencing that the animal kingdom (Metazoa), whose members are characterized by collagenous connective tissue between two sheets of epithelia, is monophyletic; its origins are closely linked with those of the kingdom Fungi (Wainright *et al.*, 1993) whose members have chitinous cell walls. Some of the most marked changes in boundary have taken place in the former plant kingdom, now slimmed down to exclude most of the former algal groups: the term algae may be useful to ecologists but it has no taxonomic significance and refers to non-embryophyte photosynthesizers. The serial endosymbiosis theory of chloroplast origin and the increasing evidence from phylogenetics that chloroplasts of all photosynthetic eukaryotes are derived from a single endosymbiotic event (chloroplast monophyly, but see Howe *et al.*, this volume) have played havoc with venerable classification schemes such as that of Christensen (1980) which used photosynthetic pigments as their main criterion. Thus Christensen's Chlorophyta (possessing chlorophylls a and b) included the Euglenophyceae (Euglenoidea) as well as the green algae (Chlorophyceae, Prasinophyceae), whereas the SSUrDNA sequencing studies have confirmed early conclusions from ultrastructure that the euglenoids have more in common with the heterotrophic trypanosomes and their allies (Kinetoplastea) and acquired their green chloroplast by secondary endosymbiosis. Similarly Christensen's Chromophyta (chlorophylls a and c present, but not b) embraced the dinoflagellates – now recognized as basically heterotrophic protozoa (Dinozoa) that acquired their chloroplasts secondarily, again probably from a eukaryote symbiont. The rest of Christensen's Chromophyta (Cryptophyceae, Rhaphidophyceae, Chrysophyceae, Haptophyceae, Bacillariophyceae, Eustigmatophyceae and Phaeophyceae) now constitute the Kingdom Chromista (Cavalier-Smith 1986), the monophyly of which is debatable.

Chromist chloroplasts, peculiarly, lie within the lumen of the rough endoplasmic reticulum; in addition, their flagella have rigid bipartite or tripartite hairs. Within this kingdom, members of the subkingdom Cryptista (Cryptophyceae, cryptomonads) retain (as the nucleomorph) the nucleus of the eukaryote symbiont from which they derived their phycobilin-containing chloroplast and have bipartite hairs on both their flagella. The other subkingdom, Chromobiota lack these features and includes the haptophytes (Haptophyceae, Prymnesiophyta) with their characteristic haptonema in addition to their two flagella, and the heterokonts – the vast assemblage of former chromophyte classes mentioned above. The heterokonts have characteristic rigid bipartite or tripartite hairs on the anterior flagellum only and they are widely referred to as the 'stramenopiles' on account of this feature. This character alone has hauled into the Chromista (or Stramenopila!) many heterotrophic former protozoa (e.g Labyrinthulida or slime nets,

Bicocoecida) and the former pseudofungi (Oomycetes and Hyphochytriomycetes; see Beakes, Chapter 20).

Christensen's last major algal group, the Rhodophyta, despite their distinctive features (no centrioles or flagella, chloroplasts with phycobilin pigments) remains within the plant kingdom. Rhodophytes share with the Chlorophyta **plus** green plants (together constituting the monophyletic Chlorobionta or Viridaeplantae) chloroplasts located in the cytosol and bounded by a simple two-membraned envelope. Phylogenetic analysis of glyceraldehyde-3-phosphate dehydrogenase genes from cytosol and chloroplasts of red algae, other photosynthetic eukaryotes and cyanobacteria suggests that chlorobiont and rhodophyte chloroplasts form a monophyletic group of cyanobacterial descent and that rhodophytes separated from chlorophytes at about the same time as animals and fungi (Liaud *et al.*, 1994).

With animal, plant and chromist kingdoms redefined we are left with unicellular eukaryotes that are largely heterotrophic, primitively phagotrophic and lacking plastids, and also lacking collagen or chitinous vegetative cell walls. These organisms constitute the redefined Kingdom Protozoa – though admittedly the definition is rather weak! And of course the kingdom is paraphyletic, because even if protozoa do include the ancestral single-celled eukaryote, all the other kingdoms must be derived from them. Whether we should separate off from the Protozoa those eukaryotes that lack mitochondria and place them in yet another kingdom, the Archezoa, is now debatable. The earliest branches on the SSUrDNA eukaryote tree – the Metamonada (e.g. the diplomonad *Giardia*) and Microspora (e.g. *Nosema*) and some other groups not yet sequenced (Figure 1.2), lack mitochondria, and indeed other membrane-bound organelles (plastids, peroxisomes, hydrogenosomes). It is therefore tempting to regard them as relicts of a stage of eukaryote evolution that preceded symbiotic incorporation of the purple photosynthetic bacteria that became mitochondria (Cavalier-Smith, 1993). As these organisms are dwellers in anaerobic environments (most are parasites or gut symbionts), however, an alternative interpretation is that the mitochondrion has been secondarily lost. This view is supported by evidence for mitochondrial heatshock proteins (HSP) in *Giardia* (HSP60; Soltys and Gupta, 1994) and *Nosema* (HSP70; Germot *et al.*, 1997). In addition, protein trees – for example those for the chaperone protein HSP70 (Germot *et al.*, 1997) and tubulins (Edlind, Chapter 5) – place the all-parasitic Microspora close to if not within the Fungi, suggesting that their lack of cytoplasmic furnishing is a consequence of long-established obligate parasitism and that they are misplaced on the SSUrDNA trees. There is now abundant evidence that among the third emergent group on that tree, the Parabasalia, *Trichomonas* spp. bear traces of mitochondrial heatshock

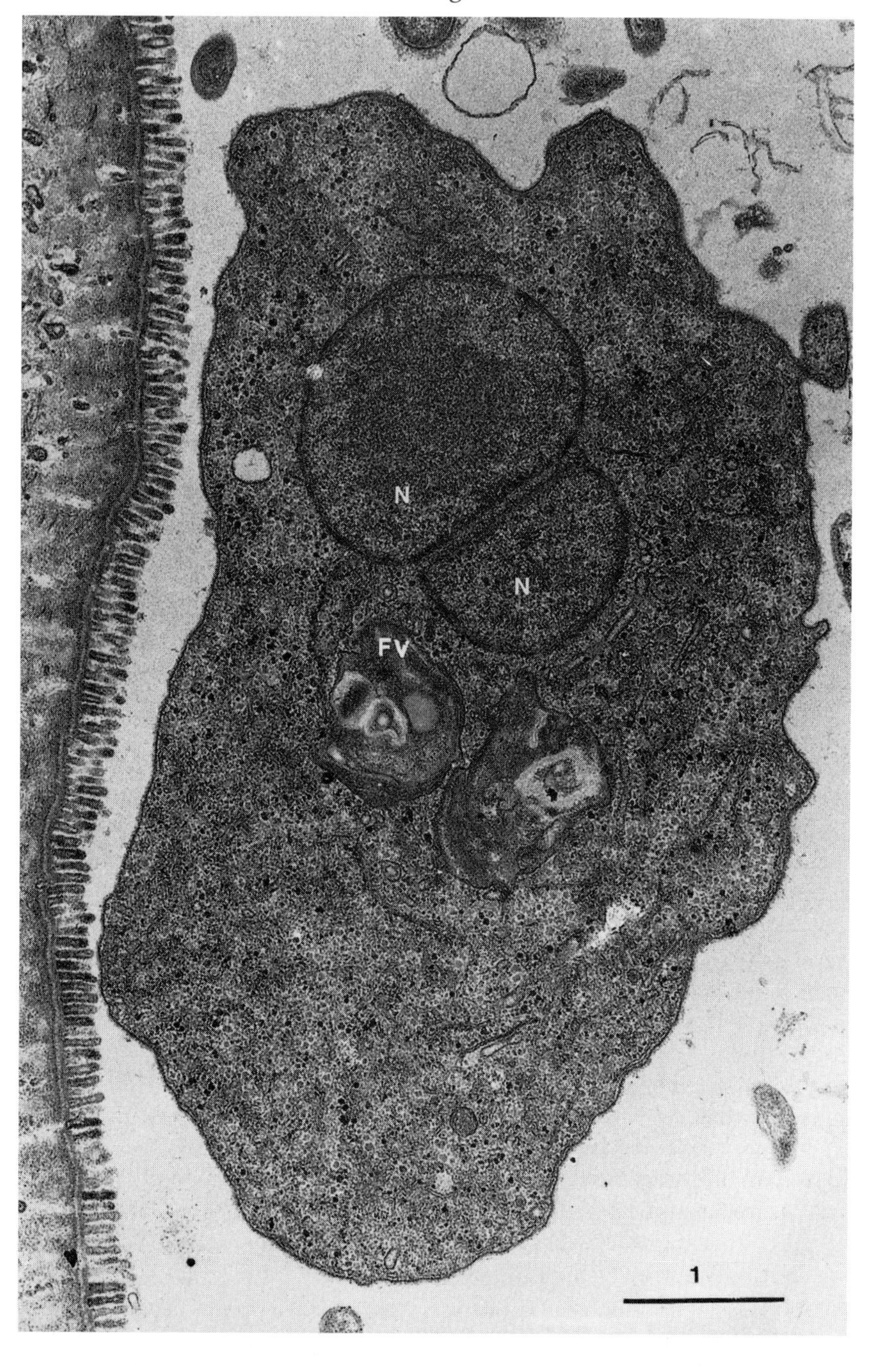
N
N
FV
1

protein genes and their products can be localized to the unusual energy-generating hydrogenosome organelles. Hydrogenosomes are in fact highly-derived mitochondria which have lost all trace of mitochondrial genome, cytochromes, the tricarboxylic acid cycle and oxidative phosphorylation as discussed by Hackstein and colleagues (Chapter 8). The Archezoa, then, are doubtfully deserving of a Kingdom of their own, and in Chapter 21 in the present book, Cavalier-Smith has returned them to the kingdom of the Protozoa.

1.3 REVISING THE SYSTEMATICS OF PROTOZOA ALONG EVOLUTIONARY LINES

Already, phylogenetic studies have toppled much of the house of cards that passed for protozoan systematics (Levine *et al.*, 1980; Cavalier-Smith, 1993, Corliss, 1994). This was really no more than a convenient filing system, its subdivisions failing to reflect evolutionary relationships. Gone is that monster group the Mastigophora (flagellates): flagellated organisms are common in chromists and chlorobionts as well as among Protozoa. And that other monster group, the Sarcodina (amoeboid organisms), although struggling on, has been redefined (Cavalier-Smith, Chapter 21). We now realize that amoeboid locomotion has been adopted in several independent lineages (Parabasalia, Percolozoa, Foraminifera, Mycetozoa, Amoebozoa, Dinozoa and so on). What we now can recognize is a series of sizeable monophyletic groups, some more robust than others (Patterson, 1994; Corliss, Chapter 22) and we may even assign them some sort of branching order on the tree of life. This widely accepted order is seriously questioned by some (Philippe and Adoutte, 1995 and Chapter 22; Khumar and Rhetsky, 1996), but its influence on classification has already become striking (Cavalier-Smith, 1993 and Chapter 21;

Figure 1.2 An archezoan? This transmission electron micrograph of an as yet unnamed parasitic amoeba browsing the vaginal epithelium (left) of the medicinal leech (*Hirudo medicinalis*) shows the basic features of archezoan organization. The cytoplasm has rough and smooth endoplasmic reticulum and abundant electron-dense glycogen particles but no mitochondria or other membrane-bound organelles, except for food vacuoles (FV) that contain endocytosed albumin. Diplokaryosis (the presence of two structurally-linked nuclei [N]) is another feature of many archezoans. Phylogenetic evidence is mounting that such organisms are secondarily amitochondriate. Scale in microns. (Micrograph: K. Vickerman.)

Corliss, 1994 and Chapter 22) and it is a recurring theme throughout this book.

The earliest-branching, amitochondriate phyla (Metamonada and Microspora) have already been discussed in relation to the concept of the Archezoa. Above them on the tree we have a middle-branching group consisting of the Parabasala, Percolozoa and Euglenozoa. The Parabasalia (which includes *Trichomonas* and *Trichonympha*), are now believed to have replaced their mitochondria with hydrogenosomes (reviewed in Müller, 1997); the other two groups contain early mitochondriate organisms with discoid mitochondrial cristae.[2] The Percolozoa are largely amoeboflagellates (such as *Naegleria* and the enigmatic hydrogenosome bearing *Psalteriomionas* – see chapters by De Jonckheere (Chapter 10) and Hackstein *et al.*, (Chapter 8) and the Euglenozoa include the euglenoids and kinetoplasteans (trypanosomes and their relatives – see Philippe, Chapter 11).

Among those protozoa with tubular cristae, perhaps the most prominent are the Alveolata and Cercozoa. Organisms of the Alveolata branch (Gajadhar *et al.*, 1991) are united by the somewhat tenuous morphological feature of membrane-bound cortical sacs present beneath the plasma membrane (Figure 1.3), and include three very different phyla – the Dinozoa (Dinoflagellata), the Apicomplexa (or Sporozoa, see Ellis *et al.*, Chapter 14), and the Ciliophora.[3] In most contemporary SSUrDNA trees, they form one of the Crown groups along with animals, fungi, plants and stramenopiles (chromists). On the SSUrDNA phylogeny, the Cercozoa form a clade, also a Crown group (Cavalier-Smith and Chao, 1997) of seemingly disparate, and until recently somewhat neglected, organisms. They embrace the common naked soil flagellates *Cercomonas* and *Heteromita*, the euglyphid (testate) amoebae, the scale bearing thaumatomonad flagellates (Figure 1.4), the cabbage club root agent (*Plasmodiophora*) and the erstwhile alga *Chlorarachnion* – now regarded as an amoeboid organism that has taken into secondary endosymbiosis a chlorophyte alga. Interestingly, molecular divergence within the genus *Cercomonas* is comparable to that existing within all fungi or land plants (Cavalier-Smith and Chao, 1997).

The Cercozoa assemblage has until now (Cavalier-Smith, Chapter 21) been referred to as the Rhizopoda (Cavalier-Smith and Chao, 1997). Their rechristening is to be welcomed, as to traditional protozoan sys-

[2] Much has been made of the change in shape of mitochondrial cristae (discoidal, tubular, sacculate, flat, branching) as indicating major transitions in evolution (Patterson, 1994), but its functional significance remains enigmatic.

[3] SSUrRNA gene sequences currently suggest that the Haplosporidia, once included in the 'Sporozoa' but not in the Apicomplexa, have an alveolate ancestry (Siddall *et al.*, 1995).

tematists the name Rhizopoda refers to that division of the Sarcodina containing all amoeboid organisms that progress via pseudopodia (as opposed to the Actinopoda which possess axopodia for this purpose). The former Rhizopoda included several groups now in an isolated or unknown position on the SSUrDNA tree, for example the possibly early-branching Archamoebae (flagellated amitochondriate amoebae, e.g. *Pelomyxa, Mastigina,* on which sequencing studies are awaited); the Amoebozoa (*Acanthamoeba, Hartmannella*); secondarily amitochondriate parasites *Entamoeba* and *Endolimax* (Clark, *et al.*, Chapter 9), slime moulds (*Physarum* and *Dictyostelium* do not appear to belong to the same clade) and Foraminifera. Sequencing genes of this last group is made particularly difficult by the presence in most species of a whole range of eukaryote symbionts (dinoflagellates, diatoms, green or red algae), parasites or prey organisms and the conclusion from the complete SSUrRNA gene sequence of the benthic *Ammonia beccarrii* that foraminiferans may represent a fourth alveolate group (Wray *et al.*, 1995) is perhaps premature (Van der Peer, Van der Auwera and De Wachter, 1996). Partial SSUrDNA sequences of five planktonic foraminifera taken from gametes (to avoid the foreign DNA problem) suggested an earlier branching of these organisms between diplomonads and kinetoplastids (Wade *et al.*, 1996).

In recent years, an appreciation of the enormous diversity of small free-living heterotrophic flagellates (Patterson and Larsen, 1991) as well as of their ecological importance has made particular demands on systematists. One of the largest and ecologically most important groups, the choanoflagellates, sequencing studies show to be undoubtedly close to animals and fungi, but despite their instantly recognizable morphology, they may be polyphyletic (Cavalier-Smith, Chapter 21) The ultrastructural features of many of the small heterotrophic flagellates do not ally them to any major group (e.g. euglenozoans, dinoflagellates, cercozoans, choanoflagellates), however, and many are doomed to spend years being transplanted from one transient dustbin group to another until respectable sequencing studies suggest a more permanent home. The converse may happen of course! Sequencing studies may confer an importance way beyond the ecological abundance or known significance of the organism – as illustrated by the strange little flagellate *Apusomonas proboscidea* (Figure 1.5) which, having had its SSUrDNA (Cavalier-Smith and Chao, 1995) sequenced, now stands at the head of a whole subphylum named in its honour (Cavalier-Smith, Chapter 21).

It will become obvious to the reader of this book that a natural classification of Protozoa is still some way off, and that for some years yet utilitarian elements will plague the systematics of these organisms, even if they no longer dominate it. But that a revolution in protozoan systematics is in full swing cannot be doubted.

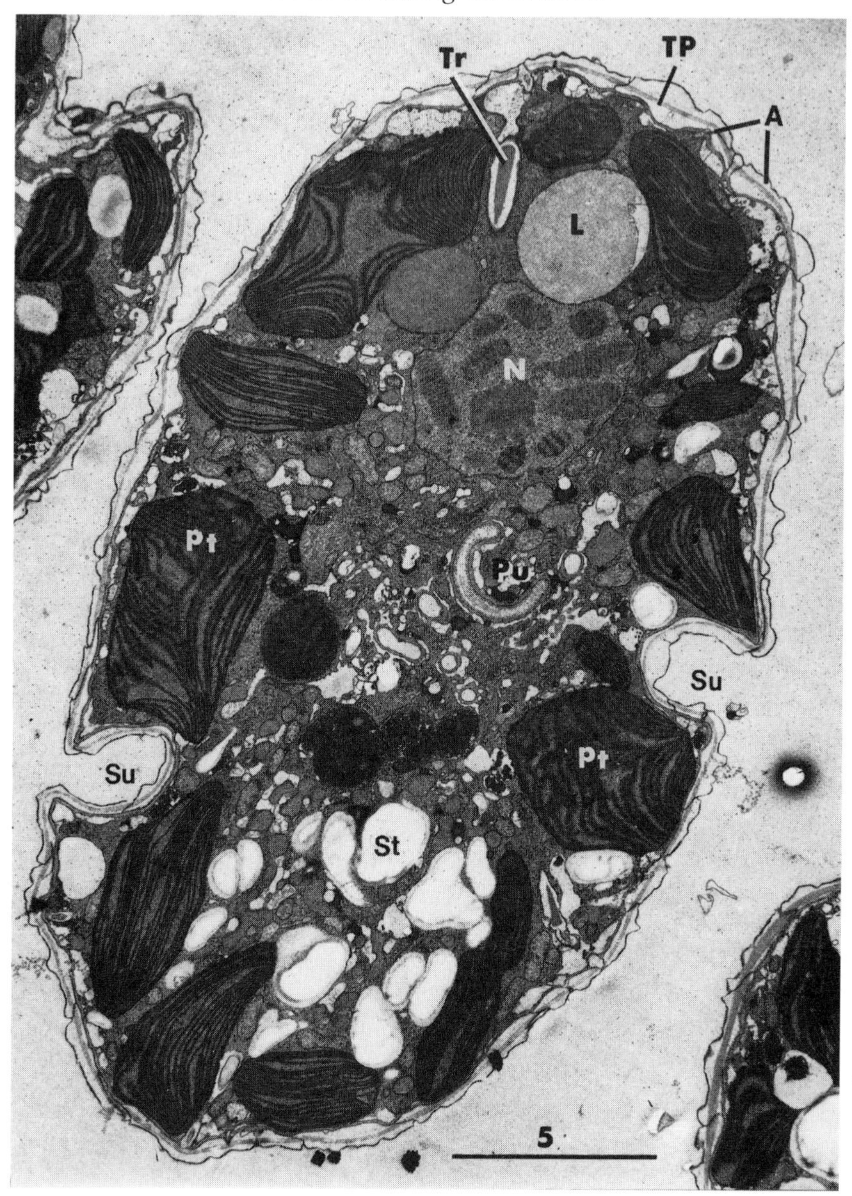

Figure 1.3 An alveolate protozoan. Transmission electron micrograph of a slightly oblique section of the thecate dinoflagellate *Woloszynskia coronata* showing the surface and equatorial flagellar groove or sulcus (Su) covered in membrane-lined alveoli (A); in the thecate dinoflagellates the alveoli contain cellulose plates (TP). Trichocysts (Tr) or other extrusive organelles are also a feature of the Alveolata. The nucleus (N) is of the typical dinokaryote type with

1.4 EVOLUTION OF THE NUCLEUS IN PROTOZOA

Phylogenetic studies have already persuaded us to revise our views on the primitive (as opposed to derived) nature of some prominent features of specific groups, especially nuclear characters. An outstanding example is the 'mesokaryotic' nature of the dinoflagellate nucleus or dinokaryon (Dodge, 1983; Sigee, 1985). As the great divide in cell organization between prokaryotes and eukaryotes rests on nuclear structure and division mechanisms, the dinokaryote dinoflagellates could be deemed mesokaryotic in that their nucleus portrayed several prokaryotic features: the absence of nucleosomal organization in their chromosomes coupled with low basic protein content (absence of histones); the narrow diameter (3nm) and ordered packing of DNA-containing fibrils in the chromosome and association of the daughter chromosomes with cell membranes during genomic segregation on an extranuclear spindle (Figure 1.6).

These suggested that dinoflagellates represented an early branch on the eukaryote tree. On the SSUrDNA tree, however, dinoflagellates are seen as part of a monophyletic assemblage, the Alveolata, which far from being basal is undoubtedly up there in the crown, moreover on a 24S rRNA tree the earliest branching dinoflagellates ('protoalveolates', e.g. *Oxyrrhis*) are those that have histones and an intranuclear mitotic spindle (Lenaers *et al.*, 1991). The supposed prokaryotic features of the dinokaryon are therefore secondarily derived (Raikov, 1995). This conclusion is astounding. If dinoflagellates can ditch their histones and abandon nucleosomal organization, where does this leave current ideas about the role of histones in gene expression (Rhodes, 1997). The dinokaryote dinoflagellates illustrate well my point about protozoa providing the exception that tests the rule, and in this case phylogenetic studies have revealed it.

The nuclear organization of the three main alveolate phyla (Dinozoa, Apicomplexa, Ciliophora) could not be more different, and only that of the Apicomplexa could be described as approaching the conventional. The Ciliophora with their nuclear dimorphism – segregation of genetic and transcriptive roles of the nucleus into separate micronucleus and macronucleus respectively – and their departure from the genetic code of other eukaryotes, clearly display another derived nuclear condition.

---

permanently condensed and banded chromosomes (see Figure 1.6). The numerous plastids (Pt, chloroplasts) of photosynthetic dinoflagellates are believed to be derived from secondary endosymbiosis on the part of a phagotrophic ancestor. L, lipid; Pu, pusule; St, starch granule. Scale in microns. (Micrograph: J.D. Dodge.)

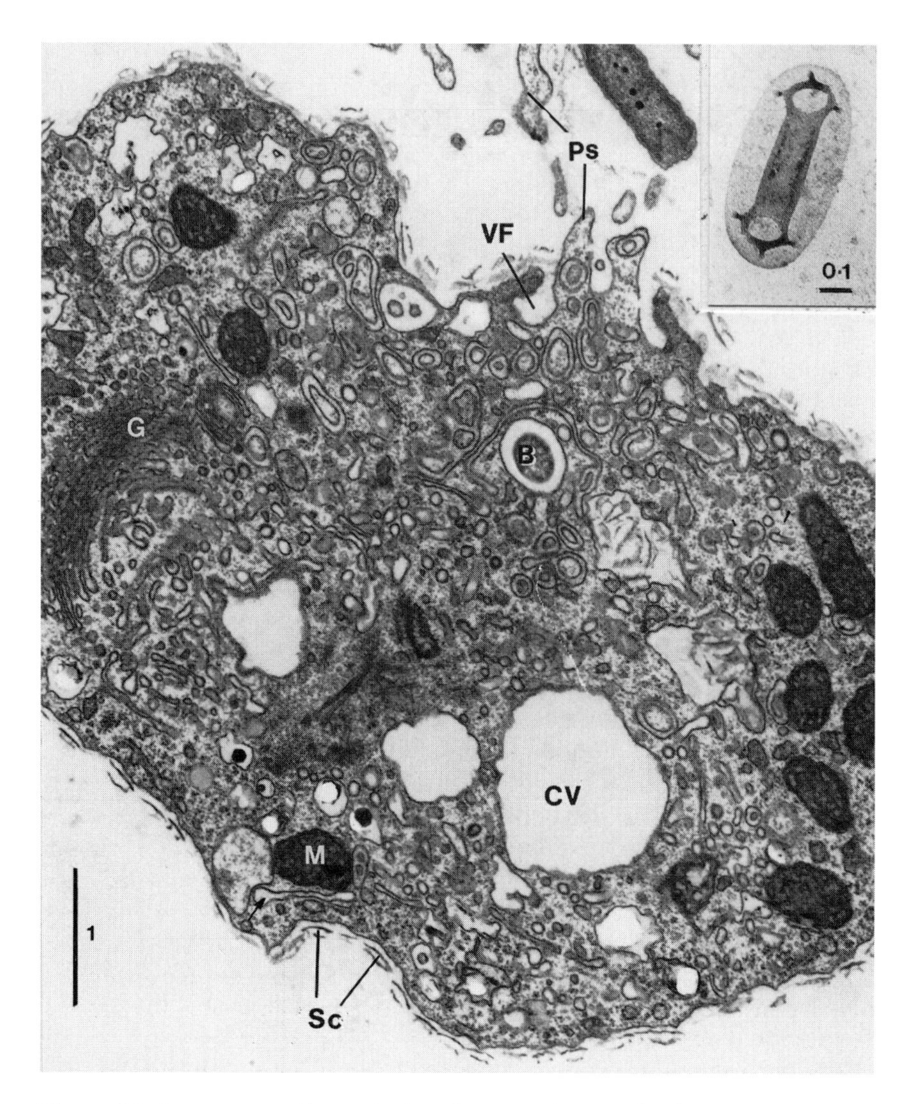

Figure 1.4 A cercozoan. Transmission electron micrograph of transverse section of the trophic flagellate phase of *Thaumatomonas lauterborni*. Under conditions of starvation, individuals of the silica scale-secreting trophic flagellate phase fuse to form symplasts in which feeding and scale secretion cease; the large Golgi apparatus (G) and the flagella regress simultaneously. Uniquely in the thaumatomonads, scales (Sc) are forged in vesicles (arrowed) associated with the mitochondria (m) and secreted over the body surface; insert shows whole mount of scale. Pseudopodia (Ps) emerging from a ventral furrow (VF, which is a posteriad extension of the anterior flagellar pocket) ingest bacteria (B) during feeding. The thaumatomonads have scale secretion and pseudopod production from a limited aperture in common with the Testaceofilosa (shelled amoebae) and use of two heterodynamic flagella and pseudopodial feeding in common with the Cercomonadidae. Scale bars in microns. (Micrographs: K. Vickerman.)

In most ciliates, both macronucleus and micronucleus divide during binary fission, a new macronucleus being regenerated only during sexual reproduction (conjugation, autogamy) when gamete nuclei derived from the micronucleus fuse to form a zygote nucleus from which both types of nucleus arise following division – and endopolyploidization in the case of the macronucleus. In members of one ciliate order, however, the Karyorelictida, the macronucleus does not divide, and a new macronucleus is generated from the micronucleus at each binary fission. From an evolutionary viewpoint, this state has long been regarded as primitive and the dividing macronucleus as derived. This view must now be overturned, however, as SSUrDNA sequence studies indicate that the Karyorelictida are derived from ancestors that possessed a dividing macronucleus (see Chapters 17 by Foissner and 18 by Hirt, Wilkinson and Embley).

Diplokaryosis, the physical association of two nuclei in the one cell (Figure 1.2) has been commented on as a feature of the Archezoa and accorded some significance in the evolution of sexual processes (Cavalier Smith, 1995). These arguments may require some revision if Microspora do turn out to be of fungal origin (Edlind, Chapter 5).

1.5 ENDOSYMBIOSIS AND EVOLUTION OF PROTOZOA

The Prokaryota/Eukaryota schism is marked not only by striking differences in lay-out and operation of the nuclear genome, but also in organization of the cytoplasm. Compartmentalization of function into distinct organelles is a feature of eukaryote organization and perhaps nowhere in evolutionary studies do protists come into their own as in the study of the origin and continuity of cell organelles, especially the mitochondrion and the plastid which contain their own DNA. The hypothesis that these organelles arose as the result of serial endosymbiosis – the incorporation of an aerobically-respiring bacterium and a photosynthesizing cyanobacterium into the early eukaryotic cell with subsequent progressive transfer of symbiont genes to the host cell nucleus (Margulis, 1970) – is now widely accepted. Evidence for the monophyly of both mitochondrion and chloroplast has been reviewed by Gray and Spencer (1996) and Howe *et al.* Chapter 15 injects a cautionary note on the latter. The widespread genetic input of symbionts into the host's genome is only just beginning to be appreciated as phylogenetic studies reveal it, and it need not always occur in the context of organelle evolution. (Martin, 1996).

One of the most striking pieces of recent evidence backing up the endosymbiotic origin of mitochondria hypothesis is the demonstration that the tiny heterotrophic flagellate *Reclinomonas americanum* contains the largest collection of genes so far identified in any mitochondrial

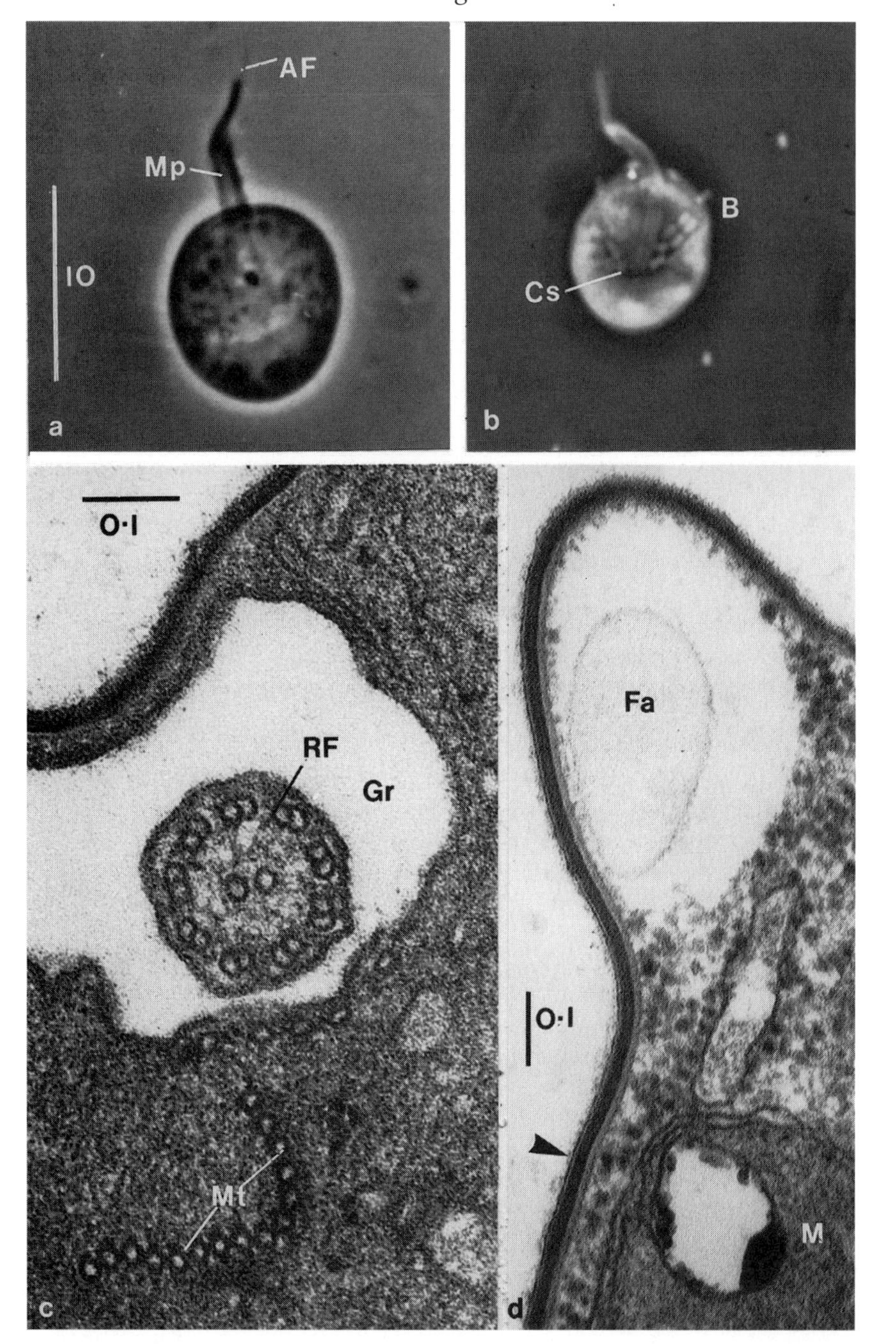
AF
Mp
IO
a
B
Cs
b
O·I
RF
Gr
Mt
c
Fa
O·I
M
d

DNA (mtDNA) (Lang *et al.*, 1997). The eubacterial characteristics of genome organization and expression, not found before in mitochondrial genomes, suggest that *R. americana* mtDNA more closely resembles the ancestral proto-mitochondrial genome than any other mtDNA investigated to date.

Phylogenetic evidence that mitochondria may have been lost by the archezoan protozoa has already been discussed. There is now mounting evidence (reviewed in Müller, 1997) that the hydrogenosomes of parabasalians such as *Trichomonas* (Figure 1.7) are derived from mitochondria as are those of some ciliates (Finlay and Fenchel, 1989) and that hydrogenosomes have arisen polyphyletically in several groups (Hackstein *et al.*, Chapter 8). Hydrogenosomes, unlike mitochondria appear to have surrendered all their DNA to the host.

Transmission electron microscopy and phylogenetic analysis of SSUrDNA sequences of the heterokont chromists suggest that chloroplast loss appears to have occurred repeatedly within this group (Cavalier-Smith *et al.*, 1996). Elsewhere reduction to a non-photosynthesizing leucoplast appears to be the rule. One of the most surprising examples of plastid persistence in this way is provided by the non-photosynthesizing plastid remnant found in malaria parasites and other members of the Apicomplexa (reviewed Wilson and Williamson, 1997 and see chapters by Ellis *et al.* (Chapter 14), Hackstein *et al.* (Chapter 8) and Wilson *et al.*, (Chapter 16)). The function of this organelle and hence the selective forces responsible for its survival are now eagerly sought, especially by those looking for chemotherapeutic leads. The question of whether this plastid complete with its relict

---

Figure 1.5 *Apusomonas proboscidea.* Described by its discoverer as 'a dug-out canoe with an elephant's trunk', this creeping soil flagellate (Phase contrast micrograph (a)) propels itself by means of a sleeved anterior flagellum (AF) borne at the end of a mastigophore (Mp) in which a recurrent flagellum lies in an open groove leading to a cytostome (Cs) into which bacteria (B) are trawled (Phase contrast micrograph (b)) and ingested by pseudopodial activity. The cytoskeleton is virtually confined to bands of microtubules running along the mastigophore and the body shape is maintained by a curved 'falx' of unknown composition and an extraordinary thick, five-layered pellicle. Transmission electron micrograph (c) shows the mastigophore in transverse section, with its recurrent flagellum (RF) in its groove (Gr) (which is lined by a simple membrane) and microtubule bands (Mt). Transmission electron micrograph (d) shows a transverse section of the falx (Fa) surrounded by thick pellicle (M, mitochondrion). SSUrDNA sequencing has shown that this singular flagellate is related to the common ancestors of animals, fungi and choanoflagellates. Scale bars in microns. (Micrographs: K. Vickerman.)

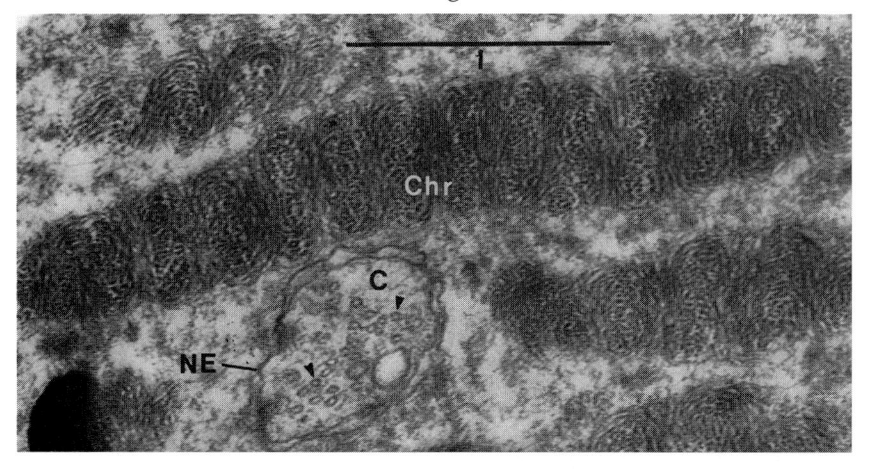

Figure 1.6 Dinokaryote chromosomes. Transmission electron micrograph showing chromosomes (Chr) of the dinoflagellate *Gymnodinium* sp. with 'screw-carpet' arrangement of DNA fibrils characteristic of the dinokaryote dinoflagellates. Also shown is a cytoplasmic channel (C) bounded by nuclear envelope (NE) passing through the dividing nucleus and bearing microtubules (at arrowheads) of the extranuclear mitotic spindle. Scale in microns. (Micrograph: B.S.C. Leadbeater.)

genome is also present in non-photosynthesizing dinoflagellates or protoalveolates should also be addressed.

What was the nature of the selection pressure that directed symbiont genes to the host cell nucleus during organelle evolution? This transfer has not been complete in mitochondria and plastids, though in the hydrogenosome it would appear to have been. The argument that transfer to the nucleus represents an economy of resources in having a single genome and a single apparatus for gene expression fails if any organellar genes are retained, though obviously transferring genes to the nucleus facilitates sexual recombination and DNA repair. Allen and Raven (1996) have come up with the ingenious suggestion that the selective advantage of transfer is decreased free radical-induced mutation: chloroplasts and mitochondria have high volume-specific rates of redox reactions, producing reactive oxygen intermediates that chemically modify DNA. The genes that remain behind in the organelle are those that encode certain key components of the photosynthetic and respiratory pathways and whose expression is controlled by the redox state of electron carriers to minimize free radical production.

Although the mitochondrion and chloroplast represent by now rock-

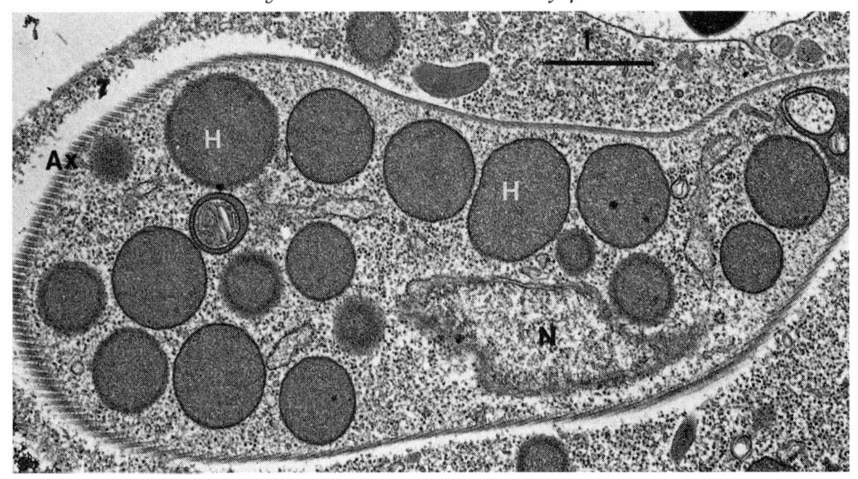

Figure 1.7 Hydrogenosomes. Transmission electron micrograph of section through the head of the axostyle of the parabasalian *Trichomonas lacertae* (from gut of lizard). Encased within the microtubular sheath of the axostyle (Ax) are the nucleus (N) and several hydrogenosomes (H). These organelles are believed to be derived from mitochondria. Scale in microns. (Micrograph: G. Brugerolle.)

solid examples of the conversion of symbionts to organelles, we should not run away with the idea that they are the only ones, nor believe that gene transfer is solely from symbiont genome to host nucleus. Symbiotic bacteria (or are they organelles?) are extremely common in many protozoa, especially ciliates where widespread occurrence of 'consortia' between ciliates and methanogenic bacteria have been reported (reviewed in Fenchel, 1996) and of course the *Caedibacter* (*kappa* particles) of *Paramecium* hold a most honourable place in the history of our understanding of genetic control of symbionts (Görtz, 1988). The bacterium-like symbionts (bipolar bodies) of the one-host trypanosomatids are known to relieve their hosts of dependence on an external source of various vitamins and amino acids, while, strangely, their presence in several genera is associated with a standard set of secular modifications of cytoskeleton and mitochondrion (Roitman and Camargo, 1985). Not all such symbiotic associations are enduring; *Paramecium* may lose its *kappa* particles and trypanosomatids may be robbed of their bipolar bodies by antibiotic treatment. Their temporary presence may leave its mark, however. Evidence that ephemeral endosymbiosis may have lasted long enough for successful symbiont gene transfer to the host nucleus to occur is impressive (Martin, 1996). Thus using sequence comparisons, Henze and colleagues (1995) demonstrated that *Euglena gracilis, Giardia lamblia* and *Entamoeba histolytica* have all acquired their cytosolic glyceraldehyde-3-phosphate dehydrogenase genes from eubacterial donors which did not ultimately stay on to become membrane-bound organelles. Further evidence for the input of bacterial genes into the eukaryotic genome is discussed by Müller and Opperdoes *et al.* in Chapters 6 and 13 in this volume.

Palmer (1995) has reviewed evidence that the nuclear encoded but chloroplast located CO_2-fixing enzyme Rubisco in photosynthetic dinoflagellates appears to have been derived from the mitochondrion. As Martin (1996) has so vividly put it, 'endosymbiotic gene transfer throws a monkey wrench into the phylogenetics of early evolution'. That most students of evolutionary biology have ignored this input through lack of interest in protists has been the continuing complaint of Lynn Margulis (Margulis and Sagan, 1997).

1.6 EVOLUTION OF PARASITISM

A parasite is one member of a two organism partnership that is dependent upon at least one gene of the other member for its own survival. One might expect loss of some of the parasite's genes as dependency on the host grows, even some simplification of parasite structure. Just how great this reduction can be is evident from phylogenetic analysis (Anderson, Chapter 19) which shows that the Myxozoa, though long classified as Protozoa, are much simplified animals, indeed with loss of cell layers, collagen and development from a blastula, one might say that they have lost all their animal characters. Similarly if the conclusion from protein trees (Canning (Chapter 4) and Edlind (Chapter 5)) is confirmed and the wholly-parasitic Microspora do turn out to be derived from Fungi, then their loss of cell structure has gone further than just mitochondria to include most cytoplasmic organelles, and rRNA gene deletions have accompanied a reversion to prokaryote-like (70S) ribosomes. The worry that all the Archezoa might be deep branching because they had undergone modifications associated with the parasitic lifestyle has been removed by sequencing of the SSUrDNA of the free-living diplomonad *Trepomonas* (Cavalier-Smith and Chao, 1996) and demonstration that it too branches as an early eukaryote.

One of the most exciting achievements of phylogenetic analysis has been to shed light on the relationship between what have been regarded as closely-related parasites or even as phases in the life of one parasite. Thus sequencing (or riboprinting) of the intestinal amoebae has cleared up that old parasitological conundrum of exactly how distinct are the different forms of *Entamoeba* living in the human colon, and especially in distinguishing between the never pathogenic *Entamoeba dispar* and the invariably pathogenic *E. histolytica* (see Clark *et al.*, Chapter 9). De

Jonckheere in Chapter 10 shows how phylogenetics has sorted out the facultatively parasitic amoeboflagellates. The success of molecular epidemiology is made plain in Hide's (Chapter 12) study of the coexistence of human-infective and human-non-infective *Trypanosoma brucei* in the same bovine host. Co-evolution of parasite and host is often assumed, but may often be proved wrong (as in Clark's amoebae!). Time will tell!

1.7 CONCLUSION

Too many facts, too much disorder, too great a burden for the brain! These criticisms of the biological sciences, first articulated by the poet Samuel Taylor Coleridge early in the last century, progressively subsided with the impact of Darwinian theory later in that century and the revelation of the universal principles of life at the molecular level in the second half of the present century. But as biologists move on from preoccupation with what living organisms have in common to an interest in what makes them differ, Coleridge's criticisms begin to bite again, and perhaps nowhere so much as in the study of Protozoa and other unicellular eukaryotes.

The new phylogenetics based on the comparison of gene sequences promises eventually to aid greatly in replacing the hitherto utilitarian systematics of Protozoa with a more natural one based on evolutionary relationships. In the meantime, however, the interested outsider may become confused by the frenetic replacement of one fly-by-night taxon with another as new sequences enter the databases and new arguments start up about what they really mean. And our increasing realization of the vast wealth of protozoa yet to be described (especially among the heterotrophic flagellates) adds to the mental burden. Nevertheless, these difficulties cannot distract from the potential contribution of protozoan phylogenetics to the biology of the next century, especially to our understanding of the evolution of the eukaryotic cell, the interplay between nuclear and cytoplasmic genomes and the nature of parasitism.

1.8 REFERENCES

Allen, J.F. and Raven, J.A. (1996) Free-radical-induced mutation *vs* redox regulation – costs and benefits of genes in organelles. *Journal of Molecular Evolution*, **42**, 482–92.

Bowler, P. (1996) *Life's Splendid Drama: Evolutionary Biology and the Reconstruction of Life's Ancestry, 1860–1940*, University of Chicago Press, Chicago.

Cavalier-Smith, T. (1986) The kingdom Chromista, origin and systematics, in *Progress in Phycological Research*, Vol.4 (eds F.E. Round and D.J. Chapman), Biopress, Bristol, pp. 309–47.

Cavalier-Smith, T. (1993) Kingdom Protozoa and its 18 phyla. *Microbiological Reviews*, **57**, 953–94.

Cavalier-Smith, T. (1995) Cell cycles, diplokaryosis and the archezoan origin of sex. *Archiv für Protistenkunde*, **145**, 189–207.

Cavalier-Smith, T. and Chao, E.E. (1995) The opalozoan *Apusomonas* is related to the common ancestor of animals, fungi and choanoflagellates. *Proceedings of the Royal Society* Series B, **261**, 1–6.

Cavalier-Smith, T. and Chao, E.E. (1996) Molecular phylogeny of the free-living archezoan *Trepomonas agilis* and the nature of the first eukaryote. *Journal of Molecular Evolution*, **43**, 551–62.

Cavalier-Smith, T. and Chao, E.E. (1997) Sarcomonad ribosomal RNA sequences, rhizopod phylogeny, and the origin of euglyphid amoebae. *Archiv für Protistenkunde*, **147**, 227–36.

Cavalier-Smith, T., Chao, E.E., Thompson, C.E. and Hourihane, S.L. (1996) *Oikomonas*, a distinct zooflagellate related to chrysomonads. *Archiv für Protistenkunde*, **146**, 273–9.

Christensen, T. (1980) *Algae. A Taxonomic Survey*, AiO Tryk, Odense, Denmark.

Corliss, J.O. (1994) An interim utilitarian ('user friendly') hierarchical classification and characterisation of the protists. *Acta Protozoologica*, **33**, 1–51.

Darwin, C. (1859) *On the Origin of Species by Means of Natural Selection or the Preservation of Favoured Races in the Struggle for Life*, John Murray, London.

Dodge, J.D. (1983) Dinoflagellates: investigations and phylogenetic speculation. *British Phycological Journal*, **18**, 335–56.

Fenchel, T. (1996) Eukaryotic life: anaerobic physiology, in *Evolution of Microbial Life* (eds D.McL. Roberts, P. Sharp, G. Alderson and M. Collins), Symposium 54, Society for General Microbiology, Cambridge University Press, Cambridge, pp. 185–203.

Finlay, B.J. and Fenchel, T. (1989) Hydrogenosomes in some anaerobic protozoa resemble mitochondria. *FEMS Microbiology Letters*, **65**, 311–14.

Gajadhar, A.A., Marquardt, W.C., Hall, R. *et al.* (1991) Ribosomal RNA sequences of *Sarcocystis muris, Theileria annulata* and *Crypthecodinium cohnii* reveal evolutionary relationship among apicomplexans, dinoflagellates and ciliates. *Molecular and Biochemical Parasitology*, **45**, 147–54.

Germot, A., Phillipe, H. and Le Guyader, H. (1997) Evidence for loss of mitochondria in microsporidia from a mitochondrial type HSP70 in *Nosema locustae. Molecular and Biochemical Parasitology*, **87**, 159–68.

Gortz, H.D. (ed.) (1988) *Paramecium*, Springer Verlag, Heidelberg.

Gray, M.W. and Spencer, D.F. (1996) Organellar evolution, in *Evolution of Microbial Life* (eds D.McL. Roberts, P. Sharp, G. Alderson and M. Collins), Symposium 54, Society for General Microbiology, Cambridge University Press, Cambridge, pp. 109–26.

Haeckel, E. (1866) *Generelle Morphologie der Organismen. Bd.II. Allgemeine Entwickelungsgesichte der Organismen*, Georg Reimer, Berlin.

Hennig, W. (1966) *Phylogenetic Systematics.*University of Illinois Press, Urbana.

Henze, K., Badr, A., Wettern, M. *et al.* (1995) A nuclear gene of eubacterial origin in *Euglena gracilis* reflects cryptic endosymbioses in protist evolution. *Proceedings of the National Academy of Sciences USA*, **92**, 9122–6.

Hillis, D.M. (1997) Biology recapitulates phylogeny. *Science*, **176**, 218–19.

Hogg, J. (1861) On the distinction between a plant and an animal, and on a fourth kingdom of Nature. *Edinburgh New Philosophical Journal (new series)*, **2**, 216–25.

Khumar, S. and Rhetsky, A. (1996) Evolutionary relationships of the eukaryotic kingdoms. *Journal of Molecular Evolution*, **42**, 183–93.

Lang, B.F., Burger, G., O'Kelly, C.J. *et al.* (1997) An ancestral mitochondrial DNA resembling a eubacterial genome in miniature. *Nature*, **387**, 493–7.

Lenaers, G., Scholin, C., Bhaud, Y. *et al.* (1991) A molecular phylogeny of dinoflagellate protists (Pyrrhophyta) inferred from the sequence of 24SrRNA divergent domains D1 and D8. *Journal of Molecular Evolution*, **32**, 53–63.

Levine, N.D, Corliss, J.O., Cox, F.E.G. *et al.* (1980) A newly-revised classification of the Protozoa. *Journal of Protozoology*, **27**, 37–58.

Liaud, M.F., Valentin, C., Martin, W. *et al.* (1994) The evolutionary origin of red algae as deduced from nuclear genes encoding cytosolic and chloroplast glyceraldehyde-3-phosphate dehydrogenases from *Chondrus crispus. Journal of Molecular Evolution*, **38**, 319–27.

Margulis, L. (1970) *Origin of Eukaryotic Cells*. Yale University Press, New Haven.

Margulis, L. and Sagan, D. (1997) *Slanted Truths. Essays on Gaia, Symbiosis and Evolution*, Springer Verlag, New York.

Martin, W.F. (1996) Is something wrong with the tree of life? *Bioessays*, **18**, 523–7.

Maynard Smith, J. and Szathmáry,E. (1995) *The Major Transitions in Evolution*, W.H. Freeman, Oxford.

Mayr, E. (1997) *This is Biology*. Harvard University Press, Cambridge, Mass.

Müller, M. (1997) Evolutionary origin of trichomonad hydrogenosomes. *Parasitology Today*, **13**, 166–7.

Palmer, J.D. (1995) Rubisco rules fail: gene transfer triumphs. *Bioessays*, **17**, 1005–8.

Patterson, D.J. (1989) Stramenopiles: chromophytes from a protistan perspective, in *The Chromophyte Algae, Problems and Perspectives* (eds J.C. Green, B.S.C. Leadbeater and W.L. Diver), Clarendon Press, Oxford, pp. 357–79.

Patterson, D.J. (1994) Protozoa: evolution and systematics. In *Progress in Protozoology* (eds K. Hausmann and N. Hulsmann), Gustav Fischer Verlag, Stuttgart-Jena, pp. 1–14.

Patterson, D.J. and Larsen, J. (eds) (1991) *The Biology of Free-Living Heterotrophic Flagellates*, Systematics Association Special Volume No.45, Clarendon Press, Oxford.

Philippe, H. and Adoutte, A. (1995) How reliable is our current view of eukaryotic phylogeny? In *Protistological Actualities* (eds G. Brugerolle and J.-P. Mignot), Blaise Pascal University, Clermont Ferrand, pp. 17–33.

Raikov, I.B. (1995) The dinoflagellate nucleus and chromosomes: mesokaryote concept reconsidered. *Acta Protozoologica*, **34**, 239–47.

Rhodes, D. (1997) Chromatin structure: the nucleosome all wrapped up. *Nature*, **389**, 231–3.

Roitman, I. and Camargo, E.P. (1985) Endosymbionts of Trypanosomatidae. *Parasitology Today*, **1**, 143–4.

Ruse, M. (1997) *Monad to Man: the Concept of Progress in Evolutionary Biology*, Harvard University Press, Cambridge, Mass.

Saville-Kent, W. (1980–1981) *A Manual of the Infusoria*, David Bogue, London.

Siddall, M.E., Stokes, N.A. and Burreson, E.M. (1995) Molecular phylogenetic evidence that the phylum Haplosporidia has an alveolate ancestry. *Molecular Biology and Evolution*, **12**, 573–81.

Sigee, D.C. (1985) The dinoflagellate chromosome. *Advances in Botanical Research*, **12**, 205–64.

Sogin, M., Silberman, J.D., Hinkle, G. and Morrison, H.G. (1996) Problems with molecular diversity in the Eukarya, in *Evolution of Microbial Life* (eds D.McL. Roberts, P. Sharp, G. Alderson and M. Collins), Symposium 54, Society for General Microbiology, Cambridge University Press, Cambridge, pp. 167–84.

Soltys, B.J. and Gupta, R.S. (1994) Presence and cellular distribution of a 60-kDa protein related to mitochondrial HSP60 in *Giardia lamblia. Journal of Parasitology*, **80**, 580–90.

Van der Peer, Y., Van der Auwera, G. and De Wachter, R. (1996) The evolution of stramenopiles and alveolates as derived by 'substitution rate calibration' of small ribosomal subunit RNA. *Journal of Molecular Evolution*, **42**, 201–10.

Wade, C.M., Darling, K.F., Kroon, D. and Leigh Brown, A.J. (1996) Early evolutionary origin of the planktonic foraminifera inferred from small subunit rDNA sequence comparisons. *Journal of Molecular Evolution*, **43**, 672–77.

Wainwright, P.O., Hinkle, G., Sogin, M.L. and Stickel, S.K. (1993) Monophyletic origins of the Metazoa, an evolutionary link with fungi. *Science*, **260**, 340–2.

Weiser, W. (1997) A major transition in Darwinism. *Trends in Ecology and Evolution*, **12**, 367–70.

Whittaker, R.H. (1969) New concepts of kingdoms of organisms. *Science*, **163**, 150–60.

Wilson, R.J.M. and Williamson, D.H. (1997) Extrachromosomal DNA in the Apicomplexa. *Microbiology and Molecular Biology Reviews*, **61**, 1–16.

Wray, C.G., Langer, M.R., De Salle, R. *et al.* (1995) Origin of the foraminifera. *Proceedings of the National Academy of Sciences USA*, **92**, 141–5.

2

The molecular phylogeny of Eukaryota: solid facts and uncertainties

Hervé Philippe and André Adoutte

Laboratoire de Biologie Cellulaire (URA 2227 du CNRS), Bâtiment 444, Université Paris-Sud; 91405 Orsay Cedex, France. E-mail: hp@bio4.bc4.u-psud.fr

ABSTRACT

Following a period of 'educated guess', the phylogeny of protists entered a new phase 10 years ago with the advent of molecular data. The order of emergence of taxa based on the comparison of 18S ribosomal RNA takes the form of a tree in three parts. A lower part contains three amitochondrial protozoan groups in an unresolved order; a middle zone contains the Euglenozoa and a diversity of amoeboid or amoebo-flagellate phyla; finally, the tip of the tree consists of a very large unresolved radiation (the 'terminal crown') comprising plants, fungi, animals and a number of major protist groups (Alveolates, Stramenopiles, Rhodophytes, etc.). This view dominates the field and has even found its way into influential textbooks. It has served as a basis to construct a scenario of progressive acquisition of the major organelles of the eukaryotic cell. However, there are reasons to question its reliability. The major concern is that of the 'clock-like' behaviour of rRNA, and this concern crystallized when strikingly incongruent trees were obtained using protein coding genes. We examine some of the molecular data critically in an attempt to account for contradictory results. We focus on a number of potential sources of artefact often neglected by workers in the field. In particular, we illustrate how inequalities in evolutionary rates can generate falsely resolved trees. By systematically comparing rates of evolution of protein coding genes with those of rRNA genes within the same set of taxa, we show that many groups of the middle zone more probably belong

Evolutionary Relationships Among Protozoa. Edited by G.H. Coombs, K. Vickerman, M.A. Sleigh and A. Warren. Published in 1998 by Chapman & Hall, London. ISBN 0 412 79800 X

to the terminal 'crown'. For example, we show that the relatively basal position of Euglenozoa in the rRNA tree is probably due to a long branch artefact. Thus, the present status of eukaryotic phylogeny can best be summarized by a vast multifurcation of the great majority of taxa, if not all of them. The fact that, in two of the presently amitochondrial phyla (microsporidia and trichomonads), there is now good evidence for secondary loss of the mitochondria, suggests that the mitochondrial endosymbiosis occurred just before this radiation. Within this broad radiation, a small number of solid monophyletic groups can nonetheless be identified (such as Metazoa+Fungi or Chlorobionta+Rhodobionta). As is the case for other probably rapid cladogenetic events in life's history, this radiation will be difficult to resolve with the use of traditional sequence data and the usefulness of more qualitative molecular or morphological characters will be stressed. It is somewhat ironic that over the last 10 years, molecular phylogeny has confirmed the monophyletic groups identified through electron microscopy but made little progress in establishing their order of emergence.

2.1 A SHORT ACCOUNT OF THE PRE-MOLECULAR ERA

Phylogenetic reconstruction starts with a good description of the organisms under study, allowing their grouping into monophyletic ensembles. Protists are no exception, as emphasized by Patterson (1994). However, the transition from taxonomy to phylogeny has been especially slow for protists because of two difficulties specific to these organisms. First, because they are small, their description relies on the use of the microscope and, more importantly, on the electron microscope (EM). The equivalents of organs, tissues, appendages, and so on, of the multicellular organisms are in fact the subcellular organelles in protists. Thus, unlike metazoans and metaphytes, detailed (i.e. ultrastructural) descriptions of protists have become available only since the 1950s and as for metazoans and metaphytes, have allowed the delineation of numerous clearly monophyletic groups. Second, it turned out that the detection of a series of nested shared derived characters, which are the basis of a sound phylogenetic reconstruction, is not easy (if not impossible) among protists on a very large scale. The reason is the lack of homologous characters present in different states among protists: either they all share a number of characters that are typically eukaryotic and are therefore simply plesiomorphies, unsuitable to reconstruct relationships, or they display specialized traits, found in only one subgroup (autapomorphies), useful for erecting monophyletic groups but again not suited to establishing inter-group relationships. The situation is in fact quite similar to that which plagues the reconstruction of a morphological phylogeny of the 30–35 metazoan phyla. Instead of recognizing this fact, both in the case of Metazoa and more recently for protists, most authors have taken the **absence** of a character as the primitive state and its presence as the derived state. This, however, can be very dangerous, as shown below.

A simple explanation for the lack of nested synapomorphies, both in protists and in metazoans, forms the main conclusion of this paper: we are dealing with rapid radiation of the lineages considered, i.e. the successive splitting of the lineages occurred during short time intervals. Since synapomorphies accumulate during the time intervals that separate cladogeneses, it can be predicted that these will be lacking in the case of evolutionary radiations.

In short, it is not surprising that, except for a few attempts (Cavalier-Smith, 1975, 1978; Ragan and Chapman, 1978; Taylor, 1978; Sleigh, 1979), phylogenetic schemes of protists have been essentially lacking during the first stage of protistology, that of light and EM description of the taxa and their grouping into monophyletic units. It is striking that most taxonomic description of protists, including some voluminous protistological treatises of the 80s and even 90s (Lee, Hunter and Bovee, 1985; de Puytorac, Grain and Mignot, 1987; Margulis, *et al.*, 1990), have only very short portions devoted to phylogeny, if any at all!

As just indicated, this state of affairs did not preclude the recognition of monophyletic ensembles on the basis of shared structural similarities. Groups such as ciliates, dinoflagellates, sporozoans (Apicomplexa), kinetoplastids, euglenids, parabasaliads, diplomonads, microsporidians, various 'algal' lineages such as rhodophytes, chlorophytes and several chlorophyl a+c bearing groups (e.g. chrysophytes, diatoms) as well as several fungal classes (e.g. ascomycetes, basidiomycetes) were recognized rather early on, as monophyletic. In 1984, Corliss recognized 45 such protist 'phyla' and this number has recently been raised to about a hundred taxa displaying distinct 'ultrastructural identity' (Patterson, 1994).

While ultrastructural analysis allowed the identification of reasonably solid and homogeneous monophyletic groups, occasionally it also suggested the existence of certain broader assemblages, identified on the basis of fewer shared characters, grouping together some seemingly more diverse organisms. For example, very early on researchers were struck by the occurrence of tubular hairs (mastigonemes) on one of the two flagella of a very broad ensemble of photosynthetic protists with a+c type chlorophyll, such as the chrysophytes. A very similar type of appendage was observed on the flagella of the gametes of brown algae (which also have chlorophyll a+c), suggesting that in spite of their huge morphological differences (brown algae can measure several metres in length!), they were related. Later, it was found that the flagella of yet two other groups of organisms, neither of which is photosynthetic, the oomycetes and the hyphochytridiomycetes, formerly classified with fungi, also bear mastigonemes. This led to the erection of a broad ensemble called heterokonts (for which stramenopiles is a junior synonym).

Another example of grouping of seemingly quite distant organisms is the association of euglenids and kinetoplastids (Kivic and Walne, 1984) into the Euglenozoa (Cavalier-Smith, 1981). Here again, the number of ultrastructural characters uniting the two phyla was rather modest but well chosen since this grouping was also confirmed by molecular phylogeny.

In addition to these super-ensembles created on the basis of the presence of a shared character, some authors used the shared absence of an organelle as a uniting feature. Most notably, Cavalier-Smith (1983, 1987a) emphasized the absence of mitochondria in three phyla of protists, the Microsporidia, Metamonada (containing *Giardia*) and Parabasalia (containing *Trichomonas*), also in some amitochondriate ameboid organisms such as *Pelomyxa*, and suggested grouping them into a paraphyletic ensemble, the Archezoa (see Brugerolle, 1991 for a description of these taxa). He placed this ensemble at the base of the eukaryotic tree, considering that they represented descendants of the 'pre-mitochondrial symbiosis' era of eukaryotic evolution (although he acknowledged that it may also have corresponded to secondary loss in some of these lineages, notably in trichomonads).

In summary, by the beginning of the 1980s, there was a consensus on the existence of most of the monophyletic groups of protists, but the relationships between them remained an open question. For example, in an insightful and prudent attempt to reconstruct relationships among the protist groups, Taylor (1978) recognized only two solid monophyletic lineages, which he called respectively the 'chromophyte' and the 'chlorophyte series'. The former grouped together all chlorophyll a+c bearing protists (i.e. the yellow, golden and brown algae), the latter the a+b bearing ones (i.e. the green algae). Other than these two strong assumptions, the rest of the approximately 30 protistan phyla were loosely attached to the mains branches with broken lines. This was especially the case for metazoans which were depicted far from any protist group, reflecting the uncertainty prevailing at that time as to their closest relative.

The development of fast methods for nucleic acids sequencing and the parallel improvement of computerized methods for treating these data and converting them into phylogenetic trees has radically changed the situation within the last ten years. A 'triumphant' period has followed, which may have given the impression that we now have solved most of the phylogenetic questions relating to protists. There are good reasons to doubt this, however, as this paper will demonstrate.

2.2 FROM THE BURST OF MOLECULAR PHYLOGENY TO THE GENESIS OF A DOGMA

Starting in the mid-sixties, the pioneering work on phylogenetic reconstruction using sequences from protein-coding genes such as haemoglobin (Zuckerkandl and Pauling, 1965) or cytochrome c (Fitch and Margoliash, 1967) convinced the biological community that a considerable amount of evolutionary information was embedded in the genomic sequences of extant species. Molecular phylogeny was born. In 1977, Woese and Fox published their first results applying this principle on a large scale and using an even more universal marker, ribosomal RNA, allowing its extension to prokaryotes and to very distant organisms. The outcome was highly encouraging: a new 'empire' of organisms was discovered, that of Archaebacteria and the tools seemed suitable to tackle the most ambitious phylogenetic reconstruction, that of the complete tree of life (Woese, 1987). With the advent of rapid sequencing techniques and, more recently, of PCR, the approach could be extended to eukaryotes on a very large scale, even to those that can not or are difficult to grow in pure culture.

In a landmark paper, Sogin, Elwood and Gunderson (1986), using the then available complete small subunit ribosomal RNA sequences of nine species of protists, obtained results that set the frame of the eukaryotic tree for following years. They first showed that the molecular diversity within protists was considerable, even exceeding that found between metazoans and metaphytes. This result is not too surprising if protists did indeed precede multicellular eukaryotes in evolution and indeed, on their distance tree, metazoans and metaphytes were located at the tip of the eukaryotic branch preceded by several protist lineages. This agreed well with the expectations of most biologists although most would probably not have predicted that the distances between protists would be so large. The second point of interest was the order of emergence of the protistan lineages: the deepest branches were those of *Euglena* followed by *Trypanosoma* then by *Dictyostelium* and then by ciliates. This finding illustrated an important point, namely that phyla which had long been thought to be primitive such as ciliates because of their peculiar genetic code were in fact of late emergence. Their peculiarities are either losses or derived features.

In the early 90s, the published phylogenetic trees showed a rather nicely-staged emergence of the various protist phyla, but there was relatively little awareness, among molecular biologists, of the reliability of these results. Trees were validated empirically by comparing the topologies obtained by distance methods to those produced by parsimony, or by comparing SSUrRNA results with LSU ones. Congruence between these approaches was taken as an indication of reliability, but insufficient attention was paid to a number of discrepancies obtained between laboratories or even within the same laboratory from the use of different tree construction methods or of different species sampling. In fact these discrepancies were hints that the corresponding portions of the trees were not robust. The bootstrap test (Felsenstein, 1985) – a powerful indicator of the robustness of a given topology (see Sanderson, 1995 and Efron, Halloran and Holmes, 1996 for recent discussions of the significance of bootstrap) – became widely used by 'tree builders' at that time, and helped to reveal the weakness of some nodes.

The major concern, in fact, was to identify the groups branching at the two extremes of the protist tree i.e. those that may have descended from the earliest eukaryotes on the one hand, and those that might constitute the sister-group of metazoans on the other hand (the sister-group of metaphytes had been rapidly identified as corresponding to the chlorophytes). For the earliest protist lineages, the two landmark papers are those of Vossbrinck *et al.* (1987) and Sogin *et al.* (1989). These showed, respectively, that Microsporidia (with *Vairimorpha* as representative) and diplomonads (*Giardia* as representative) were the first lineages to emerge among eukaryotes. These were striking results, especially as the two groups are devoid of mitochondria. Cavalier-Smith (1983) had postulated earlier that they might correspond to the earliest eukaryotic lineages, making the assumption that mitochondrial endosymbiosis occurred after their emergence. The molecular results thus seemed to confirm the notion of an initial stage devoid of mitochondria in the history of the eukaryotic cell and, in general, appeared to fit in a scenario of progressive construction of the eukaryotic cell. Indeed, several authors (Cavalier-Smith, 1987a; Leipe and Hausmann, 1993; Patterson, 1994; Schlegel, 1994; Hausmann and Hülsmann, 1996) were prompt to hypothesize stepwise acquisition of the various organelles typical of the eukaryotic cell, with present day living protists corresponding to lineages derived from these hypothetical ancestors. For example, 'Archamoebae', which lack mitochondria and were thought to have a simplified cytoskeleton, were placed at the very bottom of the tree, although no molecular data were available for them. They were followed by organisms with a more elaborate cytoskeleton, but still no mitochondria or stacked Golgi apparatus; these in turn were followed by organisms with a more conspicuous Golgi and, finally, by organisms having undergone mitochondrial symbiosis, ending in the present day eukaryotic lineages.

As to the other end of the tree, it became apparent that it is very difficult to resolve the order of emergence of the many taxa clustered at that point. The tip of the eukaryotic tree in fact takes the form of a vast radiation known as 'the terminal crown', which contains numerous lineages, including animals, plants, fungi, and various protist groups.

More recent work has suggested that the sister-group of metazoans may be the choanoflagellates, thus confirming another 19th century hypothesis (Wainright *et al.*, 1993).

In summary, by the early nineties, a fairly detailed tree of eukaryotes was available, of the type shown in Figure 2.1, with three amitochondrial protist groups at its base, followed by Euglenozoa (Kinetoplastids such as *Trypanosoma* and Euglenoids such as *Euglena*, forming together a monophyletic group), then by a number of nicely staged protist phyla most of which display an ameboid stage in their life cycle and finally surmounted by a vast terminal 'crown' comprising all the multicellular kingdoms and a number of additional protist phyla. In brief, the tree could be described as comprising three zones: an incompletely resolved base with three phyla, a middle zone with about ten phyla and a vast terminal 'crown'.

In addition to this stepwise emergence of successive phyla, molecular phylogenies also disclosed a number of higher order assemblages, grouping several phyla into monophyletic ensembles, some of which could be expected, based on previous morphological analysis, while others were surprising. Let us cite a few:

- Of the expected groups, the Heterokonta are united on the basis of pigment composition, mastigonemes and the chloroplast endoplasmic reticulum. Molecular phylogeny added to these the non photosynthetic phyla such as oomycetes (Gunderson *et al.*, 1987), hyphochytridiomycetes and labyrinthulids (Leipe *et al.*, 1994), confirming the usefulness of mastigoneme-bearing heterokont flagella as a synapomorphy, irrespective of the presence of chloroplasts. Similarly, euglenids and kinetoplastids were united by molecular data.
- More surprisingly, an ensemble containing the morphologically very diverse ciliates, dinoflagellates and apicomplexans emerged (Gajadhar *et al.*, 1991). Indeed, when looking for a derived morphological character that may be shared by these three phyla, the only one that emerged was the presence of submembranar vesicles, closely apposed to the plasma membrane and known as alveoli in ciliates.
- Another higher level assemblage is that grouping metazoans and fungi together. Here again, there were some early suggestions by Cavalier-Smith (1987b) who even wanted to include the choanoflagellates within the ensemble (Opisthokonta). His proposal was first confirmed using rRNA (Wainright *et al.*, 1993) and shortly afterwards with the use of various protein coding genes (Baldauf and Palmer, 1993).

To some extent, then, the overall picture of the eukaryotic tree simplified into about eight or ten well staged protist phyla occupying the base and the middle zone of the eukaryotic tree and a number of dense super-assemblages constituting the 'terminal crown'. This picture tended

Figure 2.1 Small subunit rRNA phylogenetic tree. About 1000 reliably aligned nucleotides of the full sequences were used. The limits of the domains are given in Philippe and Adoutte (1995). The tree was constructed through the Neighbour-joining distance program (Saitou and Nei, 1987) using only transversions. The bootstrap proportions are computed with the NJBOOT program of the MUST package (Philippe, 1993) and are indicated only when higher than 50%.

to develop into a 'dogma' and was taken by some as an established fact when discussing many aspects of cellular evolution (parallels with palaeontology, origin of cell organelles, origin of meiosis, of sex, of senescence, of introns, of RNA editing, etc.). Doubts of its validity were expressed by some workers, however, especially because changes in species sampling generated differences in topologies. Such a situation is well illustrated in the case of the identification of the earliest eukaryotic branch (Leipe *et al.*, 1993). Here, the discrepancies were attributed to widely different G+C content in the various species used for building the conflicting trees but as we shall see below, this is not the key parameter. Probably the most important factors raising doubts concerning the validity of the rRNA tree resulted from the increased use of protein sequences which showed occasionally major discrepancies with rRNA trees (for example, see Loomis and Smith, 1990). Thus, a growing awareness developed of difficulties and pitfalls in molecular phylogeny.

2.3 PITFALLS OF MOLECULAR PHYLOGENY

A number of potential pitfalls in evolutionary reconstruction are fairly obvious and have been acknowledged by workers in the field from the start. These include: (1) the use of genes that are not strictly homologous (orthologous) in the species under analysis: that is inferring a species tree from comparison of different paralogous versions of the genes, i.e. copies resulting from duplications that have occurred prior to speciation (Fitch, 1970); (2) the use of genes whose rate of evolution is inappropriate for the evolutionary interval under analysis (genes that accumulate too many mutations are difficult to align, genes that are too slow, align well but yield little signal); (3) the use of an insufficient amount of information ('too short sequences'), leading to an overwhelming effect of stochastic factors; (4) the inconsistency of the tree reconstruction method, i.e. the fact that the reconstructed tree could be incorrect even if infinitely long sequences were used (Felsenstein, 1978). However, these problems have often been underestimated probably due to an over-optimistic attitude. Meyer, Cusanovich and Kamen (1986) had shown very early on the high level of saturation of cytochrome c sequences and suggested that ribosomal RNA could be equally saturated. This important, but clearly pessimistic, paper has been almost completely ignored.

Our own work, along with that of others in the field, led us, through a quantitative approach, to focus on artefacts which have been insufficiently taken into account so far: the use of limited species samples, the mutational saturation of sequences and the inequalities in rates of molecular evolution. The importance of the last parameter had already been clearly demonstrated by Felsenstein (1978) and is known as the 'long branch attraction artefact'. This means that species with higher rates of evolution will tend to associate with each other instead of associating with their true sister-group and will also tend to emerge earlier in the tree than their true position. We shall provide below striking examples of this effect. In addition to these pitfalls, which lead to real artefacts, we have emphasized and quantified the difficulty in resolving branching patterns corresponding to short time intervals i.e. to evolutionary radiations.

2.3.1 The importance of species sampling

Practitioners of molecular phylogeny have always endeavoured to include as much information as possible in the databases analysed (i.e. the longest possible amount of sequence, obtained by concatenating sequences from as many different genes as are available). This automatically leads to a decrease in the number of species incorporated in any tree since it is quite rare that abundant data on many different genes are simultaneously available in a large array of species. The evolutionary spectrum of species analysed is also severely limited since abundant data are available only for the major 'model' organisms. A tendency has therefore developed to work on 'four species trees' and, in fact, some of the methods have been developed for such trees. We have shown, however, that such trees can be positively misleading in that they can give high bootstrap values for contradictory topologies depending on the species used as representative of a taxon irrespective of the amount of information used (Lecointre *et al.*, 1993; Philippe and Douzery, 1994). Thus, the message is clear: phylogenies containing very few species should be considered as suspect.

2.3.2 The confounding effects of mutational saturation

DNA sequences are known to undergo multiple substitutions at the same nucleotide sites during the course of evolution. This leads to the phenomenon of mutational saturation of sequences: after a given time span, homologous sequences continue to accumulate mutations in independent lineages but these changes are becoming more and more difficult to detect. Obviously, this will lead to underestimation of the divergence separating pairs of sequences and in such instances, the sequences are said to be saturated with respect to each other. The net result is that the corresponding species on the tree will appear closer to each other than they are in reality. In the extreme case, a set of species whose sequences are totally saturated with respect to each other will emerge in the tree as an unresolved multifurcation erroneously suggesting the occurrence of an evolutionary radiation. We have recently described an extreme example by using the transitions at the third base of the codons of cytochrome b (Philippe and Adoutte, 1996): all the analysed sequences of mammals are almost equally distant, unless they belong to the same species.

We have proposed a simple method to detect mutational saturation by plotting the number of actual differences observed between pairs of species as a function of the inferred number of mutations that have occurred between these sequences. The latter is deduced from parsimony or maximum likelihood reconstruction of the tree containing these sequences (Philippe *et al.*, 1994b). When the sequences under study are saturated, it can be clearly seen that, after a linear start, the curve levels off showing that, after a given threshold, the observed distances increase very little while the inferred ones continue to increase substantially. This is precisely the zone where saturation occurs i.e. the inability to detect additional mutational events. The second message is therefore also quite clear: before using a set of sequences for a given phylogenetic reconstruction, it should be checked that these sequences are not mutationally saturated.

2.3.3 Unequal rates of mutation fixation in the different taxa can lead to their erroneous positioning

As indicated above, whatever tree construction method is used, tree topologies are sensitive to inequalities in the rates at which mutations are fixed in the different taxa analysed. We have noted such effects when studying the general phylogeny of Metazoa using full SSUrRNA (Philippe, Chenuil and Adoutte, 1994a). When using all the sequences available at the time, some clearly aberrant results were obtained, such as the emergence of *Drosophila* and *Aedes* at the base of metazoans, separate from the other arthropods. These are clearly two 'long branch' species and the method used (Neighbour Joining; Saitou and Nei, 1987) was unable to correct fully for this effect and misplaced the two species, although it was able to detect that these species had evolved faster than the others. The use of a more complex model of sequence evolution, such as the maximum likelihood method (Hasegawa and Fujiwara, 1993) and a correction for variable evolutionary rates of sites within the molecule (Van de Peer *et al.* 1996), can reduce but not eliminate this artefact.

In the case of metazoans, a wealth of non-molecular evidence enables us to identify the wrong placement of certain species but that may not be the case for protists, where we have very little, if any, anatomical or palaeontological evidence that can pinpoint an aberrant positioning if whole phyla are concerned. To check for possible inequalities in mutation rate, a simple test was devised in the early days of molecular phylogeny, known as the relative rate test (Wilson, Carlson and White,

1977). This test consists of comparing the respective distances of two taxa, A and B, to a third one, C, known to be an outgroup to A and B. Any inequality in these distances will indicate that mutations have not accumulated at the same rate in the two lineages and will identify the 'fast clock' lineage. In fact, because most presently available distance matrix programs respect inequalities in evolutionary rates among different lineages in a tree, such rapidly evolving lineages will appear in the tree at the end of a 'long branch' and thus will be easily identified. Unfortunately, this test may actually be fooled by one of the artefacts we have discussed.

2.3.4 Saturation can preclude the identification of rapidly evolving sequences

When sequences of a set of species being compared are mutationally saturated, these sequences will all appear to be equidistant from each other: all differences in the real number of substitutions that have occurred between pairs of species go unnoticed because they fall in the 'plateau' of the saturation curve. The phenomenon will be all the more accentuated if the outgroup is very distant from the ingroup species since this increases the probability that the sequences are saturated with respect to those of the outgroup.

We have recently described extreme examples of this problem in the case of the mammalian phylogeny (Philippe and Adoutte, 1996). When using characters which are clearly saturated, e.g. the transitions at the third base of the codons of cytochrome b and the transversions at the same positions, all the branches in a distance tree are of similar length. But, when comparing the molecular distances with divergence dates inferred from palaeontology, the evolutionary rates display differences of at least a factor of ten. These differences begin to appear on a distance tree when using a less saturated part of cytochrome b, e.g. the transversions at the first two bases of the codon. Similarly, a difference in evolutionary rate of the complete mitochondrial genome of a factor five between rodents and carnivores can also be demonstrated (Philippe submitted), although a relative rate test was unable to detect any differences (Janke *et al.*, 1994). As a result, a clock-like behaviour is in fact very often due to the existence of a high level of saturation rather than to the existence of a real molecular clock (Philippe and Laurent, unpublished).

2.3.5 Evolutionary radiations make the determination of branching patterns very difficult

Even if one assumes that all the pitfalls listed above do not occur in a given instance, one is still left with a major and potentially insurmountable difficulty in molecular phylogeny, and one pertaining to the true historical situation: if the taxa under analysis have undergone rapid splitting, sequence data may be insufficient to reconstruct the pattern of cladogeneses. The details of the argument were provided for the case of metazoan phylogeny in relation to the Cambrian radiation (Philippe, Chenuil and Adoutte, 1994a). Briefly, if cladogenetic events have occurred at a rapid pace between the ancestors of a set of extant species, there will have been too little time available for the accumulation of specific mutations between each of the internal nodes separating the successive clades i.e. for those events that produce the molecular synapomorphies allowing the reconstruction of the branching pattern. Thus, instead of yielding a staged phylogeny consisting of successive dichotomies, one will obtain a huge unresolved 'bush', a giant multifurcation. The argument can be made more quantitative and more informative in terms of sequencing effort required to resolve any type of cladogenesis. For this, one uses the bootstrap proportion (BP), a value that is indicative of the support for a given node in a tree (Felsenstein, 1985). One calculates the values of BPs for a given node as a function of the increasing number of nucleotides considered. Supposing, for example, that one has the full 18S rRNA available for a set of species, one can first build a tree with 200 nucleotides, then with 400, 600, etc., and calculate BPs for each of these nodes. Obviously, the longer the sequence utilized, the higher the value of the BP will be, assuming that the node can indeed be resolved.

We have recently modelled the evolution of BP as a function of the sequence length (Lecointre *et al.*, 1994). We found that $BP = 100\ (1 - e^{-b\,x})$, where x is the number of nucleotides and b a parameter specific for each node. If a BP limit is chosen as significant (95%) and if a molecular clock is assumed, a simple relationship is obtained between the number of nucleotides, x, and the time between two speciations, ΔT, (Philippe *et al.*, 1994a): $x = k/\Delta T$ where k is a parameter specific for the given gene and for a given phylogenetic problem. The resolving power of molecular phylogeny can thus be estimated, but relies on the doubtful assumption of a clock-like behaviour.

In the case of the overall metazoan phylogeny (Philippe, Chenuil and Adoutte, 1994a), the complete SSUrRNA allows a reliable resolution only of speciations separated by more than 40 million years (MY). The observed lack of resolution is thus compatible with palaeontological data, which support the idea that the major metazoan phyla appeared during less than 20 MY in the early Cambrian. In general, a hyperbolic relationship is seen between time and the number of nucleotides. Schematically, three zones can be discerned: (1) a zone in which the speciations are well separated and for which few nucleotides can confidently resolve a phylogeny; (2) a zone in which the speciations are close (i.e.

corresponding to the evolutionary radiations) for which the number of nucleotides is huge and inaccessible to experiments; (3) an intermediate zone for which increasing the length of sequences increases significantly the resolving power. For example, in metazoan phylogeny, 500 nucleotides provide a resolution of 140 MY, but 80 000 are required for a resolution of 1 MY.

In summary, no foolproof 'internal' tests exist to validate a given dataset in molecular phylogeny at the moment. The only tests available are, therefore, comparisons between independent datasets and the analysis of their congruence. This is a general problem when dealing with conclusions based on comparative data in biology i.e. with conclusions that cannot be submitted to experimental testing at the bench. That is why comparison of rRNA to protein datasets is so important and, in fact, is one of the major factors that have recently led us to question the rRNA tree.

2.4 HAVE THESE PITFALLS AFFECTED THE PHYLOGENY OF PROTISTS?

With the increasingly large database of complete SSUrRNA and a substantial number of LSU sequences as well, the problem of having a large enough species sample for adequate reconstruction seemed to have been largely overcome. Similarly, an ever expanding database is becoming available for several 'universal' proteins. This has allowed, both for rRNA and for some proteins, the use of variations in species sampling as a test for the reliability of phylogeny (Lecointre *et al.*, 1993). However, all this not only did not solve protistan phylogeny but instead revealed major problems.

2.4.1 Major incongruencies are observed between rRNA and protein trees

Following the rise to prominence of rRNA, a number of protein trees began to appear, re-establishing the link with the first historical molecular trees but now, the protein sequences were deduced from the corresponding genes. These protein trees were received with great anticipation. However, due to the heavy influence that rRNA had acquired as a phylogenetic marker, only those aspects of protein trees were emphasized that were in agreement with the rRNA ones, and discrepancies were explained by *ad hoc* hypotheses. For example, an early paper dealing with actin phylogeny displayed some marked discrepancies with the rRNA one, showing, for example, that *Dictyostelium* emerged much later than on the rRNA tree (Bhattacharya, Stickel and Sogin, 1991). This was interpreted as possibly reflecting peculiarities in actin evolution. By now, more careful examination of the increasingly larger number of the trees based on different proteins has revealed some unreconciliable discrepancies. We shall briefly summarize some of these results and provide some new data.

Substantial databases are now available for several genes coding for widely distributed and well conserved proteins such as α- and β-tubulins, actin, glyceraldehyde-3-phosphate dehydrogenase (GAPDH), elongation factor-1α (EF-1α) and heat shock protein 70 (HSP 70). Before describing some results, the problem of paralogy should be raised. Indeed, in many cases, there are several paralogous genes encoding these housekeeping proteins. The inferred phylogeny will be therefore a gene tree rather than a species tree. Methods for inferring species trees from gene trees are now available (Page, 1994; Page and Charleston, 1997), but we think that their use in the case of the general phylogeny of eukaryotes should be eschewed. In fact, when several genes encode the same protein in a given species, we observed two quite different situations: (1) the sequences were very similar, suggesting a very recent duplication or the existence of an homogenisation process, that allows us to consider all these genes as a single one for purposes of phylogenetic reconstruction; this is the case for ribosomal RNA, even if more divergent copies of it can exist (Carranza *et al.*, 1996; Telford and Holland, 1997); (2) a given sequence showed a low level of identity to the others, but one can assume that this copy corresponds to a recent duplication and has undergone a very high evolutionary rate because in the phylogenetic tree it either represents a very long branch or emerges very early in the tree. That this early emergence is indeed due to the long branch attraction is reinforced by the absence of orthologous sequences of the divergent copy in other species except in closely related ones, thus refuting the hypothesis of an early duplication of the gene. In addition, this divergent gene is often if not always expressed at a very low level, suggesting that it is on its way to become a pseudogene. We thus assume that all these multigenic families evolve almost like a single gene, if one disregards some highly divergent copies.

As a first example, we will analyse a broad sample of β-tubulin genes. Tubulins constitute a multigenic family in most species, which can potentially generate serious problems of paralogy. The problem appears to be reduced for tubulin since, as shown on the tree (Figure 2.2), when all the sequences of a given species are considered, they tend to group together indicating their origin from recent duplications. One exception is the red alga *Porphyra* the sequences of which emerge at the base and at the top of the tree. We interpret this finding as follows: the early emerging sequence is misplaced due to a very high evolutionary rate thus only the late emerging ones reflect the correct evolutionary position of red algae. In view of this, we have not incorporated all

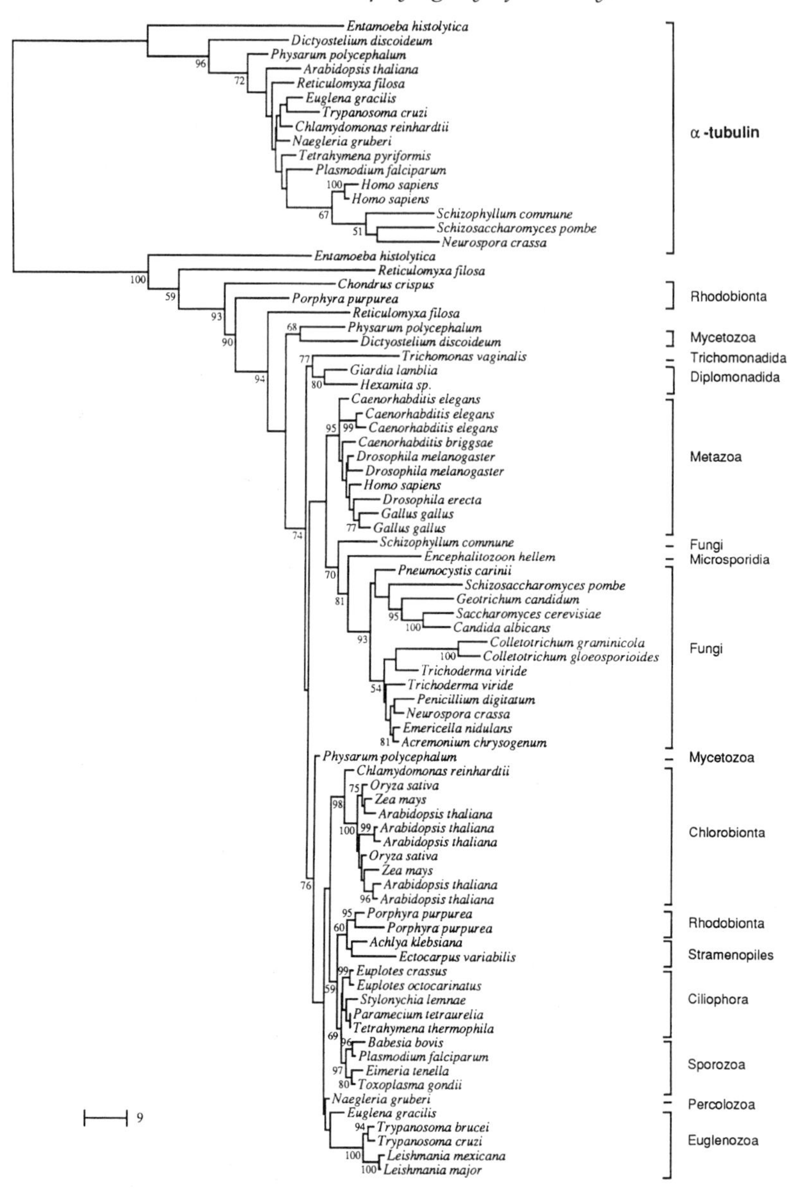

Figure 2.2 β-Tubulin phylogenetic tree. The tree was constructed as follows: the 100 most parsimonious trees were first obtained with the PROTPARS program (Felsenstein, 1993) then the one displaying the highest likelihood among them was identified, using the JTT procedure of the PROTML program in the MOLPHY package (Adachi and Hasegawa, 1992). The bootstrap proportions are computed with the NJBOOT program of the MUST package (Philippe, 1993) by using the Kimura (1983) method and are indicated only when higher than 50%.

available sequences but have tried instead to spread the sampling over as many taxa as possible. The tree was rooted using α-tubulin since α- and β-tubulin are known to derive from an ancient duplication and are still alignable over substantial parts of their sequences.

A number of major incongruencies with respect to the rRNA tree become immediately apparent. As compared with that in the rRNA tree, the order of emergence of many groups has dramatically changed and some of the reversed positions are well supported by the bootstrap test and other measures of robustness. For example, two amoebas now emerge at the base of the tree while diplomonads and trichomonads emerge much later (and as a monophyletic group). Fungi still form the sister-group to Metazoa but a microsporidian is now included among them, very far from a basal position, as already noted by Edlind *et al.* (1996) and Keeling and Doolittle (1996)! Also quite strikingly, the Euglenozoa are now among the 'crown' organisms as a sister group to a ciliate + chlorobiont clade! In short, several groups that were located at the base of the rRNA tree now emerge in the centre or at the top of the tree and, vice-versa, some late emerging groups now occupy a basal position.

Major incongruencies were also noted between rRNA and actin. Interestingly, however, these were not the same as those between rRNA and β-tubulin. An example is shown in Figure 2.3. The tree was rooted using centractin, on the basis of the same reasoning as for the rooting of the tubulin tree. Among the striking incongruencies, one can observe that ciliates are not monophyletic and most of them emerge quite low in the tree i.e. in exactly the opposite position as in the rRNA tree. Polyphyly of ciliates contradicts a wealth of morphological evidence. Obviously, since ciliates are no longer monophyletic, alveolates also have disappeared as a group: most of their representatives emerge rather low in the tree in a para/polyphyletic arrangement. In contrast, a robust monophyletic group comprising most of the 'rhizopods', ameboid organisms like *Acanthamoeba*, *Dictyostelium* and *Physarum*, emerges at the top of the tree, a feature never seen in rRNA trees leading to the conclusion that rhizopods are polyphyletic. A similar result was recently obtained by Baldauf and Doolittle (1997) using elongation factor EF-1α.

When comparing the incongruencies of the tubulin tree with those of the actin tree, some elements of rationalisation can nonetheless be proposed. It appears that in each case, organisms in which the molecule explored is little used emerge at the base of the tree. Conversely, organisms in which the protein is intensely used emerge later in the tree and form some new, unexpected monophyletic groups.

As a result, some of these incongruencies can be accounted for by unusually rapid rates of evolution of the protein within some of the

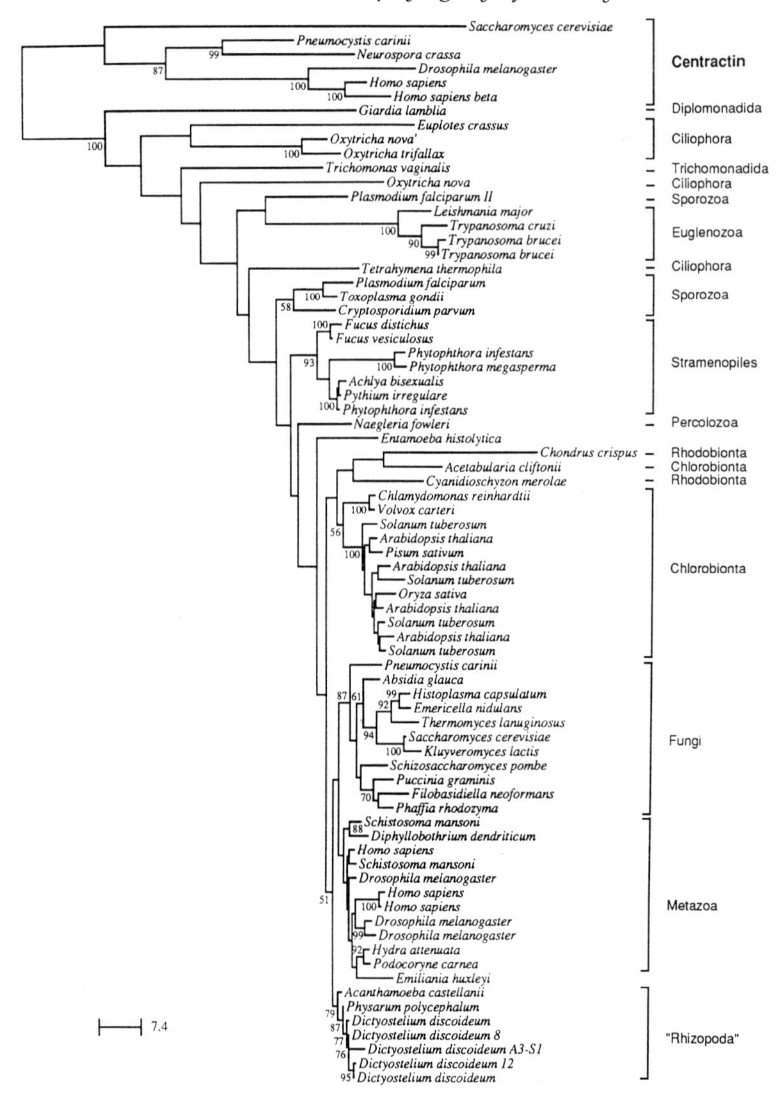

Figure 2.3 Actin phylogenetic tree. The tree was constructed as for β-tubulin (Figure 2.2).

lineages. For example, in the case of tubulin, one can hypothesize that organisms lacking cilia or flagella as well as centrioles have much fewer constraints bearing on the protein. Indeed, an axoneme is made up of over 200 proteins (Piperno *et al.*, 1977), many of which interact with tubulin, thus considerably restricting its degrees of freedom. Adoption of new devices as microtubule organizing centres, as in fungi, rhodophytes and some amoebas releases the tubulin from many of these constraints allowing it to accumulate many more substitutions than in other lineages. This in turn will lead to a well known artefact in molecular phylogeny, the long branch attraction phenomenon (Felsenstein, 1978) emphasized above.

2.4.2 Saturation of molecular markers in the case of eukaryotic phylogeny

However, the species with a suspected high evolutionary rate (actin in ciliates and tubulin in amoebas, for example) do not appear with very long branches on the actual tree. In fact, even the relative rate test does not conclusively establish significant differences in evolutionary rates between branches. As discussed above, this phenomenon could be due to mutational saturation of the genes. Such saturation is all the more likely when the overall phylogeny of eukaryotes is studied with the use of distantly related outgroup (prokaryotes, if rRNA is being used for example or by an anciently duplicated gene such as α-tubulin for a β-tubulin tree). In fact, plotting observed versus inferred number of substitutions for the three molecules analysed in this paper (Figure 2.4) shows that all are mutationally saturated. The phenomenon is more accentuated for β-tubulin and rRNA but is quite noticeable for actin also. This high level of saturation explains why the fast evolving species do not appear as long branches in the inferred tree, but instead emerge very early, due to the attraction of the long branch of the outgroup and the long branch of the fast evolving species. Some of the incongruencies observed in the β-tubulin tree (first emergence of amoebas) and in the actin tree (first emergence and polyphyly of ciliates) can thus be explained. But what is true for actin and tubulin can also be true for rRNA, explaining for example the very early emergence of microsporidia. The high level of saturation of rRNA provides a direct indication for the long suspected idea that the rRNA molecule may have undergone changes in rates of evolution depending on lineages.

2.4.3 rRNA and proteins evolve at highly different rates in different lineages

Saturation prevents the detection of difference in evolutionary rate with the use of the relative rate test because the distance to the outgroup is most subject to underestimation. To avoid this problem, one has to compare more closely related species, with less saturation. Indeed, differences in the rate of evolution of any given lineage can be observed if

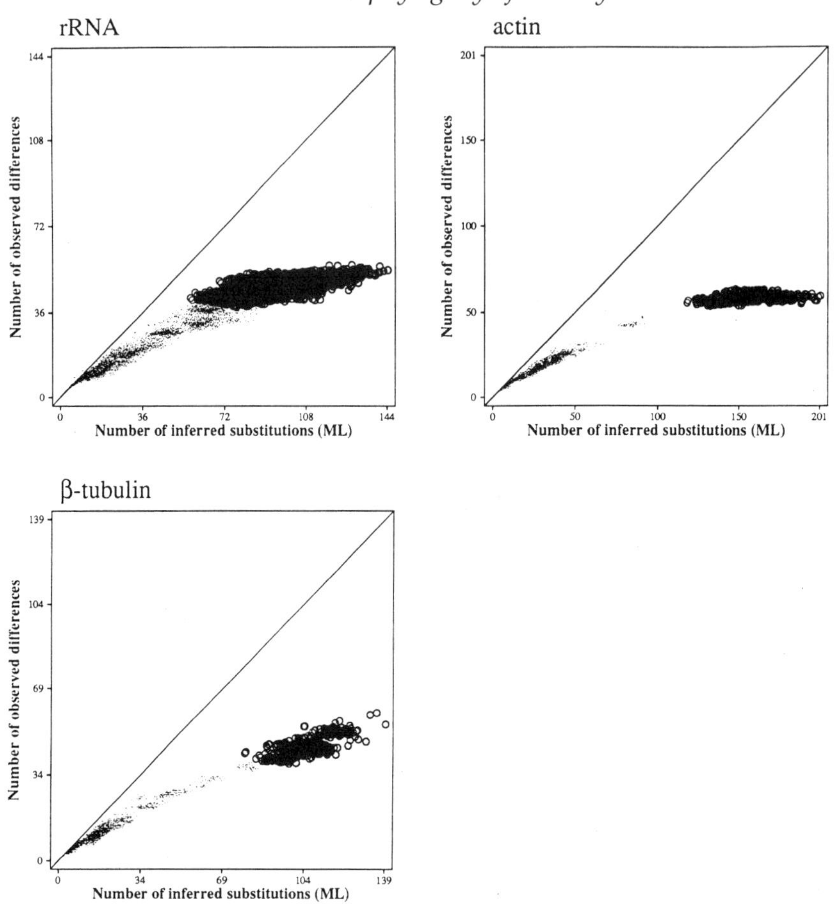

Figure 2.4 Substitution saturation curves. y-axis: the actual number of observed differences between pairs of species sequences. x- axis: the inferred number of substitutions between the same two sequences determined using a maximum likelihood program (Adachi and Hasegawa, 1992). Each dot thus defines the observed over inferred number of substitutions for a given pair of sequences. It can be seen that in the three cases, the curve levels off after a given point, indicating that while the number of inferred mutations still increases (x axis), they are no longer detected as observed differences (levelling along the y axis). Open circles indicate pairwise comparison between outgroup species and ingroup species.

one compares the ratios of distances for the same pair of taxa obtained with two different markers. For example, one can calculate the distance between *Euglena* and *Trypanosoma* (i.e. within Euglenozoa) using rRNA on one hand and β-tubulin on the other, since both sequence pairs are

Table 2.1 Relative evolutionary rates of rRNA, actin and β-tubulin

Group	*rRNA*	*Actin*	*β-tubulin*	*Actin/ rRNA*	*β-tubulin/ rRNA*
Ciliophora	2	41.5	5	21	2.5
Alveolates	5	33	7	5	1.5
Euglenozoa	10.5	23	11	2	1
Metazoa	2.8	7	11	2.5	4
Fungi	1.1	10	26	9	23.5
Chlorobionta	1.3	10	10	8	8
Physarum/Dictyostelium	11	1	21	0.1	2

The figures represent intragroup distances, expressed as per cent of differences between the sequences analysed, obtained for the same taxa using three different molecules (first three columns) and the ratios of the protein distances over the rRNA ones (last two columns).

available for these two species. Similarly, one can carry out the same calculation between *Euplotes* and *Tetrahymena* (i.e. within the Ciliophora). One can then calculate the ratio of the β-tubulin distance to the rRNA one within the two phyla.

The results are summarized in Table 2.1. First, values in the rRNA, actin and β-tubulin columns show considerable differences, both within a column and between parallel columns. For example, in the rRNA column, a distance of 11 is measured within 'Rhizopoda' while a distance of 2 is seen within Ciliophora. This result could be interpreted as reflecting an early emergence of 'Rhizopoda' combined with early divergence of the major lineages within the phylum and a converse phenomenon in ciliates. However, when one looks at the actin column, a very different picture is observed. While great differences are again observed between different phyla, the ones displaying the largest intraphylum distances are not the same as in the rRNA column: the largest distances are observed within Ciliophora and the smallest within the 'Rhizopoda'. If one compares this to the position of Ciliophora and 'Rhizopoda' in the rRNA and actin tree respectively, the phyla showing large distances emerge at the base of the tree (and the opposite for those with short distances).

The very high evolutionary rate of actin within Ciliophora explains not only their basal position in the actin tree but also the other major incongruencies that had been noted such as the lack of monophyly of Ciliophora and the ensuing disappearance of Alveolates. This deep incongruency between rRNA and actin datasets could still be rationalized by assuming that the rRNA picture is correct while actin is aberrant. That this is probably not the case is indicated by a further argument: when the ratios of the distances are examined (last two columns) even more striking differences are observed. For example, in the actin/rRNA column, the ratio is 21 for Ciliophora while it is 0.1 for 'Rhizopoda', i.e. a factor of 200! This cannot be accounted for by variations in actin evolutionary rate only; an amplifying effect of variations in rRNA rate of evolution in the reverse direction of actin must also have occurred.

Such tremendous intragroup and intergroup differences will confound any method of tree reconstruction. This raises serious doubts, for example, in the generally accepted view, based on rRNA, that microsporidians, trichomonads and diplomonads emerge at the base of the eukaryotic tree. In conclusion, therefore, it appears that none of the presently available trees can be trusted over very large evolutionary distances since the major parameters determining the topology of the tree are the lineage-dependent differences in rate of evolution of the molecule under study.

2.4.4 The results can be accounted for by a radiation followed by unequal rates of evolution

There is in fact a simple way to rationalize all the results just described. If the major phyla of eukaryotes had diversified within a relatively short time interval and if the rate of evolution of a given gene had been different among the resulting clades, one would obtain an apparently resolved tree in which rapidly evolving taxa are at the base and slowly evolving ones at the top. And a different topology would be obtained for another gene if the rates of evolution of this gene differed among the different taxa from the previous one. A schematic representation of this hypothesis is shown in Figure 2.5: the apparently resolved trees inferred for the different genes and the contradictions observed between them can now be rationalized as being the result of a broad radiation followed by lineage dependent differences in the constraints on the different molecules.

A completely independent piece of evidence suggesting the existence of strong biases in the various trees discussed up to now resides in their lack of 'balance', i.e. of symmetry. Mooers *et al.* (1995) have shown that a high level of homoplasy present in characters used for phylogenetic reconstruction leads to 'unbalanced trees', i.e. trees in which successive clades, instead of appearing as the result of successive dichotomies, yielding similar numbers of taxa on each side of successive nodes, appear in a paraphyletic disposition, all on one side of the basal node. Precisely this is observed with all the molecular markers used so far and in an especially striking manner with rRNA: the base of the tree is

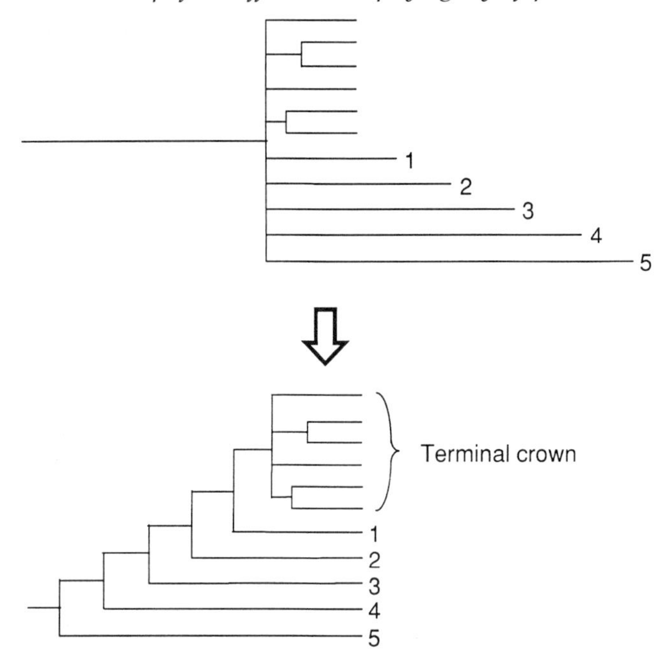

Figure 2.5 How does a pseudo-resolved phylogeny arise? This interpretative scheme shows how, starting from an evolutionary radiation (upper part), branches of unequal length (i.e. undergoing unequal rates of molecular evolution) can yield an apparently well resolved tree (lower part): lineages corresponding to the 'faster' branches will be artefactually displaced to the bottom of the tree and this will occur in proportion to their length.

very asymmetric whereas the top of the tree is rather symmetric. As a result, the bases of the three eukaryotic phylogenies discussed here (Figures 2.1–2.3) are highly questionable, in contrast to their upper parts which probably reflect reality, particularly the 'crown' of the rRNA tree.

Finally, it should be recalled that one of the 'triumphs' of the initial rRNA trees was the discovery that a number of amitochondriate taxa emerged at the base of the tree. This appeared to provide strong support for the tree since it could be easily rationalized by assuming that mitochondrial symbiosis had occurred later than the emergence of the eukaryotic cell. This conclusion was in good accordance with the notion of increasing complexity of the eukaryotic cell. However, some other amitochondriate protists, like *Entamoeba*, emerge later in the tree, suggesting a secondary loss of mitochondria in these phyla. This was confirmed when Clark and Roger (1995) found nuclear genes of clear mitochondrial origin in *Entamoeba*, confirming this hypothesis. This approach for demonstrating a secondary loss of mitochondria has now been applied to the three first emerging phyla.

Several groups (Bui, Bradley and Johnson, 1996; Germot, Philippe and Le Guyader, 1996; Horner *et al.*, 1996; Roger, Clark and Doolittle, 1996) have recently detected in trichomonads both cpn60 and HSP70 genes. Not only were eubacterial-type sequences corresponding to these proteins identified, but phylogenetic analysis showed these to branch within the mitochondrial cluster. Thus, trichomonads, although lacking mitochondria today, have had them in the past. Similar results have been obtained in diplomonads by immunological methods for detecting cpn60 (Soltys and Gupta, 1994) and by sequence analysis for triose phosphate isomerase (Keeling and Doolittle, 1997), and for HSP70 in microsporidia (Germot, Philippe and Le Guyader, 1997). That hydrogenosomes of trichomonads may be derived from mitochondria was in fact suggested by Cavalier-Smith (1987a) but the primitively amitochondrial nature of microsporidia and diplomonads was, in contrast, broadly accepted. In the framework of the radiation hypothesis, this simply means that mitochondrial symbiosis occurred prior to the radiation.

Thus, not only the available gene sequences lack support for the accepted rRNA tree, but what could have been taken as an independent evidence in favour of it must also be reconsidered.

2.5 A BIG-BANG IN EUKARYOTIC EVOLUTION?

There is no way to reconcile the protein trees with the rRNA one. In fact, the marked differences in rates of evolution shown above and the demonstration that these differences appear to be the major factor determining the placement of a phylum within the tree, show that any attempt to reconcile these datasets is futile. However, if the rRNA and protein trees are broken into two parts, one basal and asymmetric and the other apical and symmetric, a simple reconciliation appears. The basal part is artefactual due to the long branch attraction phenomenon whereas the apical part correctly reflects a huge radiation of eukaryotes (see Figure 2.5). As a result, no single gene can give a perfect picture of the eukaryotic phylogeny, which has to be inferred by a consensus approach exclusively from the symmetric parts of the trees. Obviously, some genes are better for this purpose than others. In the case studied here, if the criterion used is the discrepancy of evolutionary rates, β-tubulin is better than rRNA, itself better than actin (see Table 2.1), but if the criterion is the number of informative sites, the rRNA is better than β-tubulin, itself better than actin. In fact, our consensus phylogeny (Figure 2.6) is somewhat closer to the rRNA tree (Figure 2.1) than to the tubulin tree (Figure 2.2). The actin tree (Figure 2.3), although differs

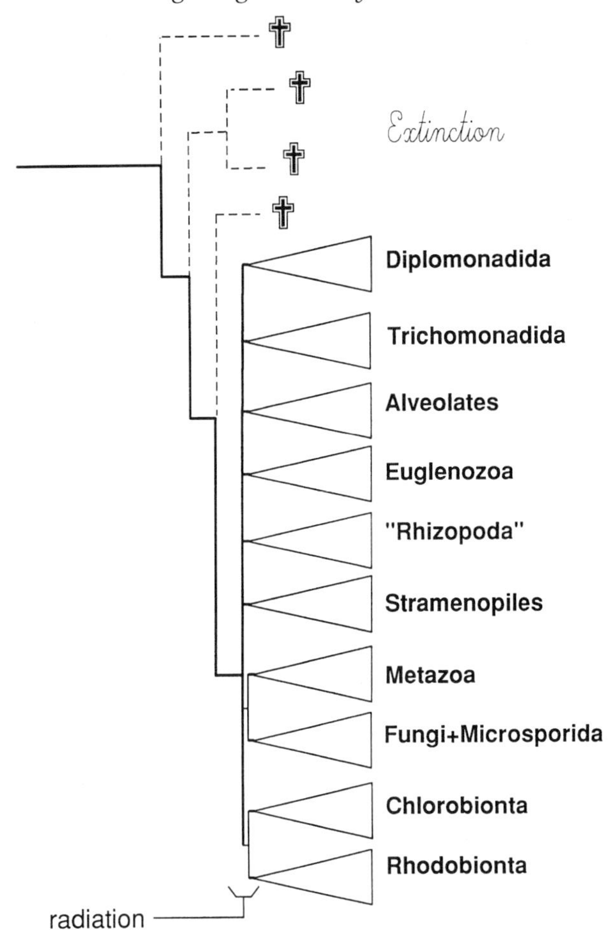

Figure 2.6 The 'big bang' hypothesis. Extant eukaryotes are assumed to have arisen during a period of rapid diversification, perhaps correlated with the establishment of mitochondrial symbiosis. Earliest (possibly anaerobic) groups have not left living representatives. A number of high level monophyletic groups can still be identified in the 'crown', such as Alveolates or Stramenopiles, but their phylogenetic relationships to the other groups are difficult to establish. Only two even higher level groups are suggested, that uniting fungi and animals and that uniting Rhodobionta and Chlorobionta

more from the consensus, as is expected from the small number of informative sites and the great variability of evolutionary rates (Table 2.1), also provides valuable insights (particularly on the monophyly of the 'rhizopods').

In summary, a number of monophyletic groups emerge on the basis of molecular data. These are usually confirmed by anatomical synapomorphies which, despite their small number, are of great quality. However, as noted in the beginning of this paper, it is very difficult, or even impossible, to find nested sets of synapomorphies, which would enable us to draw a completely resolved tree of eukaryotes in the form of successive bifurcations. It appears, therefore that neither at the level of macromolecular sequences nor at the level of ultrastructural characters can evidence for a regularly staged diversification of eukaryotes be found. One may therefore start wondering whether the evolutionary history of eukaryotes has been as gradual as we generally think evolution is. Although a number of leading evolutionists such as Simpson (1944) and Gould (1989) have, for a long time, stressed the importance of evolutionary radiations and 'quantum evolution', and although this notion has even been extended previously to the diversification of the major protist lineages (Cavalier-Smith, 1978), the prevailing view among biologists has been that of the graduality of evolution.

In their quest for 'resolved trees', biologists have avoided facing the obvious possibility that the diversification of **extant** eukaryotes may have occurred during a relatively short time interval. This is why we fail to find both morphological and molecular synapomorphies. We are not suggesting that the eukaryotic cell as a whole has been 'invented' at that time; indeed, we think that this radiation was preceded by an extended period of development but that most, if not all, descendants of this early period have disappeared. As schematized in Figure 2.6, what we identify today are only the descendants of a burst in diversification that occurred within one of the early lineages. During this burst, sufficient synapomorphies, both molecular and morphological, accumulated within some of the individual lineages to allow their identification as monophyletic groups.

Instead of taking this conclusion as a failure, we should try to understand its implications on the very mechanisms of eukaryotic evolution. If indeed a vast radiation occurred in the ancestry of extant eukaryotes, what could its basis be? One hypothesis, first suggested by Margulis (1970), is to relate it to mitochondrial symbiosis: as emphasized above, all extant protists, including amitochondriate ones, appear to descend from an ancestor that had experienced mitochondrial symbiosis. We assume, on the basis of all currently available molecular and palaeontological evidence, that this was a relatively recent event, which occurred possibly as late as 700 to 1000 million years ago, at the time when oxygen tension in the atmosphere had reached its present level (Canfield and Teske, 1996). The new respiring organism may have had one considerable advantage over its predecessors, i.e. a dramatic gain in energetic efficiency achieved from the use of oxygen as the terminal electron acceptor. We may even go one step further and assume that this advantage was so marked as to lead to the disappearance of all amitochondrial ancestors. That is why we have so far detected only descendants from the mitochondrial big bang. Alternatively, the 'big bang' may have been generated by the acquisition of phagotrophy, as first suggested by Stanier (1970) but we feel the evidence favours the mitochondrial symbiosis scenario.

Is there any hope of finding approaches that may nevertheless lead to further refinement of this tree? As described above, traditional molecular phylogenetic methods would require an unrealistic number of nucleotides to resolve this radiation confidently. We think that in fact these refinements will emerge from a few well chosen anatomical or biochemical similarities which, when analysed at the molecular level, will turn out to be true homologies. An encouraging example is provided by the epiplasm. In morphological terms, this is a semi-rigid cytoskeletal layer underlying the cell membrane in ciliates, dinoflagellates and euglenids with an important role in defining the shape and mechanical properties of the cell cortex, and seems to be immunologically related in these groups (Vigues *et al.*, 1987). According to the rRNA tree (see Figure 2.1), the two taxa bearing this membrane skeleton are very distantly related. This was first interpreted as a convergent feature imposed by the constraints on free-living single cells deprived of a rigid cell wall. Now, sequences of epiplasmic proteins have been obtained for both *Euglena* (Marrs and Bouck, 1992) and the ciliate *Pseudomicrothorax* (Huttenlauch *et al.*, 1995) and they show striking similarity. It is difficult to interpret these similarities in any way other than as reflecting evolutionary homology. This, in turn, fits well with a clustering of Euglenozoa with ciliates that is observed (on a weak basis) using β-tubulin (see Figure 2.2) and EF-1α sequences (Adoutte *et al.*, 1996). The epiplasm might therefore represent a morphological and molecular synapomorphy that unites some high level taxa (Ciliophora and Euglenozoa) into still higher ensembles. If this kind of approach is developed further, the apparently unresolved tree of eukaryotes with a high number of unrelated phyla may progressively be transformed into a partially resolved tree comprising a small number of 'super-ensembles'. It is still too early to determine how many of these large ensembles will emerge but we venture to suggest that the number will be small, possibly of the order of four or five. What is left for future work is the identification of similarly informative characters!

Acknowledgements

We are grateful to the editors for giving us the opportunity to present this work at the meeting on evolutionary relationships among Protozoa.

We thank Miklos Müller for critical reading of the manuscript and Tom Cavalier-Smith for many suggestions and corrections. The ideas expanded in this manuscript have profited from fruitful discussions with Agnès Germot, Hervé Le Guyader, Miklos Müller and Andrew Roger.

2.6 REFERENCES

Adachi, J. and Hasegawa, M. (1992) *MOLPHY: Programs for Molecular Phylogenetics. I. PROTML: Maximum Likelihood Inference of Protein Phylogeny*, Institute of Statistical Mathematics, Tokyo.

Adoutte, A., Germot, A., Le Guyader, H. and Philippe, H. (1996) Que savons-nous de l'histoire évolutive des Eucaryotes ? 2. De la diversification des protistes à la radiation des multicellulaires. *Médecine/Sciences*, **12**, 1–17.

Baldauf, S.L. and Doolittle, W.F. (1997) Origin and evolution of the slime molds (Mycetazoa). *Proceedings of the National Academy of Sciences USA*, **94**, 12007–12.

Baldauf, S.L. and Palmer, J.D. (1993) Animals and fungi are each other's closest relatives: Congruent evidence from multiple proteins. *Proceedings of the National Academy of Sciences USA*, **90**, 11558–62.

Bhattacharya, D., Stickel S.K. and Sogin M. (1991) Molecular phylogenetic analysis of actin regions from *Achlya bisexualis* (Oomycota) and *Costaria costata* (Chromophyta). *Journal of Molecular Evolution*, **33**, 525–36.

Brugerolle, G. (1991) Flagellar and cytoskeletal systems in amitochondrial flagellates: Archamoebae, Metamonada and Parabasala. *Protoplasma*, **164**, 70–90.

Bui, E.T.N., Bradley, P.J. and Johnson, P.J. (1996) A common evolutionary origin for mitochondria and hydrogenosomes. *Proceedings of the National Academy of Sciences USA*, **93**, 9651–6.

Canfield, D.E. and Teske, A. (1996) Late proterozoic rise in atmospheric oxygen concentration inferred from phylogenetic and sulphur-isotope studies. *Nature*, **382**, 127–32.

Carranza, S., Giribet, G., Ribera, C. *et al.* (1996) Evidence that two types of 18S rDNA coexist in the genome of *Dugesia (Schmidtea) mediterranea* (Platyhelminthes, Turbellaria, Tricladida). *Molecular Biology and Evolution*, **13**, 824–32.

Cavalier-Smith, T. (1975) The origin of nuclei and of eukaryotic cells. *Nature*, **256**, 463–8.

Cavalier-Smith, T. (1978) The evolutionary origin and phylogeny of microtubules, mitotic spindles and eukaryotic flagella. *BioSystems*, **10**, 93–114.

Cavalier-Smith, T. (1981) Eukaryotic kingdom: seven or nine? *BioSystems*, **14**, 461–81.

Cavalier-Smith, T. (1983) A 6-kingdom classification and a unified phylogeny, in *Endocytobiology II* (eds W. Schwemmler and H.E.A. Schenk), De Gruyter, Berlin, pp. 1027–34.

Cavalier-Smith, T. (1987a) Eukaryotes with no mitochondria. *Nature*, **326**, 332–3.

Cavalier-Smith, T. (1987b) The origin of Fungi and pseudofungi, in *Evolutionary Biology of the Fungi* (eds A.D.M. Rayner, C.M. Brasier and D. Moore), Cambridge University Press, Cambridge, pp. 339–53.

Clark, C. G. and Roger, A. J. (1995) Direct evidence for secondary loss of mitochondria in *Entamoeba histolytica. Proceedings of the National Academy of Sciences USA*, **92**, 6518-21.

Corliss, J.O. (1984) The kingdom Protista and its 45 phyla. *BioSystems*, **17**, 87–126.

Edlind, T.D., Li, J., Visvesvara, G.S. *et al.* (1996) Phylogenetic analysis of β-tubulin sequences from amitochondrial protozoa. *Molecular Phylogenetics and Evolution*, **5**, 359–67.

Efron, B., Halloran, E. and Holmes, S. (1996) Bootstrap confidence levels for phylogenetic trees. *Proceedings of the National Academy of Sciences USA*, **93**, 7085–90.

Felsenstein, J. (1978) Cases in which parsimony or compatibility methods will be positively misleading. *Systematic Zoology*, **27**, 401–10.

Felsenstein, J. (1985) Confidence limits on phylogenies: An approach using the bootstrap. *Evolution*, **39**, 783–91.

Felsenstein, J. (1993) PHYLIP Manual version 3.5. University Herbarium, University of California, Berkeley, California.

Fitch W.M. (1970) Distinguishing homologous from analogous proteins. *Systematic Zoology*, **19**, 99–113.

Fitch, W.M. and Margoliash, E. (1967) Construction of phylogenetic trees. A method based on mutation distances as estimated from cytochrome c sequences is of general applicability. *Science*, **155**, 279–84.

Gajadhar, A.A., Marquardt, W.C., Hall, R. *et al.* (1991) Ribosomal RNA sequences of *Sarcocystis muris, Theileria annulata* and *Crypthecodinium cohnii* reveal evolutionary relationship among apicomplexans, dinoflagellates, and ciliates. *Molecular and Biochemical Parasitology*, **45**, 147–54.

Germot, A., Philippe, H. and Le Guyader, H. (1996) Presence of a mitochondrial-type 70-kDa heat shock protein in *Trichomonas vaginalis* suggests a very early mitochondrial endosymbiosis in eukaryotes. *Proceedings of the National Academy of Sciences USA*, **93**, 14614–17.

Germot, A., Philippe, H. and Le Guyader, H. (1997) Evidence for loss of mitochondria in microsporidia from a mitochondrial-type HSP70 in *Nosema locustae. Molecular and Biochemical Parasitology*, **87**, 159–68.

Gould, S. J. (1989) *Wonderful Life: the Burgess Shale and the Nature of History*. Norton, New York.

Gunderson, J.H., Elwood, H., Ingold, A. *et al.* (1987) Phylogenetic relationships between chlorophytes, chrysophytes, and oomycetes. *Proceedings of the National Academy of Sciences USA*, **84**, 5823–7.

Hasegawa, M. and Fujiwara, M. (1993) Relative efficiencies of maximum likelihood, maximum parsimony, and neighbor joining methods for estimating protein phylogeny. *Molecular Phylogenetics and Evolution*, **2**, 1–5.

Hausmann, K. and Hülsmann, N. (1996) *Protozoology*, 2nd edn, Georg Thieme Verlag, Stuttgart.

Horner, D.S., Hirt, R.P., Kilvington, S. *et al.* (1996) Molecular data suggest an early acquisition of the mitochondrial endosymbiont. *Proceedings of the Royal Society of London Series B-Biological Sciences*, **263**, 1053–9.

Huttenlauch, I., Geisler, N., Plessmann, U. *et al.* (1995) Major epiplasmic proteins of ciliates are articulins: Cloning, recombinant expression, and structural characterization. *Journal of Cell Biology*, **130**, 1401–12.

Janke, A., Feldmaier-Fuchs, G., Thomas, W.K. *et al.* (1994) The marsupial mitochondrial genome and the evolution of placental mammals. *Genetics*, **137**, 243–56.

Keeling, P.J. and Doolittle, W.F. (1996) Alpha-tubulin from early-diverging eukaryotic lineages and the evolution of the tubulin family. *Molecular Biology and Evolution*, **13**, 318–38.

Keeling, P.J. and Doolittle, W.F. (1997) Evidence that eukaryotic triosephosphate isomerase is of alpha-proteobacterial origin. *Proceedings of the National Academy of Sciences USA*, **94**, 1270–5

Kimura, M. (1983) *The Neutral Theory of Molecular Evolution*. Cambridge University Press, Cambridge.

Kivic, P.A. and Walne, P.L. (1984) An evaluation of a possible phylogenetic relationship between the Euglenophyta and Kinetoplastida. *Origins of Life*, **13**, 269–88.

Lecointre, G., Philippe, H., Lê, H.L.V. and Le Guyader, H. (1993) Species sampling has a major impact on phylogenetic inference. *Molecular Phylogenetics and Evolution*, **2**, 205–24.

Lecointre, G., Philippe, H., Lê, H.L.V. and Le Guyader, H. (1994) How many nucleotides are required to resolve a phylogenetic problem? The use of a new statistical method applicable to available sequences. *Molecular Phylogenetics and Evolution*, **3**, 292–309.

Lee, J.J., Hutner, S.H. and Bovee, E.C. (1985) *An Illustrated Guide to the Protozoa*. Society of Protozoologists, Lawrence, Kansas.

Leipe, D.D., Gunderson, J.H., Nerad, T.A. and Sogin M.L. (1993) Small subunit ribosomal RNA$^+$ of *Hexamita inflata* and the quest for the first branch in the eukaryotic tree. *Molecular and Biochemical Parasitology*, **59**, 41–8.

Leipe, D. and Hausmann, K. (1993) Neue Erkenntnisse zur Stammesgeschichte der Eukaryoten. *Biologie in unserer Zeit*, **23**, 178–83.

Leipe, D.D., Wainright, P.O., Gunderson, J.H. *et al.* (1994) The stramenopiles from a molecular perspective: 16S-like rRNA sequences from *Labyrinthuloides minuta* and *Cafeteria roenbergensis*. *Phycologia*, **33**, 369–77.

Loomis, W.F. and Smith, D.W. (1990) Molecular phylogeny of *Dictyostelium discoideum* by protein sequence comparison. *Proceedings of the National Academy of Sciences USA*, **87**, 9093–7.

Margulis, L. (1970) *Origin of Eukaryotic Cells*. Yale University Press, New Haven, Connecticut.

Margulis, L., Corliss, J.O., Melkonian, M. and Chapman, D.J. (1990) *Handbook of Protoctista*. Jones & Barlett Publishers, Boston.

Marrs, J.A. and Bouck, G.B. (1992) The two major membrane skeletal proteins (articulins) of *Euglena gracilis* define a novel class of cytoskeletal proteins. *Journal of Cell Biology*, **118**, 1465–75.

Meyer, T.E., Cusanovich, M.A. and Kamen, M.D. (1986) Evidence against use of bacterial amino acid sequence data for construction of all-inclusive phylogenetic trees. *Proceedings of the National Academy of Sciences USA*, **83**, 217–20.

Mooers A.O., Page, R.D.M., Purvis A. and Harvey P.H. (1995) Phylogenetic noise leads to unbalanced cladistic tree reconstructions. *Systematic Biology*, **44**, 332–42.

Page, R.D.M. (1994) Maps between trees and cladistic analysis of historical associations among genes, organisms, and areas. *Systematic Biology*, **42**, 58–77.

Page, R.D.M. and Charleston M.A. (1997) From gene to organismal phylogeny: reconciled trees and the gene tree/species tree problem. *Molecular Phylogenetics and Evolution*, **7**, 231–40.

Patterson, D.J. (1994) Protozoa: evolution and systematics, in *Progress in Protistology* (eds. K. Hausmann and N. Hülsmann), Gustav Fischer Verlag, Stuttgart-Jena, pp. 1–14.

Philippe, H. (1993) MUST: a computer package for management utilities for sequences and trees. *Nucleic Acids Research*, **21**, 5264–72.

Philippe, H. and Adoutte, A. (1995) How reliable is our current view of eukaryotic phylogeny? in *Protistological Actualities* (eds. G. Brugerolle and J.-P. Mignot), Clermont-Ferrand, pp. 17–33.

Philippe, H. and Adoutte, A. (1996) What can phylogenetic patterns tell us about the evolutionary processes generating biodiversity? in *Aspects of the Genesis and Maintenance of Biological Diversity* (eds. M.E. Hochberg, J. Clobert and R. Barbault), Oxford University Press, Oxford, pp. 41–59.

Philippe, H., Chenuil, A. and Adoutte, A. (1994a) Can the Cambrian explosion be inferred through molecular phylogeny? *Development 1994*, suppl., 15–25.

Philippe, H. and Douzery, E. (1994) The pitfalls of molecular phylogeny based on four species, as illustrated by the Cetacea/Artiodactyla relationships. *Journal of Mammalian Evolution*, **2**, 133–52.

Philippe H., Sörhannus U., Baroin A. *et al.* (1994b) Comparison of molecular and paleontological data in diatoms suggests a major gap in the fossil record. *Journal of Evolutionary Biology*, **7**, 247–65.

Piperno, G., Huang, B. and Luck, D.J.L. (1977) Two dimensional analysis of flagellar proteins from wild-type and paralyzed mutants of *Chlamydomonas reinhardtii*. *Proceedings of the National Academy of Sciences USA*, **74**, 1600–4.

de Puytorac, P., Grain, J. and Mignot, J.P. (1987) *Précis de Protistologie*. Soc. Nouvelle Ed. Boubée, Paris.

Ragan, M.A. and Chapman, D.J. (1978) *A Biochemical Phylogeny of the Protists*. Academic Press, London, pp. 1–317.

Roger, A.J., Clark, C.G. and Doolittle, W.F. (1996) A possible mitochondrial gene in the early-branching amitochondriate protist *Trichomonas vaginalis*. *Proceedings of the National Academy of Sciences USA*, **93**, 14618–22.

Saitou, N. and Nei, M. (1987) The neighbor-joining method: a new method for reconstructing phylogenetic trees. *Molecular Biology and Evolution*, **4**, 406–25.

Sanderson, M.J. (1995) Objections to bootstrapping phylogenies: a critique. *Systematic Biology*, **44**, 299–320.

Schlegel, M. (1994) Molecular phylogeny of eukaryotes. *Trends in Ecology & Evolution*, **9**, 330–5.

Simpson, G.G. (1944) *Tempo and Mode of Evolution*. Columbia University Press, New York.

Sleigh, M.A. (1979) Radiation of the eukaryote Protista, in *The Origin of Major Invertebrate Groups* (ed. M.R. House), Systematics Association Special Vol. No 12, Academic Press, London, pp. 23–54.

Sogin, M.L., Elwood, H.J. and Gunderson, J.H. (1986) Evolutionary diversity of eukaryotic small-subunit rRNA genes. *Proceedings of the National Academy of Sciences USA*, **83**,1383–7.

Sogin, M.L., Gunderson, J.H., Elwood, H.J. *et al.* (1989) Phylogenetic significance of the kingdom concept: an unusual eukaryotic 16S-like ribosomal RNA from *Giardia lamblia. Science*, **243**, 75–7.

Soltys, B.J. and Gupta, R.S. (1994) Presence and cellular distribution of a 60-kDa protein related to mitochondrial HSP 60 in *Giardia lamblia. Journal of Parasitology*, **80**, 580–90.

Stanier, R.Y. (1970) Some aspects of the biology of cells and their possible evolutionary significance. *Symposium of the Society of General Microbiology*, **20**, 1–38.

Taylor, F.J.R. (1978) Problems in the development of an explicit hypothetical phylogeny of the lower eukaryotes. *BioSystems*, **10**, 67–89.

Telford, M. J. and Holland P.W.H. (1997) Evolution of 28S ribosomal DNA in chaetognaths: duplicate genes and molecular phylogeny. *Journal of Molecular Evolution*, **44**, 135–44.

Van de Peer, Y., Rensing, S.A., Maier, U.-G. and De Wachter, R. (1996) Substitution rate calibration of small subunit ribosomal RNA identifies chlorarachniophyte endosymbionts as remnants of green algae. *Proceedings of the National Academy of Sciences USA*, **93**, 7732–6.

Vigues, B., Bricheux, G., Metivier, C. *et al.* (1987) Evidence for common epitopes among proteins of the membrane skeleton of a ciliate, an euglenoid and a dinoflagellate. *European Journal of Protistology*, **23**, 101–10.

Vossbrinck, C.R., Maddox, J.V., Friedman, S. *et al.* (1987) Ribosomal RNA sequence suggests microsporidia are extremely ancient eukaryotes. *Nature*, **326**, 411–14.

Wainright, P.O., Hinkle, G., Sogin, M.L. and Stickel, S.K. (1993) Monophyletic origins of the Metazoa: an evolutionary link with fungi. *Science*, **260**, 340–2.

Wilson, A.C., Carlson, S.S. and White, T.J. (1977) Biochemical evolution. *Annual Review of Biochemistry*, **46**, 573–639.

Woese, C.R. (1987) Bacterial evolution. *Microbiological Reviews*, **51**, 221–71.

Woese, C.R. and Fox, G.E. (1977) Phylogenetic structure of the prokaryotic domain: the primary kingdoms. *Proceedings of the National Academy of Sciences USA*, **74**, 5088–90.

Zuckerkandl, E. and Pauling, L. (1965) Evolutionary divergence and convergence in proteins, in *Evolving Genes and Proteins* (eds V. Bryson and H.J. Vogel), Academic Press, New York, pp. 97–166.

3

Evolutionary relationships between protist phyla constructed from LSUrRNAs accounting for unequal rates of substitution among sites

Nicolas J. Tourasse and Manolo Gouy

Laboratoire de Biométrie, Génétique et Biologie des Populations - UMR CNRS 5558, Université Claude Bernard, 43 Boulevard du 11 Novembre 1918, 69622 Villeurbanne Cedex France. E-mail: mgouy@biomserv.univ-lyon1.fr

ABSTRACT

In the sequences of functional genes, the rate of substitution varies across sites. We propose here a new method for estimating evolutionary distances between nucleotide sequences that takes into account this variation and estimates it from the data. Starting from the phylogeny built from usual distances, the maximum-parsimony algorithm is used to infer the number of changes that happened at each sequence site. Then, an 'invariant + truncated negative binomial' distribution, which allows for invariant sites, is fitted to the observed distribution of changes. We show that this distribution agrees very well with real data sets. By incorporating this model into the nucleotide substitution models of Jukes and Cantor and of Kimura, two new distance estimates that detect many more substitutions than the classical ones were developed. The use of these new distances to reconstruct the phylogeny of eukaryotes from LSU rRNA sequences, gave a tree in which Euglenozoa branch within higher eukaryotes, rather than at the base of the eukaryotic tree as in phylogenies built using classical distances. Moreover, these protists appeared to have evolved at a much faster rate than did

Evolutionary Relationships Among Protozoa. Edited by G.H. Coombs, K. Vickerman, M.A. Sleigh and A. Warren. Published in 1998 by Chapman & Hall, London. ISBN 0 412 79800 X

higher eukaryotes, suggesting that a deep phylogenetic position may be artefactual and due to long branch attraction. Accounting for among-site rate variation seems to reduce this attraction and is thus a very important issue in the framework of protozoan phylogeny. These results, like those derived from recent analyses of protein sequences, challenge the currently accepted identification of early-emerging protozoan phyla.

3.1 INTRODUCTION

In functional genes such as those encoding ribosomal RNAs and proteins, the substitution rate varies over sites (Uzzell and Corbin, 1971; Holmquist *et al.*, 1983; Sullivan, Holsinger and Simon, 1995; Van de Peer, Van den Auwera and De Wachter, 1996b). This is probably due to the fact that in such molecules, the selective and functional constraints are not the same in different parts of the molecule, depending on the role of these regions in the activity of the molecule. Several studies and simulation experiments have revealed that ignorance of among-site rate variation leads to underestimation of transition/transversion rates (Wakeley, 1994) and of evolutionary distances and branch lengths, and that these underestimations are stronger for very distantly related sequences (Jin and Nei, 1990; Tateno, Takezaki and Nei, 1994). These problems can affect drastically tree reconstruction with distance matrix methods (Olsen, 1987; Van de Peer *et al.*, 1996b).

Many authors have shown that a gamma distribution seems to be an appropriate distribution to model the variation of the substitution rate over sites (Uzzell and Corbin, 1971; Yang, 1994; Yang, Goldman and Friday, 1994). Thus, this distribution has been introduced in evolutionary distance estimates (Nei and Gojobori, 1986; Jin and Nei, 1990). The gamma distribution is characterized by a shape parameter which is inversely proportional to the extent of variation. In order to compute corrected distances, one must estimate the value of this shape parameter from the data. For this, different approaches have been used (for review, see Yang, 1996). A parsimony-based approach is used here because it allows a very large number of sequences to be processed (Sullivan, Holsinger and Simon, 1995; Yang and Kumar, 1996).

However, it has been known for a long time that some sequence positions do not vary between species and that the number of substitutions inferred through sequence comparisons can be seriously underestimated if these invariable sites are ignored (Palumbi, 1989; Shoemaker and Fitch, 1989; Gu, Fu and Li, 1995; Lockhart *et al.*, 1996). In this report, we present a more realistic model in which a fraction θ of nucleotide sites are allowed to be invariable while a truncated negative binomial distribution models the distribution of the number of substitutions among variable sites. The first aim of this article is to describe this new model and the process of data fitting. Using different sets of small subunit ribosomal RNA (SSUrRNA) and protein coding sequences we will show that this model can fit real data very well. We have included in the nucleotide substitution models of Jukes and Cantor (1969) and of Kimura (1980) the 'invariant + truncated negative binomial' model of rate variation among sites and derived corresponding evolutionary distance expressions. The detailed description of how to compute these distances is the second purpose of this article.

Using progressively larger data sets (up to 1000 SSUrRNA sequences), we also show that the values of the shape parameter and of the proportion of invariant sites vary greatly with the number of species under consideration. However, the new evolutionary distance estimates are relatively stable regardless of the parameter values, except for small data sets. These new distances turned out to be useful to detect multiple changes in sequences.

By means of the new method we have reconstructed the phylogeny of the eukaryotic domain using large subunit (LSU) rRNAs and focused on the position of protozoan organisms. In the phylogeny constructed with usual distance estimates, several protozoan lineages appear to branch deeply in the eukaryotic tree with some relatively long internal branches separating early- and late-emerging phyla. Using distance estimates that account for among-site rate variation, Euglenozoa appear to branch more recently in the eukaryotic LSU rRNA tree. Moreover, most internal branches separating eukaryotic phyla are quite short showing that the eukaryotic (and particularly protozoan) phylogeny is far from being resolved.

3.2 DATA AND METHODS

3.2.1 Data sets

A comprehensive set of aligned small subunit ribosomal RNA (SSUrRNA) sequences was extracted from the rRNA database at Antwerp (Van de Peer *et al.*, 1996a). This set comprises species belonging to all three domains of life, Bacteria, Archaea and Eukarya (Woese, Kandler and Wheelis, 1990) and was used to construct universal phylogenies. A set comprising all available eukaryotic large subunit rRNA (LSU rRNA) sequences was also extracted from the same bank and served to reconstruct the phylogeny of eukaryotes.

Partial sequences, duplicate sequences from the same species, and those containing many undetermined nucleotides were removed from the data sets. We eliminated also the regions where the alignment was

Table 3.1 Analyzed sets of protein encoding genes

Gene description	*Number of sequences**	*Number of nucleotides*
H+ ATP synthase, noncatalytic subunit	74	504
Translation elongation factor EF-1α(Tu)	106	594
Translation elongation factor EF-2(G)	30	939

*All sequences were extracted from the Genbank database.

ambiguous. Gap-containing sites were generally excluded, but to avoid excessive reduction of the total number of sites, some isolated gaps were replaced by undetermined nucleotides (Ns). This treatment left 898 and 1508 positions for phylogenetic analysis for a universal set of 1000 SSUrRNA sequences and a eukaryotic set of 39 LSU rRNA sequences, respectively.

Eight subsets comprising 20, 30, 50, 100, 300, 500, 700, and 1000 SSUrRNA sequences have been constructed by randomly selecting sequences from the preceding subset. These samples were made so that representatives from each of the three domains of life were present in each subset. The same 898 positions were used whatever the subset size, so that results obtained with each subset are comparable.

Three protein genes sequenced in the three domains of life were analysed (Table 3.1). Only nonsynonymous substitutions, i.e. DNA changes which induce amino acid changes in the corresponding protein, were considered.

3.2.2 Parsimony analysis of the data sets

For each data set described above, a 4-step parsimony-based analysis of the substitution rate variation among sites was conducted: (1) Construction of a phylogeny using Kimura's two-parameter distance (that does not take into account rate variation among sites) followed by the Neighbour-Joining tree-building method; (2) Evaluation of the previous phylogeny by the maximum-parsimony method to count the minimum number of changes required at each site along this tree; (3) Fitting of the 'invariant + truncated negative binomial' model to the distribution of the number of substitutions per site computed in the second step. This gives the estimated values of the model parameters; (4) Computation of the new corrected distances (that take into account rate variation among sites). We now detail each step.

3.2.3 Construction of the starting phylogeny

For each set of sequences, evolutionary distances between sequence pairs were computed using the two-parameter distance of Kimura (1980). For protein coding genes, numbers of nonsynonymous sites and of nonsynonymous transitions and transversions were computed according to the method of Nei and Gojobori (1986). For rRNA sequences, all positions where an ambiguity character (N, R, Y) was present in one of the two sequences under consideration were ignored. A tree topology was then inferred by the Neighbour-Joining method as modified by Studier and Keppler (1988).

3.2.4 Estimation of the number of substitutions

For rRNA sequences, the minimum number of substitutions required at each site to explain the tree topology was computed using the User-tree option of the DNAPARS program of the PHYLIP Package (Felsenstein, 1993). Undetermined nucleotides were taken into account since the parsimony algorithm allows for such ambiguities. For nucleotide sequences of protein genes, the maximum-parsimony method was used to infer the number of nonsynonymous substitutions for each codon by applying the User-tree option of the PROTPARS program of the PHYLIP Package. In computing this number, nonsynonymous changes at the three positions of the codons have been considered. To do this, we have slightly modified the PROTPARS program, designed for protein sequences, so that it could read nucleotide sequences. The distribution of the number n_k of sites (codons) showing k substitutions (nonsynonymous substitutions) was then obtained.

3.2.5 Fitting of the model to the observed distribution of substitutions

The gamma distribution is characterized by a shape parameter α and a scale parameter β and its probability density is

$$f(\lambda) = \frac{\lambda^{\alpha-1} e^{-\lambda/\beta}}{\beta^{\alpha}\Gamma(\alpha)} \ (\geqslant 0,\ \alpha > 0,\ \beta > 0),$$

where l is the substitution rate per site. Assuming that, for any given site (or codon), the probability of sustaining a change remains constant over time, the number of changes occurring at this site follows a Poisson distribution with parameter $>$. Moreover, if $>$ varies according to a gamma distribution over sites (or codons), the number of substitutions per site (or codon) follows a compound Poisson $\times$ gamma distribution which is a negative binomial with shape parameter α and scale parameter β (Johnson and Kotz, 1969, pp. 124–25 and 184).

To take into account sites which have not been observed to vary between all species, our model further assumes that a proportion q of sites do not sustain substitutions. Other sites, representing a proportion 1-q, are variable and undergo substitutions that are distributed following a truncated negative binomial distribution, i.e., a negative binomial distribution from which the probability that zero substitution occurs is removed (Johnson and Kotz, 1969, p. 136). The observed distribution of the number *k* of substitutions per site (or codon) is then modelled by the following discrete distribution:

$$\begin{cases} p(0) = \theta, \\ p(k) = (1-\theta)\dfrac{1}{1-(\beta+1)^{-\alpha}}\dbinom{\alpha+k-1}{k}\left(\dfrac{\beta}{\beta+1}\right)^{k}\left(\dfrac{1}{\beta+1}\right)^{\alpha}, \quad k \geqslant 1 \end{cases} \quad (1)$$

This model contains three parameters: θ, the proportion of invariant sites, α the shape parameter and β the scale parameter of the truncated negative binomial distribution, respectively. α and β describe the variability of the substitution rate among sites. α is inversely related to the extent of variation of the substitution rate among variable sites: the lower its value, the stronger the heterogeneity. A high value of θ (i.e. a great proportion of invariable sites) indicates that substitutions are concentrated on a small number of sites. β is a scale parameter allowing the mean of the expected distribution to be adjusted to the mean of the observed one.

To fit the model to the data, we employed the maximum-likelihood method: assuming that substitutions occur independently at each site, the log-likelihood function is

$$L = \sum_{k\geqslant 0} n_k \ln[p(k)],$$

where n_k is the observed number of sites experiencing *k* changes (inferred by maximum-parsimony) and *p(k)* is the expected probability that a given site undergoes *k* substitutions (computed from Equation 1 above).

Because the observed proportion of invariant sites maximizes the likelihood function (proof not given), parameter θ was estimated by its observed value. The estimates of α and β were obtained by maximizing numerically the likelihood function using a convergent Newton-Raphson-like algorithm. Furthermore, the maximum-likelihood estimates of α and β are such that the means of the expected and observed distributions are equal (Johnson and Kotz, 1969, pp. 132 and 136-37).

3.2.6 Computation of evolutionary distances

The evolutionary distance between two sequences is the mean number of substitutions that have occurred per site since their divergence. The classical formulae of Jukes and Cantor (1969) and of Kimura (1980) rely on the hypothesis that the substitution rate does not vary over sites. We have developed two new distance estimates assuming that the number of substitutions per site follows the 'invariant + truncated negative binomial' distribution. To derive them, we have introduced this distribution into the nucleotide substitution models of Jukes and Cantor (Appendix, Equation 6) and Kimura (Equation 7), respectively.

3.3 RESULTS

3.3.1 Comparison of observed and expected distributions

Some examples of observed and expected (i.e. fitted) distributions computed for differents sets of rRNA and protein-coding sequences are presented in Figure 3.1. These graphs clearly show that, in each case, the fitted distribution agrees very well with the observed one, despite the relative heterogeneity of the latter distribution. This good fit indicates therefore that among-site rate variation does occur in the genes we have analysed. A very satisfactory point is that this agreement exists for distributions that have a great variety of shapes. Figure 3.1 shows also that the 'invariant + truncated negative binomial' model can serve to fit distributions of numbers of substitutions per site for rRNA sequences as well as distributions of numbers of substitutions per codon for protein coding genes.

3.3.2 Effect of the number of species

Parameters α and θ describe the heterogeneity of the substitution process among sites. Figure 3.2a presents the evolution of α and θ with respect to the number of sequences contained in the eight universal subsets described above. The results are very clear: the parameter values are highly dependent on the number of sequences employed to estimate them. However, when a very large number of sequences is used, the values of α and θ tend to stabilize and one can assess that each will converge towards an asymptotic value.

3.3.3 Estimates of evolutionary distances

We next studied the dependence of evolutionary distances on the number of sequences used to estimate the distribution of rates of evolution. Figure 3.2b displays the behaviour of a set of Kimura-like distances with respect to the number of species from which parameters α, β and θ were estimated. In contrast to parameter values, evolutionary distances are much more stable, though for small data sets, low dis-

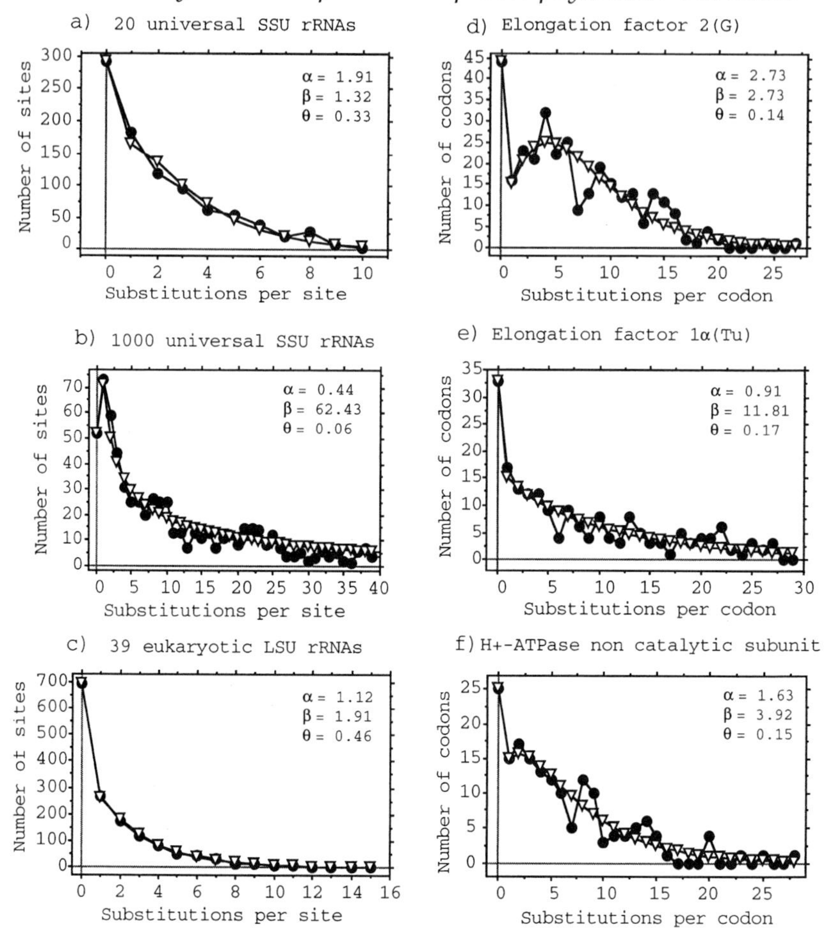

Figure 3.1 Fit of the 'invariant + truncated negative binomial' model to several data sets. Each graph displays the distribution of the number of substitutions per site (per codon for protein-coding) given by the model (open triangles) as fitted to the observed distribution inferred by parsimony (filled circles). Estimated values of the three model parameters are given in inset. Panels (a), (b) and (c) show three distinct rRNA data sets. Only the first 40 points out of the full 225-point distribution are shown in panel (b). Panels (d), (e) and (f) show three distinct protein-coding data sets. Only the first 30 points out of the full 50-point distribution are shown in panel (e).

tances are somewhat underestimated and high distances somewhat overestimated. Indeed, for a given pair of sequences, the distance estimated from 20 sequences does not differ greatly from that computed with the parameters estimated using 1000 sequences, whereas α and θ

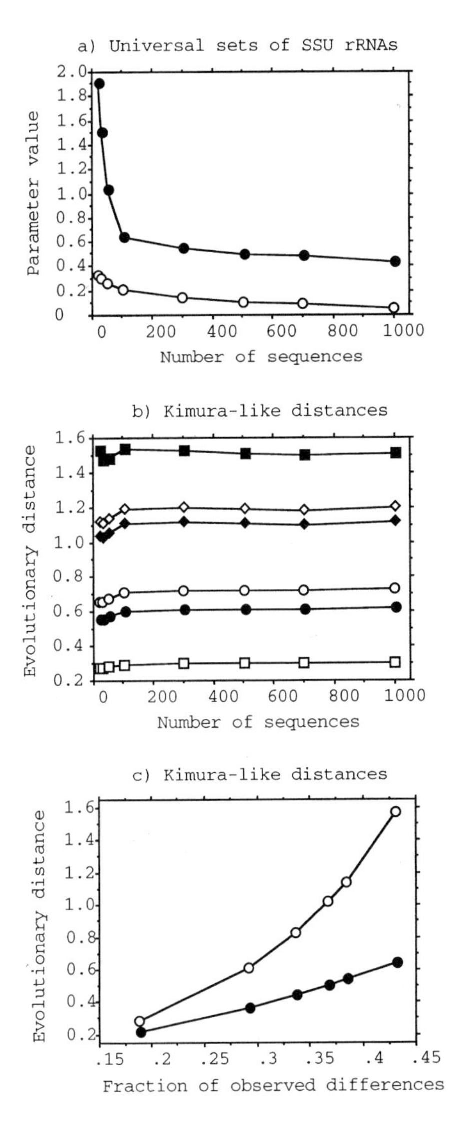

Figure 3.2 (a) Estimates of the α (filled circles) and θ (open circles) parameters as functions of the number of sequences in the universal sets of SSU rRNAs. (b) Variation of the new Kimura-like distances (Appendix, Equation 7) as functions of the number of sequences. The distances, expressed as numbers of substitutions per site, were computed on a sample of four SSU rRNA sequences belonging to all three domains of life, giving thus six pairwise distances, each represented by a different symbol. (c) Comparison of the usual distances of Kimura (filled circles) with the new distances (open circles), as functions of the observed divergence between two sequences. The distances were computed on a sample of SSU rRNAs from species belonging to all three domains of life. Corrected distances were obtained with parameter values $\alpha = 0.437$, $\beta = 62.426$, $\theta = 0.058$ estimated from a set of 1000 sequences.

values are very different (compare Figures 3.2a and 3.2b). Furthermore, this relative stability is true for low as well as for high distances. Identical findings have been obtained with the Jukes and Cantor-like estimates (results not shown).

Another important point is that the new distances are always higher than the classical estimates of Jukes and Cantor and of Kimura, as illustrated in Figure 3.2c for Kimura-like estimates only. This indicates again that ignoring among-site rate variation when it exists, leads to severe underestimation of evolutionary distances. As expected, this underestimation becomes larger as the distance becomes greater.

3.3.4 Application: phylogeny of eukaryotic LSU rRNAs

We applied the 4-step analysis described above on the set of 39 LSU rRNA sequences and reconstructed the phylogeny of Eukaryotes. For better readability, only 23 sequences are displayed in Figure 3.3. Figure 3.3a displays the 'starting' tree built by means of the Neighbour-Joining method applied to the transversional part of the classical distance of Kimura (1980). Use of transversions only allows problems due to differences in base compositions between sequences to be reduced. In this tree, Euglenozoa, as well as several other protists, branch off very early. In contrast (Figure 3.3b), Euglenozoa branch more recently and are grouped with Fungi in the 'corrected' eukaryotic tree. This tree was built using the transversional part of the new Kimura-like distances (Appendix, Equation 10) computed with parameter values estimated from the full 39-species starting tree, ($\alpha = 1.12$, $\beta = 1.91$, $\theta = 0.46$; observed and expected distributions presented in Figure 3.1c). Thus, accounting for among-site rate variation does alter reconstruction of protozoan phylogeny. The incorrect position of the two insect sequences (*Drosophila* and *Aedes*) in Figure 3.3a is repaired in Figure 3.3b where all animal sequences are clustered as expected. Finally, internal branches connecting the ancestors of Euglenozoa, Fungi, Plants, and Metazoa in the corrected tree are very short and supported by low bootstrap values (Figure 3.3b).

3.4 DISCUSSION

In this report, we used the maximum-parsimony method to infer the number of substitutions per site in sequences related by a tree and we proposed a composite model to represent the variation of this number over sites. This model was applied to the number of substitutions per site in nucleotide sequences of genes that do not code for protein, such as ribosomal RNAs, as well as to the number of amino acid-changing substitutions per codon in protein coding nucleotide sequences. This model permits that some sites (or codons) in the sequences do not change, and assumes that evolutionary changes are distributed according to a truncated negative binomial distribution.

The fitting of the model to real data has shown that it can agree very well with a great variety of observed distributions, i.e. exponential-like or bell-shaped distributions. The flexibility of the model is due to the fact that the negative binomial distribution is composed of two parameters: parameter α determines the shape of the distribution whereas β is a scaling parameter. Thus, we conclude that this model is of general use.

Furthermore, the fact that the 'invariant + truncated negative binomial' model can fit real data suggests that it is somewhat realistic. In particular, the introduction of invariant sites seems to be a very important point: the shape of some observed distributions like those presented in Figures 3.1b, 3.1d and 3.1f do not correspond at all to shapes that can be generated by a simple negative binomial. To fit these distributions, it is clearly necessary to add invariable sites. Indeed, Uzzell and Corbin (1971) and Holmquist *et al.* (1983), who fitted simple negative binomial distributions to observed data, had to exclude some invariable sites from the distributions to maximize the fit.

As shown by Wakeley (1993), the use of the parsimony method can lead to an underestimation of the actual number of changes that occurred at each site because this approach does not allow for multiple substitutions, and this underestimation is larger for the most variable sites. Wakeley demonstrated by simulations that this bias induced overestimates of the shape parameter of the negative binomial distribution. However, the most variable sites have the lowest probabilities of occurrence (Figure 3.1), so that they contribute weakly to the total likelihood and therefore they have no major influence on parameter estimation. Furthermore, we expect that the bias due to the parsimony approach becomes weaker as the number of sequences examined becomes larger because, in a large data set, a great number of intermediate sequences ensures a better estimation of the mutability of each site (see also below).

By using SSUrRNA data sets of different sizes, we have shown (Figure 3.2a) that the shape parameter α and the proportion of invariable sites θ are strongly linked to the number of species that have been used to estimate them. More precisely, α and θ decrease when the number of species increases. This can be interpreted in terms of the genetic diversity of the data set. When new sequences are added, the genetic diversity of the data set increases and one notices that (1) some sites are in fact variable whereas they appeared as invariable in the smaller data set, this explaining why θ decreases and, (2) among variable sites, some appear to have sustained more changes than inferred with a smaller number of species because of the lack of inter-

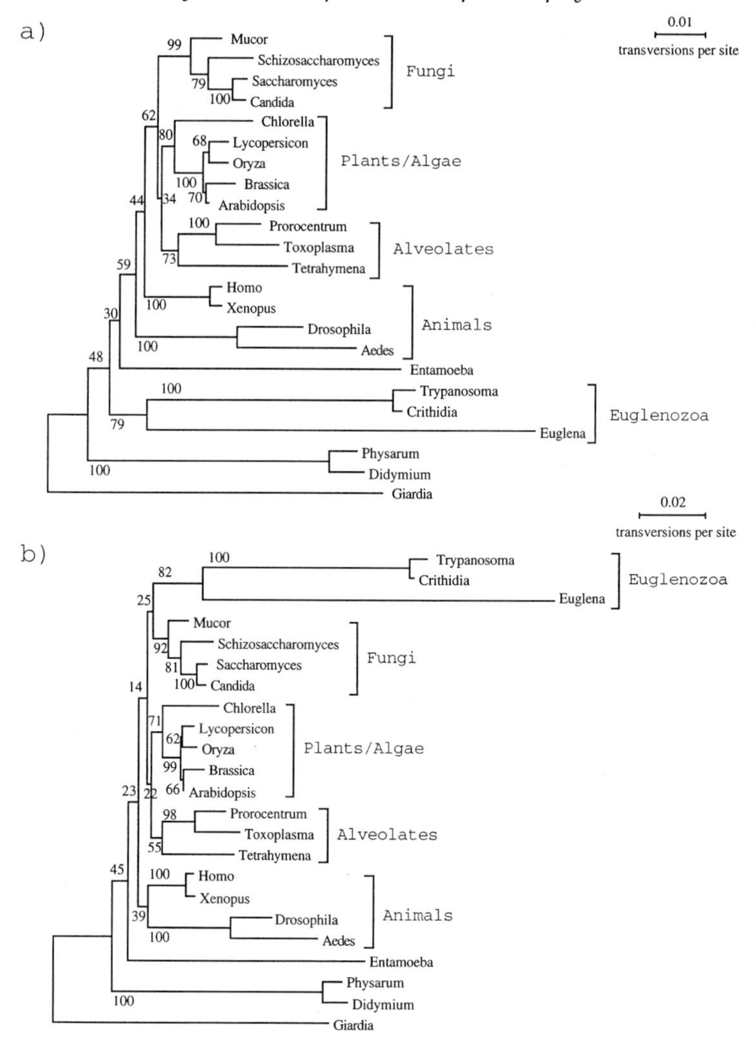

Figure 3.3 Phylogeny of Eukaryotes reconstructed with LSU rRNA sequences, using 1508 homologous sites. Trees depicted here are simplified trees constructed with a representative subset of 23 sequences chosen among the 39 currently available. The overall topologies are conserved between the 23- and 39-species trees. Tree a is the 'starting' tree built from the usual distances of Kimura. Tree b is the 'corrected' tree built using the new Kimura-like distances computed with parameter values estimated from the full 39-species starting tree, i.e., $\alpha = 1.115$, $\beta = 1.912$, $\theta = 0.458$. In each case, only transversions were considered. Both topologies have been obtained by applying the Neighbour-Joining algorithm to the corresponding distance matrices. The corrected tree can be used as a starting tree to reiterate the whole tree-building procedure (parsimony

mediate sequences. Thus, the extent of variation of the substitution rate increases, and therefore the value of parameter α decreases with the number of sequences. When a very large number of sequences is used, the variability contained in the data set gets closer to the actual variability and the parameters tend therefore to stabilize.

Given the above results, we conclude that it is not possible to use the values of α and θ estimated by parsimony to characterize the heterogeneity of the substitution process in a given molecule, unless a very large number of sequences is studied.

In contrast, the evolutionary distance estimates presented here depend only moderately on the number of sequences used (Figure 3.2b). This fact is essential in the framework of phylogenetic inferences because it allows these distances to be used to build phylogenetic trees from a moderate number of sequences. Our method works very well if a large number ($\geqslant 100$) of sequences are considered, but reasonable distance estimates can be obtained with at least 20 sequences. The two new distances have been derived by introducing the 'invariant + truncated negative binomial' model into the nucleotide substitution models of Jukes and Cantor and of Kimura. They should be particularly suitable for constructing phylogenies that include distantly related sequences because it is in such sequences that the misleading effects of rate variation among sites and multiple substitutions are strongest (Jin and Nei, 1990; Tateno, Takezaki and Nei, 1994; Yang and Roberts, 1995).

We have shown (Figure 3.2c) that the new distances detect many more substitutions than do the classical formulae of Jukes and Cantor (1969) and of Kimura (1980). The reason for this is that the new method takes into account the numerous changes that have occurred at fast-evolving sites.

Application of the new method to eukaryotic LSU rRNA sequences suggests that euglenozoan protists branch more recently in the eukaryotic radiation than in the tree reconstructed by means of classical distance estimates (Figure 3.3). This finding is in line with the results of Hashimoto *et al.* (1995) who constructed eukaryotic trees using the maximum-likelihood method applied to the amino acid sequences of elongation factors 1α and 2 and found that Euglenozoa cluster with higher eukaryotes, including Animals, Fungi, and Plants. We also built

---

computation of the number of changes at each sequence site, estimation of parameter values, distance computations, and NJ tree-building) yielding a third tree. This third tree is identical to the second, corrected tree. Numbers in front of nodes are the bootstrap proportions, out of 1000 replicates. Trees are arbitrarily rooted on the branch leading to *Giardia*

trees using SSUrRNA sequences by means of the new method and observed that the placement of Euglenozoa varies according to species samples. In some cases, Euglenozoa emerge early, that is, as in Figure 3.3a, whereas in other cases, they emerge later than Metazoa (results not shown). This phylogenetic instability may be partly due to the reduced number of sites suitable for analysis, which is twice smaller for SSU than for LSU rRNA sequences. Indeed, analysis of larger sequences constructed by concatenating the SSU and LSU molecules, gives a tree in which Euglenozoa emerge late, within the eukaryotic tree 'crown' (results not shown).

In a rooted tree, any internal node represents a single point in time. Consequently, long branches emerging from a node denote high rates of molecular evolution in the corresponding lineage. In all rRNA-derived phylogenies, most early-emerging Protozoa are connected by very long branches. Thus these lineages seem to have evolved more rapidly than other eukaryotes. Therefore, the question arises of whether these Protozoa are genuine early-emerging organisms or are placed at the base of phylogenetic trees by the long branch attraction artefact (Olsen, 1987). Application of the new method of evolutionary distance estimation changed the eukaryotic tree in two instances, for Euglenozoa and for insects. Because the true position of insects is known to be that of Figure 3.3b rather than that of Figure 3.3a, and because these species are situated on longer branches than other animals (*Homo* and *Xenopus*), we observe that the new method succeeds in avoiding the long branch attraction effect for these two sequences. The true evolutionary origin of Euglenozoa is not known *a priori*, but we conclude that their late emergence (Figure 3.3b) is more accurate than their early emergence (Figure 3.3a) as also found by Hashimoto *et al.* (1995) using elongation factors. We also note that distinct molecules (e.g. rRNA and elongation factors) may yield incongruent evolutionary trees with standard tree-building procedures not because of true differences in evolutionary histories but because of tree-building artefacts.

Accounting for among-site rate variation weakens the long branch attraction through better estimation of evolutionary distances and thus branch lengths (Figure 3.2c), as also shown by Van de Peer, Van den Auwera and De Wachter (1996b) with other eukaryotic phyla. For example, the mean distance between Euglenozoa and Physarales (*Physarum* and *Didymium*) computed on LSU rRNAs with the new method is 0.420 substitution/site while Kimura's formula gives a much smaller number, 0.261 substitution/site. Therefore, allowing for rate variation across sites is very useful in the framework of eukaryotic phylogeny, particularly for studying the phylogenetic position of Protozoa.

The bootstrap values of Figure 3.3 may suggest that both trees are equally unstable, but this is not so. If the two insect sequences are removed from Figure 3.3a, a bootstrap value of 89% separates the tree crown from early-emerging *Entamoeba*, Euglenozoa, Physarids, and *Giardia*. Moreover, 94% of bootstrap replicates of Figure 3.3a place Euglenozoa and *Giardia* outside the tree crown when the positions of insects, *Entamoeba* and Physarids are ignored. Therefore very high bootstrap values are disproved by application of the new tree-building method which accounts for rate variation among sites. The 'corrected' tree is supported by weak bootstrap values showing that the radiation of the tree 'crown' remains poorly resolved.

The late emergence of some protists in our 'corrected' LSU and combined SSU + LSU rRNA trees tends to challenge the current view of protozoan evolutionary relationships deduced from molecular phylogeny (in which several protozoan phyla are considered as early emerging in the eukaryotic radiation) as did recent analyses of protein coding genes, notably β-tubulin (Philippe and Adoutte, Chapter 2). The fact that in our LSU tree internal branches separating the origins of most eukaryotic phyla are very short and poorly supported by bootstrap resampling (Figure 3.3b) suggests that eukaryotic phyla might have emerged during a relatively short period. In conclusion, protozoan and more generally, eukaryotic phylogenies, appear currently as far from being resolved.

The different steps of the present analysis were performed using three computer programs written in the C language. They are available upon request.

3.5 APPENDIX

3.5.1 Jukes and Cantor-like distance

According to Jukes and Cantor (1969), if all substitutions occur at rate > per unit time, the probability that two homologous sites exhibit the same nucleotide in two sequences having diverged t time units ago is,

$$q(\lambda, t) = \frac{1}{4} + \frac{3}{4} e^{-\frac{8\lambda t}{3}} \qquad (2).$$

The distribution of the number of substitutions per site in the whole tree is modelled here by the distribution p(k) given in Equation 1 above. The mean of this distribution, $\overline{\lambda}_{tree}$, equals the observed average number of substitutions per site in the whole tree. For a given pair of sequences, the distribution of the number of substitutions per site between them has an unknown mean, $\overline{\lambda}_{pair}$, such that at a site where k substitutions occurred in the whole tree, the expected number of substitutions between the two members of the pair is

$$\lambda_k = k \frac{\bar{\lambda}_{pair}}{\bar{\lambda}_{tree}}$$

Therefore, the expected proportion $\bar{Q}$ of identical nucleotides between the two sequences of the pair is

$$\bar{Q} = \sum_{k=0}^{\infty} q(\lambda_k, t).\ p(k) \tag{3}$$

$$= \sum_{k=0}^{\infty} \left(\frac{1}{4} + \frac{3}{4} e^{-\frac{8k\bar{\lambda}_{pair}t}{3\bar{\lambda}_{tree}t}} \right) p(k)$$

$$= \frac{1}{4} + \frac{3}{4}\theta + \frac{3}{4} \sum_{k=1}^{\infty} p(k) e^{-\frac{8k\bar{\lambda}_{pair}t}{3\bar{\lambda}_{tree}}} \tag{4}$$

Using the generating function of the negative binomial distribution (Feller, 1962, p. 253), one can obtain:

$$\forall x, |x| \leqslant 1, \quad \sum_{k=1}^{\infty} p(k)x^k = \left(\frac{1-\theta}{V-1} \right) \left[\left(1 - \left(\frac{\beta}{1+\beta} \right) x \right)^{-\alpha} - 1 \right] \tag{5}$$

where

$$V = (1+\beta)^{\alpha}$$

By definition, the evolutionary distance d between two sequences is the expected mean number of changes that have occurred per site since divergence of the sequences, i.e., $d = 2\bar{\lambda}_{pair}t$. Therefore, replacing t by $d/(2\bar{\lambda}_{pair})$ in Equation 4 and using Equation 5, gives

$$d = -\frac{3}{4} \bar{\lambda}_{tree} \ln \left[\left(\frac{1+\beta}{\beta} \right) (1 - (Z+1)^{-1/\alpha}) \right] \tag{6}$$

with

$$Z = (V-1)\left(1 - \frac{4(1-\bar{Q})}{3(1-\theta)} \right)$$

For nucleotide sequences, $\bar{Q}$ is estimated by the observed proportion of identical bases in the two sequences. For protein-coding genes, it is the complement to 1 of the proportion of nonsynonymous substitutions at nonsynonymous sites, computed using Nei and Gojobori's (1986) method.

3.5.2 Kimura-like distance

An analogous distance based on the two-parameter model of Kimura (1980), which distinguishes transitions from transversions, can be derived. According to this model, the proportions of transitional (u) and transversional (v) differences between two sequences having diverged t time units ago are

$$u(a, b, t) = \frac{1}{4} - \frac{1}{2}e^{-4(a+b)t} + \frac{1}{4}e^{-8bt}$$

and

$$v(b, t) = \frac{1}{2} - \frac{1}{2}e^{-8bt},$$

respectively. Here, a and b are the rates of transitional and transversional substitution per site per unit time, and the total substitution rate per site is $\lambda = a + 2b$. $\bar{\lambda}_{pair}$, the mean number of substitutions that occurred in a given pair of sequences is then $\bar{\lambda}_{pair} = \bar{a}_{pair} + 2\bar{b}_{pair}$, where $\bar{a}_{pair}$ and $\bar{b}_{pair}$ are mean numbers of transitions and transversions between pair members, respectively.

Thus, by applying the same reasoning as for the Jukes and Cantor distance, we finally obtained the following overall evolutionary distance between two given sequences as

$$d = d_1 + d_2 \tag{7}$$

where

$$d_1 = -\frac{1}{4}\bar{\lambda}_{tree} \ln\left[\left(\frac{1+\beta}{\beta}\right)\left(1 - (Z_1+1)^{-1/\alpha}\right)\right], \tag{8}$$

with

$$Z_1 = (V - 1)\left(1 - \frac{2\bar{Q}}{1-\theta}\right)$$

and

$$d_2 = -\frac{1}{2}\bar{\lambda}_{tree} \ln\left[\left(\frac{1+\beta}{\beta}\right)\left(1 - (Z_2+1)^{-1/\alpha}\right)\right], \tag{9}$$

with

$$Z_2 = (V - 1)\left(1 - \frac{2\bar{P}}{1-\theta} - \frac{\bar{Q}}{1-\theta}\right).$$

In Equations (8) and (9), $\bar{P}$ and $\bar{Q}$ are estimated by the observed proportions of transitions and transversions (or nonsynonymous transitions and transversions per nonsynonymous sites in protein genes), respectively. The expected mean number of transversions that have occurred between the two sequences (i.e. the transversional part of d) is

$$d_v = 4\bar{b}_{pair}t = 2d_1 \qquad (10),$$

and the expected mean number of transitions (i.e. the transitional distance) is

$$d_u = 2\bar{a}_{pair}t = d_2 - d_1 \qquad (11).$$

3.6 REFERENCES

Feller, W. (1962) *An Introduction to Probability Theory and its Applications, Vol. I*, 2nd edn, John Wiley and Sons, Chichester.

Felsenstein, J. (1993) *PHYLIP: Phylogeny Inference Package, Version 3.5c*, University of Washington, Seattle.

Gu, X., Fu, Y.-X. and Li, W.-H. (1995) Maximum-likelihood estimation of the heterogeneity of substitution rate among nucleotide sites. *Molecular Biology and Evolution*, **12**, 546–57.

Hashimoto, T., Nakamura, Y., Kamaishi, T. *et al.* (1995) Phylogenetic place of kinetoplastid protozoa inferred from a protein phylogeny of elongation factor 1α. *Molecular and Biochemical Parasitology*, **70**, 181–5.

Holmquist, R., Goodman, M., Conroy, T. and Czelusniak, J. (1983) The spatial distribution of fixed mutations within genes coding for proteins. *Journal of Molecular Evolution*, **19**, 437–48.

Jin, L. and Nei, M. (1990) Limitations of the evolutionary parsimony method of phylogenetic analysis. *Molecular Biology and Evolution*, **7**, 82–102.

Johnson, N.L. and Kotz, S. (1969) *Distributions in Statistics: Discrete Distributions*, Houghton-Mifflin Company, Boston.

Jukes, T.H. and Cantor, C.R. (1969) Evolution of protein molecules, in *Mammalian Protein Metabolism*, (ed. H.N. Munro), Academic Press, New York, pp. 21–132.

Kimura, M. (1980) A simple method for estimating evolutionary rates of base substitutions through comparative studies of nucleotide sequences. *Journal of Molecular Evolution*, **16**, 111–20.

Lockhart, P.J., Larkum, A.W.D., Steel, M.A. *et al.* (1996) Evolution of chlorophyll and bacteriochlorophyll: the problem of invariant sites in sequence analysis. *Proceedings of the National Academy of Sciences USA*, **93**, 1930–4.

Nei, M. and Gojobori, T. (1986) Simple methods for estimating the numbers of synonymous and nonsynonymous nucleotide substitutions. *Molecular Biology and Evolution*, **3**, 418–26.

Olsen, G.J. (1987) Earliest phylogenetic branchings: comparing rRNA-based evolutionary trees inferred with various techniques. *Cold Spring Harbor Symposia on Quantitative Biology*, **52**, 825–37.

Palumbi, S.R. (1989) Rates of molecular evolution and the fraction of nucleotide positions free to vary. *Journal of Molecular Evolution*, **29**, 180–7.

Shoemaker, J.S. and Fitch, W.M. (1989) Evidence from nuclear sequences that invariable sites should be considered when sequence divergence is calculated. *Molecular Biology and Evolution*, **6**, 270–89.

Studier, J.A. and Keppler, K.J. (1988) A note on the Neighbor-Joining algorithm of Saitou and Nei. *Molecular Biology and Evolution*, **5**, 729–31.

Sullivan, J., Holsinger, K.E. and Simon, C. (1995) Among-site rate variation and phylogenetic analysis of 12S rRNA in sigmodontine rodents. *Molecular Biology and Evolution*, **12**, 988–1001.

Tateno, Y., Takezaki, N. and Nei, M. (1994) Relative efficiencies of the Maximum-Likelihood, Neighbor-Joining and Maximum-Parsimony methods when substitution rate varies with site. *Molecular Biology and Evolution*, **11**, 261–77.

Uzzell, T. and Corbin, K.W. (1971) Fitting discrete probability distributions to evolutionary events. *Science*, **172**, 1089–96.

Van de Peer, Y., Nicolaï, S., De Rijk, P. and De Wachter, R. (1996a) Database on the structure of small ribosomal subunit RNA. *Nucleic Acids Research*, **24**, 86–91.

Van de Peer, Y., Van der Auwera, G. and De Wachter, R. (1996b) The evolution of stramenopiles and alveolates as derived by 'substitution rate calibration' of small ribosomal subunit RNA. *Journal of Molecular Evolution*, **42**, 201–10.

Wakeley, J. (1993) Substitution rate variation among sites in hypervariable region 1 of human mitochondrial DNA. *Journal of Molecular Evolution*, **37**, 613–23.

Wakeley, J. (1994) Substitution-rate variation among sites and the estimation of transition bias. *Molecular Biology and Evolution*, **11**, 436–42.

Woese, C.R., Kandler, O. and Wheelis, M.L. (1990) Towards defining a natural system of organisms: proposal for the domains Archaea, Bacteria, and Eucarya. *Proceedings of the National Academy of Sciences USA*, **87**, 4576–9.

Yang, Z. (1994) Maximum-likelihood phylogenetic estimation from DNA sequences with variable rates over sites: approximate methods. *Journal of Molecular Evolution*, **39**, 306–14.

Yang, Z. (1996) Among-site rate variation and its impact on phylogenetic analyses. *Trends in Ecology and Evolution*, **11**, 367–72.

Yang, Z., Goldman, N. and Friday, A. (1994) Comparison of models for nucleotide substitution used in maximum-likelihood phylogenetic estimation. *Molecular Biology and Evolution*, **11**, 316–24.

Yang, Z. and Kumar, S. (1996) Approximate methods for estimating the pattern of nucleotide substitution and the variation of substitution rates among sites. *Molecular Biology and Evolution*, **13**, 650–9.

Yang, Z. and Roberts, D. (1995) On the use of nucleic acid sequences to infer early branchings in the tree of life. *Molecular Biology and Evolution*, **12**, 451–8.

4

Evolutionary relationships of Microsporidia

Elizabeth U. Canning

Department of Biology, Imperial College at Silwood Park, Ascot, Berkshire SL5 7PY, UK.
E-mail: e.u.canning@ic.ac.uk

ABSTRACT

Microsporidia are very small amitochondrial eukaryotic organisms, all of which are obligate intracellular parasites. They are further characterized by their 70S ribosomes with 16S and 23S rRNAs and a life cycle involving a proliferative phase (merogony) and a spore-producing phase (sporogony) culminating in specialized spores with an extrusible polar tube, through which the infective agent (sporoplasm) is inoculated into host cells. Recent attempts to resolve the evolutionary history of microsporidia on molecular data have given conflicting results. Gene sequences for the 16S rRNA and amino acid sequence for elongation factor EF-1α, when compared with a range of eukaryotes, have suggested that microsporidia are deep-branching eukaryotes, whereas analyses of amino acid sequences for β- and α-tubulin have placed them between the ascomycete and basidiomycete clusters within the animal/fungal lineage. Fungal characters which are present in or absent from microsporidia are reviewed. Attempts to determine the relationships of several microsporidian genera using 16S rRNA sequences have confirmed some previous hypotheses e.g. that at present the genus *Nosema* harbours a heterogeneous collection of species. It has also thrown up some surprising results, for example that the diplokaryotic genus *Vittaforma* is more closely related to the monokaryotic genus *Endoreticulatus* than it is to other diplokaryotic genera like *Nosema*. Many more genera will have to be added to this data bank before a realistic classification system can be constructed on molecular data.

Evolutionary Relationships Among Protozoa. Edited by G.H. Coombs, K. Vickerman, M.A. Sleigh and A. Warren. Published in 1998 by Chapman & Hall, London. ISBN 0 412 79800 X

4.1 INTRODUCTION

Microsporidia are very small unicellular organisms, all of which are intracellular parasites. An important diagnostic feature is an extrusible polar tube which serves to convey the infective agent, the sporoplasm, from the resistant spore into the cytoplasm of the host cell. Before the era of electron microscopy the microsporidian polar tube (then called a polar filament) was likened to the polar filament of the myxosporidia and the two groups have been linked in various classification schemes, notably as orders Myxosporida and Microsporida in the class Cnidosporidia Doflein. Whereas microsporidia are clearly unicellular, the spores of myxosporidia have a multicellular origin. Also, the myxosporidian polar filaments are enclosed in polar capsules and function merely to anchor the spores, playing no part in the conveyance of sporoplasms out of the spores into host cells. The two groups are now recognized as separate phyla, the phylum Microspora Sprague 1977 being one of several protistan phyla, while phylum Myxozoa Grasse, 1970 is clearly a metazoan phylum.

There are two distinct phases in the development of microsporidia, an initial proliferative phase, merogony, responsible for massive increase in numbers, and an overlapping sporogony, in which sporonts are committed to division into sporoblasts, which mature into spores. Merogonic and sporogonic phases are usually distinguished by the simple plasma membrane of the former and the secretion of a dense surface coat on the latter, the coat becoming the outer layer, exospore, of the spore wall. The spore wall has a chitin component, traditionally thought to be in the inner layer (endospore), although a recent study by freeze fracture and deep etching (Bigliardi *et al.*, 1996) suggests that the endospore is a space traversed by bridges connecting the exospore with the plasma membrane and that the chitin is primarily in the exospore.

Within the spore wall lie the sporoplasm and the extrusion apparatus. The latter consists of a polar tube coiled in the posterior half of the spore and connected via a straight section to an anchoring disc lying within a polar sac at the anterior end, beneath a somewhat thinned area of endospore. The straight section of the tube is surrounded by a series of flattened membranes and vesicles constituting the polaroplast, which plays a major part in creating an internal pressure, forcing the polar tube to evert under the correct stimuli. When fully everted the cytoplasm and nuclei of the sporoplasm pass through the tube and, if the tip of the tube has penetrated a host cell during extrusion, the sporoplasm emerging at its tip will be deposited in the host cell cytoplasm to begin development. The remaining part of the spore cavity is occupied by general cytoplasm and nuclei. At the posterior tip in the immature spore lies a collection of vesicles, likened to a primitive Golgi apparatus but lacking the characteristic flattened cisternae with forming and secreting faces. These are responsible for secretion of the polar tube, after which they de-differentiate to form the posterior vacuole in the mature spore.

The nuclear complement exists, according to genus, as a single nucleus or as two nuclei closely apposed as a diplokaryon, whose divisions are synchronized. The nuclei have a typical eukaryotic envelope of two membranes with pores. Mitosis is intranuclear with spindle microtubules arising from dense plaques in the nuclear pores. There are no centrioles. Estimates of genome size for several microsporidia range from 19.5 Mb to 2.9 Mb, the latter being among the smallest known for any eukaryote (Biderre *et al.*, 1995). The small genome size is likely to be due in part to the very high level of dependence of these organisms on their host cells. Equally, there is a case for believing that the small genome size reflects the origin of the group as an early branch in the evolution of the eukaryotes. The number of bands, possibly representing chromosomes, resolvable by pulsed field gel electrophoresis varies from 8 to 18 according to species. Little is known of the position of genes on these chromosomes although Vivarès *et al.* (1996) have established the localization of five genes of *Encephalitozoon cuniculi*, including ribosomal genes, which were found on all 11 chromosomes. In contrast Kawakami *et al.* (1994) found that ribosomal genes of *Nosema bombycis* were present on one chromosome only. The cytoplasm contains abundant ribosomes, many of which are bound to cisternae of endoplasmic reticulum, which partially encircle the nuclei. There are no mitochondria. The ribosomes are of particular interest because they are of a size more typical of prokaryotes, having sedimentation coefficients of 70S with 50S and 30S subunits (Ishihara & Hayashi, 1968) and rRNAs of 16S, 23S (Curgy, Vávra and Vivarès, 1980) and 5.0S (Kawakami *et al.*, 1992), while lacking the 5.8S rRNA typical of eukaryotes (Vossbrinck & Woese, 1986).

Today over 120 genera of microsporidia have been described, with hosts in all invertebrate phyla that have been examined, including other protists, and in all five classes of vertebrates. They form a group showing considerable diversity both in morphology and life cycle strategies. Some have simple direct life cycles with one morphological sequence, while others exhibit alternation between unikaryotic (isolated nuclei) and diplokaryotic (paired nuclei) sequences, in some cases involving an obligate alternation of hosts, fusion of gametes and subsequent reduction of chromosome number.

Microsporidia engage in a remarkable relationship with their host cells. The most obvious expression of the interaction is the orientation of host mitochondria around individual parasites in contact with the parasite plasma membrane, suggesting the utilization of host-produced

ATP by the parasites. Other expressions are the enormous enlargement or repeated duplication of host nuclei and nucleoli, envelopment of individual parasites by cisternae of host endoplasmic reticulum, gross hypertrophy of cells and formation of multiple surface coats to support the giant cells. The profound effect of parasitism on host cells suggests that the changes are under parasite control, for example by affecting the phosphorylation level of key components in signal transduction pathways, but there is no direct evidence for this. Some genera give rise to prominent branching and anastomosing excrescences which may connect with double walled tubular structures deep in the host cytoplasm, the complex possibly acting to transport enzymes for host cell degradation from the parasite or to transport metabolites towards the parasite's surface (Hollister *et al.*, 1996)

4.2 PHYLOGENETIC RELATIONSHIPS OF MICROSPORA

The defining characters of organisms belonging to the phylum Microspora are that they are: entirely parasitic; amitochondrial; possess 70S ribosomes; have a primitive Golgi; have intranuclear mitosis; form spores with extrusible polar tube through which a binucleate or uninucleate sporoplasm is ejected into host cells.

An important problem to resolve in determining the phylogenetic position of microsporidia is whether they are primitively amitochondrial (having diverged early in eukaryotic evolution before the endosymbiotic events took place, which gave rise to mitochondria) or secondarily amitochondrial (having lost mitochondria during their evolution as intracellular parasites). Consideration of this problem is suggested by the universal aggregation of host cell mitochondria at the surface of these parasites, implying that microsporidia have a requirement for ATP greater than that obtained by glycolysis alone, and perhaps indicating that microsporidia once had mitochondria but dispensed with them during their adaptation to intracellular parasitism. However, see addendum, p. 88.

The first data suggesting that microsporidia diverged early in eukaryotic evolution came from the demonstration that their ribosomal size (70S) and rRNAs (16S and 23S) are typically prokaryotic (Ishihara & Hayashi, 1968; Curgy *et al.*, 1980). An insect infecting microsporidium, *Vairimorpha necatrix*, was shown to lack the eukaryotic feature of a separate 5.8S rRNA, the corresponding sequence being present at the 5′ end of the 23S molecule (Vossbrink & Woese, 1986). The first sequence obtained of microsporidian small subunit rRNA (SSUrRNA) genes was also that of *V. necatrix* (Vossbrinck *et al.*, 1987). Comparative sequence analysis of this 16S rRNA gene (1244 nucleotides) with those of a range of prokaryotes and eukaryotes showed that the *V. necatrix* sequence was unlike the other eukaryotic sequences and led to the conclusion, based on rRNA difference as a measure of evolutionary distance, that microsporidia diverged before the evolution of mitochondria and before the 5.8S rRNA split off to form a separate molecule.

Later it was shown that other amitochondrial protists, the diplomonad *Giardia lamblia* (Sogin *et al.*, 1989) and the trichomonad *Tritrichomonas foetus* (Chakrabarti *et al.*, 1992) also have 16S rRNAs with unusual sequences but the branching order (if indeed the branch points for any of these lie at the base of the eukaryotic tree) has not been resolved on the SSUrRNA sequences. *Entamoeba histolytica* lacks mitochondria but has typical eukaryotic rRNAs. Many more SSUrRNA sequences have been obtained for microsporidia, all confirming their unusual nature for eukaryotes and supporting the view that they branch deeply in the eukaryotic tree (Figure 4.1).

Very recently attempts have been made to resolve the evolutionary history of microsporidia using other sequences: the deduced amino acid sequence from part of the gene encoding elongation factor 1α (EF-1α) (Kamaishi *et al.*, 1996); the sequence of a small nuclear rRNA (U2 RNA) homologue (DeMaria *et al.*, 1996); and sequences of β-tubulin (Edlind *et al.*, 1996) and α-tubulin (Li *et al.*, 1996). Kamaishi *et al.* (1996) argued that amino acid (aa) sequences are free from the bias conferred by drastic G + C content differences in SSUrRNA genes and would give a more robust estimation of the early divergence of eukaryotes. The aa sequences, within the GTP binding domain and within a putative actin-binding site of EF-1α and its eubacterial homologue EF-Tu, were compared for the microsporidium *Glugea plecoglossae* and a range of eukaryotes, archaebacteria and eubacteria. At the GTP-binding site the *G. plecoglossae* aa sequence was typically eukaryotic, having 10 of the 11 aa at positions 121–131 (*Homo sapiens* sequence positions), whereas there were only four or seven in the bacteria. They proposed that the one residue deletion could have occurred along the microsporidian line, with the common eukaryotic ancestor containing the 11-aa segment. At the actin-binding site 12 aa present in animal (human) and fungal (*S. ceresvisiae*) sequences at positions 214–225 were absent from all except *G. plecoglossae* which had 11 aa, although these did not correspond closely to those in the human and fungal insertion. They proposed that this sequence could have been inserted independently in the branches leading to the animal/fungal lineage and microsporidian lineage and their phylogenetic tree (Figure 4.2) shows *G. plecoglossi* as a deep branch in the eukaryotic lineage.

DeMaria *et al.* (1996) characterized a U2 RNA homologue isolated from *V. necatrix*. The sequence of 165 nucleotides represents the smallest known U2 homologue, other than those of the trypanosomes. Align-

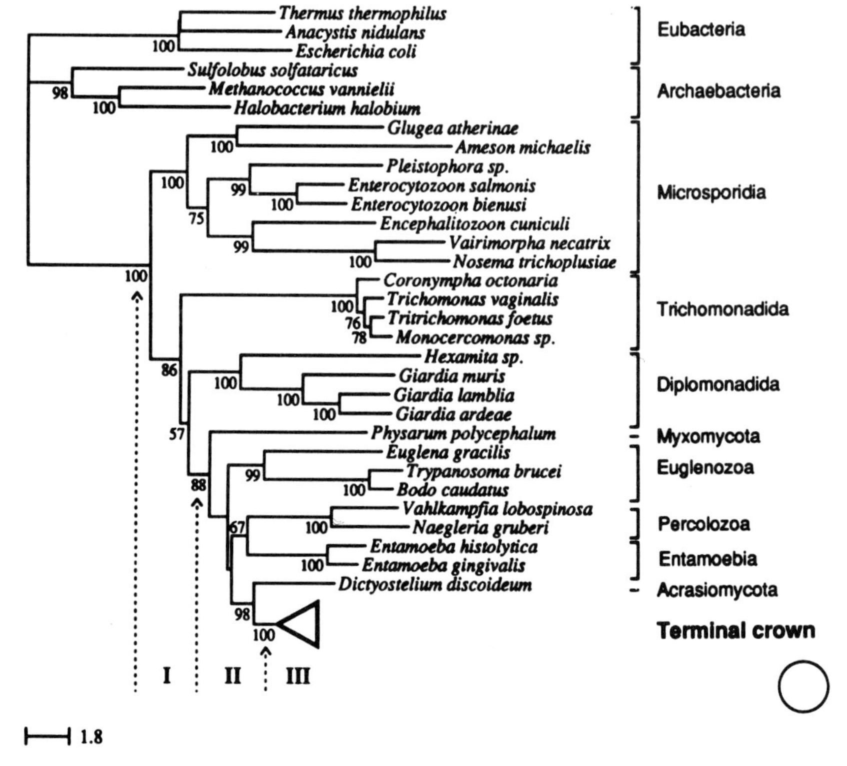

Figure 4.1 Small subunit rRNA phylogenetic tree constructed by Neighbour-Joining method using Eubacteria and Archaebacteria as outgroups, showing the microsporidia as the first emerging group of eukaryotes. Reproduced from Philippe and Adoutte (1995).

ment of the *V. necatrix* DNA sequence with human and consensus U2 RNA sequences showed that there was significant sequence identity in the 5′ region, including almost exact matching with the highly invariant 24–46 consensus region, which contains the GTAGTA branch point, which is involved in base pairing with pre-m RNA. However, five individual nucleotides were found to differ at positions previously found to be invariant, and an Sm binding site was lacking. There was a greater degree of divergence in the 3′ region. The secondary structure showed three of the four principal stem loop structures of consensus U2 RNA but, among other differences, there was a substituted U at *V. necatrix* position 21 which is A in all other U2 RNAs, forming part of a sequence involved in base pairing with U6 RNA. A major difference was a cis-diol cap structure at the 5′ end of *V. necatrix* U2 RNA which

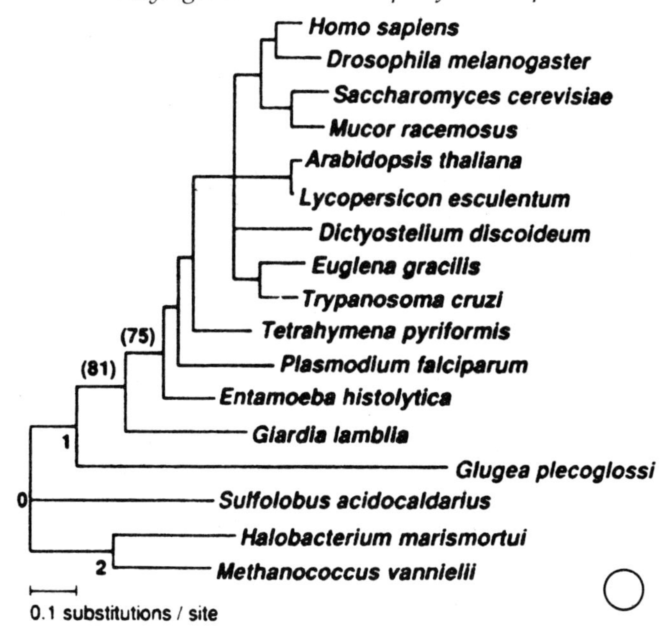

Figure 4.2 Phylogenetic tree based on the amino acid sequence of EF-1α, inferred by maximum likelihood method, using Archaebacteria as outgroup, showing the microsporidian *Glugea plecoglossi* as the earliest branch of the eukaryotic line. Reproduced from Kamaishi *et al.*, 1996.

was not 2.7.7 trimethylguanosine (present in all sn RNAs except U6) or y-monomethyl phosphate (present in U6).

On the assumption that microsporidia are very primitive eukaryotes, DeMaria *et al.* (1996) reasoned that the existence of a U2 RNA homologue in microsporidia indicated that a pre-mRNA splicing machinery arose very early in eukaryotic evolution before mitochondrial symbiosis. On the other hand Cavalier-Smith (1991), having proposed that spliceosomal introns originated from group II introns *after* the latter had been introduced into the nucleus as a result of mitochondrial symbiosis, argued that microsporidia must once have had mitochondria if they had U2 RNA. These arguments being circular, the demonstration of U2 RNA in microsporidia does not settle whether they are primitive or secondarily amitochondrial.

Edlind *et al.* (1996) proposed that ambiguities regarding phylogeny of amitochondrial protists might be clarified by analysing genes for β-tubulin, a component of microtubules which are found universally in eukaryotes. In microsporidia, microtubules are known only in intranuclear spindles. The sequences for the four groups of amitochondrial protists (*Giardia, Trichomonas, Entamoeba* and the microsporidium *Encephalitozoon hellem*) were aligned with previously reported sequences representing plants, animals, a chromist and eight 'protozoan' phyla. These were analysed, using various α and γ- tubulin sequences as outgroups, by parsimony and distance matrix methods. Nearly identical trees were obtained by parsimony using different outgroups. All trees had two major lineages (fungal/animal and plant/protozoa), while *Giardia, Trichomonas* and *Entamoeba* branched independently before the major lineages. *E. hellem* β- tubulin branched within the fungal/animal lineage, usually between the basidiomycete and ascomycete clusters. Three branch points sustained this position, although bootstrap support was not high (see Figure 5.3 in the chapter by Edlind, this volume). A nearly identical tree was generated by the distance matrix method but with only modest bootstrap support (57%) for the basidiomycete – *E. hellem* – ascomycete cluster. Similar results were obtained when partial β-tubulin sequences of *Encephalitozoon (Septata) intestinalis, E. cuniculi* and *Nosema locustae,* as well as that of *E. hellem,* and the α-tubulin sequence of *N. locustae,* were aligned with the corresponding sequences from representative animals, fungi, plants and protozoa (Li *et al.*, 1996).

The relationship of microsporidia to the animal/fungal lineage is also suggested by the data for the aa sequence downstream of the putative actin binding site of EF-1α, where an 11 (microsporidian) or 12 (animal and fungal) aa insertion was uniquely shared by these organisms (Kamaishi *et al.*, 1996; see p. 81 this volume). However it can be argued either that the segment was inserted only once, into the branch which led to the animal/fungal lineage or twice independently leading to the microsporidia on the one hand and to the common ancestor of animals and fungi on the other.

Various non-molecular features of microsporidia add weight to the concept of their relatedness to fungi. Paired, synchronously dividing (diplokaryotic) nuclei occur only in microsporidia and ascomycete fungi. However, in ascomycetes, the paired nuclei do not take the form of two closely appressed hemispheres, as is characteristic of the microsporidian diplokaryon. Spores of both groups contain chitin. Intranuclear mitoses with centriolar plaques occur in microsporidia and in at least some fungal groups, including the ascomycetes but it must be noted that the presence of chitin and these nuclear features also occur in several protistan groups. Microsporidia show drug sensitivity typical of fungi: paromomycin sensitivity depends on a binding site towards the 3′ end of the SSUrRNA, which is present in bacteria, plants and many 'protozoa', including *Giardia, Entamoeba* and *Trichomonas* (paromomycin sensitive) but absent in fungi, other metazoa and microsporidia (paromomycin resistant) (Katiyar, Visvesvara and Edlind, 1995); analysis of *Encephalitozoon* β-tubulin residues demonstrated an additional residue His 6, as well as 5 previously identified residues, associated with sensitivity to the anti-fungal benzimidazoles (Edlind *et al.*, 1996; Li *et al.*, 1996), one of which, albendazole, has been shown to ameliorate or cure microsporidian infections *in vivo* (Blanshard *et al.*, 1993) and *in vitro* (Colbourn *et al.*, 1994). Fusion of gametes has been documented in some microsporidian life cycles but not in the other amitochondrial protists. It is worth mentioning, also, that some microsporidia (*Amblyospora, Parathelohania*) have complex life cycles involving two generations of mosquito hosts and an obligate copepod intermediate host (Sweeney, Hazard and Graham, 1985). Similar alternations of mosquito and copepod hosts occur in the life cycle of species of *Coelomomyces*, a chytridiomycete fungus (Whisler, Zebold and Shemanchuk, 1974). However, not too much emphasis should be placed on this, as gametogenesis and fusion of gametes occur in different hosts (copepods for *Coelomomyces* and mosquitoes for the microsporidia). Characters of microsporidia, which differ from fungi, are their typically protistan cellular architecture and absence of hyphae, mitochondria, cell walls and ergosterol-containing membranes.

4.3 RELATIONSHIPS OF MICROSPORIDIAN GENERA

The earliest attempts to classify microsporidia were based on morphological features such as spore shape and number of spores derived from a sporont. Tuzet *et al.* (1971) introduced a primary dichotomy based on the presence or absence of an envelope (pansporoblast membrane = sporophorous vesicle) around the spores. Weiser (1977) introduced nuclear differences as the fundamental character, distinguishing those microsporidia with unpaired nuclei (unikaryotic) from those with paired nuclei (diplokaryotic). Unfortunately this did not take account of the many genera that exhibit both arrangements of nuclei at different stages of the life cycle. The most recently proposed classification (Sprague, Becnel and Hazard, 1992) incorporated the chromosome cycle as a basic taxonomic character. Sequences of rRNA genes have provided new insights into microsporidian systematics and indicated some unexpected generic relationships.

The largest number of named species of microsporidia, described from a wide range of invertebrate hosts and from vertebrates, including man, have been assigned to the genus *Nosema*. The genus is defined as diplokaryotic throughout development and producing two free sporoblasts (and later two free spores) per sporont (disporoblastic). The genus *Vairimorpha* has one sporogonic sequence like that of *Nosema* and a second sequence, in which a diplokaryotic sporont undergoes meiosis to give rise to eight uninucleate (haploid) sporoblasts in a sporophorous vesicle.

Baker *et al.* (1994) used a 350 nucleotide segment of the 23S rRNA to examine the affinity of several species of *Nosema* (from Lepidoptera, Diptera and Orthoptera) and several species of *Vairimorpha* (all from Lepidoptera). Using as outgroups *Amblyosoma* and *Parathelohania* from mosquitoes, trees were generated by parsimony (Figure 4.3). The *Vairimorpha* spp. fell into two groups which correlate well with morphological and ecological characters. The *Nosema* spp. fell into a lepidopteran group, which included the type species *N. bombycis*, while the others were not closely related to each other or to the lepidopteran group. The results were in accord with the previous concept that the genus *Nosema* is a heterogeneous assembly of unrelated genera (Sprague, 1978).

Baker *et al.* (1995) examined a range of microsporidia, using the complete sequences of the 16S rRNA, which varied from 1234 bases in *N. bombycis* to 1352 in *Ichthyosporidium* sp.. The hosts were widely different, embracing Lepidoptera, Hymenoptera, a decapod crustacean, a

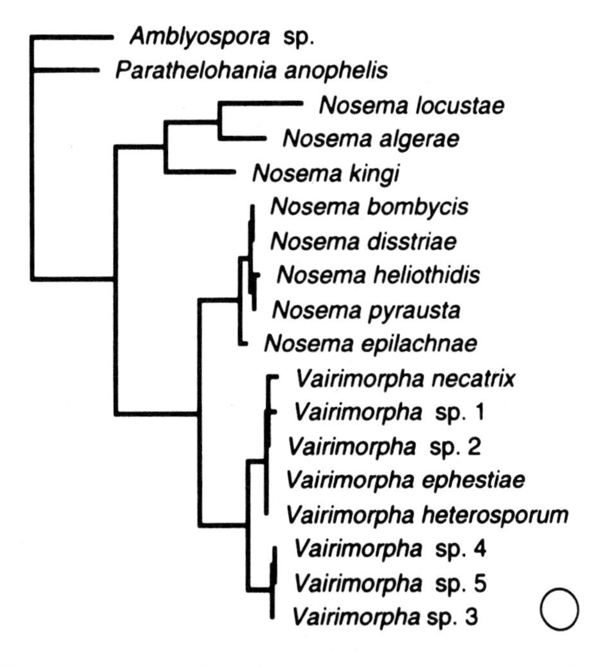

Figure 4.3 Phylogenetic tree based on parsimony analysis using the polymorphic microsporidia *Amblyospora* and *Parathelohania* as outgroups, showing the relationships of *Vairimorpha* spp. and *Nosema* spp. Reproduced from Baker *et al.*, 1994.

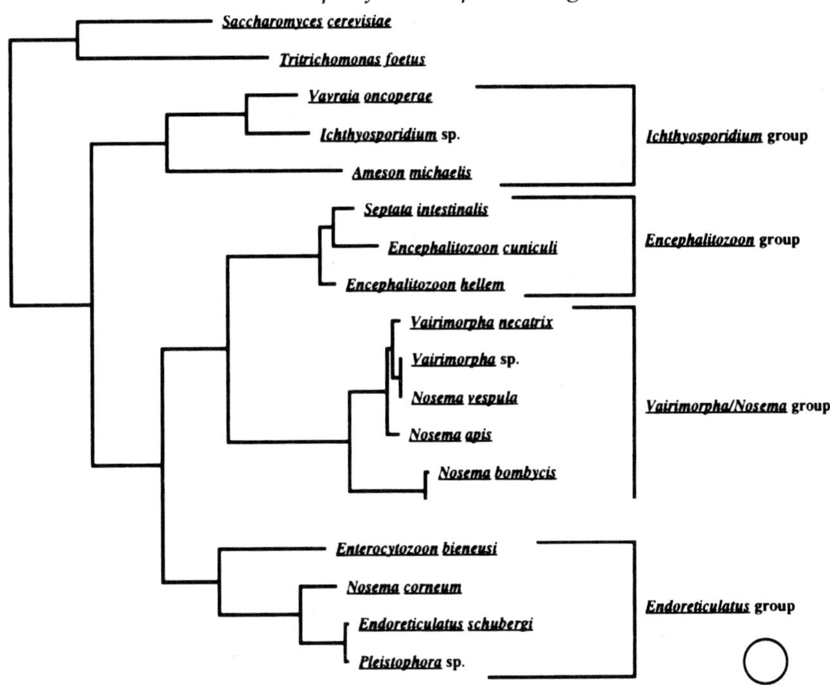

Figure 4.4 Phylogenetic tree based on parsimony analysis using *Saccharomyces* and *Tritrichomonas* as outgroups, showing the relationships of various microsporidian genera. Of particular note are the *Ichthyosporidium* and *Endoreticulatus* groups, both of which include unikaryotic species (unpaired nuclei) and diplokaryotic species (paired nuclei). Reproduced from Baker *et al.*, 1995.

fish, and rabbit; and four species that had been isolated from man. Four groups were obtained by parsimony analysis with strong bootstrap support (Figure 4.4).

The *Encephalitozoon* group, as expected and as previously found (Vossbrinck *et al.*, 1993; Zhu *et al.*, 1994), contained three closely related unikaryotic species from mammals. The *Vairimorpha* and *Nosema* spp. confirmed the pattern found previously (Baker *et al.*, 1994) of a lepidopteran group of true *Nosema* spp. with the other so-called *Nosema* spp. being more closely related to *Vairimorpha*. The most surprising results came in the other clusters. The unikaryotic species *Vavraia oncoperae* from Lepidoptera and diplokaryotic *Ichthyosporidium* sp. from a fish formed a clade with the diplokaryotic *Ameson michaelis* as a sister group. Another cluster included the unikaryotic *Enterocytozoon bieneusi* and diplokaryotic *Nosema corneum* (since transferred from *Nosema* to a new genus *Vittaforma* by Silveira & Canning, (1995) both from man and unikaryotic *Endoreticulatus schubergi* and *Pleistophora* sp. (since identified as another species of *Endoreticulatus*) both from Lepidoptera. A morphological feature uniting *Vittaforma* and *Endoreticulatus* is that every stage is surrounded by a complete cisterna of host endoplasmic reticulum. The clustering of species, which are either wholly unikaryotic or wholly diplokaryotic, has been a major taxonomic surprise.

In summary, strong support for the hypothesis that microsporidia arose as an early branch of the eukaryotic line has been provided by SSUrRNA gene sequences and this has been reinforced by same EF-1α data. Microsporidia have several developmental and physiological features in common with fungi and a possible relationship with the fungi has been suggested by the aa sequences of β-tubulin. However, an explanation would be required for the 16S ribosomes and absence of a separate 5.8S rRNA in microsporidia, if these parasites had evolved from the fungi. Although the absence of typical fungal structures in microsporidia may be explained by secondary loss due to their obligate parasitic habit, no evidence has been forthcoming for the past presence of mitochondria in microsporidia (but see addendum). Further investigation of the possible fungal origin is needed. Meanwhile the rRNA data is being used to advantage in determining generic relationships within the phylum Microspora.

Addendum

Since submission of this manuscript two articles have been published which weaken the view that microsporidia diverged from the main eukaryotic stock before the mitochondrial symbiotic event. Symbiont-derived Hsp genes have been identified in two microsporidia (Germot, Philippe and Le Guyader, 1997; Hirt *et al.*, 1997). As the protein products of these genes normally function in mitochondria, the likely explanation is that microsporidia no longer have functional mitochondria but retain evidence of their past presence. This strengthens the view that microsporidia are not primitively amitochondrial but their relationship with the fungi still requires clarification.

4.4 REFERENCES

Baker, M.D., Vossbrinck, C.R., Maddox, J.V. and Undeen, A.H. (1994) Phylogenetic relationships among *Vairimorpha* and *Nosema* species (Microspora) based on ribosomal RNA sequence data. *Journal of Invertebrate Pathology*, **64**, 100–6.

Baker, M.D., Vossbrinck, C.R., Didier, E.S. *et al.* (1995) Small subunit ribosomal DNA phylogeny of various microsporidia with emphasis on AIDS related forms. *Journal of Eukaryotic Microbiology*, **42**, 564–70.

Biderre, C., Pagès, M., Méténier, G. *et al.* (1995) Evidence for the smallest nuclear genome (2.9 Mb) in the microsporidium *Encephalitozoon cuniculi. Molecular and Biochemical Parasitology*, **24,** 239–41.

Bigliardi, E., Selmi, M.G., Lupetti, P. *et al.* (1996) Microsporidian spore wall: ultrastructural findings on *Encephalitozoon hellem* exospore. *Journal of Eukaryotic Microbiology*, **43,** 181-6.

Blanshard, C., Ellis, D.S., Dowell, S.P. *et al.* (1993) Electron microscopic changes in *Enterocytozoon bieneusi* following treatment with albendazole. *Journal of Clinical Pathology*, **46,** 898–902.

Cavalier-Smith, T. (1991) Kingdom Protozoa and its 18 phyla. *Microbiological Reviews*, **57,** 953-94.

Chakrabarti, D., Dame, J.B., Gutell, R.R. and Yowell, C.A. (1992) Characterization of the rDNA unit and sequence analysis of the small subunit rRNA and 5.8S rRNA genes from *Tritrichomonas foetus. Molecular and Biochemical Parasitology*, **52,** 75–83.

Colbourn, N.I., Hollister, W.S., Curry, A. and Canning, E.U. (1994) Activity of albendazole against *Encephalitozoon cuniculi* in vitro. *European Journal of Protistology*, **30,** 211–20.

Curgy, J.-J., Vávra, J. and Vivarès, C. (1980) Presence of ribosomal RNAs with prokaryotic properties in *Microsporidia*, eukaryotic organisms. *Biologie Cellulaire*, **38,** 49–52.

DeMaria, P., Palic, B., Debrunner-Vossbrinck, B.A. *et al.* (1996) Characterization of the highly divergent U2 RNA homolog in the microsporidian *Vairimorpha necatrix. Nucleic Acids Research*, **24,** 515–22.

Edlind, T., Katiyar, S., Visvesvara, G. and Li, J. (1996) *Evolutionary Origins of Microsporidia and Basis for Benzimidazole Sensitivity: an Update.* Proceedings of the Joint Meeting of the American Society of Parasitologists and the The Society of Protozoologists, Tucson, Arizona, Abstract **90**, 101–2.

Edlind, T.D., Li, J., Visvesvara, G.S., *et al.* (1996) Phylogenetic analysis of β-tubulin sequences from amitochondrial Protozoa. *Molecular Phylogenetics and Evolution*, **5,** 359–67.

Germot, A., Philippe, H. and Le Guyader, H. (1997) Evidence for loss of mitochondria in Microsporidia from a mitochondrial-type HSP 70 in *Nosema locustae. Molecular and Biochemical Parasitology*, **87**, 159–168.

Hirt, R.P., Healy, B., Vossbrinck, C.R., *et al.* (1997) A mitochondrial Hsp 70 orthologue in *Vairimorpha necatrix:* molecular evidence that microsporidia once contained mitochondria. *Current Biology*, **7**, 995–998.

Hollister, W.S., Canning, E.U., Weidner, E. *et al.* (1996) Development and ultrastructure of *Trachipleistophora hominis* n.g., n.sp. after *in vitro* isolation from an AIDS patient and inoculation into athymic mice. *Parasitology*, **122,** 143–54.

Ishihara, R. and Hayashi, Y. (1968) Some properties of ribosomes from the sporoplasm of *Nosema bombycis. Journal of Invertebrate Pathology*, **11,** 377–85.

Kamaishi, T., Hashimoto, T., Nakamura, Y. *et al.* (1996) Protein phylogeny of translation Elongation Factor EF-1α suggests microsporidians are extremely ancient eukaryotes. *Journal of Molecular Evolution*, **42**, 257–63.

Katiyar, S.K., Visvesvara, G.S. and Edlind, T.D. (1995). Comparisons of ribosomal RNA sequences from amitochondrial protozoa: implications for processing, mRNA binding and paromomycin susceptibility. *Gene*, **152,** 27–33.

Kawakami, Y., Inoue, T., Kikuchi, M. *et al.* (1992) Primary and secondary structures of 5S ribosomal RNA of *Nosema bombycis* (Nosematidae, Microsporidia). *Journal of Sericultural Science of Japan*, **61,** 321–7.

Kawakami, Y., Inoue, T., Ito, K. *et al.* (1994) Identification of a chromosome harboring the small subunit ribosomal RNA gene of *Nosema bombycis. Journal of Invertebrate Pathology*, **64**, 147–8.

Li, J., Katyar, S.K., Hamelin, A. *et al.* (1996) Tubulin genes from AIDS-associated microsporidia and implications for phylogeny and benzimidazole sensitivity. *Molecular and Biochemical Parasitology*, **78**, 289–95.

Philippe, H. and Adoutte, A. (1995) How reliable is our current view of eukaryotic phlogeny? in *Protistological Actualities,* Proceedings of the Second European Congress of Protistology, Clermont-Ferrand, 1995.

Silveira, H. and Canning, E.U. (1995) *Vittaforma corneae* N. Comb. for the human microsporidium *Nosema corneum* Shadduck, Meccoli, Davis and Font, 1990 based on its ultrastructure in the liver of experimentally infected athymic mice. *Journal of Eukaryotic Microbiology*, **42,** 158-65.

Sogin, M.L., Gunderson, J.H., Elwood, H.J. *et al.* (1989) Phylogenetic meaning of the kingdom concept: an unusual ribosomal RNA from *Giardia lamblia. Science*, **243,** 75–7.

Sprague, V. (1978) Characterization and composition of the genus *Nosema. Miscellaneous Publications of the Entomological Society of America*, **11,** 5–16.

Sprague, V., Becnel, J.J. and Hazard, E.I. (1992) Taxonomy of the phylum Microspora. *Critical Reviews in Microbiology*, **18,** 285–395.

Sweeney, A.W., Hazard, E.I. and Graham, M.F. (1985) Intermediate host for an *Amblyospora* sp. (Microspora) infecting the mosquito *Culex annulirostris. Journal of Invertebrate Pathology*, **46,** 98–102.

Tuzet, O., Maurand, J., Fize, A. *et al.* (1971) Proposition d'un nouveau cadre systématique pour les genres de Microsporidies. *Compte Rendu hebdomadaire des séances de l'Académie des Sciences. Paris*, **2723,** 1268–71.

Vivarès, C., Biderre, C., Duffieux, F. *et al.* (1996) Chromosomal localization of five genes in *Encephalitozoon cuniculi* (Microsporidia). *Journal of Eukaryotic Microbiology*, **43**, 97S.

Vossbrinck, C.R. and Woese, C. (1986) Eukaryotic ribosomes that lack a 5.8S RNA. *Nature*, **320,** 287–8.

Vossbrinck, C.R., Maddox, J.V., Friedman, S. *et al.* (1987) Ribosomal RNA sequence suggests microsporidia are extremely ancient eukaryotes. *Nature*, **326,** 411–14.

Vossbrinck, C.R., Baker, M.D., Didier, E.S. *et al.* (1993) Ribosomal DNA sequences of *Encephalitozoon hellem* and *Encephalitozoon cuniculi*: species identification and phylogenetic construction. *Journal of Eukaryotic Microbiology*, **40,** 345–62.

Weiser, J. (1977). Contribution to the classification of microsporidia. *Vestnik Ceskoslovenske Spolecnosti Zoologicke*, **41,** 308–20.

Whisler, H.C., Zebold, S.L. and Shemanchuk, J.A. (1974). Life history of *Coelomomyces psorophorae. Proceedings of the National Academy of Sciences USA*, **72,** 693–6.

Zhu, X., Wittner, M, Tanowitz, H.B. *et al.*(1994) Ribosomal RNA sequences of *Enterocytozoon bieneusi, Septata intestinalis* and *Ameson michaelis*: phylogenetic construction and structural correspondence. *Journal of Eukaryotic Microbiology*, **41,** 204–9.

5

Phylogenetics of protozoan tubulin with reference to the amitochondriate eukaryotes

Thomas D. Edlind

Department of Microbiology and Immunology, MCP-Hahnemann School of Medicine, Allegheny University of the Health Sciences, 2900 Queen Lane, Philadelphia, PA 19129, USA. E-mail: edlind@auhs.edn

ABSTRACT

The anaerobic protozoa include several phyla of clinically important parasites, and furthermore provide potential insight into early eukaryotic evolution. Organisms representing three of these phyla (Microsporidia, Metamonada and Parabasalia) have unusually small and atypical rRNA. Although this could reflect an early branch in evolution, there is a clear need for additional phylogenetic markers, preferably unrelated to the translational machinery. A defining feature of eukaryotic cells is the microtubule, formed by polymerization of α and β-tubulin. Over 100 full-length β-tubulin sequences have been determined from a wide range of organisms, including the amitochondriate protozoa *Giardia lamblia* (Metamonada), *Trichomonas vaginalis* (Parabasalia) *Encephalitozoon hellem* (Microsporidia) and *Entamoeba histolytica* (Archamoeba). These sequences (approximate length, 450 amino acids) are highly conserved, with >50% identity and few if any insertions or deletions between any pair of sequences. Moreover, comparing β-tubulin with α- or γ-tubulin, there is 30–40% identity and only 4–9 insertions or deletions of 1–3 residues. These properties permit relatively unambiguous sequence alignments. In addition to full-length sequences, PCR with primers corresponding to conserved regions of tubulin has been used to generate a large number of partial sequences from diverse protists.

Evolutionary Relationships Among Protozoa. Edited by G.H. Coombs, K. Vickerman, M.A. Sleigh and A. Warren. Published in 1998 by Chapman & Hall, London. ISBN 0 412 79800 X

In our initial studies of β-tubulin as a phylogenetic marker, residues 1–430 from the amitochondriate protozoa cited above were aligned with representative animal, plant, fungal and protozoan β-tubulins. Aligned sequences were analysed by the distance matrix and parsimony methods within the PHYLIP package. Both α- and γ-tubulin were tested as outgroups. Nearly all analyses generated the same tree, and key elements within this tree were supported by bootstrapping. Expanded datasets of partial sequences (residues 108–259) generated consistent and in some cases more definitive results. At the base of the β-tubulin-based tree are *Entamoeba* spp. and *T. vaginalis*. These are followed, less reproducibly, by *G. lamblia* and the slime mould *Physarum polycephalum*. A major division then occurs leading to animal–fungus and protozoan–plant lineages. Within the latter are the widely accepted clusters of Kinetoplasta/Euglenoida, Cilophora/Apicomplexa and algae/higher plants. Furthermore, the Mycetozoa and Oomycetes are within the protozoan–plant lineage and not within fungi, and in the basidiomycetes are clearly divided. Surprisingly, the amitochondriate microsporidia branch within the fungi. This result is obtained with both complete *E. hellem* and partial *Nosema locustae* β-tubulins as well as with a partial *N. locustae* α-tubulin. Although several examples illustrate that the potential remains for artefactual branching, overall tubulin appears to be an excellent molecule for phylogenetic analysis of diverse protozoa.

Is there a protozoan, mired in the mud or lurking in our lumens, that can reveal to us the early evolutionary history of eukaryotes? The studies reviewed in this chapter have examined this basic question, focusing on the anaerobic, mitochondria-lacking protozoa. While definitive answers are not yet in hand, recent studies of one particular class of molecules, the tubulins, have been particularly enlightening.

5.1 AMITOCHONDRIATE PROTOZOA: OUR EUKARYOTIC ROOTS?

As defined by Cavalier-Smith (1987, 1993), the kingdom Archezoa includes organisms that are primitively without mitochondria; that is, diverged from the eukaryotic lineage prior to the endosymbiotic acquisition of these organelles. The original proposal for the Archezoa included four phyla: Archamoeba, Metamonada, Parabasalia, and Microsporidia. With the exception of the latter, these phyla share several key features. First, consistent with their lack of mitochondria, they have an anaerobic metabolism with ferredoxin-based electron transport, and consequent high sensitivity to metronidazole. Also, their most well studied representatives are extracellular, lumen-dwelling parasites: *Entamoeba histolytica* (Archamoeba) in the large intestine, *Giardia lamblia* (Metamonada) in the small intestine, and *Trichomonas vaginalis* (Parabasalia) in the genital tract. The microsporidia, in contrast, are obligate intracellular parasites found in a wide variety of eukaryotic hosts; several species (e.g. *Encephalitozoon hellem*) have been identified recently in humans with AIDS. They are not truly anaerobic as they associate closely with host cell mitochondria (see chapter 4) and are resistant to metronidazole (Beauvais *et al.*, 1994).

About ten years ago the first rRNA sequences from these phyla were reported. These sequences are remarkable for their brevity: rRNAs from *T. vaginalis, G. lamblia,* and *E. hellem* are about 90, 80, and 70%, respectively, the lengths of other protist rRNAs (for compilation, see Neefs *et al.*, 1993). Metamonada and microsporidial rRNAs are significantly shorter even than bacterial rRNAs (and hence not 'bacterial-like' as often stated). Furthermore, these rRNA sequences are highly divergent, from other organisms and from each other. Not too surprisingly then, in phylogenetic trees their rRNAs form long branches near the eukaryotic base (Sogin, 1989; Leipe *et al.*, 1993; Cavalier-Smith, 1993). *E. histolytica* rRNA, on the other hand, is longer than most rRNAs and branches among mitochondria-containing protozoa.

For several reasons, however, we should not be content as yet with these rRNA-based trees, especially with regards to the amitochondriate protozoa. First of all, phylogenetic analysis of elongation factor 1α (EF-1α) sequences suggests a significantly lower branch for *E. histolytica* than suggested by rRNA analysis (Hasegawa *et al.*, 1993). Second, the branching order of rRNAs from the amitochondriate protozoa was shown to vary with the dataset or outgroup used (Leipe *et al.*, 1993). Third, if an organism has a skewed base content in its genome and consequently in its rRNA, as with the GC-rich *G. lamblia,* this may generate artefactual branching (Hasegawa and Hashimoto, 1993). Fourth, the alignment of sequences that are significantly different in length, as is the case with amitochondriate rRNAs, becomes more ambiguous and hence the analysis suspect. Finally, there is a clear possibility that the abbreviated rRNAs of microsporidia are due to deletion events rather than reflecting a primitive condition (see below). Deletions are likely to be accompanied by compensating changes in adjacent or even distant regions of the remaining rRNA.

Ideally, phylogenetic analysis is insensitive to these random sequence changes associated with skewed base compositions or deletions and additions. However, all phylogenetic methods are variably sensitive to 'long branch' artefacts when dealing with highly divergent sequences. The early branching of one or more of the amitochondriate protozoa could represent such an artefact.

5.2 PROTEIN SEQUENCES AS PHYLOGENETIC MARKERS

Some of the potential problems associated with rRNA can be minimized by using protein sequences. In particular, skewed genome base compositions may have little effect due to the degeneracy of the genetic code (Hashimoto *et al.*, 1994). Also, some proteins may be more resistant to deletions or additions, hence facilitating alignment and reducing random sequence variation. The most important advantage of using proteins, however, may simply be that there are many to choose from. Specific proteins are more useful than others in answering specific phylogenetic questions.

With regards to the amitochondriate protozoa and the phylogeny of early branching eukaryotes, a number of different proteins have already been used. There is a large database of EF-1α and to a lesser extent EF-2 sequences, including sequences from Metamonada, Microsporidia, and *E. histolytica*. EF-based trees have been generated that place *G. lamblia* and the microsporidian *Glugea plecoglossi* at the base, while *E. histolytica* is relatively low but among mitochondrial-containing protozoa (Bauldauf and Palmer, 1993; Hasegawa *et al.*, 1993; Hashimoto *et al.*, 1994; Kamaishi *et al.*, 1996). In addition to their extensive databases, an important advantage of EF sequences relates to outgroup. To root early eukaryotes, either or both eubacterial and archaebacterial EF sequences have been used. However, there are limitations to the use of EF sequences in phylogeny as well. Primarily, it is likely that EF proteins (and other components of the translational machinery such as amino acyl-tRNA synthetases) co-evolve with rRNA. If, for example, microsporidial rRNA has suffered deletions and extensively diverged, this might explain the unusual sequence of *G. plecoglossi* EF-1α (Kamaishi *et al.*, 1996) and its early branching could be artefactual. Another limitation is the current lack of a representative from phylum Parabasalia.

With a focus on *T. vaginalis*, Müller and colleagues have phylogenetically analysed the sequences of several enzymes of 'core metabolism' common to all free-living organisms. Intriguingly, for glyceraldehyde-3-phosphate dehydrogenase, there is evidence for horizontal gene transfer: the *T. vaginalis* sequence is more closely related to eubacterial than to other eukaryotic or archaebacterial sequences (Markos, Miretsky and Muller, 1993). On the other hand, other enzymes from this organism are most closely related to other eukaryotic sequences (Markos *et al.*, 1996). Thus, these enzymes provide an interesting but somewhat ambiguous picture of early eukaryotic evolution, apparently complicated by genetic exchange between nucleus and bacterial endosymbiont. A further limitation to using these enzymes for phylogeny is that the database currently lacks microsporidial sequences; indeed, these enzymes may be absent in these obligate parasites.

In addition to the above, limited studies have been done with sequences of the cytoskeletal protein actin (Bauldauf and Palmer, 1993; Drouin, Moniz de Sa and Zuter, 1995). Proteins such as actin which are characteristic of eukaryotes but not bacteria could potentially provide a more accurate picture of eukaryotic evolution than proteins (or rRNA) which are shared by all organisms. The reasons for this are: (1) such proteins would not be affected by horizontal gene transfer from bacterial endosymbionts, and (2) the evolution of eukaryote-specific proteins is intuitively more likely to parallel the evolution of the nucleus, the defining eukaryotic feature. However, the absence of a bacterial homologue means there is no bacterial outgroup for eukaryotic actins. Alternatively, if there were two or more types of actin, they could be used to root each other. While there are indeed several types of actin in higher eukaryotes, there appears to be only one type in lower eukaryotes (Drouin, Moniz de Sa and Zuker, 1995). A potentially useful alternative, however, is contraction (Philippe and Adoutte, 1995).

5.3 TUBULIN: STRUCTURE, FUNCTION AND DIVERSITY

A defining feature of eukaryotic cells is the microtubule, the main component of the mitotic spindle responsible for chromosome segregation. In many eukaryotes, microtubules have additional functions as major components of the flagella (or cilia) and cytoskeleton. Microtubules are formed by the polymerization of tubulin, in turn formed by the dimerization of α-tubulin and β-tubulin subunits. In the cell, microtubule formation initiates within a microtubule-organizing center or centrosome which varies considerably between different phyla but appears to always include a third type of tubulin, γ-tubulin. All tubulins have similar lengths (440–460 amino acids) and their sequences are conserved: within each tubulin type there is generally >70% identity between any two species, while between types identity is 30 to 40% (for review, see Burns, 1991). Clearly, α-, β-, and γ-tubulins evolved by gene duplication events, early in or prior to eukaryotic evolution. Additional gene duplication events have also occurred more recently (i.e. within genera and species), generating two or more copies of a single tubulin type in many organisms.

An early study (Little, 1985) concluded that tubulins were of limited use in phylogenetic analysis since rates of divergence in some species (yeasts) were higher than in others (animals). Clearly, this study suffered from the very limited and non-representative sequence database available at the time. In recent years, the database has expanded to include full or nearly full length sequences from nearly 100 diverse species, including representatives from at least 12 protozoan phyla. The α-tubulin database represents about 50 species and the γ-tubulin database about 20. Nevertheless, the tubulin databases remain somewhat biased towards animals, ascomycetous fungi, and higher plants. There are many sequences from these three groups that vary little from one another, while the protozoa, non-green algae, and non-ascomycetous fungi with fewer representatives contribute much more diversity.

Until recently, the tubulin database was inadequate for the purpose of examining early eukaryote evolution. Of the four phyla of amitochondriate protozoa, only Metamonada was represented, by *G. lamblia* β-tubulin (Kirk-Mason, Turner and Chakraborty, 1988). In fact, this sequence represented only one of three β-tubulin gene copies present in this organism (Kirk-Mason, Turner and Chakraborty, 1989). However, this major deficiency in the tubulin database has now been addressed with the sequencing of full length β-tubulins from *T. vaginalis, E. histolytica*, and *E. hellem* (Katiyar and Edlind, 1994, 1996; Li *et al.*, 1996). In addition, partial β-tubulin sequences have been determined for the two additional gene copies found in *G. lamblia* and five to six additional copies found in *T. vaginalis* (Katiyar and Edlind, 1994; Edlind *et al.*, 1996b). Partial sequences were also determined for a second *Entamoeba* species (*E. invadens*) and three additional microsporidia: *Encephalitozoon cuniculi, Encephalitozoon (Septata) intestinalis*, and *Nosema locustae* (Li *et al.*, 1996; Edlind *et al.*, 1996a). Complete *E. histolytica* and partial *N. locustae* α-tubulin sequences have also been reported (Sanchez *et al.*, 1994; Li *et al.*, 1996).

Much of this sequence determination relied on the use of PCR with primers based on conserved tubulin sequences. The locations of conserved β-tubulin residues (Katiyar and Edlind, 1996) are shown in Figure 5.1.

Both universally conserved residues and those that vary only once in the database are shown, since the latter may have trivial origins such as sequencing errors. For PCR, we have had success with primers based on β-tubulin residues 8–15 (QA/TGQCGNQ), 99–107 (NNWAKGHYT), 203–210 (DNEALYDI), 259–266 (PFPRLHFF), 396–403 (HWYTGEGM), and 421–428 (EYQQYQD/EA). Degenerate positions are best dealt with by using I (inosine) when the degeneracy is 4, but either G or T alone when the degeneracy is 2 (taking advantage of the G-T 'wobble' pair). Partial gene sequences have been completed by inverse PCR, RACE techniques when cDNA could be prepared, or by conventional cloning from a genomic DNA library. Gene copy number was evaluated by Southern blotting with at least three enzymes, and multiple clones were sequenced to ensure that each copy was represented.

```
beta    M-REIVHIQAGQCGNQIGAKFWEVISDEHGIDPSGTYSGDSDLQ--LERVDVFYNEATGGRYVPRAILMDLEPGTMDSVRAGPFGQLFRPDN-FVFGQTG-AGNN 100
alpha   .-..VIS.HV..A.I...NAC..LFCL....Q.D.QMPS.QVVAGGDDAFNT.FS.TGA.KH...CVFV....TVV.E..T.TYR...H.EQ-LIS.KED-.A.. 102
gamma   .P...ITL.C........VE..KQLCN..N..QE.ILKNNNF.N--ED.K.I.FYQ.DDEHFI.G.L.F....RVIN.IQTSEYRN.YN.E.M.ISKEG.G.... 103

beta    WAKGHYTEGAELIDAVLDVLRKEAEGCDCLQGFQITHSLGGGTGSGMGTLLISKIREEYPDRIMETFSVFPSP-KV-SDTVVEPYNATLSVHQLVENADEVQ 200
alpha   F.R....I.K.IV.VC..R...L.DN.TG....LMFNAV.......L.C..LERLAID.GKKSKLN.CSW...-Q.-.TA......SV..T.S.L.HT.VAI 202
gamma   .GC.-.SQ.HKVEEEII.MIDR.VDNS.N.E..ILS..IA........SY.LELLNDN.SKKMIQ......LLTNES..V..Q...SI.TLKR.ILST.S.V 204

beta    VIDNEALYDICFRTLKLTTPTYGDLNHLVSAAMSGVTCSLRFPGQLNSDLRKLAVNLIPFPRLHFFMYGFAPLTSRGSQQYRA-LTVPELTQQMFDAKNMM 300
alpha   ML....I....KKN.DIER...TN..R.IAQVI.SL.A....D.A..V.VTEFQT..V.Y..I..MLSSY..II.AEKAYHEQ-.S.S.I.NSA.EPAS.. 302
gamma   ....TS.NR.FVER...NN..FQQT.TII.NV..AS.TT..Y..SM.N.MIS.ISS..IN.KC..LITSYT.I.IDKHISNVQKT..LDVMKRLLHT..I. 305

beta    CTSDPRHGRYLTACAMFRGRMSTKEVDEQMLNVQNKNSSYFVEWIPHNTKSSVCDIPPL---G--L-K-M-AVTFVGNSTAIQEMFKRVSDQFTAMFRRKAFLHWYTG 400
alpha   AKC.....K.MAC.L.Y..DVVP.D.NAAVATIKT.R.IQ..D.C.TGF.CGINYQ..TVVP.GD.A.V.R..CMIS.....A.V.S.MDQK.DL.YAKR..V...V. 410
gamma   VSAPV.R.M.ISILNII..ETDPTQ.HKGLQRIRDRKLVN.IK.N.ASIQVTLAKQS.HVV-S--QH.V--CGLMMA.H.S.STL.E.CVT..DRLYK.R...EN.KK 408

beta    EGMDEMEFTEAESNMNDLVSEYQQYQDATAEEEGEFEEEEGDVEA  445
alpha   ...E.G..S..REDLAA.EKD.EEVGIESNDG...D.GY.  450
gamma   .S.FSSADGQGNFEEMESSK.IT.NLIDEYKSAERDDYFTNTYI 452
```

Figure 5.1 Alignment of β-, α-, and γ-tubulin sequences from *Plasmodium falciparum*. Dots represent identity to the β-tubulin sequence; dashes represent gaps introduced to maximize alignment. Double underline, conserved in all β-tubulins; single underline, conserved with one exception (adapted from Katiyar and Edlind, 1996).

5.4 VARIABLES IN PHYLOGENETIC ANALYSIS OF TUBULINS

There are a number of variables in phylogenetic analysis that can affect the results, some of which are specific to tubulin. The first of these is the dataset used. While including all or nearly all available sequences might seem to be the least biased approach, as mentioned above the tubulin sequence database is clearly biased towards 'higher' eukaryotes. It is best therefore to select a diverse, representative set of sequences; for example, two or three sequences per phylum, where no two sequences are >95% identical. Many organisms have multiple tubulin gene copies, in which case the most conserved copy is selected (the divergent copies presumably have specialized functions, or may not even be expressed). Some reported β-tubulin sequences are unusually divergent (compared to related species) for unknown reasons, such as those from *Bombyx mori* (silkworm), *Euplotes crassus* (ciliate), and even the common yeast (*Saccharomyces cerevisiae* and *Candida albicans*). While divergent sequences should not be excluded from the dataset just for this reason, it is nevertheless important to compare results with and without these sequences to see what effect their inclusion has. In one analysis, we observed that the inclusion of *C. albicans* β-tubulin generated an anomalous result (Figure 2B in Edlind *et al.*, 1996b).

A second variable is sequence alignment. The tubulins have a clear advantage over rRNAs and many proteins in this regard. Up to residue 430, no more than one or two single residue insertions or deletions are required to align any two β-tubulin sequences. The 10 to 30 residues following 430 are highly variable, although generally acidic, and are excluded from the analysis. Furthermore, α, β, and γ-tubulins can be aligned with little ambiguity as well. As shown in Figure 5.1 which uses *Plasmodium falciparum* sequences as an example, there are only two regions, involving insertions of 2 and 8 residues in α-tubulin, where alignment between β- and α-tubulin is ambiguous. Similarly, there are only three regions where alignment between β- and γ-tubulin is ambiguous.

A third variable in phylogenetic analysis is the number of informative sites. While smaller than rRNA, the tubulins are larger than most proteins that have been used. Within residues 1 to 430, there are 349 residues that vary at least once, and 273 that vary twice or more (Figure 5.1). If partial tubulin sequences derived from PCR products are used, the number of informative sites is obviously reduced and the results should be interpreted more cautiously.

The outgroup chosen represents a fourth important variable. As mentioned above, for rRNAs, EFs, and enzymes of core metabolism, outgroups from the eubacteria or archaebacteria are used. For tubulin, on the other hand, the α, β, and γ-tubulins provide appropriate outgroups for each other. Indeed, comparing the results obtained with two different outgroups (e.g. α- and γ-tubulin with a β-tubulin dataset) provides one measure of robustness.

A final variable is certainly the methods of phylogenetic analysis employed. As with selection of dataset and outgroup, the best option is to use at least two distinct methods and compare results. For our studies, we have used the distance and parsimony-based methods available in the PHYLIP package (Felsenstein, 1993).

5.5 TREES BASED ON NEARLY COMPLETE TUBULIN SEQUENCES

A parsimony-based tree of representative β-tubulin sequences encompassing residues 1 to 430 is shown in Figure 5.2. Along with animals, plants, and fungi, this tree includes sequences from the four phyla of amitochondriate protozoa (Archezoa?) and eight additional protozoan phyla (nine if the oomycete *Achlya klebsiana* from kingdom Chromista (Cavalier-Smith, 1993) is included). The tree is rooted with both α- and γ-tubulin sequences; various individual outgroups tested did not alter the result.

The reproducibility of specific branches was evaluated by 100 bootstrap resamplings (Felsenstein, 1993); the percentages of trees in which a given branch (encompassing exclusively those sequences shown to the right of the branch) was obtained are shown (where no number is shown, reproducibility was $<50\%$). When this same dataset was analysed by the distance method, an anomalous result lacking bootstrap support was obtained: the ascomycetous yeasts *C. albicans* and *Schizosaccharomyces pombe* branched near the base of the tree, separate from other fungi. However, as mentioned above, the *C. albicans* sequence is relatively divergent, and when it was omitted a distance-based tree was obtained that was almost identical to the parsimony-based tree (Edlind *et al.*, 1996b).

In both trees, the earliest branch is *E. histolytica* β-tubulin, followed reproducibly by *T. vaginalis* β-tubulin. Similarly, in both trees these are soon followed by a split leading to two major lineages: a fungus–animal lineage and a protozoan–plant lineage. Depending on the method used, the *G. lamblia* and *Physarum polycephalum* (slime mould) sequences branch either just before or just after this split, but bootstrap support is weak in both cases. Regardless, the two lineages are well defined. The fungus–animal lineage includes an ascomycete–basidiomycete cluster (unfortunately, no full length zygomycete sequences are available as yet) which, quite surprisingly, includes the sequence from the microsporidian *E. hellem*. The protozoan–plant lineage includes a number of previously recognized relationships (Cavalier-Smith, 1993), as well as a few new

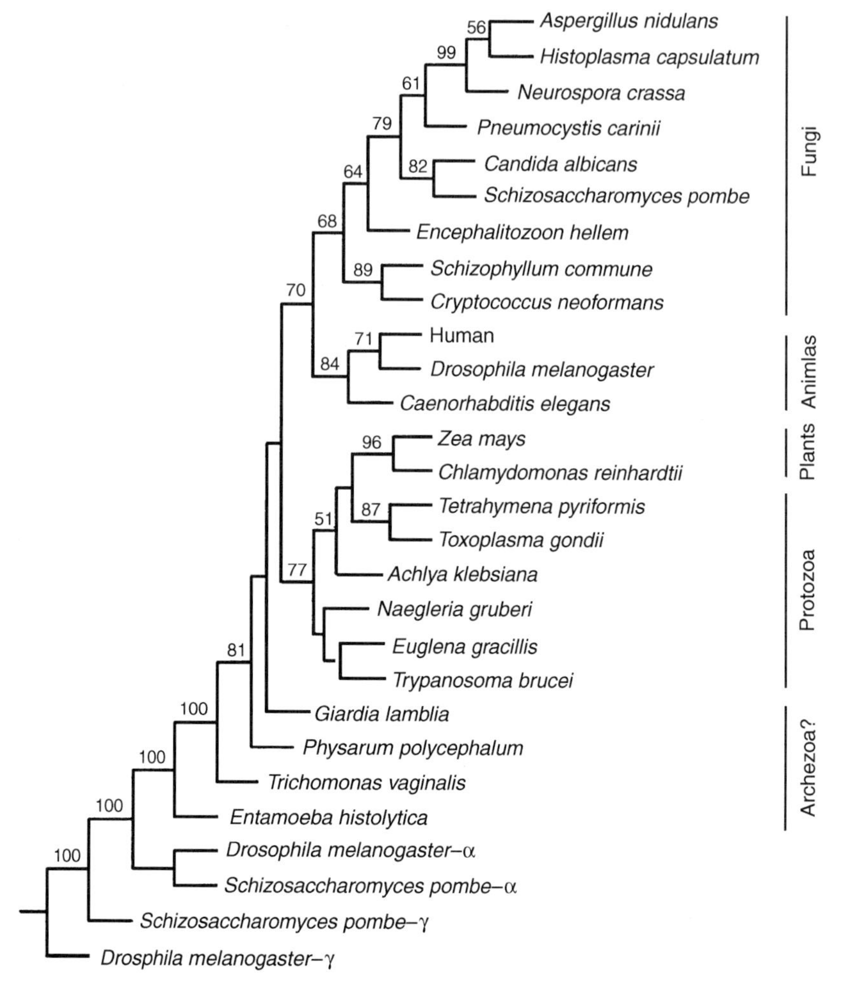

Figure 5.2 Parsimony-based phylogenetic tree of representative β-tubulin sequences (residues 1 to 430). Selected α- and γ-tubulin sequences were also included; *D. melanogaster* γ-tubulin was the designated outgroup. Tree shown is one of two shortest trees; the second shortest tree was identical except that the *G. lamblia* and *P. polycephalum* branches were switched. Numbers indicate percentage of trees in which the group of species to the right of that branch was found in 100 bootstrap resamplings (from Edlind *et al.*, 1996b).

ones. Specifically, there are reproducible clusters of (1) higher plant (*Zea mays*) with unicellular alga (*Chlamydomonas reinhardtii*), (2) ciliate (*Tetrahymena pyriformis*) with apicomplexan (*Toxoplasma gondii*), and (3) *Euglena gracilis* with kinetoplastid (*Trypanosoma brucei*). The inclusion of the oomycete (*A. klebsiana*) sequence among the protozoa rather than fungi is similarly consistent with current opinion. To our knowledge, the clustering (in both parsimony and distance-based trees) of *Naegleria gruberi* (phylum Percolozoa) with *E. gracilis* and *T. brucei* has not been previously described. A more fundamental observation may be that the plant kingdom is a branch within the protozoan–plant lineage rather than an independent branch off the main line of descent, as implied by most trees (Sogin, 1989; Bauldauf and Palmer, 1993; Drouin, Moniz de Sa and Zuker, 1995; Doolittle *et al.*, 1996; Kamaishi *et al.*, 1996). Evolutionary relationships between plants and the Ciliophora–Apicomplexa group (which includes Dinozoa, for which tubulin sequences are lacking) have been proposed (based on, for example, the presence of plastid DNA in Apicomplexa and many Dinozoa), but to our knowledge not previously supported by phylogenetic analysis.

Do α-tubulin and β-tubulin give consistent trees? This has been examined by analysing paired sequences from 14 species (Edlind *et al.*, 1996b). The dataset however included only one amitochondriate (*E. histolytica*) and hence is not particularly useful for examining early eukaryotic evolution. Nevertheless, with both tubulins *E. histolytica* branched first, followed by fungi and animals (as separate lineages) and then protozoa and plants. In some respects, the two molecules give complementary results: while there is significant bootstrap support (84%) for a *T. gondii–T. pyriformis* branch in the β but not α-tubulin tree, the *E. gracilis–T. brucei* branch receives significant support (75%) in the α but not β-tubulin tree. Also, as with β-tubulin (see above), α-tubulins support a relationship between *N. gruberi* and the *E. gracilis–T. brucei* group.

5.6 TREES BASED ON PARTIAL TUBULIN SEQUENCES

Starting with relatively little or relatively impure DNA, partial tubulin sequences can be rapidly generated by PCR using various primer pairs (see above). These partial sequences can be used to answer specific phylogenetic questions. As shown in Figure 5.3A, parsimony analysis of β-tubulin residues 108 to 259 generated a tree very similar to the tree generated with residues 1 to 430 (Figure 5.2), although as expected there is less bootstrap support for most branches.

More importantly, this tree may be used to examine two questions: first, do all amoebae have highly atypical β-tubulins that branch early, as with *E. histolytica* tubulin? If yes, this would suggest that the *E. histolytica* early branch is an artefact. Partial β-tubulin sequences were determined from *Acanthamoeba polyphaga* (phylum Rhizopoda), an aerobic amoeba. In contrast to *E. histolytica, A. polyphaga* β-tubulin branched within the protozoan–plant lineage (Figure 5.3A). Indeed, the inclusion

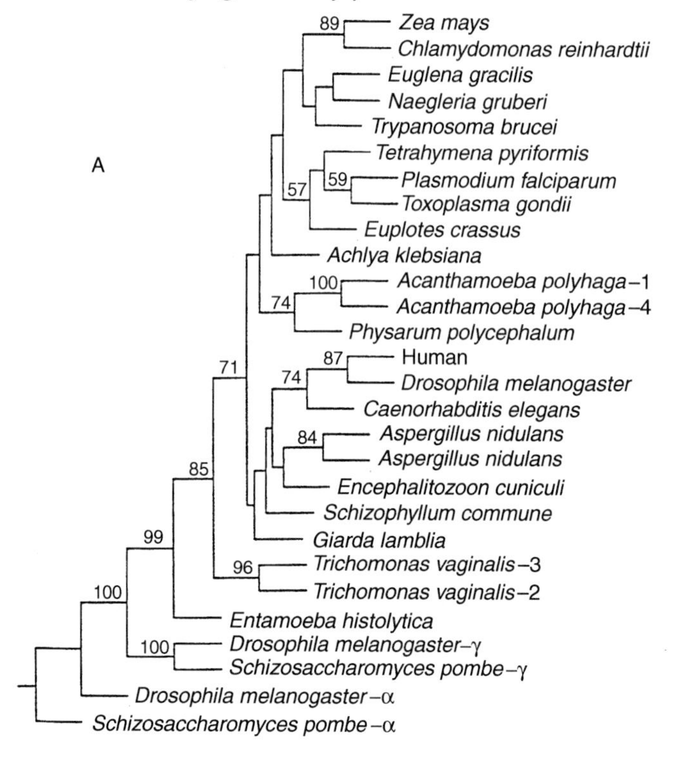

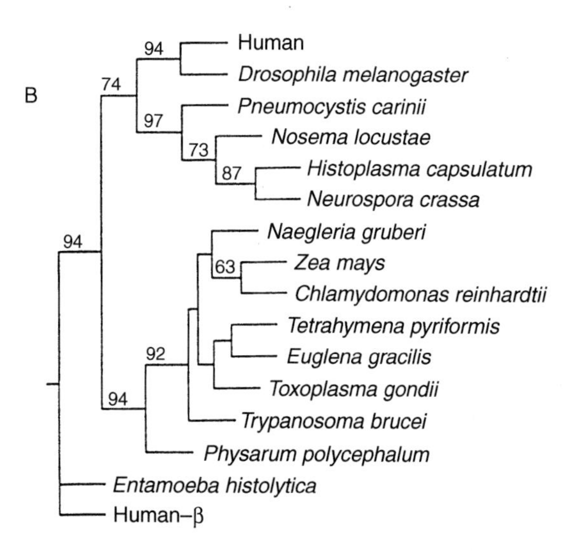

Figure 5.3 Phylogenetic trees of representative partial tubulin sequences. (A) Residues 108 to 259 of β-tubulin analysed by parsimony; selected α- and γ-tubulin sequences were included as outgroups (from Edlind *et al.*, 1996b). (B)

of *A. polyphaga* sequences resolves the location of the *P. polycephalum* branch which was ambiguous in the prior analysis (Figure 5.2): *P. polycephalum,* and the related *Dictyostelium discoideum* (not shown) cluster within the protozoan–plant lineage. These three genera also form a tight cluster in an actin-based phylogeny (Drouin, Moniz de Sa and Zuker, 1995).

The second question addressed with partial sequences was: do the conserved and divergent copies of β-tubulin branch independently? Again, if yes, this would suggest a potential source of artefacts. In the three cases examined, involving conserved and divergent sequences from *Aspergillus nidulans, A. polyphaga,* and *T. vaginalis,* the sequences branched together. For the latter two organisms which have from four to seven β-tubulin gene copies (Katiyar and Edlind, 1994; Edlind *et al.,* 1996b), determination of these sequences was greatly facilitated by PCR.

Partial tubulin sequences have also been used to further investigate the phylogeny of microsporidia. Using PCR, partial β-tubulin sequences were determined for two additional species related to *E. hellem, E. cuniculi* and *E. (Septata) intestinalis,* and a fourth unrelated species parasitic to lepidopterian insects, *Nosema locustae* (see chapter 4, this volume). The microsporidian sequences branched together, and furthermore they continued to branch with fungal sequences, although this relationship had only weak (60%) bootstrap support (Li *et al.,* 1996). However, phylogenetic analysis of a partial α-tubulin sequence from *N. locustae* generated a more convincing result, with 97% bootstrap support for a microsporidia–fungus connection (Figure 5.3B).

Finally, partial β-tubulin sequences were used to determine if *E. histolytica* is a true representative of its genus. It is: partial sequences from the reptilian parasite *Entamoeba invadens* were as unusual as those from *E. histolytica* (Katiyar and Edlind, 1996).

5.7 CURRENT TUBULIN-BASED MODELS FOR EARLY EUKARYOTIC EVOLUTION

Recent studies involving rRNA (Wainright *et al.,* 1993) and several different proteins (Bauldauf and Palmer, 1993; Doolittle *et al.,* 1996) have demonstrated that animals and fungi share a common lineage. Our

---

Residues 110 to 259 of α-tubulin analysed by distance method; human β-tubulin was the outgroup (from Li *et al.,* 1996). Numbers indicate percentage of trees in which the group of species to the right of that branch was found in 100 bootstrap resamplings.

studies reviewed here using both α- and β-tubulin sequences have confirmed this relationship. Furthermore, with an expanded database that included sequences from the four phyla of amitochondriate protozoa, and with the use of appropriate outgroups, the tubulin-based phylogenies have resolved a second major lineage encompassing plants and most protozoa.

Where do the amitochondriate protozoa fall relative to the 'big split' leading to these two major lineages? With the notable exception of the microsporidia (see below), the amitochondriate β-tubulins branch either clearly before (*E. histolytica* and *T. vaginalis*) or right around (*G. lamblia*) this split. This relationship suggests another way to define the Archezoa, in addition to 'primitively without mitochondria': organisms that branched before this major split. Further, it suggests that the acquisition of aerobic metabolism (or perhaps sexual reproduction) was closely associated with the dramatic radiation of eukaryotic species that comprise the two major lineages. The Archezoa survived by and large only as parasites or symbionts of these aerobic eukaryotes.

While rRNA and tubulin-based trees agree with regards to the early branching of *T. vaginalis* and *G. lamblia*, these two types of molecules give very different results for *E. histolytica* and microsporidial species. Specifically, with tubulin, *E. histolytica* branches very early while microsporidia branch within the fungi, while with rRNA, microsporidia branch early and *E. histolytica* branches with mitochondria-containing protozoa.

Since several protein sequences also place *E. histolytica* among the mitochondria-containing protozoa (Hasegawa *et al.*, 1993; Clark and Roger, 1995; Drouin, Moniz de Sa and Zuker, 1995), the simplest explanation for the observed discrepancy is that *Entamoeba* tubulins are extremely divergent rather than ancient, and their early branching is artefactual. This could have resulted from a reduced or altered role of microtubules in this organism, presumably related to its amoeboid nature. However, β-tubulin from an aerobic amoeba, *A. polyphaga*, was not similarly divergent, so *Entamoeba* may be unique in this regard. Indeed, *E. histolytica* forms an atypical mitotic spindle lacking a metaphase stage (Orozco *et al.*, 1988; Meza, 1992). Thus, the potential artefactual branching of *Entamoeba* species does not invalidate tubulin sequences as phylogenetic markers. Rather, it indicates that specific exceptions are to be expected, as with any marker. In this case, there is a logical basis for the exception: an amoeboid cell with minimal and apparently atypical use of microtubules. *T. vaginalis* and *G. lamblia*, in contrast, make extensive use of microtubules.

Microsporidia do not outwardly resemble fungi, but this is true as well for the AIDS-associated pathogen *Pneumocystis carinii*, a former protozoan reclassified as a fungus. As with *P. carinii* (Li and Edlind, 1994), tubulin-based phylogenetic analyses reviewed above strongly support a fungal origin for microsporidia. Analyses that place organisms within an established group are more reliable than analyses that place organisms at the extremities of a tree. What reasons then might explain an artefactual placement of microsporidia at the base of the rRNA (Sogin, 1989; Leipe *et al.*, 1993) and EF-1α (Kamaishi *et al.*, 1996) eukaryotic trees? In retrospect, phylogenetic analysis of microsporidial rRNA was questionable since secondary structure analysis indicates it has suffered multiple deletions. That is, there are many otherwise universally conserved helices that are absent in only two types of rRNA: microsporidial and mitochondrial (Gutell, Gray and Schnare, 1993; Neefs *et al.*, 1993; Katiyar, Visvesvara and Edlind, 1995), and for the latter there is no dispute that deletions are responsible. (This analogy to mitochondria may go even further, since both in their own ways are obligate intracellular parasites.) To preserve ribosome function, compensating changes would likely occur in adjacent and possibly even distant rRNA sequences in response to deletion events. Thus, deletions have an unpredictable effect on rRNA sequences and phylogenetic analysis of those sequences. With regards to EF-1α, since it is similarly involved in protein synthesis, it very likely coevolves with rRNA. Thus, the early branching obtained with both markers can be explained as long branch artefacts resulting from their highly divergent sequences.

Although microsporidial rRNA and EF-1α are potentially unsuitable for phylogenetic analysis, there may still be clues within their sequences as to the origins of these organisms. For example, within the 'decoding region' near the 3′ end of small subunit rRNA is a base pair that is conserved in eubacteria, archaebacteria, plants, and nearly all protozoa (including *G. lamblia*, *T. vaginalis*, and *E. histolytica*), but is absent in fungi, animals, and microsporidia (Katiyar, Visvesvara and Edlind, 1995). Although only one base pair, its functional importance has been demonstrated by mutational analysis (De Stasio and Dahlberg, 1990). Perhaps more convincingly, there is an 11 residue insertion in EF-1α from the microsporidian *G. plecoglossi* (Kamaishi *et al.*, 1996) that correlates with a 12 residue insertion specific to the fungus-animal lineage (Bauldauf and Palmer, 1993). Sequence divergence within the insertion is high but consistent with the overall divergence of *G. plecoglossi* EF-1α (Figure 5.4). Additional observations in support of a fungal affinity for microsporidia, in particular their capacity for sexual reproduction, have been reviewed in Chapter 4.

5.8 PROSPECTS

As with any molecule (including rRNA), the use of tubulin in phylogeny might be compromised by its variable use in cells: while the mitotic spindle appears to be universal, cytoskeletal and flagellar micro-

```
               GDNM E S NMPWYKGW  E KAG   G TLLEALD   PP RP  consensus
Xenopus        ....L.P.P....F...KITR.E.SGS.T.......C.L..S..  Animals
Drosophila     ....L.A.DRL......NI.R.E.KAD.K...D...A.L..S..
Saccharomyces  ....I.ATT.A......EK.T...VVK.K.....I.A.EQ.S..  Fungi (Ascomycete)
Puccinia       ....L.E.T..G.F...TK.T...VSK.K...D.I.A.E..S..        (Basidiomycete)
Mucor          ....LDE.T....F...NK.T...SKT.K.....I.A.E..V..        (Zygomycete)
Glugea         .I.IV.KGDKFE.F...KPVSG..D-SIF..EG..NSQI..P..  Microsporidia
Giardia        ...IM.K.DK....E.------------PC.ID.I.GLKA.K..  Archezoa (Metamonada)
Hexamita       ...IM.A.PKT....------------KC.I.CI.GLKA.K..
Entamoeba      ....I.P.T......------------P..IG...SVT..E..  Protozoa
Blastocystis   ....I.H.A......------------P.......NVH..K..
Dictyostelium  ....L.R.DK.E...------------P.......A.VE.K..
Euglena        ....I.A.E..G...------------L..IG...NLE..K..
Aridopsis      ....I.R.T.LD...------------P.......Q.NE.K..  Plants
Sulfolobus     ...VTHK.TK....N.------------P..E.L..QLEI.PK.  Archaebacteria
Halobacterium  ...IA.E.EHTG..D.------------EI.....NELPA.EP.
```

Figure 5.4 Alignment of EF-1α residues (approximate positions 200 to 240) from representative organisms. The top line is a consensus sequence (based only on the sequences shown). Dots indicate identity to the consensus sequence; dashes indicate gaps introduced to maximize alignment.

tubules are not. As discussed, this is the most likely explanation for the incongruence between *Entamoeba* phylogenies based on tubulin versus rRNA and other proteins. Given the diversity of life that evolution has generated, no single molecule will prove unambiguous for every organism. However, tubulins appear to be particularly useful for looking at the 'big picture' of eukaryotic evolution. Their use is relatively straightforward owing to the ease with which they can be aligned, the insensitivity to choice of outgroup (β, α, or γ-tubulin), and relative ease of cloning tubulin genes using PCR with primers based on conserved sequences. There are however, significant gaps in the tubulin databases that need to be filled; for example, dinoflagellates, diatoms, and zygomycetes are currently unrepresented. To better resolve the evolutionary events leading to the major split, additional amitochondriate genera need to be studied, such as *Hexamita* (Metamonada) and *Dientamoeba* (Parabasalia). Analysing other amoebae may explain the apparently anomalous results obtained for *Entamoeba*. Finally, since microtubules are required for meiosis as well as mitosis, tubulin sequences may hold clues to the evolution of sexual reproduction.

5.9 REFERENCES

Bauldauf, S.L. and Palmer, J.D. (1993) Animals and fungi are each other's closest relatives: congruent evidence from multiple proteins. *Proceedings of the National Academy of Sciences USA*, **90**, 11558–62.

Beauvais, B., Sarfati, C., Challier, S. and Derouin, F. (1994) In vitro model to assess effect of antimicrobial agents on *Encephalitozoon cuniculi*. *Antimicrobial Agents and Chemotherapy*, **38**, 2440–8.

Burns, R.G. (1991) α-, β-, and γ-Tubulins: Sequence comparisons and structural constraints. *Cell Motility and Cytoskeleton*, **20**, 181–9.

Cavalier-Smith, T. (1987) Eukaryotes with no mitochondria. *Nature*, **326**, 332–3.

Cavalier-Smith, T. (1993) Kingdom protozoa and its 18 phyla. *Microbiology Reviews*, **57**, 953–94.

Clark, C.G. and Roger, A.J. (1995) Direct evidence for secondary loss of mitochondria in *Entamoeba histolytica*. *Proceedings of the National Academy of Sciences USA*, **92**, 6518–21.

De Stasio, E.A. and Dahlberg, A.E. (1990) Effects of mutagenesis of a conserved base-paired site near the decoding region of *Escherichia coli* 16S ribosomal RNA. *Journal of Molecular Biology*, **212**, 127–33.

Doolittle, R.F., Feng, D.F., Tsang, S. *et al.* (1996) Determining divergence times of the major kingdoms of living organisms with a protein clock. *Science*, **271**, 470–1.

Drouin, G., Moniz de Sa, M. and Zuker, M. (1995) The *Giardia lamblia* actin gene and the phylogeny of eukaryotes. *Molecular Evolution*, **41**, 841–9.

Edlind, T.D., Katiyar, S., Visvesvara, G. and Li, J. (1996a) Evolutionary origins of microsporidia and basis for benzimidazole sensitivity: an update. *Journal of Eukaryotic Microbiology*, in press.

Edlind, T.D., Li, J., Visvesvara, G.S. *et al.* (1996b) Phylogenetic analysis of β-tubulin sequences from amitochondriate protozoa. *Molecular and Phylogenetic Evolution*, **5**, 359–67.

Felsenstein, J. (1993) PHYLIP Version 3.5c. Distributed by the author, Department of Genetics, University of Washington, Seattle.

Gutell, R.R., Gray, M.W. and Schnare, M.N. (1993) A compilation of large subunit (23S and 23S-like) ribosomal RNA structures: *Nucleic Acids Research*, **21**, 3055–74.

Hasegawa, M., Hashimoto, T., Adachi, J. *et al.* (1993) Early branchings in the evolution of eukaryotes: Ancient divergence of *Entamoeba* that lacks mitochondria revealed by protein sequence data. *Journal of Molecular Evolution*, **36**, 380–8.

Hasegawa, M. and Hashimoto, T. (1993) Ribosomal RNA trees misleading? *Nature*, **361**, 23.

Hashimoto, T., Nakamura, Y., Nakamura, F. *et al.* (1994) Protein phylogeny gives a robust estimation for early divergences of eukaryotes: phylogenetic place of a mitochondria-lacking protozoan, *Giardia lamblia*. *Molecular and Biological Evolution*, **11**, 65–71.

Kamaishi, T., Hashimoto, T., Nakamura, Y. *et al.* (1996) Protein phylogeny of translation elongation factor EF-1α suggests microsporidians are extremely ancient eukaryotes. *Journal of Molecular Evolution*, **42**, 257–63.

Katiyar, S.K. and Edlind, T.D. (1994) β-Tubulin genes of *Trichomonas vaginalis*. *Molecular and Biochemical Parasitology*, **64**, 33–42.

Katiyar, S.K. and Edlind, T.D. (1996) *Entamoeba histolytica* encodes a highly divergent β-tubulin. *Journal of Eukaryotic Microbiology*, **43**, 31–4.

Katiyar, S.K., Visvesvara, G.S. and Edlind, T.D. (1995) Comparisons of ribosomal RNA sequences from amitochondriate protozoa: implications for processing, mRNA binding and paromomycin susceptibility. *Gene*, **152**, 27–33.

Kirk-Mason, K.E., Turner, M.J. and Chakraborty, P.R. (1988) Cloning and sequence of β-tubulin cDNA from *Giardia lamblia. Nucleic Acids Research*, **16**, 2733.

Kirk-Mason, K.E., Turner, M.J. and Chakraborty, P.R. (1989) Evidence for unusually short tubulin mRNA leaders and characterization of tubulin genes in *Giardia lamblia. Molecular Biochemistry and Parasitology*, **36**, 87–100.

Leipe, D.D., Gunderson, J.H., Nerad, T.A. and Sogin, M.L. (1993) Small ribosomal RNA of *Hexamita inflata* and the quest for the first branch in the eukaryotic tree. *Molecular Biochemistry and Parasitology*, **59**, 41–8.

Li, J. and Edlind, T.D. (1994) Phylogeny of *Pneumocystis carinii* based on β-tubulin sequence. *Journal of Eukaryotic Microbiology*, **41**, 97S.

Li, J., Katiyar, S.K., Hamelin, A. *et al.* (1996) Tubulin genes from AIDS-associated microsporidia and implications for phylogeny and benzimidazole sensitivity. *Molecular Biochemistry and Parasitology*, **78**, 289–95.

Little, M. (1985) An evaluation of tubulin as a molecular clock. *Biosystems*, **18**, 241–7.

Markos, A., Miretsky, A. and Müller, M. (1993) A glyceraldehyde 3-phosphate dehydrogenase with eubacterial features in the amitochondriate eukaryote, *Trichomonas vaginalis. Journal of Molecular Evolution*, **37**, 631–43.

Markos, A., Morris, A., Rozario, C. and Müller, M. (1996) Primary structure of a cytosolic malate dehydrogenase of the amitochondriate eukaryote, *Trichomonas vaginalis. FEMS Microbiology Letters*, **135**, 259–64.

Meza, I. (1992) *Entamoeba histolytica*: phylogenetic considerations. A review. *Archives of Medical Research*, **23**, 1–5.

Neefs, J.M., Van de Peer, Y., De Rijk, P. *et al.* (1993) Compilation of small ribosomal subunit RNA structures. *Nucleic Acids Research*, **21**, 3025–49.

Orozco, E., Solis, F.J., Dominquez, J. *et al.* (1988) *Entamoeba histolytica*: cell cycle and nuclear division. *Experimental Parasitology*, **67**, 85–95.

Philippe, H. and Adoutte, A. (1995) How reliable is our current view of eukaryotic phylogeny? in *Protistological Actualities* (eds G. Brugerole and J.-P. Mignot) Clermont-Ferrand, pp. 17–33.

Sanchez, M.A., Peattie, D.A., Wirth, D. and Orozco, E. (1994) Cloning, genomic organization and transcription of the *Entamoeba histolytica* α-tubulin-encoding gene. *Gene*, **146**, 239–44.

Sogin, M.L. (1989) Evolution of eukaryotic microorganisms and their small subunit ribosomal RNAs. *American Zoologist*, **29**, 487–99.

Wainright, P.O., Hinkle, G., Sogin, M.L. and Stickel, S.K. (1993) Monophyletic origins of the metazoa: an evolutionary link with fungi. *Science*, **260**, 340–2.

6

Enzymes and compartmentation of core energy metabolism of anaerobic protists – a special case in eukaryotic evolution?

Miklós Müller

The Rockefeller University, 1230 York Avenue, New York, NY 10021, USA. E-mail: mmuller@rockvax.rockefeller.edu

ABSTRACT

Protists display a great variety in their morphological and biochemical organization, from relatively simple levels to high complexity. The enzymatic and organellar organization of the core energy metabolism of protists also varies. Of the few metabolic patterns recognized so far two are amitochondriate: Type I without metabolic compartmentation (seen in *Giardia lamblia* and *Entamoeba histolytica*) and Type II with cytosolic/hydrogenosomal compartmentation (seen in *Trichomonas vaginalis* and other parabasalids, anaerobic ciliates and fungi). All amitochondriate eukaryotes lack biochemically recognizable mitochondrial functions (tricarboxylic acid cycle, cytochrome mediated electron transport and oxidative phosphorylation), but they also display positive distinguishing features. They oxidize pyruvate with pyruvate:ferredoxin oxidoreductase, an enzyme restricted to prokaryotes and amitochondriate eukaryotes, and not with the non-homologous pyruvate dehydrogenase complex. Type I organisms have a number of other enzymes of extended glycolysis which differ from those found in Type II organisms and in mitochondriate protists, e.g. pyruvate dikinase instead of pyruvate kinase, acetate thiokinase (ADP-forming) instead of succinate thiokinase, bifunctional acetyl-CoA reductase/aldehyde reductase instead of alcohol dehydrogenase. Type II organisms differ from the mito-

Evolutionary Relationships Among Protozoa. Edited by G.H. Coombs, K. Vickerman, M.A. Sleigh and A. Warren. Published in 1998 by Chapman & Hall, London. ISBN 0 412 79800 X

chondriate group by lacking mitochondria but containing hydrogenosomes with a pathway producing H_2. Since the acquisition of mitochondria has been regarded as a later event in eukaryotic evolution, amitochondriate groups, or at least some of them, were assumed to be premitochondrial. This conclusion has been integrated into proposals to divide the eukaryotes into premitochondriate Archaezoa and postmitochondriate Protozoa. A confrontation of the metabolic types with the rRNA derived phylogenetic position of amitochondriate organisms shows, however, that either type is present in more than one lineage. The data argue for the polyphyletic origin of the amitochondriate condition by convergent evolution and also indicate that most, or probably all, amitochondriate lineages arose from mitochondriate ones. Energy metabolism thus does not display a clear directional evolution. The results pose the challenging task of defining the phylogenetic history of the enzymes and organelles participating in protist energy metabolism. Accumulating sequence data on the metabolic enzymes of amitochondriate protists indicate that the history of individual enzymes is complex and rarely congruent with the rRNA tree. All enzymes of amitochondriate core metabolism studied so far are more closely related to their eubacterial homologs than to archaebacterial ones. Elucidation of the paths whereby similar metabolic systems were acquired by diverse protists is a goal of future studies.

6.1 INTRODUCTION

The biological and morphological diversity of unicellular eukaryotes, protists, has fascinated scientists and laymen for a long time. The notion that a comparison of diverse protists should give insight into the evolution of the eukaryotic cell resulted in numerous proposals for phylogenetic reconstruction of the protist world and for establishing the relationships of protists with multicellular organisms. A pervasive thought was that during evolution morphological and biochemical complexity of the protist cell increases (Cavalier-Smith, 1993; Hülsmann and Hausmann, 1994), though proposals for evolution by functional losses have also been put forward (Lwoff, 1951).

These considerations have been applied to various aspects of the protist cell. One of these, core metabolism, has attracted significant interest, since it was recognized relatively early that the metabolic characteristics of protists are more diverse than those of multicellular organisms. Certain protists were found to lack mitochondria (Cavalier-Smith, 1987b; Cavalier-Smith, 1993; Fenchel and Finlay, 1995) and to rely on less complex metabolic machinery than others (Müller, 1988; Fairlamb, 1989; Coombs and Müller, 1995), raising the hope that through extensive comparisons one will be able to reconstruct the evolutionary paths leading to the characteristic metabolic patterns of the multicellular organisms. Two quotes demonstrate this: 'The shapeless, amoeba-like, multinucleate freshwater protist, *Pelomyxa palustris* is obviously a most ancestral animal, since it contains real nuclei but no mitochondria ... It is likely that such an animal will display ancestral traits in its fermentation and respiration' (Leiner *et al.*, 1968, my translation from the German original); '*Entamoeba histolytica*, grown axenically may be one of the most primitive organisms, a 'metabolic fossil'' (Reeves, 1984). The 'ancestral' status of such organisms is by no means a foregone conclusion, since they could have equally well arisen from more 'advanced' ancestors through the loss of their mitochondrial enzyme systems and the organelle itself. Since the amitochondriate groups inhabit anoxic and hypoxic habitats (Fenchel and Finlay, 1995), such functional losses could represent adaptations to the lack of O_2 as terminal electron acceptor. This chapter revisits this problem by looking at the core metabolism of such 'simple' protists and testing the notions of their 'primitivity' and 'ancestral nature'.

After an outline of the metabolic features of amitochondriate organisms, three problems will be examined. First, the systematic distribution of amitochondriate protists will be described. Second, the idea that certain metabolic characteristics of amitochondriate metabolism represent ancestral features and argue for an ancestral nature of these organisms will be revisited. Third, selected enzymes of amitochondrial protists will be examined to evaluate their history and to trace the origins of amitochondriate metabolism in protists.

6.2 AMITOCHONDRIATE METABOLISM

Morphological and ecological criteria strongly indicate that numerous protists living in anoxic and hypoxic environments are amitochondriate (Fenchel and Finlay, 1995). Supporting biochemical data, however, exist only for a few organisms that are almost all parasites or symbionts of mammals (Table 6.1).

These limited data allow the metabolic characterization of organisms belonging to at least five separate lineages. Based on the subcellular organization of their core metabolism these can be grouped into two metabolic types. Species without compartmentalized energy metabolism, *Giardia lamblia* (Adam, 1991; Jarroll and Paget, 1995) and *Entamoeba histolytica* (Reeves, 1984; McLaughlin and Aley, 1985), are Type I organisms. The other group is characterized by the presence of a membrane-bounded organelle of core metabolism, the hydrogenosome (Müller, 1993). This represents Type II core metabolism with the parabasalids, *Trichomonas vaginalis* (Müller, 1989) and *Tritrichomonas foetus* (Müller, 1976), the ciliates *Daystricha ruminantium* from sheep rumen (Yarlett, Lloyd and Williams, 1985; Williams, 1986), *Trimyema com-*

Table 6.1 Amitochondriate organisms studied in detail with biochemical methods

Group and species[a]	*Type*	*Compartmentation*	*Major end products*[b]
Diplomonads (early branch)			
Giardia lamblia	I	none	et, ac[c]
Parabasalids (early branch)			
Trichomonas vaginalis	II	hydrogenosome	lact, ac, H_2[d]
Tritrichomonas foetus	II	hydrogenosome	succ, ac, H_2[e]
Entamoebids (intermediate branch)			
Entamoeba histolytica	I	none	et, ac[f]
Ciliates (late branch)			
Dasytricha ruminantium	II	hydrogenosome	lact, but, ac, H_2[g]
Trimyema compressum	II	hydrogenosome	form, ac, H_2[h]
Fungi (late branch)			
Neocallimastix spp.	II	hydrogenosome	form, ac, lact, et, H_2[i]

[a] In view of lack of consensus on the higher level taxonomy of protists, only vernacular names are used for higher taxa. For phylogenetic positions see Figure 6.2.
[b] ac, acetate; but, butyrate; et, ethanol; form, formate; lact, lactate; succ, succinate
References: [c]Jarroll and Paget, 1995, [d]Müller, 1988, [e]Müller, 1976; [f]Reeves, 1984; [g]Yarlett *et al.*, 1985, Ellis *et al.*, 1991; [h]Goosen *et al.*, 1990; [i]Marvin-Sikkema *et al.*, 1990; Yarlett, 1994.

pressum, a free-living species (Goosen *et al.*, 1990), and other ciliates, as well as chytrid fungi of the rumen (Yarlett *et al.*, 1986; Marvin-Sikkema *et al.*, 1993; Yarlett, 1994) as its major representatives. These two types differ significantly, although both are functionally amitochondriate and share a number of properties. A brief overview of their energy metabolism will provide a background for the subsequent discussion. For additional details see Müller (1988), Coombs and Müller (1995) and Fenchel and Finlay (1995).

The organisms mentioned above do not require oxygen, are resistant to inhibitors of mitochondrial respiration and lack mitochondrial enzyme systems, e.g. pyruvate dehydrogenase complex, tricarboxylic acid cycle, cytochrome mediated electron transport and electron transport linked oxidative phosphorylation (Müller, 1988). They are fermentative, and oxidize their energy substrates incompletely to organic endproducts and CO_2. The hydrogenosome-containing Type II species produce also H_2, a property also noted in certain green algae (Schulz, 1996), but not in other eukaryotes. The efficiency of such metabolism, since O_2 is not used as terminal electron acceptor, is much lower than that of mitochondrion-containing organisms.

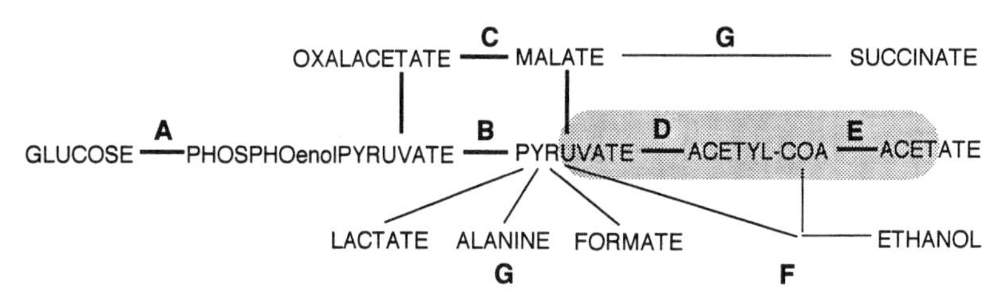

Figure 6.1 Main steps in carbohydrate catabolism of amitochondriate protists. A – Glycolytic formation of phosphoenolpyruvate from glucose; B – direct conversion of phosphoenolpyruvate into pyruvate; C – malate bypass; D – oxidative decarboxylation of pyruvate to acetyl-CoA; E – liberation of acetate from acetyl-CoA; F – ethanol formation; G – formation of other end products. Heavy lines denote conversions known from all species studied, light lines those found only in certain organisms, or representing alternative pathways. Shading covers the part of the pathway that is hydrogenosomal in Type II organisms.

Figure 6.1 provides a scheme of the core metabolism of amitochondriate protists. They share with the mitochondriate organisms a classical Embden-Meyerhof glycolytic pathway, from glucose and glycogen to phosphoenolpyruvate (A), which is subsequently converted to pyruvate, either directly (B) or via a malate bypass (C). In amitochondriate protists, as in other organisms, pyruvate is a branchpoint where pathways leading to different endproducts separate. A pathway present in all amitochondriates leads to acetate with acetyl-CoA as the key intermediate (D and E). Ethanol, another endproduct, is produced either from pyruvate or acetyl-CoA (F). The formation of succinate, lactate, formate and alanine (G), also produced by some but not all amitochondriates, will not be discussed.

Reducing equivalents generated in the pyruvate to acetyl-CoA conversion are transferred to either organic acceptors, with the formation of organic endproducts or to protons with the production of H_2. In the nature of some of these reactions, and of the enzymes catalysing them, amitochondriate organisms differ from the organisms belonging to the mitochondrial type characteristic of most eukaryotes (Table 6.2). The two amitochondriate types also differ from each other in their metabolic characteristics.

The main biochemical feature that separates amitochondriate organisms from mitochondriate ones is the presence of pyruvate:ferredoxin oxidoreductase (PFO), an iron-sulphur protein, instead of pyruvate dehydrogenase (PDH) complex, as the enzyme catalysing oxidative decarboxylation of pyruvate to acetyl-CoA (D). The differences between

Table 6.2 Characteristics of major types of eukaryotic core carbohydrate metabolism

Block	*Property*	*Amitochondriate*		*Mitochondriate*
		Type I	*Type II*	
	Metabolic organelle	none	hydrogenosome	mitochondrion
A-B	Hexose → Pyruvate			
	Number of PPi-linked enzymes	2-3[a]	1	none
	Formation of pyruvate	dikinase	kinase	kinase
D	Pyruvate → Acetyl CoA			
	Pyruvate oxidase	PFO[b]	PFO[b]	PDH[b]
	Electron acceptor	ferredoxin	ferredoxin	NAD
	Terminal acceptor and product	organic → organic	H^+ → H_2	O_2 → H_2O
E	Acetyl CoA →			
	Product	acetate	acetate	CO_2
	Pathway	acetate thiokinase	CoA-transferase	TCA cycle
	ATP formation			
	Substrate level	acetate thiokinase	succinate thiokinase	succinate thiokinase
	Electron transport linked	none	none	present
F	Pyruvate → ethanol via	acetyl-CoA[c]	acetaldehyde	acetaldehyde

[a]From *G. lamblia*, two known. PPi-linked PEP carboxyphosphotransferase present in *E. histolytica* was not tested.
[b]PFO – pyruvate:ferredoxin oxidoreductase; PDH – pyruvate dehydrogenase complex.
[c]Based on *E. histolytica*. Pathway of ethanol formation in *G. lamblia* not elucidated yet.

these two enzymes are manifold, though both perform a similar overall reaction.

Both amitochondriate types also differ from most, but not all, mitochondriate organisms in having a inorganic pyrophosphate (PPi)-dependent and not an ATP-dependent glycolytic phosphofructokinase (in A). This indicates that PPi plays an important role in amitochondriate organisms.

Type I amitochondriate species differ from the rest of the eukaryotes, i.e. Type II amitochondriates and mitochondriates, in three major aspects. First, the enzyme responsible for the conversion of phosphoenolpyruvate to pyruvate (B) is not pyruvate kinase but a second PPi-linked enzyme, pyruvate dikinase. Second, the conversion of acetyl-CoA to acetate, accompanied by substrate level phosphorylation of ADP to ATP (E), is catalysed by a single enzyme, acetyl-CoA synthetase (ADP-forming), otherwise found only in a few prokaryotes. Third, the pathway leading to ethanol formation (F) is not the widely distributed pyruvate decarboxylase-alcohol dehydrogenase system but, at least in *E. histolytica*, it proceeds via acetyl-CoA and is catalysed by the bifunctional acetyl-CoA reductase/aldehyde reductase enzyme, again known only from certain prokaryotes.

Type II amitochondriate species differ from all other eukaryotes in having the pathway from pyruvate to acetate localized in a membrane-bounded organelle, the hydrogenosome (Figure 6.1, shaded area). Acetate formation and conservation as ATP of the energy of the thioester bond of acetyl-CoA formed by PFO proceeds via succinyl-CoA as an intermediate and is effected by a two-enzyme system comprising an acyl-CoA transferase and succinate thiokinase. This organelle produces H_2 by disposing of the electrons generated in the PFO reaction via a simple electron transfer chain consisting of ferredoxin and hydrogenase.

The differences between Type I and Type II organisms cannot be sufficiently emphasized. Though both are functionally amitochondriate and share PFO as a key enzyme, they represent markedly different structural and enzymatic organization of metabolism. It is likely that their evolutionary histories have been quite different.

6.3 TAXONOMIC DISTRIBUTION OF AMITOCHONDRIATE PROTISTS

The unusual, amitochondriate types of eukaryotic metabolism were recognized during the 1970s, but each one only in a single, well defined group of organisms; Type I in *Entamoeba histolytica* (Reeves, 1984) and Type II in parabasalids (Müller, 1980). These findings were of real significance in pointing out that eukaryotic cells are more diverse in the core metabolism than previously thought (Müller, 1988). Since this recognition coincided with the general acceptance of the endosymbiotic origin of mitochondria, the potential evolutionary implications of these findings did not remain unnoticed. Based on them a simple scenario was proposed (Müller, 1992), which, however, quickly turned out to be erroneous. According to this, the metabolic machinery of Type I amitochondriate organisms descended directly from the system that functioned in 'premitochondrial' eukaryotes. Subsequently, with the emergence of organelles of core metabolism, this system became remodelled in two separate directions. One led to the appearance of Type II metabolism in parabasalids, characterized by the presence of hydrogenosomes, while the other was the development of the aerobic eukaryotic cell through the acquisition of mitochondria.

From the early 1980s, many additional protists were found to be amitochondriate, as confirmed biochemically for some species (Table 6.1), but still to be verified for many others. The absence or presence of mitochondria was soon adopted as a major taxonomic character. In combination with other features defining cellular complexity, it was used to separate the unicellular eukaryotic world into two main groups, Archaezoa and Protozoa. Archaezoa encompass organisms 'that (allegedly) primitively lack mitochondria, plastids, typical Golgi bodies, hydrogenosomes and peroxisomes...' while the Protozoa have 'tubular (with a few notable exceptions) cristate mitochondria (when absent, replaced by hydrogenosomes), Golgi bodies, and peroxisomes' and in photosynthetic species also chloroplasts (Cavalier-Smith, 1993). In brief, mitochondrial acquisition was regarded as the watershed dividing two primary groups.

If we accept the much quoted rRNA trees as representing the overall phylogeny of eukaryotes and superimpose over them the distributions of the main metabolic types, both Type I and Type II organisms turn out be polyphyletic (Figure 6.2). In addition, in some lineages amitochondriate species coexist with mitochondriate ones as seen most clearly among the ciliates (see chapter 18) and fungi (see chapter 8). Even if the overall phylogeny of protists will be revised (see chapter 2) (Palmer and Delwiche, 1996), the polyphyletic distribution of various metabolic types remains a valid conclusion, which shows that the amitochondriate condition had to emerge repeatedly and independently in several protist lineages (Müller, 1996).

The most important aspect of this phylogenetic distribution, however, is that both Type I and Type II organisms are present in groups that are regarded as early branches of the rRNA tree and also in branches of late emergence (Müller, 1996). If certain amitochondriate protists indeed reflect an ancestral 'premitochondrial' state, this implies that similar metabolic organizational types arose in some lineages by the conservation of the ancestral state and in others by a loss of a more complex machinery and acquisition of certain 'ancestral' features. This would represent a striking case of convergent evolution by two highly different routes. The alternative possibility is that the acquisition of mitochondria has preceded the divergence of all extant eukaryotes, thus we have no living example of a 'premitochondriate' protist. While most organisms then retained the mitochondriate condition, the amitochondriate ones lost their mitochondrial machinery, partially or completely, together with the mitochondrial genome. More importantly during this remodel-

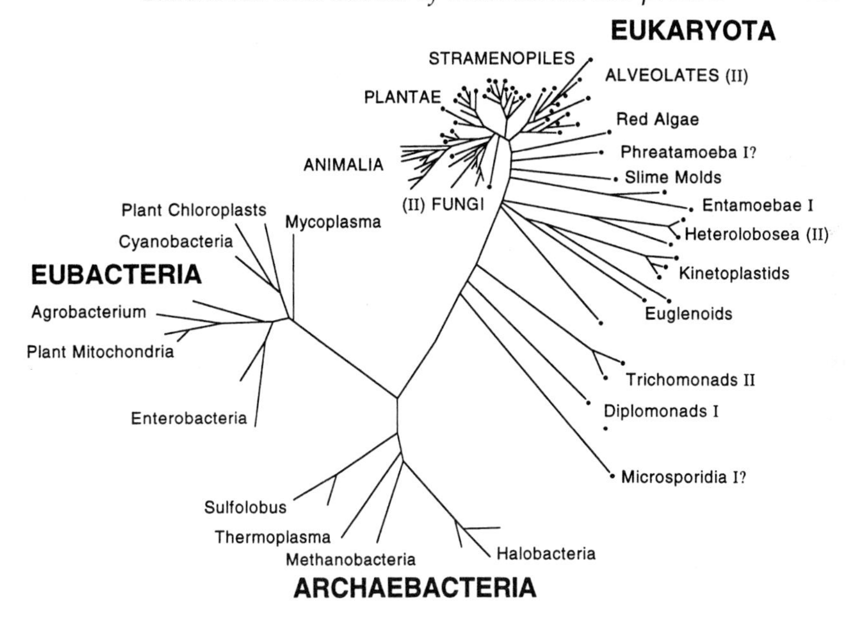

Figure 6.2 Broad polyphyletic distribution of amitochondriate metabolic types among lineages as revealed on a schematic small subunit rRNA phylogenetic tree. The overall reconstruction is statistically robust, but has been questioned recently (see chapter 2). The detailed order of emergence of early branches is not established (Leipe *et al.*, 1993; Philippe and Adoutte, 1995). 'I' and 'II' designate the two amitochondriate metabolic types. Without parenthesis they denote putative metabolic organization of whole branches, in parenthesis they indicate that species with given metabolic type are present together with mitochondriate species. Putative Type I lineages are noted 'I?'. Groups without comment are not known to have amitochondriate species. The phylogenetic position of several putative Type I archaezoans remains to be established. Based on trees presented by Sogin, 1991; Leipe *et al.*, 1993; Hinkle *et al.*, 1994. Modified from Müller, 1996.

ling of their energy metabolism they also acquired the enzymes characteristic of amitochondriate metabolic types. Recent data on two amitochondriate protists, *E. histolytica* (Clark and Roger, 1995) and *T. vaginalis* (see chapter 7) (Bui, Bradley and Johnson, 1996; Germot, Philippe and Le Guyader, 1996; Roger, Clark and Doolittle, 1996; Horner *et al.*, 1996) strongly indicate that their ancestors contained mitochondria. It remains to be seen if similar evidence will be found for other putative Archaezoa or all of them will turn out to have mitochondrion-containing ancestors.

6.4 PUTATIVE ANCESTRAL FEATURES OF AMITOCHONDRIATE METABOLISM

As compared with mitochondriate organisms, amitochondriate protists exhibit two remarkable metabolic characteristics. These are the prominent roles of inorganic pyrophosphate (PPi) as energy carrier and of nonheme iron-sulphur centres as redox carriers. PPi and iron-sulphur centres are thought to have played major roles during early evolution, and to antedate the appearance of adenine nucleotides and heme iron (see Daniel and Danson, 1995). This has led to proposals to regard their functional significance to amitochondriate protists as relics of the early evolutionary history of eukaryotic cells (Reeves, 1984; Müller, 1988, 1992). Based on the unusual metabolism of *E. histolytica*, a complete 'ancestral' pathway from glucose to acetate and ethanol, in which only PPi and iron-sulphur centres participate, has been designed (Reeves, 1984). Such proposals, if verified, could have a major impact on our notions on the evolution of eukaryotic metabolism. However, no extant eukaryote is known which would rely exclusively on PPi and nonheme iron for its central metabolism. These compounds are just parts of a more complex metabolic machinery in all eukaryotes.

The free energy of the pyrophosphate bond is relatively high (Davies, Poole and Saunders, 1993), thus PPi can replace ATP in certain reactions, as frequently seen in amitochondriate organisms. The most studied example is the phosphorylation of fructose-6-phosphate to fructose-1,6-bisphosphate. In these organisms this key step in glycolysis (in A) is catalysed by PPi-linked phosphofructokinase (PPi-PFK) (Mertens, 1991, 1993) instead of the ATP-linked enzyme (ATP-PFK) of most mitochondriate eukaryotes. Amitochondriate fungi, however, are an exception and have ATP-linked activity (Marvin-Sikkema *et al.*, 1993). Type I amitochondriates have additional PPi-linked enzymes. Both well studied species contain pyruvate dikinase (B) (Reeves, 1984; Hrdý, Mertens and Nohynková, 1993), which is found also in eubacteria and plants. *E. histolytica* contains at least two more PPi-linked enzymes, PEP-carboxyphosphotransferase and PPi-acetate kinase (Reeves, 1984). It is to be emphasized, however, that these enzymes are still the exceptions and ATP remains the main high energy phosphate compound in these organisms.

The presence of enzymes that utilize the free energy of the pyrophosphate bond is not bound to the amitochondrial condition. PPi-PFK has also been detected in some mitochondriate protist species (Mertens, 1993; Denton, Thong and Coombs, 1994). PPi-linked glycolytic enzymes, PPi-PFK (Mertens, 1991) and pyruvate dikinase (Mertens, 1993), are present also in plants along with ATP-PFK and pyruvate kinase. Plants also contain inorganic pyrophosphatases which couple

PPi hydrolysis to proton translocation across the vacuolar membrane (Rea *et al.*, 1992).

The significance of the PPi-linked enzymes probably lies in the fact that by using PPi instead of ATP in glycolysis, the input of ATP into the pathway decreases, thus the overall efficiency of the process is higher. The occurrence of PPi-dependent glycolytic enzymes in amitochondriate organisms is probably not an ancestral character but an adaptation to the relatively inefficient core metabolism of these organisms (Reeves, 1976; Mertens, 1991, 1993).

The role of iron-sulphur proteins presents a different problem. Such proteins are widely distributed both in anaerobic and aerobic organisms, including in mitochondria, and play a major role in redox reactions. They coexist in all organisms with other redox components, of which flavins and pyridine nucleotides are almost ubiquitous, and quinones are present in most organisms including amitochondriate eukaryotes. The only real difference is the absence in the latter of one class of redox components, the cytochromes (Müller, 1988), thus their redox machinery is characterized by a negative character. This by itself is obviously not diagnostic of their ancestral nature.

The role of iron-sulphur proteins in amitochondriate protists, however, is restricted to one specific pathway, the oxidative decarboxylation of pyruvate and the disposal of the electrons generated in this process. Instead of a lipoamide- and pyridine nucleotide-dependent pathway, characteristic of mitochondria (Wieland, 1983), the iron-sulphur protein PFO is its main catalyst. This is linked to ferredoxins and in Type II amitochondriates to hydrogenase, also iron-sulphur proteins. Such ferredoxin linked reactions of relatively low midpoint potentials are largely restricted to anaerobic, photosynthetic and N_2-fixing organisms. These reactions and electron carriers have been regarded as having arisen early in evolution (Eck and Dayhoff, 1966; Kerscher and Oesterhelt, 1982; Daniel and Danson, 1995; Schönheit and Schäfer, 1995). Their presence in amitochondriate protists is well correlated with the anoxic environments of these organisms.

In conclusion, the presence of PPi-linked enzymes is unlikely to represent an ancestral feature in amitochondriate eukaryotes. While iron-sulphur proteins by themselves could be regarded as ancestral, their presence and significant role in amitochondriates do not argue compellingly for the ancestral nature of their metabolism.

6.5 HISTORY OF INDIVIDUAL ENZYMES

The taxonomic sample is highly biased and too small to even begin the reconstruction of the history of the individual enzymes of core metabolism, let alone of their assembly into the operating metabolic machineries. Instead of making a futile attempt, only a few enzymes will be discussed, to illustrate the problems of reconstructing their evolutionary history.

First PFO will be examined, the hallmark of amitochondriate organisms, an enzyme not known from mitochondriate species with the curious exception of mitochondria of *Euglena gracilis*. These mitochondria do not contain a pyruvate dehydrogenase complex, but have pyruvate:NADPH oxidoreductase (Watanabe *et al.*, 1993), a fusion protein consisting of PFO and ferredoxin:NADH oxidoreductase (see Hrdý and Müller, 1995).

PFO is present in various eubacteria and archaebacteria. It is closely related to pyruvate:flavodoxin oxidoreductase, an enzyme involved in N_2 fixation (Wahl and Orme-Johnson, 1987), encoded by *nifJ*. Members of this enzyme family are of two types, one consists of four or of two different subunits, while the other is a homomer. All archaebacterial PFOs (Kletzin and Adams, 1996), as well as some eubacterial ones (Smith, Blamey and Adams, 1994; Hughes *et al.*, 1995), belong to the first type. PFOs of *Clostridium pasteurianum* and several other eubacteria, together with the *nifJ* products, are members of the single subunit group.

The PFO sequence is known from *G. lamblia* (GenBank L27221), *E. histolytica* (Rodriguez *et al.*, 1996) and *T. vaginalis* (Hrdý and Müller, 1995). No sequences are available for the eubacterial homomeric PFOs but a number of *nifJ* genes of eubacteria have been sequenced. The protist PFO sequences display about 40% positional identity of amino acids, and are also closely related to the *nifJ* sequences. No homology can be detected between these enzymes and the components of PDH dehydrogenase complex. In phylogenetic reconstructions, the three protist sequences form a single clade and the eubacterial *nifJ* sequences form another with a short node connecting these two clades. The data are insufficient to decide whether one single event led to the presence of PFO in the three protists or there were several independent events of acquisition.

Comparison of sequences of the multi-subunit enzymes of prokaryotes with those of the homomeric enzymes reveal their homology but show that the two groups are quite distant from each other. The closest known relatives of protist PFOs are eubacterial and not archaebacterial (Figure 6.3). The single subunit represents a fusion product of the separate subunits of multi-subunit enzymes (Kletzin and Adams, 1996). The enigmatic history of this enzyme remains one of the most fascinating problems of the origin of amitochondriate metabolism.

Electron acceptors for PFOs are iron-sulphur proteins, ferredoxins. These proteins are broadly distributed in the living world. Although all have one or two low redox potential iron-sulphur centres, in fact they

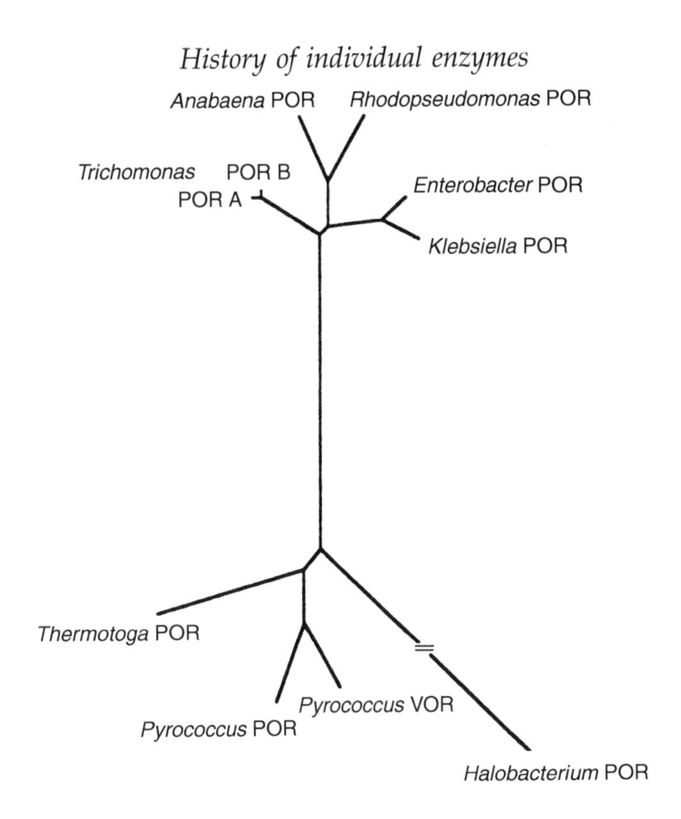

Figure 6.3 Phylogenetic reconstruction of the evolutionary relationships of ketoacid:ferredoxin and pyruvate:flavodoxin oxidoreductases showing the close relationship of the *Trichomonas vaginalis* enzyme to the eubacterial ones and the great divergence between these and the archaebacterial ones. Branch-lengths denote relative distances. The *H. halobium* POR branch is approximately double of what is shown. POR – pyruvate:acceptor (ferredoxin or flavodoxin) oxidoreductase; VOR – 2-ketoisovalerate:ferredoxin oxidoreductase. With permission from Kletzin and Adams, 1996.

are highly diverse, constituting several protein families (Matsubara and Saeki, 1992). The major ferredoxins of *G. lamblia* (Townson *et al.*, 1994), *T. vaginalis* (Gorrell, Yarlett and Müller, 1984) and *E. histolytica* (Reeves, Guthrie and Lobelle-Rich, 1980) have been characterized biochemically. For the first only an amino-terminal sequence is available (Townson *et al.*, 1994) while the other two have been sequenced (Huber *et al.*, 1989; Johnson *et al.*, 1990). The results show that ferredoxins of these species belong to three separate protein families. *G. lamblia* contains a 6 kDa protein, having probably a [3Fe-4S] and a [4Fe-4S] centre, *T. vaginalis* a 9 kDa [2Fe-2S] ferredoxin and *E. histolytica* a 6 kDa 2[4Fe-4S] protein.

The *T. vaginalis* ferredoxin (Johnson *et al.*, 1990) is related to proteins that participate in mixed-function oxidase systems in mitochondria and aerobic bacteria, while the *E. histolytica* ferredoxin (Huber *et al.*, 1989) is related to proteins found in strict anaerobic bacteria (e.g. *Clostridium* species). These findings do not reveal the history of these proteins, but strongly indicate their separate origins.

The reactions in which the reduced ferredoxin is reoxidized are not elucidated in Type I organisms, though it is clear that the electrons are finally transferred to organic acceptors. In hydrogenosome-containing Type II organisms, the acceptors are protons, which are reduced to H_2 by the action of a hydrogenase. This enzyme of *T. vaginalis* is an Fe-only hydrogenase related to the few eubacterial (*Clostridium* and *Desulfovibrio* species) Fe-only hydrogenases sequenced so far (Bui and Johnson, 1996). The simple pyruvate to H_2 electron transport of *T. vaginalis* hydrogenosomes thus comprises two enzymes related to those found in strict anaerobic bacteria, which are linked by a ferredoxin characteristic of aerobic eubacteria and mitochondria.

The proteins discussed so far are characteristic of the amitochondriate protists only and are usually absent from mitochondriate organisms. Now two steps of glycolysis proper will be examined, which are catalysed by enzymes present in all eukaryotes for which significant databases exist.

Phosphofructokinase catalyses the conversion of fructose-6-phosphate into fructose-1,6-bisphosphate. As mentioned above, the enzyme of almost all amitochondriate organisms studied uses PPi as phosphoryl donor instead of ATP. The PPi- and ATP-linked phosphofructokinases derive from a common ancestor but diverged from each other early (Fothergill-Gilmore and Michels, 1993) (Figure 6.4). Their amino-terminal halves can be convincingly aligned, while the carboxyl-terminal halves are significantly more divergent and can be aligned only for closely related enzymes. Phylogenetic reconstruction based on the amino-terminal half of the molecule reveals significant divergence of the protist PPi-PFKs (Figure 6.4). No clear evolutionary trends can be discerned. Interestingly PPi-PFKs of the two Type I organisms do not form a clade but have separate histories. The unexpected observation that the *G. lamblia* enzyme is within the plant PPi-PFK clade is supported also by the fact that in contrast to all other protist enzymes it can be convincingly aligned also in its carboxyl-terminal half with both subunits of the plant enzymes.

Glyceraldehyde 3-phosphate dehydrogenase (GAPDH), another glycolytic enzyme, also presents a complex picture. Phylogenetic reconstructions based on the available extensive database indicate the existence of several paralogous protein lineages and a number of horizontal gene transfer events (Fothergill-Gilmore and Michels, 1993;

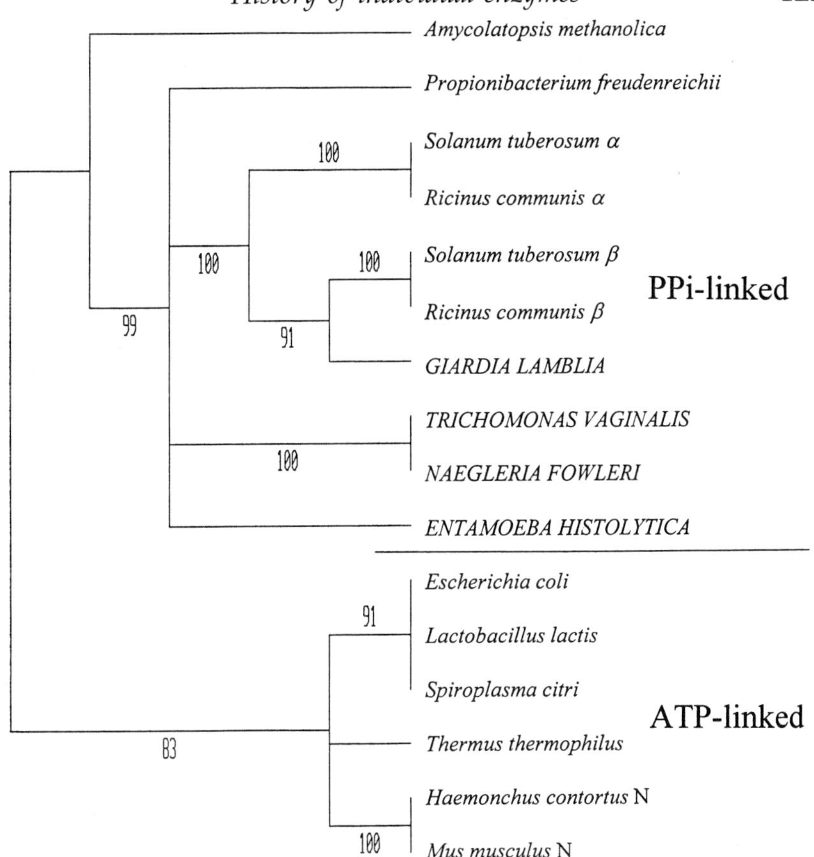

Figure 6.4 Collapsed phylogenetic tree of PPi- and ATP-linked phosphofructokinases showing that the enzymes in amitochondriate protists are not monophyletic. Neighbour-joining method, based on an amino acid distance matrix calculated from the more conserved amino-terminal half of the molecules. Bootstrap proportions (BP) based on 500 replicates. Nodes with BP values over 75 are depicted, nodes with lower BP values are shown as unresolved multifurcations. Protist sequences from Rozario *et al.*, 1995; Wessberg *et al.*, 1995; Bruchhaus *et al.*, 1996, other sequences from GenBank. The *Trichomonas vaginalis* sequence and the reconstruction shown are unpublished.

Henze *et al.*, 1995). GAPDH of eukaryotes displays a characteristic signature in the so-called S-loop region that differentiates it from eubacterial and chloroplast enzymes. Of the organisms discussed here, the *T. vaginalis* enzyme has an eubacterial signature (Markoš, Miretsky and Müller, 1993), while GAPDH of *G. lamblia* and other diplomonads as

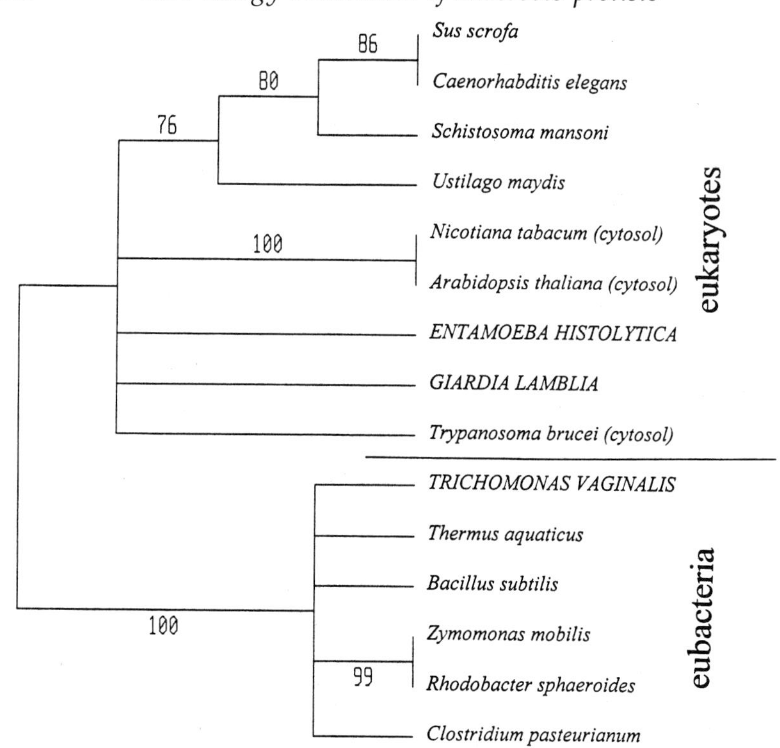

Figure 6.5 Collapsed phylogenetic tree of glyceraldehyde-3-phosphate dehydrogenase showing the eubacterial affinities of the *Trichomonas vaginalis* enzyme and the eukaryotic affinities of the enzymes from *Giardia lamblia* and *Entamoeba histolytica*. Neighbour-joining method, based on an amino acid distance matrix. For details of representation, see Figure 6.4. Based on Markoš *et al.*, 1993 and Rozario *et al.*, 1996.

well as of *E. histolytica* display a eukaryotic signature (Rozario *et al.*, 1996). The respective position of these enzymes in a phylogenetic reconstruction is firmly in the eubacterial and eukaryotic branches (Figure 6.5), showing that this enzyme has different origins in Type I and Type II amitochondriate organisms.

The reliability of the conclusions made above needs further comment, in view of the relatively limited information content and the possible mutational saturation of the sequences analysed. Although this is a real concern, it should be noted that the discussed phylogenetic relationships are based on highly conservative interpretation of the trees obtained. They are also consistently supported by specific sequence properties, i.e. the lack of separate subunits in PFO, the position of the cysteine-residues coordinating the iron-sulphur centres in the ferredoxins, the similarity or divergence of the carboxyl-terminal half of the PFK molecules and the eubacterial or eukaryotic type of S-loop of GAPDH. None of these characteristics is likely to have arisen by convergent evolution.

Altogether, the data show a remarkable diversity of the enzymes so far studied. Although clearly there is much more to discover, the results so far indicate that the machinery of core metabolism of the few amitochondriate protists studied is a complex mosaic of pieces which had separate histories. The surprising metabolic similarity of various amitochondriate organisms possibly does not reflect a common origin but is a result of convergence.

Concerning the overall relationship of the enzymes of the eukaryotic core metabolism to those in the two other domains, sequence data so far consistently indicate a closer similarity to eubacterial proteins than to archaebacterial ones, as documented above for PFO. In view of the still limited information on archaebacterial enzymes, it would be premature to assume that this conclusion is carved in stone. Attempts made to establish given eubacterial groups as the sources of individual core enzymes are yet to be successful, possibly due to a lack of sufficient eubacterial data bases. This situation will undoubtedly change in the near future, with the current explosive growth of sequence information.

6.6 METABOLIC COMPARTMENTATION

The absence or presence of a metabolic compartment is, as discussed above, a crucial difference between Type I and Type II amitochondriate organisms. The key metabolic step, the oxidative decarboxylation of pyruvate by PFO, takes place in different subcellular compartments in the two metabolic types. The acquisition of this enzyme by the ancestors of the extant amitochondriate organisms thus resulted in two different metabolic systems, possibly though separate evolutionary paths.

The establishment of the Type I metabolic organization is a completely open question. Some lineages with Type I metabolism clearly derive from mitochondrion-bearing ancestors as shown for *E. histolytica* (Clark and Roger, 1995). It remains to be seen whether this is true for all Type I organisms or some will turn out to be genuine Archaezoa.

Hydrogenosomes, characterizing Type II organisms are present in a number of independent lineages, indicating that they arose repeatedly in separate events (see chapter 18). Their evolutionary origin remains obscure but there is increasing evidence in favour of the notion that probably in all cases they arose from mitochondria (Finlay and Fenchel, 1989; Bui, Bradley and Johnson, 1996; Germot, Philippe and Le

Guyader, 1996; Horner *et al.*, 1996; Roger, Clark and Doolittle, 1996) (see chapter 7) though for the fungal organelle a peroxisomal origin is also being considered (see chapter 8). The earlier notion that certain hydrogenosomes descended from integrated endosymbionts different from the ancestors of mitochondria (Müller, 1980) probably will have to be abandoned. The mechanisms of this convergent conversion and the origins of the unique enzymes present in hydrogenosomes remain to be elucidated. It seems to be significant that the hydrogenosomal machinery differs significantly from the metabolism of Type I amitochondriate species both in its electron transport leading to H_2 formation and in the mechanism of acetate formation accompanied by substrate level phosphorylation (Müller, 1988). This implies that hydrogenosomes cannot be regarded as the products of a simple remodelling of mitochondria by losses of a number of constituents and retargeting of cytosolic enzymes of Type I amitochondriates, as it has been suggested earlier (Cavalier-Smith, 1987a). Further research will have to shed light on the origin of targeting signals for those hydrogenosomal proteins that are not found in mitochondria and on the processes of wholesale functional losses during the conversion.

6.7 OUTLOOK

The findings considered above confirm that the well known diversity of unicellular eukaryotes is also reflected in their core energy metabolism. They also show that this metabolic diversity does not represent a linear evolutionary trend but is probably due to separate reorganizations of the enzymatic machinery, processes that involve the acquisition and deletion of single genes or sets of genes as well as the enzymatic modifications of existing organelles of core metabolism (Müller, 1996). The eukaryotic energy generating system is clearly much more flexible than is commonly accepted. Other biochemical aspects of amitochondriate protists also display marked diversity, as seen in amino acid (North and Lockwood, 1995), purine and pyrimidine metabolism (Berens, Krug and Marr, 1995), and in defense mechanisms against oxygen toxicity (Bacchi and Yarlett, 1995; Docampo, 1995; Brown, Upcroft and Upcroft, 1995).

All major components of eukaryotic core metabolism have their homologues in all three domains, indicating that classical glycolysis (Kengen, Stams and de Vos, 1996), various modes of pyruvate catabolism (Kengen, Stams and de Vos, 1996) and heme protein dependent electron transport (Castresana and Saraste, 1995; Schäfer, Purschke and Schmidt, 1996) arose before the separation of the three domains of life. If this tenet withstands the test of future studies, its corollary is that few principally new building blocks will be found in protist energy metabolism.

Sequence data on enzymes of glycolysis and its extensions in both Type I and Type II amitochondriate protists reveal consistently a closer relationship with their eubacterial than archaebacterial homologs. This is in contrast to many other enzymes, primarily of DNA replication, transcription and translation, showing a close relationship between archaebacteria and eukaryotes, (Baldauf, Palmer and Doolittle, 1996; Hashimoto and Hasegawa, 1996), in agreement with the interpretation of results of rRNA sequence comparisons (Sogin, 1991; Olsen and Woese, 1993). Eubacterial similarities of the enzymes of amitochondriate core metabolism suggest that sometime during early eukaryotic evolution the energy generating machinery inherited from their archaebacterial ancestors was lost and replaced by a new set of enzymes. The event responsible for such bulk replacement and the source of the new metabolic system are still unknown. The acquisition of mitochondria is a plausible candidate, but remains to be verified. The data also indicate that after a new core metabolism of eubacterial origin became established in eukaryotes, further replacements of enzymes and pathways led to its further divergence. Our task is now to map the paths leading to the metabolic diversity of protists and, more importantly, to elucidate the benefits the divergent organisms derive from their specific metabolic makeups.

Acknowledgements

Members, past and present, of the author's laboratory contributed many of the experimental data and ideas discussed in this paper. In addition the author has benefited from stimulating discussions and shared preliminary information from numerous colleagues over the years. Lack of space does not permit to list these individuals but the author wishes to express his indebtedness here. Original research in the author's laboratory received generous support for over a quarter of a century through grant AI 11942 from the USPHS National Institutes of Health.

6.8 REFERENCES

Adam, R.D. (1991) The biology of *Giardia* spp. *Microbiological Reviews*, **55**, 706–32.

Bacchi, C.J. and Yarlett, N. (1995) Polyamine metabolism, in *Biochemistry and Molecular Biology of Parasites* (eds J.J. Marr and M. Müller), Academic Press, London. pp. 119–31.

Baldauf, S.L., Palmer, J.D. and Doolittle, R.F. (1996) The root of the universal tree and the origin of eukaryotes based on elongation factor phylogeny. *Proceedings of the National Academy of Sciences USA*, **93**, 7749–54.

Berens, R.L., Krug, E.C. and Marr, J.J. (1995) Purine and pyrimidine metabolism, in *Biochemistry and Molecular Biology of Parasites* (eds J.J. Marr and M. Müller), Academic Press, London. pp. 89–117.

Brown, D.M., Upcroft, J.A. and Upcroft, P. (1995) Free radical detoxification in *Giardia duodenalis*. *Molecular and Biochemical Parasitology*, **72**, 47–56.

Bruchhaus, I., Jacobs, T., Denart, M. and Tannich, E. (1996) Pyrophosphate–dependent phosphofructokinase of *Entamoeba histolytica*: molecular cloning, recombinant expression and inhibition of pyrophosphate analogues. *Biochemical Journal*, **316**, 57–63.

Bui, E.T.N. and Johnson, P.J. (1996) Identification of characterization of [Fe]-hydrogenases in the hydrogenosome of *Trichomonas vaginalis*. *Molecular and Biochemical Parasitology*, **76**, 305–10.

Bui, E.T.N., Bradley, P.J. and Johnson, P.J. (1996) Chaperonins in hydrogenosomes of *Trichomonas vaginalis*. *Proceedings of the National Academy of Sciences USA*, **93**, 9651–6.

Castresana, J. and Saraste, M. (1995) Evolution of energetic metabolism: the respiration-early hypothesis. *Trends in Biochemical Sciences*, **20**, 443–8.

Cavalier–Smith, T. (1987a) The simultaneous symbiotic origin of mitochondria, chloroplasts, and microbodies. *Annals of the New York Academy of Science*, **503**, 55–72.

Cavalier–Smith, T. (1987b) Eukaryotes with no mitochondria. *Nature*, **326**, 332–3.

Cavalier–Smith, T. (1993) Kingdom protozoa and its 18 phyla. *Microbiological Reviews*, **57**, 953–94.

Clark, C.G. and Roger, A.J. (1995) Direct evidence for secondary loss of mitochondria in *Entamoeba histolytica*. *Proceedings of the National Academy of Sciences USA*, **92**, 6518–21.

Coombs, G.H. and Müller, M. (1995) Energy metabolism in anaerobic protozoa, in *Biochemistry and Molecular Biology of Parasites* (eds J.J. Marr and M. Müller), Academic Press, London. pp. 33–47.

Daniel, R.M. and Danson, M.J. (1995) Did primitive microorganisms use nonheme iron proteins in place of NAD/P? *Journal of Molecular Evolution*, **40**, 559–63.

Davies, J.M., Poole, R.J. and Sanders, D. (1993) The computed free energy change of hydrolysis of inorganic pyrophosphate and ATP: apparent significance for inorganic-pyrophosphate-driven reactions of intermediary metabolism. *Biochimica et Biophysica Acta*, **1141**, 29–36.

Denton, H., Thong, K.-W. and Coombs, G.H. (1994) *Eimeria tenella* contains a pyrophosphate-dependent phosphofructokinase and a pyruvate kinase with unusual allosteric regulators. *FEMS Microbiology Letters*, **115**, 87–91.

Docampo, R. (1995) Antioxidant mechanisms, in *Biochemistry and Molecular Biology of Parasites* (eds J.J. Marr and M. Müller), Academic Press, London. pp. 147–60.

Eck, R. and Dayhoff, M.O. (1966) Evolution of the structure of ferredoxin based on living relics of primitive amino acid sequences. *Science*, **152**, 363–6.

Ellis, J.E., McIntyre, P.S., Saleh, M. *et al.* (1991) Influence of CO_2 and low concentrations of O_2 on fermentative metabolism of the rumen ciliate *Dasytricha ruminantium*. *Journal of General Microbiology*, **137**, 1409–17.

Fairlamb, A.H. (1989) Novel biochemical pathways in parasitic protozoa. *Parasitology*, **99**, S93–S112.

Fenchel, T. and Finlay, B.J. (1995) *Ecology and Evolution in Anoxic Worlds*, Oxford University Press, Oxford, pp. ix-276.

Finlay, B.J. and Fenchel, T. (1989) Hydrogenosomes in some anaerobic protozoa resemble mitochondria. *FEMS Microbiology Letters*, **65**, 311–4.

Fothergill-Gilmore, L.A. and Michels, P.A.M. (1993) Evolution of glycolysis. *Progress in Biophysics and Molecular Biology*, **59**, 105–235.

Germot, A., Philippe, H. and Le Guyader, H. (1996) Presence of a mitochondrial-type 70-kDA heat shock protein in *Trichomonas vaginalis* suggests a very early mitochondrial endosymbiosis in eukaryotes. *Proceedings of the National Academy of Sciences USA*, **93**, 14614–17.

Goosen, N.K., van der Drift, C., Stumm, C.K. and Vogels, G.D. (1990) End products of metabolism in the anaerobic ciliate *Trimyema compressum*. *FEMS Microbiology Letters*, **69**, 171–6.

Gorrell, T.E., Yarlett, N. and Müller, M. (1984) Isolation and characterization of *Trichomonas vaginalis* ferredoxin. *Carlsberg Research Communications*, **49**, 259–68.

Hashimoto, T. and Hasegawa, M. (1996) Origin and early evolution of eukaryotes inferred from the amino acid sequences of translation elongation factors 1α/Tu and 2/G. *Advances in Biophysics*, **32**, 73–120.

Henze, K., Badr, A., Wettern, M. *et al.* (1995) A nuclear gene of eubacterial origin in *Euglena gracilis* reflects cryptic endosymbioses during protist evolution. *Proceedings of the National Academy of Sciences USA*, **92**, 9122–6.

Hinkle, G., Leipe, D.D., Nerad, T.A. and Sogin, M.L. (1994) The unusually long small subunit ribosomal RNA of *Phreatamoeba balamuthi*. *Nucleic Acids Research*, **22**, 465–9.

Horner, D.S., Hirt, R.P., Klivington, D. *et al.* (1996) Molecular data suggest an early acquisition of the mitochondrion endosymbiont. *Proceedings of the Royal Society of London, Series B*, **263**, 1053–9.

Hrdý, I., Mertens, E. and Nohynková, E. (1993) *Giardia intestinalis*: Detection and characterization of a pyruvate phosphate dikinase. *Experimental Parasitology*, **76**, 438–41.

Hrdý, I. and Müller, M. (1995) Primary structure and eubacterial relationships of the pyruvate:ferredoxin oxidoreductase of the amitochondriate eukaryote, *Trichomonas vaginalis*. *Journal of Molecular Evolution*, **41**, 388–96.

Huber, M., Garfinkel, L., Gitler, C. *et al.* (1989) Nucleotide sequence analysis of an *Entamoeba histolytica* ferredoxin gene. *Molecular and Biochemical Parasitology*, **31**, 27–34.

Hughes, N.J., Chalk, P.A., Clayton, C.L. and Kelly, D.J. (1995) Identification of carboxylation enzymes and characterization of a novel four-subunit pyruvate:flavodoxin oxidoreductase from *Helicobacter pylori*. *Journal of Bacteriology*, **177**, 3953–9.

Hülsmann, N. and Hausmann, K. (1994) Towards a new perspective in protozoan evolution. *European Journal of Protistology*, **30**, 365–71.

Jarroll, E.L. and Paget, T.A. (1995) Carbohydrate and amino acid metabolism in *Giardia*: a review. *Folia Parasitologica (Praha)*, **42**, 81–9.

Johnson, P.J., d'Oliveira, C.E., Gorrell, T.E. and Müller, M. (1990) Molecular analysis of the hydrogenosomal ferredoxin of the anaerobic protist, *Trichomonas vaginalis*. *Proceedings of the National Academy of Sciences USA*, **87**, 6097–101.

Kengen, S.W.M., Stams, A.J.M. and de Vos, W.M. (1996) Sugar metabolism of hyperthermophiles. *FEMS Microbiological Reviews*, **18**, 119–37.

Kerscher, L. and Oesterhelt, D. (1982) Pyruvate:ferredoxin oxidoreductase – new findings on an ancient enzyme. *Trends in Biochemical Sciences*, **7**, 371–4.

Kletzin, A. and Adams, M.W.W. (1996) Molecular and phylogenetic characterization of pyruvate and 2-ketoiosvalerate ferredoxin oxidoreductases from *Pyrococcus furiosus* and pyruvate ferredoxin oxidoreductase from *Thermotoga maritima*. *Journal of Bacteriology*, **178**, 248–57.

Leiner, M., Schweikhardt, F., Blascke, G. *et al.* (1968) Die gärung und atmung von *Pelomyxa palustris* Greeff. *Biologisches Zentralblatt*, **87**, 567–91.

Leipe, D.D., Gunderson, J.H., Nerad, T.A. and Sogin, M.L. (1993) Small subunit ribosomal RNA+ of *Hexamita inflata* and the quest for the first branch in the eukaryotic tree. *Molecular and Biochemical Parasitology*, **59**, 41–8.

Lwoff, A. (1951) Introduction to biochemistry of protozoa, in *Biochemistry and Physiology of Protozoa*, vol. 1 (ed. A. Lwoff), Academic Press, New York, pp. 1–26.

Markoš, A., Miretsky, A. and Müller, M. (1993) A glyceraldehyde 3-phosphate dehydrogenase with eubacterial features in the amitochondriate eukaryote, *Trichomonas vaginalis*. *Journal of Molecular Evolution*, **37**, 631–43.

Marvin–Sikkema, F.D., Richardson, A.J., Stewart, C.S. *et al.* (1990) Influence of hydrogen-consuming bacteria on cellulose degradation by anaerobic fungi. *Applied and Environmental Microbiology*, **56**, 3793–7.

Marvin-Sikkema, F.D., Pedro Gomes, T.M., Grivet, J.-P. *et al.* (1993) Characterization of hydrogenosomes and their role in glucose metabolism of *Neocallimastix* sp. L2. *Archives of Microbiology*, **160**, 388–96.

Matsubara, H. and Saeki, K. (1992) Structural and functional diversity of ferredoxins and related proteins, in *Advances in Inorganic Chemistry. Volume 38. Iron-Sulphur Proteins* (ed. R. Cammack), Academic Press, San Diego. pp. 223–80.

McLaughlin, J. and Aley, S. (1985) The biochemistry and functional morphology of *Entamoeba*. *Journal of Protozoology*, **32**, 221–40.

Mertens, E. (1991) Pyrophosphate-dependent phosphofructokinase, an anaerobic glycolytic enzyme? *FEBS Letters*, **285**, 1–5.

Mertens, E. (1993) ATP versus pyrophosphate: Glycolysis revisited in parasitic protists. *Parasitology Today*, **9**, 122–6.

Müller, M. (1976) Carbohydrate and energy metabolism of *Tritrichomonas foetus*, in *Biochemistry of Parasites and Host-parasite Relationships*, (ed. H. Van den Bossche), Elsevier, Amsterdam, pp. 3–14.

Müller, M. (1980) The hydrogenosome. *Symposia of the Society for General Microbiology*, **30**, 127–42.

Müller, M. (1988) Energy metabolism of protozoa without mitochondria. *Annual Review of Microbiology*, **42**, 465–88.

Müller, M. (1989) Biochemistry of *Trichomonas vaginalis*, in *Trichomonads Parasitic in Humans* (ed. B.M. Honigberg), Springer, New York, pp. 53–83.

Müller, M. (1992) Energy metabolism of ancestral eukaryotes: a hypothesis based on the biochemistry of amitochondrate parasitic protists. *BioSystems*, **28**, 33–40.

Müller, M. (1993) The hydrogenosome. *Journal of General Microbiology*, **139**, 2879–89.

Müller, M. (1996) Energy metabolism of amitochondriate protists, an evolutionary puzzle, in *Christian Gottfried Ehrenberg–Festschrift* (eds M. Schlegel and K. Hausmann), Leipziger Universitätsverlag, Leipzig, pp. 63–76.

North, M.J. and Lockwood, B.C. (1995) Amino acid and protein metabolism, in *Biochemistry and Molecular Biology of Parasites* (eds J.J. Marr and M. Müller), Academic Press, London, pp. 67–88.

Olsen, G.J. and Woese, C.R. (1993) Ribosomal RNA: a key to phylogeny. *The FASEB Journal*, **7**, 113–23.

Palmer, J.D. and Delwiche, C.F. (1996) Second-hand chloroplasts and the case of the disappearing nucleus. *Proceedings of the National Academy of Sciences USA*, **93**, 7432–5.

Philippe, H. and Adoutte, A. (1995) How reliable is our current view of eukaryotic phylogeny? in *Protistological Actualities (Proceedings of the Second European Congress of Protistology Clermont-Ferrand, 1995)* (eds G. Brugerolle and J.-P. Mignot), pp. 17–33.

Rea, P.A, Kim, Y., Sarafian, V. *et al.* (1992) Vacuolar H^+-translocating pyrophosphatases: a new category of ion translocase. *Trends in Biochemical Sciences*, **17**, 348–53.

Reeves, R.E. (1976) How useful is the energy in inorganic pyrophosphate? *Trends in Biochemical Sciences*, **1**, 53–5.

Reeves, R.E. (1984) Metabolism of *Entamoeba histolytica* Schaudinn, 1903. *Advances in Parasitology*, **23**, 105–42.

Reeves, R.E., Guthrie, J.D. and Lobelle–Rich, P. (1980) *Entamoeba histolytica*: Isolation of ferredoxin. *Experimental Parasitology*, **49**, 83–8.

Rodriguez, M.A., Hidalgo, M.E., Sanchez, T. and Orozco, E. (1996) Cloning and characterization of the *Entamoeba histolytica* pyruvate:ferredoxin oxidoreductase gene. *Molecular and Biochemical Parasitology*, **78**, 273–7.

Roger, A.J., Clark, C.G. and Doolittle, W.F. (1996) A possible mitochondrial gene in the early-branching amitochondriate protist *Trichomonas vaginalis*. *Proceedings of the National Academy of Sciences USA*, **93**, 14618–22.

Rozario, C., Smith, M.W. and Müller, M. (1995) Primary sequence of a putative pyrophosphate-linked phosphofructokinase gene of *Giardia lamblia*. *Biochimica et Biophysica Acta*, **1260**, 218–22.

Rozario, C., Morin, L., Roger, A.J. *et al.* (1996) Primary structure and phylogenetic relationships of glyceraldehyde-3-phosphate dehydrogenase genes of free-living and parasitic diplomonad flagellates. *Journal of Eukaryotic Microbiology*, **43**, 330–40.

Schäfer, G., Purschke, W. and Schmidt, C.L. (1996) On the origin of respiration: electron transport proteins from archaea to man. *FEMS Microbiological Reviews*, **18**, 173–88.

Schönheit, P. and Schäfer, T. (1995) Metabolism of hyperthermophiles. *World Journal of Microbiology and Biotechnology*, **11**, 26–57.

Schulz, R. (1996) Hydrogenases and hydrogen production in eukaryotic organisms and cyanobacteria. *Journal of Marine Biotechnology*, **4**, 16–22.

Smith, E.T., Blamey, J.M. and Adams, M.W.W. (1994) Pyruvate ferredoxin oxidoreductases of the hyperthermophilic archaeon, *Pyrococcus furiosus*, and the hyperthermophilic bacterium, *Thermotoga maritima*, have different catalytic mechanisms. *Biochemistry*, **33**, 1008–16.

Sogin, M.L. (1991) Early evolution and the origin of eukaryotes. *Current Opinion in Genetics and Development*, **1**, 457–63.

Townson, S.M., Hanson, G.R., Upcroft, J.A. and Upcroft, P. (1994) A purified ferredoxin from *Giardia duodenalis. European Journal of Biochemistry*, **220**, 439–46.

Wahl, R.C. and Orme-Johnson, W.H. (1987) Clostridial pyruvate oxidoreductase and the pyruvate-oxidizing enzyme specific for nitrogen fixation in *Klebsiella pneumoniae* are similar enzymes. *Journal of Biological Chemistry*, **262**, 10489–96.

Watanabe, F., Yamaji, R., Isegawa, Y. *et al.* (1993) Characterization of aquacobalamin reductase (NADPH) from *Euglena gracilis. Archives of Biochemistry and Biophysics*, **305**, 421–7.

Wessberg, K.L., Skolnick, S., Xu, J. *et al.* (1995) Cloning, sequencing and expression of the pyrophosphate-dependent phosphofructo-1-kinase from *Naegleria fowleri. Biochemical Journal*. **307**, 143–9.

Wieland, O.H. (1983) The mammalian pyruvate dehydrogenase complex: structure and regulation. *Reviews of Physiology, Biochemistry and Pharmacology*, **96**, 123–70.

Williams, A.G. (1986) Rumen holotrich ciliate protozoa. *Microbiological Reviews*, **50**, 25–49.

Yarlett, N., Lloyd, D. and Williams, A.G. (1985) Butyrate formation from glucose by the rumen protozoon *Dasytricha ruminantium. Biochemical Journal*, **228**, 187–92.

Yarlett, N., Orpin, C.G., Munn, E.A. *et al.* (1986) Hydrogenosomes in the rumen fungus *Neocallimastix patriciarum. Biochemical Journal*. **236**, 729–39.

Yarlett, N. (1994) Fermentation product generation in rumen chytridiomycetes, in *The Anaerobic Fungi* (eds D.O. Mountfort and C.G. Orpin), Marcel Dekker, New York, pp. 129–46.

7

Molecular phylogeny of *Trichomonas* and *Naegleria*: implications for the relative timing of the mitochondrion endosymbiosis

David S. Horner, Robert P. Hirt and T. Martin Embley

Department of Zoology, The Natural History Museum, Cromwell Road, London SW7 5BD, UK. E-mail: dsh@nhm.ac.uk

ABSTRACT

We have investigated support from protein sequence data for alternative relationships for *Naegleria* and *Trichomonas* using constrained maximum likelihood analyses. In particular we have investigated support for the published hypothesis that Percolozoa, represented by *Naegleria,* represents an earlier branch than *Trypanosoma* or *Trichomonas*. Resolution of this question is of particular importance because of its implications for the relative timing of the mitochondrion endosymbiosis. Our observation that *Trichomonas* contains a mitochondrial chaperonin (cpn-60), and is thus unlikely to be primitively amitochondriate, makes the position of *Trichomonas* relevant to this debate. None of our analyses, or any published phylogenies, provide strong support for a 'Percolozoa early' topology. Most protein phylogenies suggest that *Naegleria* is part of the eukaryotic 'crown' and that *Trichomonas* and *Trypanosoma* branch prior to it.

7.1 THE SMALL SUBUNIT RIBOSOMAL RNA GENE TREE FOR DEEP BRANCHING EUKARYOTES

Examination of small subunit ribosomal RNA (SSUrRNA) phylogenies (Cavalier-Smith, 1993a, Hinkle *et al.*, 1994, Sogin *et al.*, 1996) reveals several features. The first is the so called 'crown group' containing

Evolutionary Relationships Among Protozoa. Edited by G.H. Coombs, K. Vickerman, M.A. Sleigh and A. Warren. Published in 1998 by Chapman & Hall, London. ISBN 0 412 79800 X

Plantae, Fungi, Metazoa, Alveolata and Stramenopila. This group contains the great majority of eukaryotic taxa whose SSUrRNA genes have been sampled. The middle region of the tree includes the kinetoplastids and euglenoids (which constitute a monophyletic group, the Euglenozoa; for taxonomy see Corliss, 1994), the Percolozoa, the Mycetozoa, the Entamoebidae and several other amoeboid taxa (Hinkle *et al.*, 1994, Philippe and Adoutte, 1995, Sogin *et al.*, 1996). At the bottom of the SSUrRNA tree are three amitochondriate groups, the Metamonada, Microspora and Parabasala.

Strong support for the relative branching order of Microspora, Metamonada and Parabasala has not been obtained, with inference method and outgroup choice affecting branching order (Leipe *et al.*, 1993). A similar situation prevails in the middle region of the tree, with published analyses suggesting almost all possible permutations of relative branching order between Euglenozoa, Percolozoa, Mycetozoa and Entamoebidae (e.g. Baverstock *et al.*, 1989, Berchtold, Breunig and Konig, 1995, Branke *et al.*, 1996, Cavalier-Smith, 1993a, Hinkle *et al.*, 1994, Leipe *et al.*, 1993).

Recent studies have illustrated the potential limitations of standard distance and parsimony methods for constructing trees based on nucleotide datasets (Galtier and Gouy, 1995, Lockhart *et al.*, 1994). Unequal base composition of SSUrRNA sequences violates most models of nucleotide substitution (Galtier and Gouy, 1995). This may be particularly important in inferring eukaryotic phylogeny (Hasegawa and Hashimoto, 1993) since some of the SSUrRNA sequences near the base of the tree are particularly heterogeneous in composition (G + C = 74.7% for *Giardia lamblia* (Sogin *et al.*, 1989) and G + C = 37.4% for *Vairimorpha necatrix* (Vossbrinck *et al.*, 1987). Furthermore, outgroup sequences among the Archaea and Bacteria may also show significantly higher proportions of G+C%. Using an algorithm designed to estimate evolutionary distance without assumptions of homogeneity or stationarity of nucleotide substitution, Galtier and Gouy (1995) recently recovered microsporidia as the first eukaryotic branch from the main trunk of the tree. A recent SSUrRNA phylogeny using distances calculated from transversions only (see Fitch, 1971 for discussion) provides moderately strong bootstrap support for the Microspora as the earliest branch with Parabasala and Metamonada as the next two branches (Philippe and Adoutte, 1995).

Some aspects of the topology of SSUrRNA trees are consistent with inferences suggested from ultrastructural observations of certain protists. For example, in SSUrRNA trees the Microspora, Metamonada and Parabasala, all of which lack functional mitochondria, are often the earliest diverging protist lineages. This is consistent with the hypothesis that some or all of these taxa might be primitively amitochondriate

(Cavalier-Smith, 1993a, Patterson and Sogin, 1993). The presence of 70S ribosomes in Microspora and Parabasala (Ishihara and Hayashi, 1968; Champney, Chittum and Samuels, 1992; van Keulen *et al.*, 1993), and the lack of discernible golgi dictyosomes in Metamonada and Microspora have also been cited as evidence for a position near the base of the SSUrRNA tree (Cavalier-Smith, 1991, 1993a,b). The apparent lack of recognizable dictyosomes in Percolozoa has been interpreted as a primitive condition which it shares with Metamonada and Microspora, and as evidence that Percolozoa diverged before Parabasala and Euglenozoa which possess them (Cavalier-Smith, 1991, 1993a,b). Under this evolutionary scenario the Percolozoa assume significance as the first protists to contain functional mitochondria, and the Parabasala must be considered as secondarily amitochondriate (Cavalier-Smith, 1991, 1993b). However, there are now data to show that the metamonad *Giardia lamblia* is capable of golgi-mediated sorting of proteins during encystation and that it possesses a developmentally induced, morphologically identifiable golgi apparatus (Luján *et al.*, 1995 and references therein). N-acetylgalactosamine transferase and galactosyltransferase activities, which are normally associated with the medial and trans-golgi respectively, were also shown to be present. As in mammalian cells, the cellular localization of these activities is sensitive to the drug Brefeldin A (Luján *et al.*, 1995 and references therein). Taken together, these results suggest that *Giardia lamblia* possesses a golgi system with a similar level of complexity to those found in higher eukaryotes. In the light of these data the apparent absence of golgi dictyosomes in Percolozoa may be better interpreted as a derived feature rather than a primitive one.

7.2 PROTEIN PHYLOGENIES FOR PROTISTS

Given the lack of resolution for the branching order of deep branching eukaryotes in published SSUrRNA trees, it is desirable to seek support for relationships from independent data sets. The amino acid composition of protein sequences is relatively unaffected by genomic base composition (Hashimoto *et al.*, 1994, 1995a). Thus, protein sequences may be less susceptible to systematic errors of this particular type than nucleotide sequences. Relatively few protein coding genes have been used to generate hypotheses of relationships between protists. These include translation elongation factors EF-1α and EF-2 (Hashimoto *et al.*, 1995a,b); β-tubulin (Edlind *et al.*, 1996, Philippe and Adoutte, 1995); actin (Drouin, Moniz de Sa and Zucker, 1995) and the chaperonin cpn-60 (Clark and Roger, 1995). While supporting some relationships inferred from SSUrRNA analyses, the trees generated from these proteins have also suggested different groupings.

7.3 ARE ALL EARLY BRANCHING EUKARYOTES PRIMITIVELY AMITOCHONDRIATE?

It has been suggested that the shared absence of functional mitochondria in the Parabasala, Metamonada and Microspora might be a primitive trait, because they diverged from the eukaryotic stock prior to the mitochondrion endosymbiosis (Patterson and Sogin, 1993). Interestingly, whereas Metamonada and Microspora lack discernable membrane-bound respiratory organelles, trichomonads and other members of the Parabasala possess another double membraned organelle, the hydrogenosome (Lindmark and Müller, 1973). The trichomonad hydrogenosome oxidizes pyruvate and produces hydrogen (Müller, 1993). One hypothesis for its origins, based upon biochemical considerations, is that it is derived from an endosymbiosis involving an anaerobic *Clostridium*-like Gram-positive bacterium (Müller, 1980). Others have suggested that the trichomonad hydrogenosome might have evolved from former mitochondria when their ancestors first colonized animal guts (Cavalier-Smith, 1987). This last view is inextricably linked to Cavalier-Smith's later hypothesis that Percolozoa, which do have mitochondria, diverged before Parabasala (1991, 1993a,b). Resolution of hydrogenosome origins is complicated by the apparent absence of an associated organelle genome which could be analysed independently for its evolutionary origins (Müller, 1992).

Mitochondrion to nucleus gene transfer in eukaryotes is supported by the nuclear location of many genes which encode mitochondrial proteins (Gray, 1992). Thus it is possible that even in eukaryotes which have lost functional mitochondria there might be the retention, depending on their functions, of mitochondrion-derived genes on the host nuclear genome. This reasoning was recently used by Clark and Roger (1995) who identified and analysed a mitochondrial chaperonin (cpn-60) sequence left on the genome of the amitochondriate protist *Entamoeba histolytica*. The cpn-60 protein plays a key role in protein import in both mitochondria and chloroplasts (Hartl, Hlodan and Langer, 1994), each of which has a particular homologue derived from its ancestral bacterium (Viale and Arakaki, 1994). Cpn-60 is involved in the mediation of unfolding and correct refolding of peptides imported through the mitochondrial/plastid membranes and of peptides synthesized within each organelle.

Cpn-60 sequences are useful for phylogenetic inference because they are highly conserved and well sampled in terms of bacterial taxa (where the homologous protein is called groEL) and organelle sequences (Viale and Arakaki, 1994). There is also a strongly supported mitochondrial clade, which like the SSUrRNA of mitochondria (Yang *et al.*, 1985), roots in the alpha-subdivision of the proteobacteria (Viale and Arakaki, 1994).

We have recently investigated the presence of cpn-60 on the genomes of *Trichomonas vaginalis* and *Naegleria fowleri* (Horner *et al.*, 1996). Our aim was twofold, firstly to explore the possibility that the genome of *Trichomonas* might retain mitochondrion genes in line with the hypothesis that it once contained mitochondria (Cavalier-Smith, 1987). It is also worth mentioning that clues relevant to alternative symbiotic origins might also be gained as long as the organelle retains its ancestral protein import functions. For example, all bacteria have groEL (cpn-60) and specifically there are *Clostridium* sequences for comparison. Our second aim was to infer the phylogenetic position of *Naegleria*, as a member of the Percolozoa, and thus investigate its relationships to other mitochondrion-containing eukaryotes.

The maximum likelihood (ML) method (Kishino, Miyata and Hasegawa, 1990; Adachi and Hasegawa, 1992) was chosen as the main analytical method to investigate our data because it has a number of desirable properties over other methods of phylogenetic inference (for discussion see Swofford *et al.*, 1996). Maximum likelihood also provides a convenient way, through likelihood difference tests (Kishino and Hasegawa, 1989), of evaluating how well alternative hypotheses of relationships i.e. tree topologies, fit the data under a particular model.

Maximum likelihood analysis recovered the cpn-60 sequences from *Trichomonas vaginalis* and *Naegleria fowleri* within a clade otherwise defined by mitochondrion-derived sequences (Figure 7.1).

This topology was supported by high bootstrap probabilities in distance matrix analyses of both DNA and protein sequences (for details see Horner *et al.*, 1996). Constrained likelihood analyses demonstrated that placement of the *N. fowleri* or *T. vaginalis* sequences outside of the mitochondrial clade constituted a significantly worse explanation of the data at the 95% confidence level. Bootstrap partitions revealed that most of the decay at the node defining the mitochondrial clade in parsimony and distance analyses, was caused by the behaviour of the highly divergent *Entamoeba histolytica* sequence, as previously observed by Clark and Roger (1995). Removal of this sequence increased bootstrap support for the *Trichomonas*/mitochondrial clade (Horner *et al.*, 1996).

The recovery of a cpn-60 sequence from *T. vaginalis* within a well supported mitochondrial clade strongly suggests that ancestors of this organism once contained the mitochondrion endosymbiont. Similar results have recently been reported by independent laboratories (Bui, Bradley and Johnson, 1996; Germot, Philippe and Le Guyader, 1996; Roger, Clark and Doolittle, 1996). This result is also compatible with (but does not prove) the mitochondrial conversion hypothesis for the origin of hydrogenosomes in trichomonads (Cavalier-Smith, 1987). It has recently been demonstrated that a protein reacting with GroEL

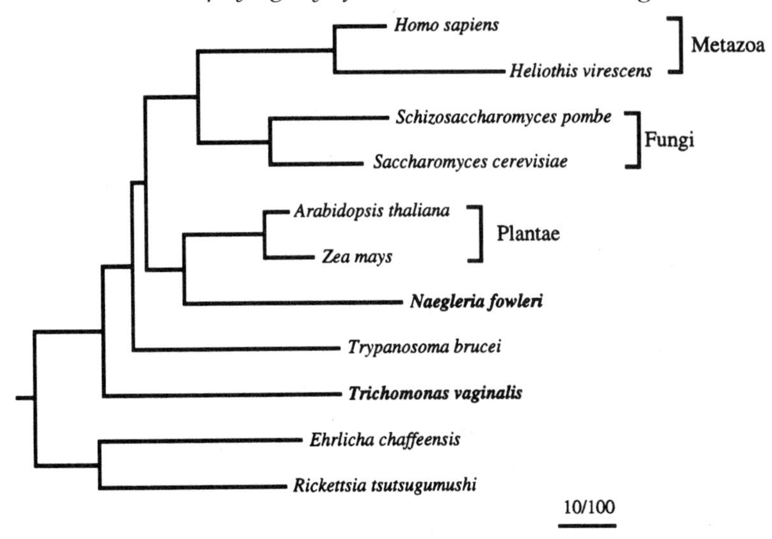

Figure 7.1 Protein maximum likelihood analyses of a cpn-60 dataset. Maximum likelihood tree for the 33 taxa dataset described in Horner *et al.* (1996) (33 taxa and 349 amino acids positions, lnL = -11852.3). Only eukaryotic taxa and two alpha proteobacteria are shown, a more detailed tree showing additional bacterial taxa has been presented (Horner *et al.* 1996). The scale bar represents 10 estimated changes per 100 positions.

antisera localizes within hydrogenosomes of *Tritrichomonas mobilensis* (Bozner, 1996) and *Trichomonas vaginalis* (Bui, Bradley and Johnson, 1996). While this is also consistent with the mitochondrial conversion hypothesis, these experiments did not conclusively resolve this issue as they depended on broad specificity antibodies.

The positioning of *T. vaginalis* at the base of the mitochondrial cpn-60 clade is consistent with inferences drawn from SSUrRNA analyses that *Trichomonas* is an early eukaryotic divergence (Leipe *et al.*, 1993). If this is indeed the case, it would suggest that the (presumed unique) endosymbiotic event which gave rise to mitochondria, occurred earlier in eukaryotic history than previously presumed (Gray, 1992, but see Cavalier-Smith, 1987). In addition, since the mitochondrial cpn-60 clade roots within the alpha-proteobacteria, it follows that the origin of the alpha-proteobacteria must predate that of the *Trichomonas* divergence among eukaryotes. Interestingly, most published rooted phylogenies based upon SSUrRNA or proteins place alpha-proteobacteria as an internal branch of the bacterial tree (van De Peer *et al.*, 1994; Eisen,

1995). A scenario of significant bacterial diversification before eukaryote diversification is consistent with data from comparison of paralogous protein sequences which suggest that the tree of life roots on the bacterial branch, and that eukaryotes and Archaea are more recent taxa (Iwabe *et al.*, 1989; Brown and Doolittle, 1995). Additional work is now needed to see if traces of a past mitochondrion presence can be detected among Metamonada and Microspora.

The association of the *Naegleria fowleri* sequence with those of higher plants (Figure 7.1) in the maximum likelihood tree (and distance and parsimony trees see Horner *et al.*, 1996) was surprising given the much deeper position of *Naegleria* in published SSUrRNA trees (e.g. Sogin *et al.*, 1996). As a test of the robustness of this relationship we used likelihood difference tests to assign confidence limits on different topologies representing alternative positions for *Naegleria* within the cpn-60 tree. We also investigated if the cpn-60 dataset allowed *Naegleria* to branch before *Trypanosoma* or *Trichomonas* in accordance with Cavalier-Smith's (1991, 1993a,b) hypothesis that Percolozoa evolved before either of these taxa.

Analysis of a dataset omitting *Entamoeba histolytica* revealed that the *Trypanosoma brucei* or *Trichomonas vaginalis* sequence could be placed as the first branch, without a significant reduction in log likelihood for the tree. In contrast, placing *Naegleria* before either of these taxa produced trees which had significantly worse log likelihoods. Thus, the current cpn-60 dataset does not support the published 'Percolozoa-early' scenario (Cavalier-Smith, 1991, 1993a,b). The very interesting relationship of *Naegleria* to plant mitochondrion sequences was not robust to perturbation. Thus, constrained trees placing *Naegleria fowleri* as an earlier branch than plants, as in SSUrRNA trees, was not significantly worse than the maximum likelihood tree.

7.4 RELATIONSHIPS OF *NAEGLERIA* AND OTHER PROTISTS BASED UPON ACTIN SEQUENCES

The highly conserved structural protein actin is implicated in muscle contraction as well as cytoskeletal formation, intracellular transport and some forms of movement in protists and Fungi (for review see Reece, McElroy and Wu, 1992). Actin genes have been cloned and sequenced from over 100 species and used to infer eukaryotic phylogenies (Reece, McElroy and Wu, 1992; Baldauf and Palmer, 1993, Drouin, Moniz de Sa and Zucker, 1995).

The inferred amino acid sequences of over 60 actin coding regions were recently analysed by the neighbour-joining method (Drouin, Moniz de Sa and Zucker, 1995). The resulting tree was rooted on *Giardia lamblia* and shares many features with the SSUrRNA tree including the emergence of trypanosomatids as a deeper branch than *Naegleria*. We have analysed the position of *Naegleria* using maximum likelihood analysis and an alignment containing comparable taxa to those available for the cpn-60 gene. Rooting was on *G. lamblia* whose position was assumed to be non-controversial. In the maximum likelihood tree (Figure 7.2A) *Naegleria* was recovered as an internal branch to *Trypanosoma* and *Leishmania* as in published actin trees and as suggested by the SSUrRNA tree. Forcing *Naegleria* to branch before these taxa in line with the 'Percolozoa early' hypothesis (Cavalier-Smith, 1991, 1993a,b) was significantly less likely. However, enforcing the intriguing relationship between *Naegleria* and plants, which was detected in the cpn-60 tree, was not a significantly worse explanation of the actin dataset analysed. *Entamoeba histolytica* was recovered as the sister taxon to the Metazoa and Fungi in the maximum likelihood tree (Figure 7.2A). However this was not a robust relationship, and the SSUrRNA placement for *Entamoeba histolytica* could be enforced without a significantly worse log likelihood score (Figure 7.2B) for the tree.

7.5 RELATIONSHIPS OF *NAEGLERIA* AND OTHER PROTISTS BASED UPON β-TUBULIN SEQUENCES

β-tubulin is one of the principal constituents of microtubular structures (cytoskeleton, axonemes and the mitotic spindle) which are characteristic of eukaryotic cells. It is among the most highly conserved genes known in eukaryotes (Bermudes, Hinkle and Margulis, 1994; Doolittle, 1995) which suggests that it may be suitable for phylogenetic reconstruction of ancient divergences. Over 90 β-tubulin sequences are now available for comparison, including representatives of several putatively deep branching phyla. β-tubulin phylogenies, like those derived from SSUrRNA, actin and cpn-60, support a sister group relationship between Fungi and Metazoa as well as the monophyly of higher plants (Baldauf and Palmer, 1993). Tubulin phylogenies calculated with distance, parsimony and likelihood methods also suggest a specific relationship between *Naegleria*, the Euglenozoa and higher plants but without strong bootstrap support in distance and parsimony methods (Philippe and Adoutte, 1995; Edlind *et al.*, 1996).

We have considered a subset of β-tubulin protein sequences within the ML framework and the JTT-f substitution matrix. The maximum likelihood tree, which was arbitrarily rooted on *Giardia lamblia*, contained a clade comprising Euglenozoa, *Naegleria* and Plantae (Figure 7.3).

Unambiguous clades (Euglenozoa, Metazoa, Fungi and Plantae) were subsequently constrained and alternate hypotheses of relationships for *Naegleria* were investigated. Trees in which the Plantae / Euglenozoa /

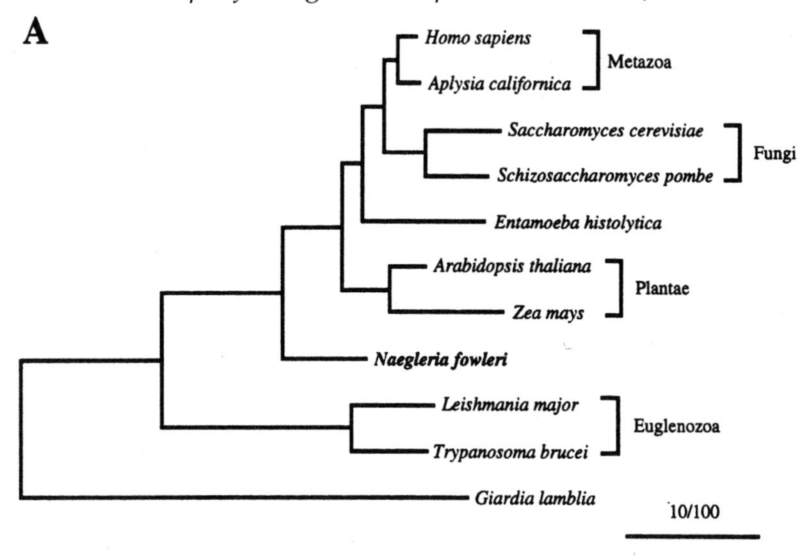

B

	Tree	ln L	D ln L ± S.D.	Significantly worse?
1.	((((M,F),Ent),P),NAE),E,Gia)	-3394.7	0 -> best	–
2.	((((M,F),P),Ent),NAE),E,Gia)	-3398.9	- 4.2 ± 6.4	NO
3.	((((M,F),P),NAE),Ent),E,Gia)	-3409.7	-15.0 ± 11.9	NO
4.	(((M,F),(P,NAE)),Ent),E,Gia)	-3412.0	-18.2 ± 11.4	NO
5.	(((M,F),Ent),(P,NAE)),E,Gia)	-3416.0	-21.3 ± 9.1	YES
6.	((((M,F),Ent),P),E),NAE,Gia)	-3409.7	-15.0 ± 6.7	YES
7.	((((M,F),P),NAE),E),Ent,Gia)	-3423.8	-29.1 ± 13.9	YES

Figure 7.2 Protein maximum likelihood analyses of an actin dataset. A. Maximum Likelihood Tree. A collection of actin protein sequences were aligned and analysed as for the cpn-60 dataset (Figure 7.1) (11 taxa and 373 positions, lnL for ML tree = -33947.7) and rooted on *Giardia*. B. Comparison of lnL for a selection of tree topologies with the taxa shown in A. Log-likelihood (lnL) and attached lnL difference with standard deviations are shown (DlnL ± SD), see text. Seven trees are shown illustrating the lack of support for the position of, among other taxa, *Entamoeba* and *Naegleria* , see text. M: Metazoa, F: Fungi, P: Plantae, E: Euglenozoa, NAE: *Naegleria*, Ent: *Entamoeba*, Gia: *Giardia*.

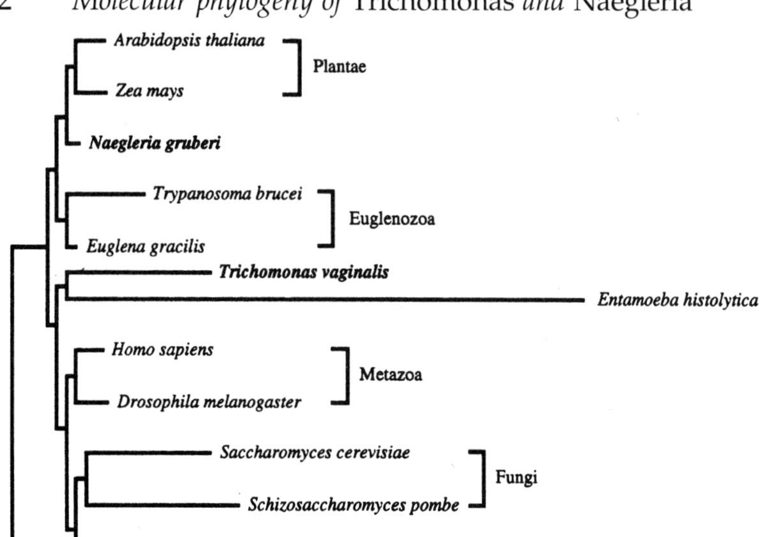

Figure 7.3 Protein maximum likelihood analysis of a β-tubulin dataset. Maximum Likelihood tree. A collection of β-tubulin protein sequences were aligned and analysed as for the cpn-60 dataset (Figure 7.1) (13 taxa and 430 positions, lnL = -4648.37) and rooted on *Giardia*. The long branch for the *Entamoeba* sequence is striking, compare with tree in Edlind *et al.* (1996).

Naegleria clade was disrupted were unacceptable at the 95% confidence interval. However, user defined trees which depicted an 'SSUrRNA-like' topology except with respect to the Microspora (see below) in which *Naegleria* and Euglenozoa diverge prior to the 'crown taxa' were acceptable explanations of the data as long as the *Entamoeba histolytica* sequence, which has an extremely long branch length and has previously been noted to be highly divergent (Katiyar and Edlind, 1996), was excluded. Unlike the cpn-60 and actin datasets, the β-tubulin analysis generally yielded no strong support to a particular relative branching order of Percolozoa and Euglenozoa. Interestingly, the topology indicated by ML analysis of cpn-60 sequences placing Euglenozoa as an early divergence and with *Naegleria* forming a monophyletic group with Plantae, was significantly worse than the ML tree. When the tree is rooted on *Giardia lamblia, Trichomonas vaginalis* must branch before Euglenozoa or *Naegleria gruberi* to constitute an acceptable explaination of the β-tubulin dataset within the likelihood framework.

A recent analysis of β-tubulin sequences recovered a relationship between *Encephalitozoon hellem* (Microspora) and Fungi (Edlind *et al.*, 1996). This topology led Edlind and co-workers (Edlind *et al.*, 1996) to speculate that the Microspora may be highly derived Fungi rather than deep branching protists. We also recovered this association in our ML analysis and evaluated support for the *Encephalitozoon*/Fungi grouping. All tested topologies in which *Encephalitozoon* and fungi were split were unacceptable at the 95% confidence level, regardless of the fungal taxa considered. Interestingly, removal of the fungi from the analysis resulted in an association between *Encephalitozoon* and the Metazoan sequences.

7.6 THE POSITION OF *TRICHOMONAS* AND *NAEGLERIA* INFERRED FROM MOLECULAR DATA: ABSENCE OF SUPPORT FOR THE 'PERCOLOZOA EARLY' HYPOTHESIS

We have investigated support from protein sequence data for alternative relationships for *Naegleria, Trypanosoma* and *Trichomonas.* In particular we have investigated support for the hypothesis that Percolozoa, represented by *Naegleria,* represents an earlier branch than either of these two taxa as suggested by Cavalier-Smith (1991, 1993a,b). Resolution of this question has particular interest because of its implications for the dating of mitochondrion endosymbiosis. Our observation that *Trichomonas* contains a mitochondrial chaperonin, although it lacks functional mitochondria, makes the position of *Trichomonas* relevant to this debate. None of our analyses, or indeed any published unconstrained protein trees, provide strong support for a 'Percolozoa early' topology. Our analyses were based upon a small selection of taxa accommodated by maximum likelihood, and it has been demonstrated that inferences are frequently influenced by taxon sampling and may change with more data (Lecointre et al., 1993; van De Peer et al., 1994). Nevertheless, there is, as far as we are aware, no strong positive support for the 'Percolozoa-early' hypothesis. The absence of golgi dictyosomes in *Naegleria* (Cavalier-Smith, 1993a) is a negative character which can also be interpreted as a derived trait (see earlier).

We have shown by constrained ML analysis that the branching order for *T vaginalis* and *N. fowleri* suggested by SSUrRNA sequences can be accommodated by restricted cpn-60, actin and β-tubulin datasets. However, contrary to SSUrRNA analyses, unconstrained phylogenies derived from actin, β-tubulin and cpn-60 sequences suggest that *Naegleria* emerges within the eukaryotic crown. Another protein coding gene for which extensive phylogenetic analysis of protist taxa has been performed is the translation elongation factor EF-1α (Hashimoto *et al.*, 1994, 1995a). At this time, sequences for representatives of the Percolozoa and

Parabasala are not available. However, it is interesting to note that analysis of this gene, like β-tubulin suggest that the Euglenozoa form part of the eukaryotic crown, although a specific relationship with the Plantae is not indicated (Hashimoto *et al.*, 1995a). Thus the relative position of the Percolozoa in EF-1α phylogenies is of great interest.

Acknowledgements

DSH is supported by the an IRF grant from the Natural History Museum. RPH is supported by a Wellcome Trust Fellowship (number 045702/Z/95/Z).

NOTE ADDED IN PROOF

Subsequent to the preparation of this manuscript, several results relevant to the timing of the mitochondrion endosymbiosis have been published. Independent data from this and another laboratory show that the two microsporidia *Vairimorpha necatrix* and *Nosema locustae* encode mitochondrial orthogues of hsp-70 (Hirt *et al.*, 1997, Germot *et al.*, 1997). In addition, Andrew Roger and co-workers have demonstrated that a mitochondrial type cpn-60 is encoded and expressed by the metamonad *Giardia lamblia* (Roger *et al.*, 1998). These data are consistent with the speculation raised in section 7.3 that all extant eukaryotic lineages have been exposed to and received a genetic contribution from the mitochondrion symbiont.

7.7 REFERENCES

Adachi, J. and Hasegawa, M. (1992) MOLPHY, in *Computer Science Monographs*, Institute of Statistical Mathematics, Tokyo.

Baldauf, S.L. and Palmer, J.D. (1993) Animals and fungi are each other's closest relatives: congruent evidence from multiple proteins. *Proceedings of the National Academy of Sciences USA*, **90**, 11558–62.

Baverstock, P.R., Illana, S., Christy, P.E. *et al.* (1989) srRNA evolution and phylogenetic relationships of the genus *Naegleria* (Protista:Rhizopoda). *Molecular Biology and Evolution*, **6**, 243–57.

Berchtold, M., Breunig, A. and Konig, H. (1995) Culture and phylogenetic characterisation of *Tritrichomonas trypanoides* Duboscq & Grasse 1924, n. comb.: a trichomonad flagellate isolated from the hindgut of the termite *Reticulitermes santonensis* Feytaud. *Journal of Eukaryotic Microbiology*, **42**, 388–91.

Bermudes, D., Hinkle, G. and Margulis, L. (1994) Do prokaryotes contain microtubules? *Microbiological Reviews*, **58**, 387–400.

Bozner, P. (1996) The heat shock response and major heat shock proteins of *Tritrichomonas mobilensis* and *Tritrichomonas augusta. Journal of Parasitology*, **82**, 103–11.

Branke, J., Berchtold, M., Breunig, A. *et al.* (1996) 16S-like rDNA sequence and phylogenetic position of the diplomonad *Spironucleus muris* (Lavier 1936). *European Journal of Protistology*, **32**, 227–33.

Brown, J.R. and Doolittle, W.F. (1995) Root of the universal tree of life based upon ancient aminoacyl-tRNA synthetase gene duplications. *Proceedings of the National Academy of Sciences USA*, **92**, 2441–5.

Bui, E.T.N., Bradley, P.J. and Johnson, P.J. (1996) A common evolutionary origin for mitochondria and hydrogenosomes. *Proceedings of the National Academy of Sciences USA*, **93**, 9651–6.

Cavalier-Smith, T. (1987) Eukaryotes with no mitochondria. *Nature*, **326**, 332–3.

Cavalier-Smith, T. (1991) Cell diversification in heterotrophic flagellates, in *The Biology of Free-living Heterotrophic Flagellates* (eds. D.J. Patterson, and J. Larsen), The Systematics Association and Clarendon Press, Oxford, pp. 113–32.

Cavalier-Smith, T. (1993a) Kingdom Protozoa and its 18 phyla. *Microbiological Reviews*, **57**, 953–94.

Cavalier-Smith, T. (1993b) Percolozoa and the symbiotic origin of the metakaryotic cell, in *Endocytobiology V* (eds S. Sato, M. Ishida and H. Ishikawa), Tübingen University Press, Tübingen, pp. 399–406.

Champney, W.S., Chittum, H.S. and Samuels, R. (1992) Ribosomes from trichomonad protozoa have prokaryotic characteristics. *International Journal of Biochemistry*, **24**, 1125–33.

Clark, C.G. and Roger, A.J. (1995) Direct evidence for secondary loss of mitochondria in *Entamoeba histolytica*. *Proceedings of the National Academy of Sciences USA*, **92**, 6518–21.

Corliss, J.O. (1994) An interim utilitarian (user friendly) hierarchical classification and characterisation of the protists. *Acta Protozoologica*, **33**, 1–51.

de Peer, Y. V., Neefs, J.-M., de Rijk, P. *et al.* (1994) About the order of divergence of the major bacterial taxa during evolution. *Systematic and Applied Bacteriology*, **17**, 32–8.

Doolittle, R.F. (1995) The origins and evolution of eukaryotic proteins. *Philosophical Transactions of the Royal Society B.*, **349**, 235–40.

Drouin, G., Moniz de Sa, M. and Zucker, M. (1995) The *Giardia lamblia* actin gene and the phylogeny of eukaryotes. *Journal of Molecular Evolution*, **41**, 841–9.

Edlind, T.D., Li, J., Visvesvara, G.S. *et al.* (1996) Phylogenetic analysis of β-tubulin sequences from amitochondrial protozoa. *Molecular Phylogenetics and Evolution*, **5**, 359–67.

Eisen, J.A. (1995) The RecA protein as a model molecule for molecular systematic studies of bacteria: comparison of trees of RecAs and 16S rRNAs from the same species. *Journal of Molecular Evolution*, **41**, 1105–23.

Fitch, W.M. (1971) Toward defining the course of evolution: minimum change for a specific tree topology. *Systematic Zoology*, **20**, 406–16.

Galtier, N. and Gouy, M. (1995) Inferring phylogenies from DNA sequences of unequal base compositions. *Proceedings of the National Academy of Sciences USA*, **92**, 11317–21.

Germot, A., Philippe, H. and Le Guyader, H. (1996) Presence of a mitochondrial-type HSP70 in *Trichomonas* suggests a very early mitochodrial endosymbiosis in eukaryotes. *Proceedings of the National Academy of Sciences USA*, **93**, 14614–17.

Germot, A., Philippe, H. and Le Guyader, H. (1997) Evidence for loss of mitochondria in Microsporidia from a mitochondrial-type HSP70 in *Nosema locustae. Molecular and Biochemical Parasitology*, **7**, 159–68.

Gray, M.W. (1992) The endosymbiont hypothesis revisited. *International Review of Cytology*, **141**, 233–57.

Hartl, F., Hlodan, R. and Langer, T. (1994) Molecular chaperones in protein folding: the art of avoiding sticky situations. *Trends in Biochemical Sciences*, **19**, 20–5.

Hasegawa, M. and Hashimoto, T. (1993) Ribosomal RNA trees misleading? *Nature*, **361**, 23.

Hashimoto, T., Nakamura, Y., Kamaishi, T. *et al.* (1995a) Phylogenetic place of kinetoplastid protozoa inferred from protein phylogeny of elongation factor 1α. *Molecular and Biochemical Parasitology*, **70**, 181–5.

Hashimoto, T., Nakamura, Y., Kamaishi, T. *et al.* (1995b) Phylogenetic place of mitochondrion-lacking protozoan, *Giardia lamblia*, inferred from amino acid sequences of elongation factor 2. *Molecular Biology and Evolution*, **12**, 782–93.

Hashimoto, T., Nakamura, Y., Nakamura, F. *et al.* (1994) Protein phylogeny gives a robust estimation for early divergences of eukaryotes: phylogenetic place of a mitochondria-lacking protozoan *Giardia lamblia*. *Molecular Biology and Evolution*, **11**, 65–71.

Hinkle, G., Leipe, D.D., Nerad, T.A. and Sogin, M.L. (1994) The unusually long small subunit ribosomal RNA of *Phreatamoeba balamuthi*. *Nucleic Acids Research*, **22**, 465–9.

Hirt, R.P., Healy, B., Vossbrinck, C.R. *et al.* (1997) A mitochondrial hsp70 orthologue in *Vairimorpha necatrix*: molecular evidence that microsporidia once contained mitochondria. *Current Biology*, **7**, 995–8.

Horner, D.S., Hirt, R.P., Kilvington, S. *et al.* (1996) Molecular data suggest an early acquisition of the mitochondrion endosymbiont. *Proceedings of the Royal Society of London (B)*, **263**, 1053–9.

Ishihara, R. and Hayashi, Y. (1968) Some properties of ribosomes of the sporoplasm of *Nosema bombycis*. *Journal of Invertebrate Pathology*, **11**, 377–85.

Iwabe, N., Kuma, K., Hasegawa, M. *et al.* (1989) Evolutionary relationship of archaebacteria, eubacteria, and eukaryotes inferred from phylogenetic trees of duplicated genes. *Proceedings of the National Academy of Sciences USA*, **86**, 9355–9.

Katiyar, S.K. and Edlind, T.D. (1996) *Entamoeba histolytica* encodes a highly divergent β-tubulin. *Journal of Eukaryotic Microbiology*, **43**, 31–4.

Kishino, H. and Hasegawa, M. (1989) Evaluation of the maximum likelihood estimate of the evolutionary tree topologies from DNA sequence data, and the branching order in Hominoidea. *Journal of Molecular Evolution*, **29**, 170–9.

Kishino, H., Miyata, T. and Hasegawa, M.. (1990) Maximum likelihood inference of protein phylogeny and the origin of chloroplasts. *Journal of Molecular Evolution*, **30**, 151–60.

Lecointre, G., Philippe, H., Lê, H.L.V. and Le Guyader, H. (1993) Species sampling has a major impact on phylogenetic inference. *Molecular Phylogenetics and Evolution*, **2**, 205–24.

Leipe, D.D., Gunderson, J.H., Nerad, T.A. and Sogin, M.L. (1993) Small subunit ribosomal RNA of *Hexamita inflata* and the quest for the first branch in the eukaryotic tree. *Molecular and Biochemical Parasitology*, **59**, 41–8.

Lindmark, D.G. and Müller, M. (1973) Hydrogenosome, a cytoplasmic organelle of the anaerobic flagellate, *Tritrichomonas foetus*, and its role in pyruvate metabolism. *Journal of Biological Chemistry*, **248**, 7724–8.

Lockhart, P.J., Steel, M.A., Hendy, M.D. and Penny, D. (1994) Recovering evolutionary trees under a more realistic model of sequence evolution. *Molecular Biology and Evolution*, **11**, 605–12.

Lujan, H.D., Marotta, A., Mowatt, M.R. *et al.* (1995) Developmental induction of golgi structure and function in the primitive eukaryote *Giardia lamblia. Journal of Biological Chemistry*, **270**, 4612–18.

Müller, M. (1980) The hydrogenosome, in *The Eukaryotic Microbial Cell* (eds. G.W. Gooday, D. Lloyd and A.P.J. Trinci), Cambridge University Press, Cambridge, pp. 127–43,

Müller, M. (1992) Energy metabolism of ancestral eukaryotes: a hypothesis based on the biochemistry of amitochondrial parasitic protists. *Biosystems*, **28**, 33–40.

Müller, M. (1993) The hydrogenosome. *Journal of General Microbiology*, **139**, 2879–89.

Patterson, D.J. and Sogin, M.L.(1993) Eukaryotic origins and protistan diversity, in *The Origin and Evolution of the Cell* (eds. H. Hartman and K. Matsuno), World Scientific, Singapore, pp. 13–46.

Philippe, H. and Adoutte, A. (1995) How reliable is our current view of eukaryotic phylogeny?, in *Protistological Actualities: Proceedings of the Second European Congress of Protistology* (eds G. Brugerolle and J.-P. Mignot), Clermont-Ferrand, pp. 17–32.

Reece, K.S., McElroy, D. and Wu, R. (1992) Function and evolution of actins, in *Evolutionary Biology* (ed. M.K. Hecht), Plenum Press, New York, pp. 1–34.

Roger, A.J., Clark, C.G. and Doolittle, W.F. (1996) A possible mitochondrial gene in the early-branching protist *Trichomonas vaginalis. Proceedings of the National Academy of Sciences USA*, **93**, 14618–22.

Roger, A.J., Svard, S.G., Tover, J. *et al.* (1998) A mitochondrial-like chaperonin 60 gene in *Giardia lamblia*: Evidence that diplomonads once harbored an endosymbiont related to the progenitor of mitochondria. *Proceedings of the National Academy of Science USA*, **95**, 229–34.

Sogin, M.L., Gunderson, J.H., Elwood, H.J. *et al.* (1989) Phylogenetic meaning of the kingdom concept: an unusual ribosomal RNA from *Giardia lamblia. Science*, **243**, 75–7.

Sogin, M.L., Morrison, H.G., Hinkle, G. and Silberman, J.D. (1996) Ancestral relationships of the major eukaryotic lineages. *Microbiología SEM*, **12**, 17–28.

Swofford, D.L., Olsen, G.J., Waddell, P.J. and Hillis, D.M. (1996) Phylogenetic inference, in *Molecular Systematics* 2nd edn (eds D.M. Hillis, C. Moritz and B.K. Mable), Sinauer Associates, Sunderland, pp 407–514.

van Keulen, H., Gutell, R.R., Gates, M.A. *et al.* (1993) Unique phylogenetic position of Diplomonadida based on the complete small subunit ribosomal RNA sequence of *Giardia ardeae, G. muris, G. duodenalis* and *Hexamita sp. FASEB Journal*, **7**, 223–31.

Viale, A.M. and Arakaki, A.K. (1994) The chaperone connection to the origins of the eukaryotic organelles. *Federation of European Biochemical Societies Letters*, **341**, 146–51.

Vossbrinck, C.R., Maddox, J.V., Friedman, S. *et al.* (1987) Ribosomal RNA sequence suggests microsporidia are extremely ancient eukaryotes. *Nature*, **326**, 411–14.

Yang, D, Oyaizau, Y., Oyaizu, H. *et al.* (1985) Mitochondrial origins. *Proceedings of the National Academy of Sciences USA*, **82**, 4443–7.

8

Hydrogenosomes and plastid-like organelles in amoeboflagellates, chytrids, and apicomplexan parasites

J.H.P. Hackstein, F.G.J. Voncken, G.D. Vogels, J. Rosenberg and U. Mackenstedt*

Department of Microbiology, Faculty of Science, University of Nijmegen, Toernooiveld, NL-6525 ED Nijmegen, The Netherlands. E-mail: hack@sci.kun.nl

ABSTRACT

The evolution of the eukaryotic cell is characterized by a sequence of endosymbiotic events and cytoplasmic compartmentalizations. Various anaerobic protists and fungi evolved hydrogenosomes – membrane-bound organelles that are involved in energy metabolism. These organisms do not share a recent common ancestry, and evidence is presented for a non-endosymbiotic and polyphyletic origin of hydrogenosomes. Distinct morphological and molecular differences discriminate the various hydrogenosomes. Those of *Trichomonas* and *Psalteriomonas* are devoid of nucleic acids and protein-synthesizing machinery and so they depend completely on the import of nuclear-encoded proteins. It is argued that the 'ease' of their evolution, the absence of nucleic acids, and the mode of their multiplication make it unlikely that hydrogenosomes were derived from recent endosymbiotic ancestors. This suggestion is also supported by molecular data that reveal substantial DNA sequence differences between genes encoding putative hydrogenosomal proteins of *Trichomonas* and *Psalteriomonas* on the one hand, and anaerobic Chytridiomycetes on the other. Electron microscopy reveals that hydrogenosomes from chytrids resemble peroxisomes of aerobic yeasts and fungi. However, evidence is presented that these

Evolutionary Relationships Among Protozoa. Edited by G.H. Coombs, K. Vickerman, M.A. Sleigh and A. Warren. Published in 1998 by Chapman & Hall, London. ISBN 0 412 79800 X

organelles have a chimeric origin; putative hydrogenosomal adenylate kinases from chytrids combine a mitochondrial ancestry with a peroxisomal C-terminal PTS1 ('SKL') targeting signal. Evidence for the functional significance of the presence of both hydrogenosomes and plastid-like organelles in *P. lanterna* is also presented.

8.1 INTRODUCTION

One of the most fascinating aspects of the eukaryotic cell is its structural complexity and high degree of functional compartmentalization. The hypothesis that this complexity evolved through several endosymbiotic events provided an invaluable concept for studying the evolution of the eukaryotic cell (Margulis, 1993, 1996; Gupta and Golding, 1996). It could be shown that mitochondria and plastids possess a genome and a protein synthesizing machinery. Phylogenetic analysis of the DNA sequences of their genomes pointed to their descent from α proteobacteria and cyanobacteria, respectively (Gray, 1989, 1992; Gupta and Golding, 1996; Melkonian, 1996).

Other cellular compartments such as microbodies (peroxisomes), hydrogenosomes, Golgi-apparatus, and endoplasmic reticulum are devoid of DNA. All the protein constituents of these organelles are encoded by nuclear genes, synthesized in the cytoplasm, and imported into the organelles. Therefore, the evolutionary origin of these cellular compartments cannot be reconstructed easily: there is no organellar DNA left that could allow a phylogenetic analysis, and the discovery of the ancestry of these organelles has to rely on DNA sequence analysis of nuclear genes that encode organellar proteins, a search for potential targeting signals and a genetic and molecular dissection of the particular import machineries (Margulis, 1993, 1996; Palmer, 1995; Thorsness and Weber, 1996; Hackstein *et al.*, 1997a).

8.2 HYDROGENOSOMES

Hydrogenosomes evolved in a number of taxa of anaerobic protists and fungi, i.e. Parabasalia, Heterolobosea, Ciliata and Chytridiomycetes (Müller, 1993; Brul and Stumm, 1994). They are membrane-bounded organelles that lack a genome – at least in *Trichomonas* and *Psalteriomonas.* They are involved in anaerobic energy metabolism and are characterized by the presence of hydrogenase and pyruvate:ferredoxin oxidoreductase (PFO) (Müller, 1993). The evolutionary descent of the hydrogenosome is controversial: mitochondrial, endosymbiotic, or endogenous (microbody) origins have been postulated (Cavalier-Smith, 1987). Currently a mitochondrial origin is favoured because of the discovery of a set of chaperonins that share a common ancestry with mitochondrial homologues (see Palmer, 1997). However, morphological and biochemical evidence suggests that the various hydrogenosomes are not the same (Müller, 1993; Marvin-Sikkema *et al.*, 1992, 1993a; Coombs and Hackstein, 1995). The hydrogenosomes of *Trichomonas* and *Psalteriomonas* show no structural affinities with mitochondria apart from two closely opposed membranes, whereas the hydrogenosomes of certain ciliates resemble mitochondria in various aspects: electron microscopy not only reveals the presence of internal structures such as tubuli or cristae, it also suggests that they may even house a genome and protein-synthesizing machinery (Yarlett *et al.*, 1984; Finlay and Fenchel, 1989; Paul, Williams and Butler, 1990). The number of membranes that bound hydrogenosomes seems to be variable, for instance even *Trichomonas* hydrogenosomes have been reported to be bounded by a single membrane (Kulda, Nohynková and Ludvik, 1986). Chytrid hydrogenosomes appear to be single membrane-bound (Yarlett *et al.*,1986; Marvin-Sikkema *et al.*, 1992, 1993a, b). This characteristic has been used to demonstrate their relatedness to peroxisomes and other microbodies, whereas the presence of two bounding membranes had been regarded as a mitochondrial feature. However, peroxisomes can also become bounded by a double membrane (Elgersma *et al.*, 1993), and the morphogenesis of the hydrogenosomes of *Trichomonas* and *Psalteriomonas* involves close contacts or even continuity with the endoplasmatic reticulum similar to developing peroxisomes (Hruban and Rechcigl, 1969; Kulda *et al.*, 1986; Coombs and Hackstein, 1995; Benchimol *et al.*, 1996a, b; Hackstein *et al.*, 1997a; c.f. Figure 8.1). Multiplication of the hydrogenosomes of *Trichomonas* seems to occur by segmentation and partition, not unlike mitochondria – but also similar to peroxisomes (Hruban and Rechcigl, 1969; Benchimol *et al.*, 1996b). In addition, the observed stacking of hydrogenosomes in *Psalteriomonas* (Figure 8.1) and the presence of Ca^{2+}-accumulating structures that have been called 'marginal plates' are peroxisomal rather than mitochondrial traits (Kulda *et al.*, 1986; Marvin-Sikkema *et al.*, 1993a, b; Makita, 1995; Benchimol *et al.*, 1996a, b). Thus, electron microscopy fails to provide unequivocal evidence for either a mitochondrial or a peroxisomal ancestry; rather the observations are consistent with the hydrogenosomes of *Trichomonas, Psalteriomonas* and chytrid fungi being organelles that possesses unique characteristics, different in certain aspects from both mitochondria and peroxisomes.

Several nuclear genes of *Trichomonas* that encode hydrogenosomal enzymes have been isolated and cloned. DNA sequence analysis reveals a considerable divergence with respect to comparable enzymes from non-hydrogenosomal sources, and the rate of divergence seems to be different for the various genes (Länge, Rozario and Müller, 1994; Hrdy

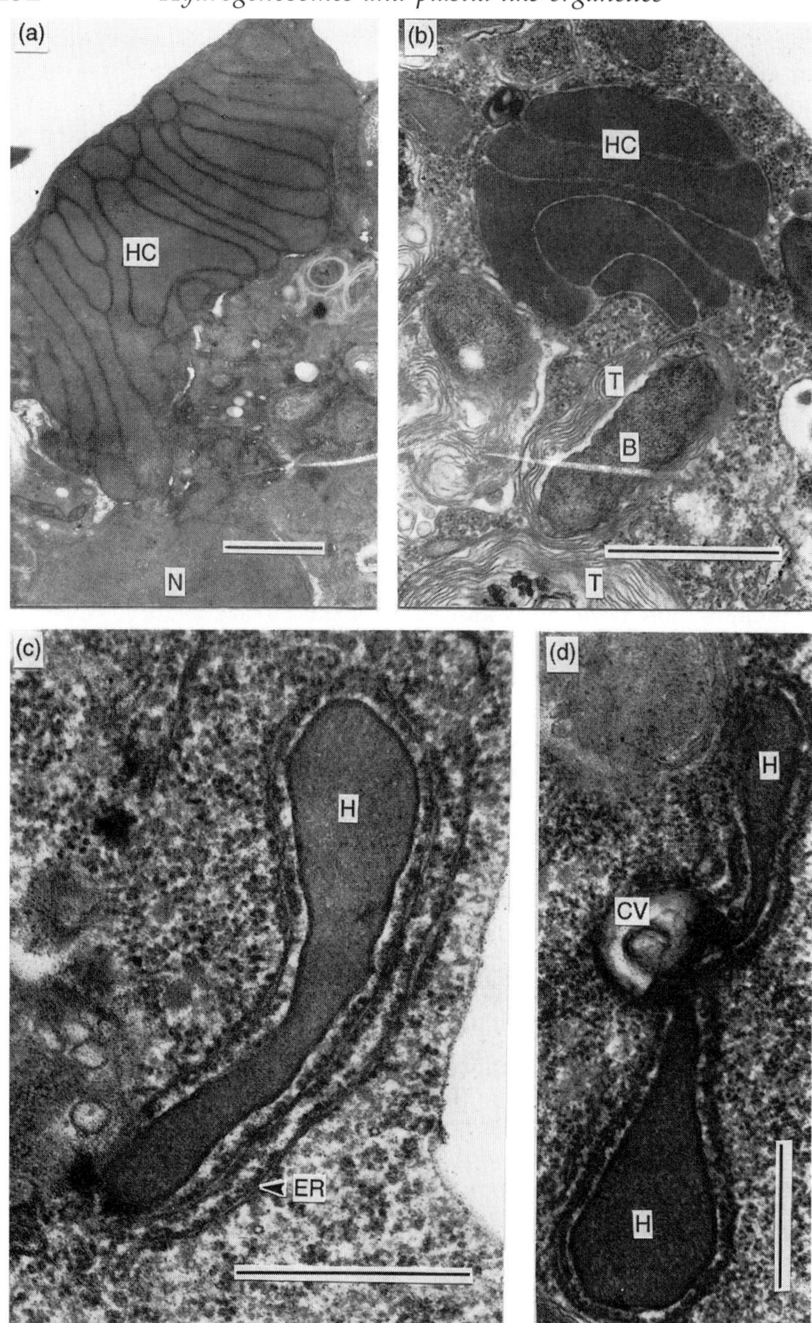

Figure 8.1 Hydrogenosomes of *Psalteriomonas lanterna.* (a) Large stack of hydrogenosomes (HC); nucleus (N). Bar, 1 μm. (b) Small, partially assembled stack of

and Müller, 1995a, b, and references therein). The data seem not to support the hypothesis of a common – possibly mitochondrial – origin. However, since a rigid phylogenetic analysis has not yet been published, it remains unclear whether the genes used for the sequence comparisons are orthologous or not. Most of the enzymes analyzed so far carry N-terminal extensions of moderate length. These have been interpreted as import signal peptides, in analogy to the N-terminal mitochondrial targeting sequences (Lahti and Johnson, 1991; Johnson, Lahti and Bradley, 1993; Bui, Bradley and Johnson, 1996; Hackstein *et al.*, 1997a). However, confirmatory experimental evidence in favour of a 'mitochondrial' targeting mechanism is still lacking (but see Bradley *et al.*, 1997), and there are no reports of components characteristic of mitochondrial import machinery in *Trichomonas* or any other organism that hosts hydrogenosomes. A functional dissection of the import machinery is essential, since it has been shown that in humans suffering from PH1 primary hyperoxaluria the 'mistargeting' of the peroxisomal alanine:glyoxylate aminotransferase 1 to mitochondria does not simply depend on the generation of a mutant N-terminal mitochondrial targeting sequence. A second mutation seems to be required before efficient mistargeting can occur (Leiper, Birdsey and Oatey, 1995). Moreover, the cytoskeletal centrins also possess a N-terminal extension that forms an amphipathic α helix. Nevertheless, there are no indications that any centrins are imported into mitochondria or plastids (Steinkötter, 1997). Therefore, it seems questionable whether the presence of N-terminal extensions and the discovery of chaperonins with a presumed mitochondrial/α proteobacterial ancestry in *Trichomonas* (Horner *et al.* 1996; Bui, Bradley and Johnson, 1996; Germot, Philippe and Le Guyader, 1996; Roger, Clark and Doolittle, 1996; Palmer, 1997) are sufficient to conclude that there was a mitochondrial ancestry of the hydrogenosome.

8.2.1 Multiple origins of hydrogenosomes

Hydrogenosomes evolved in very divergent taxa of anaerobic protists and fungi (Müller, 1993; Brul and Stumm, 1994; Coombs and Hackstein, 1995; Hackstein and Vogels, 1997). These taxa are likely to be polyphyletic, and neither phylogenetic analysis of the small subunit rDNA-

---

hydrogenosomes (HC); thylakosomes (T); food bacterium (B). Bar, 1 μm. (c) Developing hydrogenosome (H), surrounded by two layers of rough endoplasmic reticulum (ER). Bar, 0.5 μm. (d) Multiplication stage of a hydrogenosome; central vesicle (CV); hydrogenosome (H). Bar, 0.5 μm.

sequence data nor electron microscopy allows an unequivocal positioning of the various deep branching protists (Sogin 1991; Schlegel 1994) (see chapter 2). Chytridiomycete fungi belong to the 'crown' group of predominantly aerobic eukaryotes. It appears that they acquired an anaerobic lifestyle secondarily; phylogenetic analysis of their rDNA genes reveals that chytrids cluster with aerobic representatives of related taxa (Sogin, 1991; Dore and Stahl, 1991; Mennim, 1997) (see chapter 20). Chytridiomycete fungi evolved hydrogenosomes, and one might speculate whether their mitochondria were transformed into hydrogenosomes. However, ultrastructural studies have failed to reveal characteristics that support a mitochondrial descent of their hydrogenosomes. Instead, they showed that substantial morphological affinities exist between hydrogenosomes and peroxisomes (Yarlett *et al.*, 1986; Marvin-Sikkema *et al.*, 1992, 1993a, b; Müller, 1993).

Surprisingly, the genes encoding hydrogenosomal adenylate kinases of the chytrids *Piromyces* sp. and *Neocallimastix frontalis* are of mitochondrial descent (Voncken, Hochstenbach and Hackstein, 1996). But, consistent with the presence of the adenylate kinase in a peroxisome-like hydrogenosome, the 'mitochondrial' enzymes possess putative C-terminal peroxisomal PTS1 targeting signals (Voncken, Hochstenbach and Hackstein, 1996; compare with Gould *et al.*, 1990; Walton, Hill and Subramani, 1995; Subramani, 1996). This C-terminal extension is absent from all adenylate kinase 2 (AK2)-encoding genes of aerobic yeasts and fungi studied so far, but shared by *Piromyces* sp. and *Neocallimastix frontalis* – which represent two divergent taxa of anaerobic chytrids (Mennim, 1997). The presence of SKL-termini of hydrogenosomal proteins in *N. frontalis* had been postulated earlier because of the labelling of the hydrogenosomes and of putative hydrogenosomal enzymes by an antiserum directed against the SKL-epitope (Marvin-Sikkema, 1993a). It seems possible that one step in the evolution of the hydrogenosome of the chytrids has been the forwarding of a nuclear-encoded mitochondrial adenylate kinase to the evolving organelle by the addition of a new targeting signal – a phenomenon that happened several times in the evolution of the peroxisomes and other organelles (Stabenau, 1992; Brennicke *et al.*, 1993; Olsen and Harada, 1995; Long *et al.*, 1996; Voncken *et al.*, in preparation). Preliminary evidence seems to confirm that the peroxisomal targeting signal of the hydrogenosomal AK of *N. frontalis* is functional in *Saccharomyces cerevisiae*. Moreover, there is some evidence for the presence of components of a peroxisomal import machinery and key enzymes of peroxisomal β-oxidation (Voncken and Hackstein 1997; compare with Subramani, 1996).

Furthermore, a gene encoding a putative hydrogenosomal malic enzyme variant of *N. frontalis* shares a common ancestry with mitochondrial malic enzymes (van der Giezen *et al.*, 1997). It encodes a N-terminal extension similar to a mitochondrial targeting signal that seems to be sufficient to facilitate mitochondrial import in a heterologous system. However, it has also a putative PTS1-like C-terminal extension (KNL) that might be responsible for the localization in the hydrogenosomes (see Elgersma, 1995; Makita, 1995). The analysis of a putative hydrogenosomal β subunit of a succinyl-CoA synthetase from *N. frontalis* revealed similar characteristics, although a potential C-terminal signal could not be identified (Brondijk *et al.*, 1996). These observations do not, however, preclude import into a peroxisome-like organelle for the protein: it has been shown that peroxisomal import can occur via a variety of mechanisms, and even an apparent lack of targeting signals (Subramani, 1996).

Thus hydrogenosomes of chytrids have both mitochondrial and peroxisomal traits at the molecular level. It remains to be shown whether any component of a mitochondrial import machinery has been retained, or alternatively, how many 'foreign' enzymes were needed to convert the peroxisome-like precursor organelle into a hydrogenosome. It is also to be established whether peroxisomal enzymes have been retained. Perhaps the most compelling evidence will come from analysis of the components of the import machinery. In conclusion, there is strong evidence that the hydrogenosomes of chytridiomycete fungi have a chimeric origin: their morphology and the targeting signals indicate a descent completely different from that inferred from the coding sequences of enzymes. The idea that peroxisomes can contain different enzymes is not without precedent. It has been known for a long time that peroxisomes can be very different, even in the various tissues of one organism (de Duve, 1983; Stabenau, 1992; Olsen and Harada, 1995). Retargeting seems to be a frequent phenomenon in peroxisomes (Clayton and Michels, 1996) (see chapter 13). Therefore, it is reasonable to speculate that hydrogenosomes, at least in trichomonads, psalteriomonads and chytrids, evolved by the supplementing of a peroxisome-like subcellular compartment with enzymes from various sources that are essential for a hydrogenosomal function.

8.2.2 The hydrogenosomes of *Psalteriomonas lanterna*

Psalteriomonas lanterna is a free-living, heterotrophic, microaerophilic amoeboflagellate that lacks mitochondria, classical plastids, microbodies, and well developed dictyosomes (Broers *et al.*, 1990; Broers, 1992; Hackstein *et al.*, 1994). However, it possesses hydrogenosomes and unusual photosynthetic organelles (thylakosomes, see below) (Hackstein *et al.*, 1997a, b). Light microscopy established the presence of a prominent, globular structure in both the flagellate and amoeba stages of the organism. Cytochemical studies revealed hydrogenase activity in this structure (Broers, 1992), and electron microscopy confirmed that the globular complex was built up from stacked individual hydrogenosomes (Figure 8.1).

Electron microscopy on exponentially growing flagellate cells of *P. lanterna* revealed the presence of 1–4 stacked hydrogenosomal complexes per cell (Figure 8.1a, b). These complexes vary considerably in the number of hydrogenosomes per stack. The larger consist of 10–20 densely packed hydrogenosomes, the smaller ones contain only around five of these organelles. The individual hydrogenosomes are sausage- or disc-shaped; they are surrounded by two closely opposed membranes and possess a rather uniform matrix. There are no indications of the presence of internal cristae or tubular structures, ribosomes, or any other structural differentiations inside the matrix of the hydrogenosomes. Electron microscopy has not provided any evidence that the big, central complexes divide. Rather it suggests that the small stacks seem to assemble *de novo* from individual hydrogenosomes prior to cell division. Individual hydrogenosomes lie scattered in the cytoplasm predominantly at the periphery of the cells as sausage- and dumb-bell-shaped organelles. Their ultrastructural appearance identifies them as hydrogenosomes, and a careful inspection of the preparations obtained by indirect immunofluorescence labelling with the hydrogenase antiserum revealed that these organelles were labelled similar to the hydrogenosomal complexes (Broers, 1992). The peripheral hydrogenosomes are enwrapped by 1–2 cisternae of endoplasmic reticulum (ER) that is densely covered with ribosomes (Figure 8.1c). In the matrix of the organelles, neither ribosomes nor membraneous structures could be detected. It is likely therefore that the hydrogenosomal proteins are synthesized on the ER-bound ribosomes and imported into the developing hydrogenosome. These ER-enclosed organelles had been misinterpreted earlier as modified mitochondria (Broers *et al.*, 1990; Broers, 1992). Their ultrastructure, their reaction with the anti-hydrogenase serum, and the absence of cytochrome oxidase labelling reveal, however, that they are hydrogenosomes (Brul *et al.*, 1993; the title of the abstract reflects the misinterpretation of the thylakosome as a mitochondrion).

The dumb-bell-shaped peripheral hydrogenosomes are similar to the putative 'fission'- stages described from *Trichomonas* (see Nielsen and Diemer, 1976; Kulda, Nohynková and Ludvik, 1986; Benchimol *et al.*, 1996a, b). However, quite a number of them in *P. lanterna* are connected at their narrowest (median) part to a vesicle that is in intimate contact with the ER (Figure 8.1d). The lumina of both lobes of the hydrogenosome seem to open into the central vesicle. The lobes have the appearance of a partially inflated balloon, but a connection of the surrounding ER cisterns with the lumen of the central vesicle is unlikely. The membrane of the central vesicle is morphologically different from the

ER membranes, and similar in contrast to the membranes that surround the immature hydrogenosomes (Figure 8.1).

A morphometric analysis has been performed in order to analyse whether the various types of hydrogenosomes might represent different developmental stages. The analysis of 167 different hydrogenosome profiles revealed that the longest (and broadest, respectively) hydrogenosomes are found in the stacked complexes; the others showed a nearly bell-shaped length distribution that covered both the sausage- and dumb-bell stages. Apparently, the free, peripheral hydrogenosomes belong to one and the same distribution, regardless of whether they are connected with the median vesicle or not. There is no indication of a bimodal distribution: the lack of significant numbers of 'half-sized' hydrogenosomes and the absence of a separate collective of long, 'double-sized', dumb-bell-shaped profiles suggests that the dumb-bell-shaped hydrogenosomes were not undergoing fission. Rather, the frequency distribution – like the electron micrographs – favours the interpretation that the two lobes of a developing hydrogenosome are budding from the central vesicle that is in close contact with the endoplasmic reticulum. Thus, electron microscopy reveals that hydrogenosomes of *P. lanterna* have a unique mode of multiplication. It is different from the multiplication stages in *Trichomonas* and unlike the division of plastids and mitochondria (Schnepf, 1980; Benchimol *et al.*, 1996a, b).

8.3 INTERACTIONS OF THE HYDROGENOSOME WITH OTHER ORGANELLES OR ENDOSYMBIONTS

In many though not all anaerobic ciliates with hydrogenosomes, endosymbiotic or episymbiotic methanogenic bacteria are found in close association with the hydrogen producing organelles (Brul and Stumm, 1994; Fenchel and Finlay, 1995; Hackstein and Vogels, 1997). They were also present in *P. lanterna* when it was established in culture (Broers *et al.*, 1990; Broers, 1992). The methanogens of *P. lanterna* were lost during prolonged culturing. However, their loss did not result in a release of H_2 into the headspace of the culture vials in approximately equivalent amounts to the methane released earlier. This suggests the presence of an alternative hydrogen sink.

There is evidence that the anaerobic amoeboflagellate *P. lanterna* hosts membranous, plastid-like organelles that potentially provide a photosynthetic hydrogen sink (Figure 8.2) (Hackstein *et al.*, 1994, 1997b). High light intensities stimulate a consumption of hydrogen and carbon dioxide injected into the headspace of the culture vials, suggesting a light dependent reduction of carbon dioxide by hydrogen. If methanogens are co-cultured with *Psalteriomonas* cells, hydrogenosome-derived hydrogen is converted into methane that cannot be degraded by the

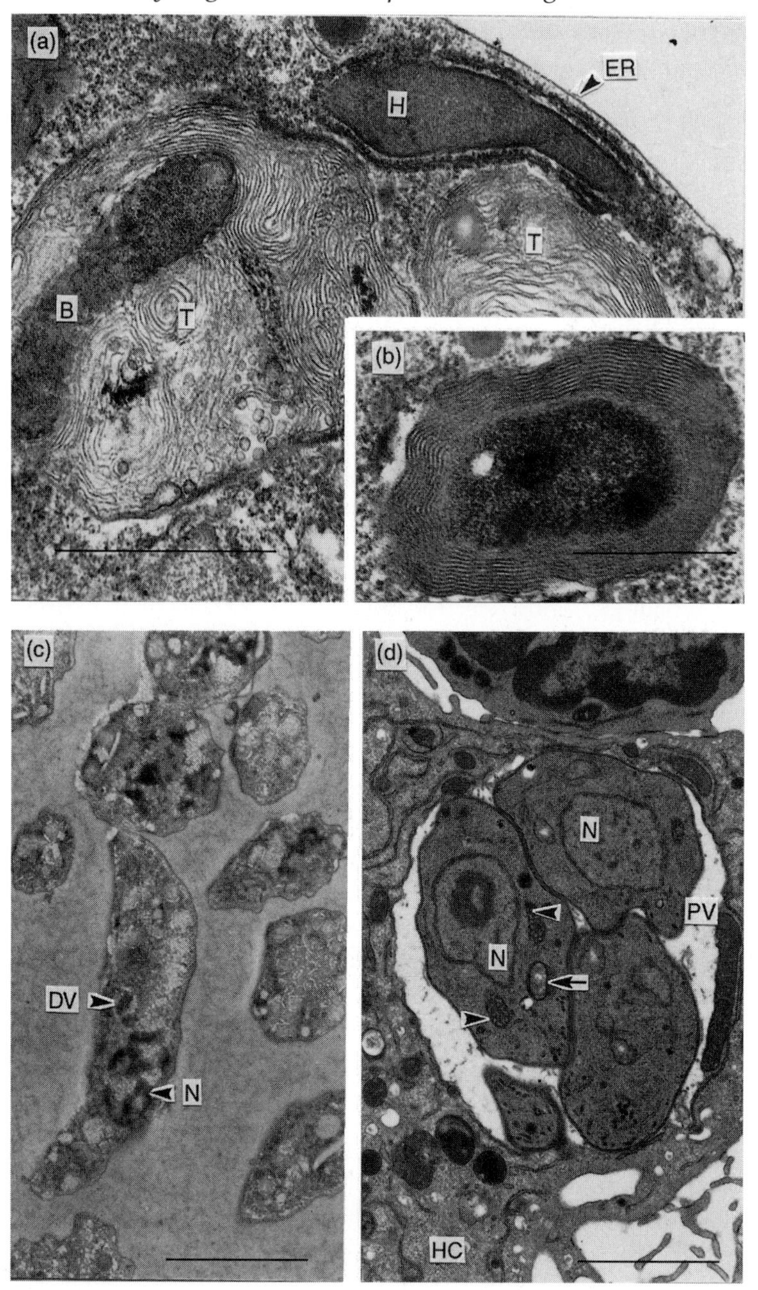

Figure 8.2 Thylakosomes and hydrogenosomes of *Psalteriomonas lanterna.* (a) Fully developed thylakosomes (T) with engulfed bacterium (B). The developing hydrogenosomes (H) surrounded by cisterns of rough endoplasmic reticulum

microbiota present in the culture vials. As methane production in the light and in the dark was significantly different, it can be tentatively concluded that the hydrogenosomes in the light either release lower amounts of hydrogen, or, more likely, that *Psalteriomonas* cells host a light-dependent sink for hydrogen (Table 8.1). There are no indications that, under these conditions, light inhibits the growth or metabolic activities of methanogens.

Thus, it is likely that hydrogenosomes can support the function of plastid-like organelles such as thylakosomes and not only of photosynthetic endosymbionts such as described for the ciliate *Strombidium purpureum* (Fenchel and Bernard, 1993; Bernard and Fenchel, 1994; Hackstein and Vogels, 1997). Apparently, hydrogenosomes can functionally replace mitochondria in photosynthetic anaerobes. Mitochondria are essential for the function of present-day plastids, and this need might limit the potential of photosynthetic eukaryotes to cope with anaerobiosis (Lawlor, 1993).

Table 8.1 Formation of methane (nmol) during the growth of *Psalteriomonas lanterna* under varying light conditions

	Dark cultures	*Light cultures*
methane	227 ± 44	83 ± 29

Polyxenic cultures of *P. lanterna* (containing free methanogenic bacteria) were incubated under anaerobic conditions in the light or in the dark. After one week, gas samples of the headspace were analysed by gas chromatography. The number of *P. lanterna* cells at the end of the experiment was approximately 3500/ml. Methane concentrations (mean and SD) are based on four independent assays. In the dark, three times more methane was formed than in the light during the incubation period. Methanogenic bacteria apparently convert the released hydrogenosomal H_2 to methane that cannot be degraded under anaerobic conditions.

(ER) were misinterpreted by Broers (1992) as mitochondria. Bar, 1 μm. (b) 'Young' developmental stage of a thylakosome. Bar, 1 μm. (c) *Toxoplasma gondii* tachyzoites, probed with an antiserum against double-stranded DNA. The nucleus (N) is heavily labelled. There is also indication for the presence of DNA in the plastid-like organelle, the double-walled versicle (DV). Bar, 1 μm. (d) *Toxoplasma gondii* in conventional electron microscopic preparation. A double walled vesicle is arrowed; rhoptries are indicated by arrowheads; N, nucleus; HC, host cell; PV, parasitophorous vacuole. Bar, 1 μm.

8.4 APICOMPLEXAN PLASTIDS

Apicomplexan parasites possess three genomes: a nuclear, a mitochondrial, and a plastidic (Feagin, 1994; Williamson *et al.*, 1994; Wilson *et al.*, 1996; Köhler *et al.*, 1997) (see chapter 16). The mitochondrion hosts a genome of about 6 kb – one of the smallest, functional mitochondrial genomes known. The plastid-like genome is also extremely reduced to a size of about 35 kb. Yet there is ample evidence that it is transcribed and supports a functional translation machinery (Preiser, Williamson and Wilson, 1995). The localization of the plastid genome, however, has been subject to many speculations. Siddal (1992) postulated that it might be hosted by an organelle variously called *Hohlzylinder*, double-walled vesicle, Golgi-adjunct, and *vesicules plurimembranaires*. Many though not all apicomplexan parasites harbour these elusive organelles (van der Zypen and Piekarski, 1967; Mehlhorn, 1988; Siddal, 1992). The organelles are distributed to the daughter cells during cell division (Hackstein *et al.*, 1995), and infrequently dumb-bell-shaped organelles are seen in the cytoplasm of the host cell suggesting a division of the organelles. Using an antiserum against dsDNA, the nucleus and the double-walled vesicle is labelled (Figure 8.2). By *in situ* transcript hybridization, McFadden *et al.* (1996) showed that a plastidic 16S RNA is transcribed inside the double-walled vesicles.

Thus, all evidence accumulated so far suggests that the double membrane-bounded vesicles of the apicomplexan parasites are highly derived, albeit functional plastids. Substantial differences in transcript abundance and inhibitor sensitivity indicate that the importance of mitochondria and plastids varies in the different developmental stages (Tomavo and Boothroyd, 1995; Feagin and Drew, 1995). Denton *et al.* (1996) showed that in *T. gondii* there are highly significant differences between the enzyme activities in tachyzoites and bradyzoites, including phosphofructokinase, pyruvate kinase, lactate dehydrogenase, NAD(P)-linked isocitrate dehydrogenase, and succinate dehydrogenase. These observations support suggestions that bradyzoites lack a functional TCA cycle and suggest that these developmental stages rely on an anaerobic metabolism. At present we can only speculate on the role and function of the plastid-like organelle in these stages, indeed there remain many unknowns about the functions of the plastid in apicomplexan parasites.

8.5 CONCLUSIONS

8.5.1 Hydrogenosomes have a polyphyletic origin

Molecular and ultrastructural evidence suggests that hydrogenosomes are polyphyletic. The hydrogenosomes of *Trichomonas*, *Psalteriomonas*, anaerobic fungi, and several ciliate species are clearly morphologically different (Yarlett *et al.*, 1984, 1986; Finlay and Fenchel, 1989; Paul, Williams and Butler, 1990; Marvin-Sikkema *et al.*, 1992, 1993; Benchimol *et al.*, 1996a, b). Phylogenetic analysis of the derived amino acid sequences of hydrogenosomal proteins has confirmed that hydrogenosomes are different (Brul *et al.*, 1994; Voncken *et al.*, submitted) (see chapter 6). Even within an apparently monophyletic taxon such as the ciliates, anaerobiosis and hydrogenosomes evolved at least three or four times (Embley *et al.*, 1995) (see chapter 18). Apparently the eukaryotic cell has the potential to evolve hydrogenosomes with ease in response to anaerobiosis.

8.5.2 Hydrogenosomes are chimeric organelles

An anaerobic eukaryote needs a subcellular compartment in order to evolve a hydrogenosome. This compartment has to be equipped with enzymes that are essential for hydrogenosomal functions (Müller, 1993). It is likely that genes encoding suitable proteins were recruited from various sources, since after integration into the nuclear genome and the addition of suitable targeting- and import-signals the acquired enzymes can easily be used to equip a hydrogenosome if the ancestral compartment possesses a suitable import machinery. Apparently, the ancestral cellular compartments that were used to evolve a hydrogenosome in the different taxa were different. The single membrane and the targeting signal of the adenylate kinase of the chytrid hydrogenosome reveal similarities to those of peroxisomes of aerobic organisms, whereas the closely opposed double membranes of the other hydrogenosomes and their chaperonines have been used as an argument to support the hypothesis of an endosymbiotic or mitochondrial ancestry of the hydrogenosome of the trichomonads and *Psalteriomonas* (see Palmer, 1997). For the hydrogenosomes of *Trichomonas* and *Psalteriomonas,* however, a chimeric origin is more likely because both peroxisomal and mitochondrial traits can be found. The lack of an organellar genome precludes a straightforward discovery of the descent of the various hydrogenosomes, and the problem of how an organelle that has lost its genome completely is maintained and propagated is still to be fully elucidated.

Our studies on the hydrogenosomes of chytrids have revealed that peroxisomes are good candidates to provide the basis for the evolution of organelles with novel functions because peroxisomes are highly flexible cellular compartments that can fulfil a broad spectrum of different functions (Lazarow and Fujiki, 1985; Cavalier-Smith, 1987; Gross, 1993; Makita, 1995). Peroxisomes can also evolve into glycosomes and glyoxysomes (Michels and Hannaert, 1994; Wiemer *et al.*, 1995; Clayton and Michels, 1996; Clayton, Häusler and Blattner, 1995) (see chapter 13).

In those species that lack peroxisomes, e.g. *Trichomonas* and *Psalteriomonas*, the ancestral, double membraned, cellular compartment has still to be identified, but it can be argued that it was a kind of peroxisome and not a mitochondrion. Molecular dissection of the hydrogenosomal import machinery should provide better evidence for the organelle's evolutionary origin.

8.5.3 Are the plastid-like organelles in apicomplexan parasites functional equivalents of hydrogenosomes?

The complex life cycle of apicomplexan parasites involves developmental stages that seem to be more or less anaerobic whereas other stages are clearly aerobic (Coombs and Hackstein, 1995; Denton *et al.*, 1996). Stage-specific functions for mitochondria and plastids can be postulated, and one might speculate whether a functional coupling between mitochondria and plastids might have provided the evolutionary constraint that is responsible for the maintenance of both organelles. Alternatively, one might propose that the regular encountering of phasic aerobiosis by the apicomplexan parasites might have prevented the evolution of hydrogenosomes. Under anaerobic conditions, a specialized plastid might be able to fulfil equivalent functions: plastids can provide a subcellular compartment that allows the generation of ATP, that hosts an electron transport chain and potentially also a hydrogenase (Schulz, 1996). These possibilities are hypothetical at the moment and have to be analysed carefully. Mitochondrial functions of the parasite are similar to the mitochondrial functions of the host cell, but the apicomplexan's plastidic functions might be unique to the parasite. Since a number of plastidic functions are known to be targets of rather specific inhibitors, the apicomplexan plastid could be a highly specific target for new classes of chemotherapeutic agents (Hackstein *et al.*, 1995).

In conclusion, the evolution of protists provides evidence for incredible flexibility in the invention of metabolic variants, the differentiation of suitable organelles and the acquisition of highly specialized endosymbionts. Eukaryotic cells are capable of reacting flexibly to changing environmental conditions by modifying pre-existing cellular compartments. The evolution of hydrogenosomes and plastid-like organelles in protists that cope with either permanent or phasic anaerobiosis illustrates once again one of the most exciting and successful evolutionary strategies – tinkering (Jacob, 1977).

Acknowledgements

We thank Holger Schlierenkamp and Theo van Alen for excellent technical support. The discussions with Ron Hochstenbach and Brigitte

Boxma are gratefully acknowledged, also the proof-reading by Urs Hackstein.

8.6 REFERENCES

Benchimol, M., Almeida, J. d. C.A. and de Souza, W. (1996a) Further studies on the organization of the hydrogenosome of *Tritrichomonas foetus*. *Tissue and Cell*, **28**, 287–99.

Benchimol, M., Johnson, P.J. and de Souza, W. (1996b) Morphogenesis of the hydrogenosome: an ultrastructural study. *Biology of the Cell*, **87**, 197–205.

Bernard, C. and Fenchel, T. (1994) Chemosensory behaviour of *Strombidium purpureum*, an anaerobic oligotrich with endosymbiotic purple non-sulphur bacteria. *Journal of Eukaryotic Microbiology*, **41**, 391–6.

Bradley, P.J., Lathi, C.J., Plümper, E. and Johnson, P.J. (1997) Targeting and translocation of proteins into the hydrogenosome of the protist *Trichomonas*: similarities with mitochondrial protein import. *EMBO Journal*, **16**, 3484–93.

Brennicke, A., Grohmann, L., Hiesel, R. *et al.* (1993) The mitochondrial genome on its way to the nucleus: different stages of gene transfer in higher plants. *FEBS Letters*, **325**, 140–5.

Broers, C.A.M. (1992) Anaerobic psalteriomonad amoeboflagellates. Thesis, University of Nijmegen, The Netherlands.

Broers, C.A.M., Stumm, C.K., Vogels, G.D. and Brugerolle, G. (1990) *Psalteriomonas lanterna* gen. nov., sp. nov., a free-living amoeboflagellate isolated from freshwater anaerobic sediments. *European Journal of Protistology*, **25**, 369–80.

Brondijk, T.H.C., Durand, R., van der Giezen, M. *et al.* (1996) *scsB*, a cDNA encoding the hydrogenosomal β subunit of succinyl-CoA synthetase from the anaerobic fungus *Neocallimastix frontalis*. *Molecular and General Genetics*, **253**, 315–23.

Brul, S., Kleijn, J., Lombardo, M.C.P. *et al.* (1993) Coexistence of hydrogenosomes and mitochondrial structures in the amoeboflagellate *Psalteriomonas lanterna*. *Journal of Eukaryotic Microbiology*, **40**, 17A.

Brul, S., Veltman, R.H., Lombardo, M.C.P. and Vogels, G.D. (1994) Molecular cloning of hydrogenosomal ferredoxin cDNA from the anaerobic amoeboflagellate *Psalteriomonas lanterna*. *Biochimica et Biophysica Acta*, **1183**, 544–6.

Brul, S. and Stumm, C.K. (1994) Symbionts and organelles in anaerobic protozoa and fungi. *Trends in Ecology and Evolution*, **9**, 319–24.

Bui, E.T.N., Bradley, P.J. and Johnson, P.J. (1996) A common evolutionary origin for mitochondria and hydrogenosomes. *Proceedings of the National Academy of Sciences USA*, **93**, 9651–6.

Cavalier-Smith, T. (1987) The simultaneous symbiotic origin of mitochondria, chloroplasts, and microbodies. *Annals of the New York Academy of Sciences*, **503**, 55–71.

Clayton, Ch., Häusler, T. and Blattner, J. (1995) Protein trafficking in kinetoplastid protozoa. *Microbiological Reviews*, **59**, 325–44.

Clayton, C.E. and Michels, P. (1996) Metabolic compartmentalization in African trypanosomes. *Parasitology Today*, **12**, 465–71.

Coombs, G.H. and Hackstein, J.H.P. (1995) Anaerobic protists and anaerobic ecosystems, in *Protistological Actualities. Proceedings of the Second European Congress of Protistology* (eds G. Brugerolle and J.-P. Mignot), Clermont-Ferrand, France, pp. 90–101.

Denton, H., Roberts, C.W., Alexander, J. *et al.* (1996) Enzymes of energy metabolism in the bradyzoites and tachyzoites of *Toxoplasma gondii. FEMS Microbiology Letters*, **137**, 103–8.

Dore, J. and Stahl, D.A. (1991) Phylogeny of anaerobic rumen chytridiomycetes inferred from small subunit ribosomal RNA sequence comparisons. *Canadian Journal of Botany*, **69**, 1964–71.

de Duve, Ch. (1983) Microbodies in the living cell. *Scientific American*, **248/5**, 52–62.

Elgersma, Y. (1995) Transport of proteins and metabolites across the peroxisomal membrane in *Saccharomyces cerevisiae*. Thesis, University of Amsterdam, The Netherlands.

Elgersma, Y., van den Berg, M., Tabak, H.F. and Distel, B. (1993) An efficient positive selection procedure for the isolation of peroxisomal import and peroxisomal assembly mutants of *Saccharomyces cerevisiae. Genetics*, **135**, 731–40.

Embley, T.M., Finlay, B.J., Dyal, P.L. *et al.* (1995) Multiple origins of anaerobic ciliates with hydrogenosomes within the radiation of aerobic ciliates. *Proceedings of the Royal Society London, Series B*, **262**, 87–93.

Feagin, J.E. (1994) The extrachromosomal DNAs of apicomplexan parasites. *Annual Review of Microbiology*, **48**, 81–104.

Feagin, J.E. and Drew, M.E. (1995) *Plasmodium falciparum*: Alterations in organelle transcript abundance during the erythrocytic cycle. *Experimental Parasitology*, **80**, 430–40.

Fenchel, T. and Bernard, C. (1993) A purple protist. *Nature*, **362**, 300.

Fenchel, T. and Finlay, B.J. (1995) *Ecology and Evolution in Anoxic Worlds*. Oxford University Press, Oxford.

Finlay, B.J. and Fenchel, T. (1989) Hydrogenosomes in some anaerobic protozoa resemble mitochondria. *FEMS Microbiology Letters*, **65**, 311–14.

Germot, A., Philippe, H. and Le Guyader, H. (1996) Presence of a mitochondrial-type HSP70 in *Trichomonas vaginalis* suggests a very early mitochondrial endosymbiosis in eukaryotes. *Proceedings of the National Academy of Sciences USA*, **93**, 14614–17.

Giezen, M. van der, Rechinger, K.B., Svendsen, I. *et al.* (1997) A mitochondrial-like targeting signal on the hydrogenosomal malic enzyme from the anaerobic fungus *Neocallimastix frontalis:* support for the hypothesis that hydrogenosomes are modified mitochondria. *Molecular Microbiology*, **23**, 11–21.

Gould, S.J., Keller, G.-A., Schneider, M. *et al.* (1990) Peroxisomal protein import is conserved between yeast, plants, insects and mammals. *EMBO Journal*, **9**, 85–90.

Gray, M.W. (1989) The evolutionary origins of organelles. *Trends in Genetics*, **5**, 294–9.

Gray, M.W. (1992) The endosymbiont hypothesis revisited. *International Review of Cytology*, **141**, 233–357.

Gross, W. (1993) Peroxisomes in algae: their distribution, biochemical function and phylogenetic importance, in *Progress in Phycological Research*, Vol. 9 (eds F.E. Round and D.J. Chapman), Biopress, pp. 47–78.

Gupta, R.S. and Golding, G.B. (1996) The origin of the eukaryotic cell. *Trends in Biochemical Sciences*, **21**, 166–71.

Hackstein, J.H.P., Schubert, H., v.d. Berg, M. *et al.* (1994) A novel photosynthetic organelle in anaerobic mastigotes. *Endocytobiology and Cell Research*, **10**, 261.

Hackstein, J.H.P., Mackenstedt, U., Mehlhorn, H. *et al.* (1995) Parasitic apicomplexans harbour a chlorophyll a-D1 complex, the potential target for therapeutic triazines. *Parasitology Research*, **81**, 207–16.

Hackstein, J.H.P. and Vogels, G.D. (1997) Endosymbiotic interactions in anaerobic protozoa. *Antonie van Leeuwenhoek*, **71,** 151–8.

Hackstein, J.H.P., Rosenberg, J., Broers, C.A.M. *et al.* (1997a) Biogenesis of hydrogenosomes in *Psalteriomonas lanterna*: no evidence for an exogenosomal ancestry, in *Eukocyotism and Symbiosis. Intertaxonic combination* versus *symbiotic adaptation*. (eds H.E.A. Schenk, R.G. Herrmann, K.W. Jeon *et al.*), Springer Verlag, Berlin, pp. 63–70.

Hackstein, J.H.P., Schubert, H., Rosenberg, J. *et al.* (1997b) Plastid-like organelles in anaerobic mastigotes and parasitic apicomplexans, in *Eukocyotism and Symbiosis. Intertaxonic combination* versus *symbiotic adaptation.* (eds H.E.A. Schenk, R.G. Herrmann, K.W. Jeon *et al.*), Springer Verlag, Berlin, pp. 49–58.

Horner, D.S., Hirt, R.P., Kilvington, S. *et al.* (1996) Molecular data suggest an early acquisition of the mitochondrion endosymbiont. *Proceedings of the Royal Society London, Series B*, **263**, 1053–9.

Hruban, Z. and Rechcigl, M. Jr. (1969). *Microbodies and Related Particles. Morphology, Biochemistry, and Physiology*. Academic Press, New York and London.

Hrdy, I. and Müller, M. (1995a) Primary structure and eubacterial relationships of the pyruvate:ferredoxin oxidoreductase of the amitochondriate eukaryote *Trichomonas vaginalis*. *Journal of Molecular Evolution*, **41**, 388–96.

Hrdy, I. and Müller, M. (1995b) Primary structure of the hydrogenosomal malic enzyme of *Trichomonas vaginalis* and its relationship to homologous enzymes. *Journal of Eukaryotic Microbiology*, **42**, 593–603.

Jacob, F. (1977) Evolution and tinkering. *Science*, **196**, 1161–6.

Johnson, P.J., Lahti, C.J. and Bradley, P.J. (1993) Biogenesis of the hydrogenosome in the anaerobic protists, *Trichomonas vaginalis*. *Journal of Parasitology*, **79**, 664–70.

Köhler, S., Delwiche, Ch.F., Denny, P.W. *et al.* (1997) A plastid of probable green algal origin in apicomplexan parasites. *Science*, **275,** 1485–9.

Kulda, J., Nohynková, E. and Ludvik, J. (1986) Basic structure and function of the trichomonad cell. *Acta Universitatis Carolinae – Biologica*, **30**, 181–98.

Länge, S., Rozario, C. and Müller, M. (1994) Primary structure of the hydrogenosomal adenylate kinase of *Trichomonas vaginalis* and its phylogenetic relationships. *Molecular and Biochemical Parasitology*, **66**, 297–308.

Lahti, C.J. and Johnson, P.J. (1991) *Trichomonas vaginalis* hydrogenosomal proteins are synthesized on free polyribosomes and may undergo processing upon maturation. *Molecular and Biochemical Parasitology*, **46**, 307–10.

Lawlor, D.W. (1993) *Photosynthesis: Molecular, Physiological and Environmental Processes.* Longman Scientific and Technical, Harlow, Essex, England.

Lazarow, P.B. and Fujiki, Y. (1985) Biogenesis of peroxisomes. *Annual Review of Cell Biology*, **1**, 489–530.

Leiper, J.M., Birdsey, G.M. and Oatey, P.B. (1995) Peroxisomes proliferate. *Trends in Cell Biology*, **5**, 435–7.

Long, M., de Souza, S.J., Rosenberg, C. and Gilbert, W. (1996) Exon shuffling and the origin of the mitochondrial targeting function in plant cytochrome C1 precursor. *Proceedings of the National Academy of Sciences USA*, **93**, 7727–31.

Makita, T. (1995) Molecular organization of hepatocyte peroxisomes. *International Review of Cytology*, **160**, 303–52.

Margulis, L. (1993) *Symbiosis in Cell Evolution. Microbial Communities in the Arcane and Proterozoic Eons*, 2nd edn, W.H. Freeman and Company, New York.

Margulis, L. (1996) Archaeal-eubacterial mergers in the origin of eukarya: phylogenetic classification of life. *Proceedings of the National Academy of Sciences USA*, **93**, 1071–6.

Marvin-Sikkema, F.D., Lahpor, G.A., Kraak, M.N. *et al.* (1992) Characterization of an anaerobic fungus from llama faeces. *Journal of General Microbiology*, **138**, 2235–41.

Marvin-Sikkema, F.D., Kraak, M.N., Veenhuis, M. *et al.* (1993a) The hydrogenosomal enzyme hydrogenase from the anaerobic fungus *Neocallimastix sp. L2* is recognized by antibodies directed against the C-terminal microbody protein targeting signal SKL. *European Journal of Cell Biology*, **61**, 86–91.

Marvin-Sikkema, F.D., Pedro Gomes, T.M., Grivet, J.-P. *et al.* (1993b) Characterization of hydrogenosomes and their role in glucose metabolism of *Neocallimastix* sp. L2. *Archives of Microbiology*, **160**, 388–96.

McFadden, G.I., Reith, M.E., Munholland, J. *et al.* (1996) Plastid in human parasites. *Nature*, **381**, 482.

Mehlhorn, H. (ed.) (1988) *Parasitology in Focus*. Springer Verlag, Berlin.

Melkonian, M. (1996) Systematics and evolution of the algae: Endocytobiosis and evolution of the major algal lineages. *Progress in Botany*, **57**, 281–11.

Mennim, G. (1997) The application of ribosomal DNA sequence data and other molecular approaches to the study of anaerobic gut fungi. PhD Thesis, University of Manchester, UK.

Michels, P.A.M. and Hannaert, V. (1994) The evolution of kinetoplastid glycosomes. *Journal of Bioenergetics and Biomembranes*, **26**, 213–19.

Müller, M. (1993) The hydrogenosome. *Journal of General Microbiology*, **139**, 2879–89

Nielsen, M.H. and Diemer, N.H. (1976) The size, density, and relative area of chromatic granules ("hydrogenosomes") in *Trichomonas vaginalis* Donne from cultures in logarithmic and stationary growth. *Cell and Tissue Research*, **167**, 461–5.

Olsen, L.J. and Harada, J.J. (1995) Peroxisomes and their assembly in higher plants. *Annual Review of Plant Physiology and Molecular Biology*, **46**, 123–46.

Palmer, J.D. (1995) Rubisco rules fall; gene transfer triumphs. *Bioessays*, **17**, 1005–12.

Palmer, J.D. (1997) Organelle genomes: Going, going, gone. *Science*, **275**, 790–1.

Paul, R.G., Williams, A.G. and Butler, R.D. (1990) Hydrogenosomes in the rumen entodiniomorphid ciliate *Polyplastron multivesiculatum*. *Journal of General Microbiology*, **136**, 1981–9.

Preiser, P., Williamson, D.H. and Wilson, R.J.M. (1995) tRNA genes transcribed from the plastid-like DNA of *Plasmodium falciparum*. *Nucleic Acids Research*, **23**, 4329–36.

Roger, A.J., Clark, C.G. and Doolittle, W.F. (1996) A possible mitochondrial gene in the early-branching amitochondriate protist *Trichomonas vaginalis*. *Proceedings of the National Academy of Sciences USA*, **93**, 14618–22.

Schlegel, M. (1994) Molecular phylogeny of eukaryotes. *Trends in Ecology and Evolution*, **9**, 330–5.

Schnepf, E. (1980) Types of plastids: their development and interconversions, in *Chloroplasts* (ed. J. Reinert), Springer Verlag, Berlin, pp. 1–27.

Schulz, R. (1996) Hydrogenases and hydrogen production in eukaryotic organisms and cyanobacteria. *Journal of Marine Biotechnology*, **4**, 16–22.

Siddal, M.E. (1992) Hohlzylinder. *Parasitology Today*, **8**, 90–1.

Sogin, M. (1991) Early evolution and the origin of eukaryotes. *Current Opinion in Genetics and Development*, **1**, 457–63.

Stabenau, H. (ed.) (1992) *Phylogenetic Changes in Peroxisomes of Algae – Phylogeny of Plant Peroxisomes*, University Press, Oldenburg.

Steinkötter, J. (1997) Centrin und centrinbindende Proteine in Grünalgen. A. Phylogenetische Analysen centrinkodierender Regionen aus Grünalgen und anderen Eukaryoten. B. Biochemische und molekulare Identifikation potenieller centrinbindender Proteine aus *Spermatozopsis similis* (Chlorophyceae). Thesis, University of Cologne, Germany.

Subramani, S. (1996) Convergence of model systems for peroxisome biogenesis. *Current Opinion in Cell Biology*, **8**, 513–18.

Tomavo, S. and Boothroyd, J.C. (1995) Interconnection between organellar functions, development and drug resistance in the protozoan parasite *Toxoplasma gondii*. *International Journal for Parasitology*, **25**, 1293–9.

Thorsness, P.E. and Weber, E.R. (1996) Escape and migration of nucleic acids between chloroplasts, mitochondria, and the nucleus. *International Review of Cytology*, **165**, 207–34

Voncken, F.G.J., Hochstenbach, R. and Hackstein, J.H.P. (1996) Evolution of hydrogenosomes in anaerobic protoctista. *Verhandlungen der Deutschen Zoologischen Gesellschaft*, **81**, 26.

Voncken, F.G.J. and Hackstein, J.H.P. (1997) New evidence for a peroxisomal ancestry of fungal hydrogenosomes. Society for General Microbiology, 137th ordinary meeting. Heriot-Watt University, Edinburgh, p. 41.

Voncken, F.G.J., Boxma, B., Verhagen, E. *et al.* (1997) A chimeric origin of fungal hydrogenosomes, submitted.

Walton, P.A., Hill, P.E. and Subramani, S. (1995) Import of stably folded proteins into peroxisomes. *Molecular Biology of the Cell*, **6**, 675–83.

Wiemer, E.A.C., Hannaert, V., v.d. Ijssel, P.R.L.A. *et al.* (1995) Molecular analysis of glyceraldehyde-3-phosphate dehydrogenase in *Trypanosoma borelli*: an evolutionary scenario of subcellular compartmentalization in kinetoplastida. *Journal of Molecular Evolution*, **40**, 443–54.

Williamson, D.H., Gardner, M.J., Preiser, P. *et al.* (1994) The evolutionary origin of the 35 kB circular DNA of *Plasmodium falciparum*: new evidence supports a possible rhodophyte ancestry. *Molecular and General Genetics*, **243**, 249–52.

Wilson, R.J.M., Denny, P.W., Preiser, P.R. *et al.* (1996) Complete gene map of the plastid-like DNA of the malaria parasite *Plasmodium falciparum*. *Journal of Molecular Biology*, **261**, 155–72.

Yarlett, N., Colemen, G.S., Williams, A.G. and Lloyd, D. (1984) Hydrogenosomes in known species of rumen entodiniomorphid protozoa. *FEMS Microbiology Letters*, **21,** 15–19.

Yarlett, N., Orpin, C.G., Munn, E.A. *et al.* (1986) Hydrogenosomes in the rumen fungus *Neocallimastix patriciarum*. *Biochemical Journal*, **236**, 729–39.

v. d. Zypen, E. and Piekarski, G. (1967) Ultrastrukturelle unterschiede von *Toxoplasma gondii*. *Zeitschrift für Bakteriologie, Parasitologie, Infektion und Hygiene*, **203**, 495–517.

9

Molecular systematics of the intestinal amoebae

C. Graham Clark, Jeffrey D. Silberman, Louis S. Diamond and Mitchell L. Sogin*

*Department of Medical Parasitology, London School of Hygiene and Tropical Medicine, Keppel Street, London WC1E 7HT, UK. E-mail: g.clark@lshtm.ac.uk

ABSTRACT

Humans are host to four genera of intestinal amoebae – *Entamoeba, Endolimax, Dientamoeba* and *Iodamoeba*. Ribosomal RNA gene data are now available for the first three genera. Earlier antigen and electron microscopy studies recognizing *Dientamoeba fragilis* as an aberrant trichomonad have been upheld by rDNA sequence analysis. However, it could not be proven unequivocally that *D. fragilis* has secondarily lost the cytoskeletal and other features normally associated with trichomonads. Within the genus *Entamoeba*, species descriptions have classically been based on a limited range of morphological characters, the validity of which has been uncertain. Over 80 isolates assigned to this genus have been studied by riboprinting (restriction site polymorphism analysis of PCR amplified small subunit ribosomal RNA genes) and over a dozen small subunit rRNA gene (rDNA) sequences are now available. The classical subdivision of the genus into those species that produce cysts with one, four and eight nuclei when mature is upheld by the rDNA data. *Entamoeba gingivalis*, a species that does not encyst, branches with the group that produces four-nucleated cysts indicating a secondary loss of this ability. Species infecting mammals are interspersed on phylogenetic trees with those that infect reptiles, indicating multiple independent colonizations rather than co-evolution of host and parasite. The species *Ent. moshkovskii* appears to be a free-living descendant of parasitic ancestry. Significant cryptic variation was detected in three species (*Ent. coli, Ent. gingivalis* and *Ent.*

Evolutionary Relationships Among Protozoa. Edited by G.H. Coombs, K. Vickerman, M.A. Sleigh and A. Warren. Published in 1998 by Chapman & Hall, London. ISBN 0 412 79800 X

moshkovskii). *Endolimax nana* has always been grouped with *Entamoeba* on morphological grounds, although the justification for doing so has been somewhat tenuous. The monophyly of *Entamoeba* and *Endolimax* is supported in some analyses but not in others, with the divergence point being very deep when they form a single clade. The placement of *Entamoeba histolytica* within the eukaryotic tree has been controversial. rDNA based analyses have consistently placed it on a comparatively late branch. The inclusion of sequences from additional *Entamoeba* species plus *Endolimax nana* only strengthens the support for a late branch point, implying that both these genera are secondarily amitochondriate. Secondary loss of morphological, cellular and/or biochemical characteristics would appear to be a common feature in intestinal amoebae, as indeed it is in parasitic protists in general. As a result, differentiation of primitive and derived traits may be difficult and cautious interpretations are warranted.

9.1 INTRODUCTION

Four genera of amoebae are recognized as colonizing the human large intestine. (In this manuscript the word amoeba will be used to describe the morphology and mode of locomotion of an organism only and should not be taken to have any taxonomic significance.) Additional genera have been described from other hosts, but little is known about those organisms. The most widely studied genus of intestinal amoebae is *Entamoeba* which has been described from hosts in all classes of vertebrates and in some invertebrates. With few exceptions, all have a simple life cycle consisting of an infective cyst stage and a multiplying trophozoite stage. There is no cyst stage in some species. The cyst is ingested with faecally contaminated food or water and it passes through the stomach and small intestine. The trophozoite emerges as the cyst enters the large intestine and establishes a new infection. In response to unknown stimuli, amoebae re-encyst and cysts are shed periodically in the stool thus completing the life cycle (Martínez-Palomo, 1993). This life cycle is also characteristic of the genera *Endolimax* and *Iodamoeba*. The fourth genus, *Dientamoeba*, has no known cyst stage and its mode of transmission is poorly understood.

Intestinal amoebae do not cause disease in their hosts, with two notable exceptions. The agent of human amoebic dysentery and amoebic liver abscess is *Ent. histolytica*, which is responsible for 50 000 to 100 000 deaths annually worldwide (Walsh, 1986). A second species, *Ent. invadens*, causes a fatal disease in snakes and lizards (Ratcliffe and Geiman, 1934). The other amoeba that has been implicated in human disease is *D. fragilis* although not all authors agree on its role as a pathogen (Grendon, DiGiacomo and Frost, 1995). Infection with this amoeba certainly is not implicated in mortality of the host.

While a single species of the genera *Dientamoeba* (*D. fragilis*), *Endolimax* (*End. nana*) and *Iodamoeba* (*I. bütschlii*) have been reported to infect humans, as many as eight species of *Entamoeba* have been isolated with varying frequency from this host. Not all such reports are universally accepted, in part because species identification is not straightforward. As with all naked amoebae there are relatively few morphological criteria on which to base a diagnosis or a species description. In the genus *Entamoeba*, classical descriptions have relied on: 1. the host species; 2. the sizes of the cyst and trophozoite; 3. the appearance of the nucleus and of the 'chromatoid bars' (crystalline arrays of ribosomes observed in immature cysts); 4. the number of nuclei in the mature cyst (either one, four or eight). Clearly, some of these criteria are subjective, while others are potentially variable and of uncertain specificity. In such cases there is a risk of creating invalid species through use of misleading criteria, such as the host species. Likewise, it is possible that significant differences may be overlooked if genetic divergence is not reflected in morphological change. Molecular tools can be used to provide an independent evaluation of the morphological criteria currently used in the identification of the intestinal amoebae.

In classical taxonomic schemes the four genera of intestinal amoebae were initially all placed in the Family Entamoebidae. Three of them remain there today. With the advent of electron microscopy, *D. fragilis* was unequivocally identified as an aberrant amoeboid trichomonad and removed from the Entamoebidae (Camp, Mattern and Honigberg, 1974). As far as we have been able to determine, no electron microscopy studies of either *Iodamoeba* or *Endolimax* have been published. Other than sharing a niche and producing cysts there is little to unite the remaining genera in one family. Again, the validity of grouping these organisms needs independent evaluation. Therefore, in addition to the relationships among species of *Entamoeba*, the relatedness of the four genera of intestinal amoebae also needs to be re-examined.

The final outstanding question is where these amoebae belong within the eukaryotes. It is known that *Entamoeba* and *Dientamoeba* lack classical mitochondria and it is suspected that these organelles are also absent from the other two genera. This absence was for many years taken as an indication of primitiveness and *Entamoeba* in particular was often held up as a paradigm for the primordial eukaryotic cell (Bakker-Grunwald and Wöstmann, 1993). Unexpectedly, phylogenetic analyses using the small subunit ribosomal RNA sequence of *Ent. histolytica* indicated that it was a relatively late branching organism and therefore that it must have secondarily lost mitochondria and other organelles. This placement has not been universally accepted, however, and additional data are needed to confirm the branch-point of the *Entamoeba* lineage and to place the other genera in the eukaryotic tree.

9.2 METHODS

We have used small subunit ribosomal RNA (SSUrRNA) genes to address each of the three areas of uncertainty outlined above. The diversity among species of the genus *Entamoeba* was initially evaluated using riboprinting (Clark and Diamond, 1991) – restriction fragment length polymorphism analysis of polymerase chain reaction amplified SSUrRNA genes – while SSUrRNA gene sequences were determined for selected *Entamoeba* species, *End. nana* and *D. fragilis* to address the relationships among the genera and to place them on the eukaryotic tree. Unfortunately, to date we have been unable to obtain material for analysis of *I. bütschlii.*

Riboprinting was performed using 12 restriction enzymes which are estimated to give between 10% and 15% indirect coverage of the gene sequence. Riboprint data were evaluated using the proportion of co-migrating fragments to estimate genetic distances or by using each DNA fragment as a character in parsimony analysis. The two methods gave congruent results. Sequence data were evaluated using a variety of distance, parsimony and likelihood analyses. Bootstrap analyses were performed and various combinations of species were included as out-groups to evaluate the robustness of the sequence-based trees generated.

9.3 RESULTS

9.3.1 Relationships within *Entamoeba*

Over 80 isolates of *Entamoeba* were riboprinted (Clark and Diamond, 1997). All the classically defined species were upheld in that they had distinct patterns. However, three species showed intraspecific variation. In the case of *Ent. gingivalis* variation was observed only with a single restriction enzyme. In the other two cases, *Ent. coli* and *Ent. moshkovskii,* the intraspecific variation was much more significant, exceeding that observed between several other pairs of species. With one exception, all the variants formed a monophyletic clade within the genus (Figure 9.1). Since riboprint variation is in most cases the only indication of differences within the two species we have decided not to recognize the variants with specific names at this point. However, a means of recognizing the existence of variation and identifying the variants is needed and we use the term 'ribodeme' (Clark and Pung, 1994) to describe a

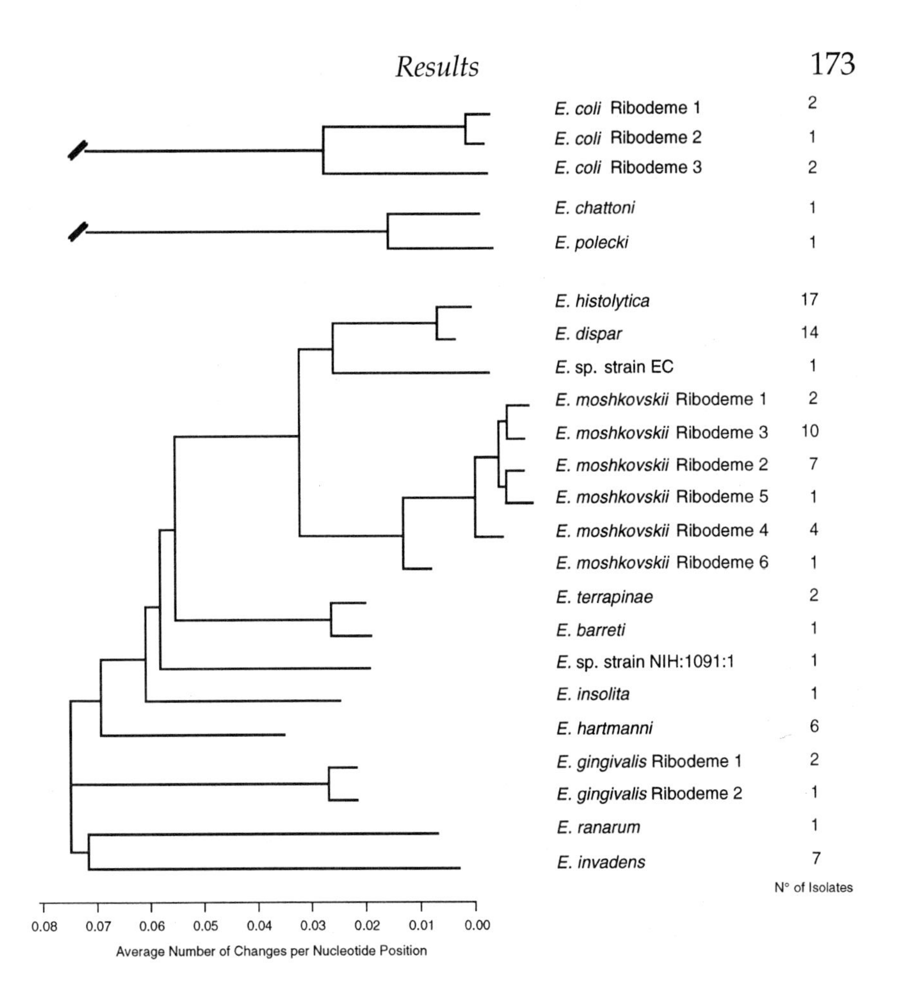

Figure 9.1 Phylogenetic relationships among *Entamoeba* species and ribodemes by riboprint analysis. The fragment co-migration data derived from the riboprint patterns were converted into estimated genetic distances by the method of Nei and Li (1979) and the tree was constructed by the method of Fitch and Margoliash (1967). The total tree standard deviation for the four-nucleated cyst producing species is 10.9%, one-third of which is due to *E. moshkovskii* Ribodeme 4. The number of isolates studied that gave each riboprint pattern is also listed. The normal host range for the species are: *E. coli* (primates); *E. chattoni* (primates); *E. polecki* (pigs); *E. histolytica* (primates); *E. dispar* (primates); *E. sp.* EC (free-living?); *E. moshkovskii* (free-living, rarely in humans); *E. terrapinae* (turtles); *E. barreti* (snapping turtle); *E. sp.* NIH:1091:1 (iguana); *E. insolita* (tortoises); *E. hartmanni* (primates); *E. gingivalis* (primates); *E. ranarum* (frogs/toads); *E. invadens* (reptiles).

population of organisms within a species that share the same riboprint pattern. All species other than the three mentioned above have a single ribodeme while in *Ent. gingivalis*, *Ent. coli* and *Ent. moshkovskii* we have identified two, three and six ribodemes, respectively. The one exception to the monophyly mentioned above is *Entamoeba* sp. isolate EC which was originally described as *Ent. moshkovskii* but based on isoenzymes and riboprinting is clearly distinct and appears to be more closely related to *Ent. histolytica*. This isolate will be redescribed as representing a new species.

As previously mentioned, trees based on distance and parsimony analysis of the riboprint data were largely congruent, differing only in the degree of resolution obtained and in the placement of one ribodeme within *Ent. moshkovskii*. If the relationships depicted by the tree are valid, a number of conclusions can be drawn. The first is that the classical grouping of *Entamoeba* species based on the number of nuclei observed in mature cysts is upheld. The distance among groups of species that produce cysts with eight (*Ent. coli*), one (*Ent. chattoni* and *Ent. polecki*) or four nuclei (the remaining species) is very large. Secondly, *Ent. gingivalis*, an oral amoeba that does not encyst, clearly clusters with the species producing cysts with four nuclei indicating it is likely descended from an ancestor with that characteristic. Thirdly, within the group producing cysts with four nuclei species derived from mammalian sources are interspersed with those from reptiles. This indicates that there have been multiple colonizations of the host rather than descent and co-speciation.

The estimated genetic distances among species of *Entamoeba* sometimes exceeded the resolving power of fragment co-migration based analysis (Nei, 1987). Therefore, the SSUrRNA genes representative of each of the clades detected by riboprinting were cloned and sequenced (Silberman Clark, Diamond and Sogin in preparation). In conjunction with those already deposited in GenBank, a total of 13 complete sequences are now available for *Entamoeba* species. Phylogenetic analyses result in trees that are largely consistent with those derived from riboprint data, differing only in the position of one branch (Figures 9.2, 9.3). In addition, however, more resolution is obtained in the deeper branches within the genus. The reptilian species *Ent. invadens* and the amphibian species *Ent. ranarum* are revealed as being on the earliest branch within the group producing cysts with four nuclei, confirming the secondary loss of encystation in an ancestor of *Ent. gingivalis*.

9.3.2 Relationships among genera

Phylogenetic analyses of the *D. fragilis* SSUrDNA sequence confirmed this organism's affinity with the trichomonads, as had been predicted

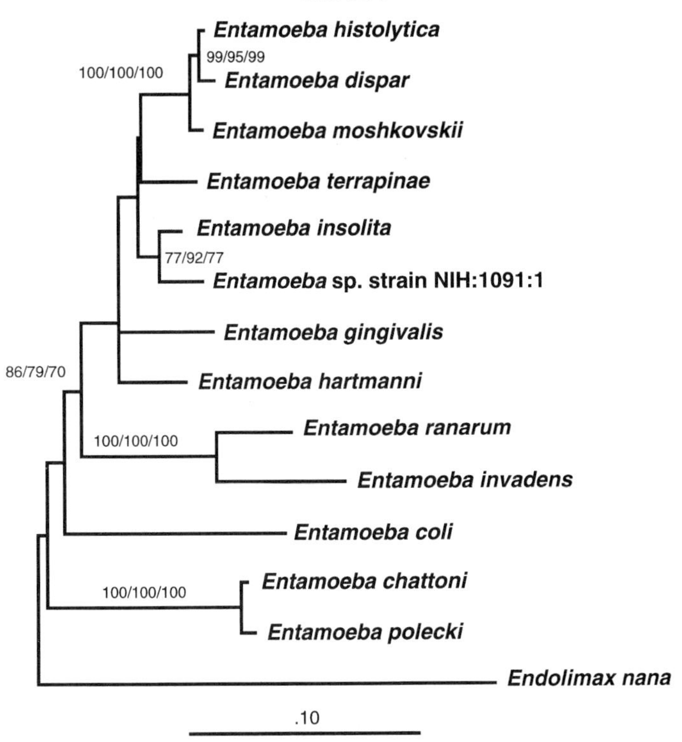

Figure 9.2 Phylogenetic analysis of *Entamoeba* SSUrRNA genes with *Endolimax nana* as the outgroup organism; 1342 unambiguously aligned positions were used in the phylogenetic reconstruction. Significant values from 100 bootstrap replicates from Least-squares distance (Kimura two parameter correction), Maximum likelihood and Maximum parsimony (branch and bound search) are shown on a tree derived from Neighbour-Joining analyses (as LS/ML/MP). If any one of the three methods gave less than 50% support for a node the values were omitted from the figure. The scale bar corresponds to 10 changes per 100 positions.

by electron microscopy (Silberman, Clark and Sogin, 1996). Although it appears to diverge near the base of the trichomonad radiation, it is unlikely that *D. fragilis* represents the ancestral form from which the better known trichomonads, with their flagella and highly structured cytoskeletons, evolved. Rather, we feel it is likely that, as in the case of other apparently primitive organisms, *D. fragilis* has secondarily lost the characteristics of a 'typical' trichomonad. Identification of the earlier diverging termite gut symbionts on the trichomonad branch may help settle this issue.

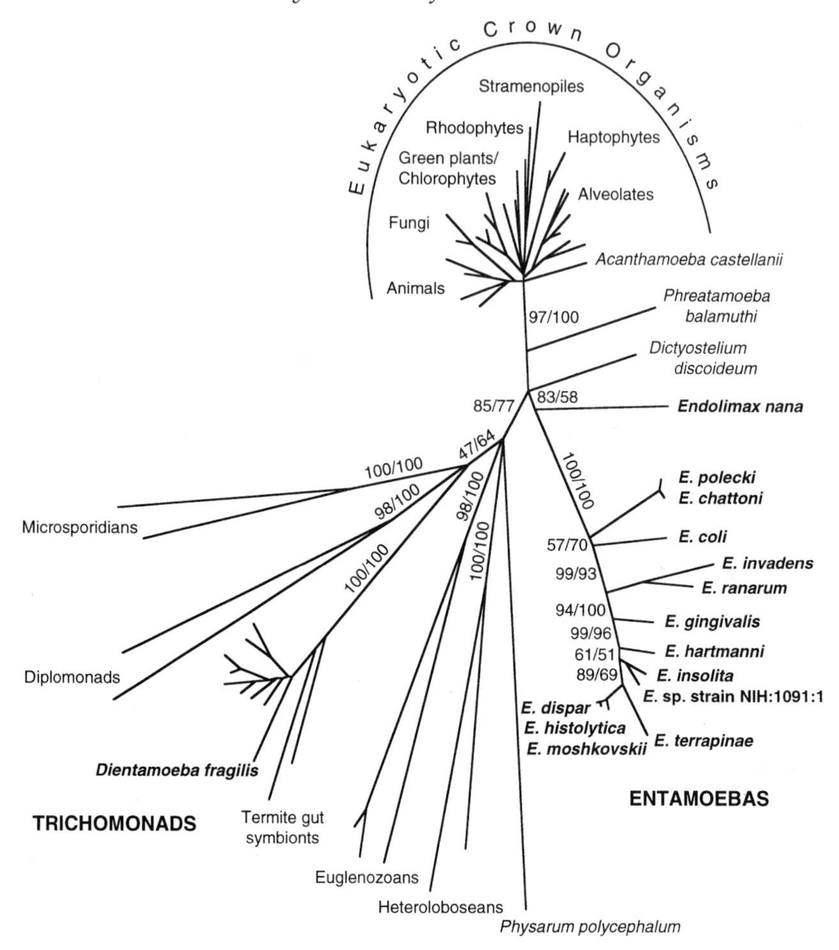

Figure 9.3 Unrooted molecular phylogeny of eukaryotes using SSUrRNA sequences. DNA sequences were aligned based on conserved primary and secondary structures; 1226 unambiguously aligned positions were used for phylogenetic inferences. A Neighbour-Joining tree is shown with selected values from 100 bootstrap replicates of Least-squares distance and Neighbour-Joining analyses (Kimura two parameter correction), respectively. Most species names were omitted for clarity; only amoeboid species are specifically identified. The three nodes supported in 100% of all bootstrap replicates performed for Figure 9.2 were also supported at that level in this analysis but are not labelled here for clarity.

The order of divergence among the three major lineages of *Entamoeba* could only be determined when the *Entamoeba* tree was rooted. Unfortunately, it is clear that the three lineages diverged from one another in a comparatively short period of time and, because of low bootstrap support, the order of divergence remains unresolved. The position of *End. nana* in the eukaryotic tree is tantalisingly close to the base of the *Entamoeba* branch. In some analyses *Endolimax* and *Entamoeba* form a single clade while in others *Endolimax* is on a distinct, but adjacent, branch. It is possible, therefore, that the classification of these two genera in the Family Entamoebidae will ultimately be upheld.

9.3.3 Relationships with other eukaryotes

The position of the branch(es) leading to *Endolimax* and *Entamoeba* indicates that both genera are descended from an ancestor that had mitochondria with the classic cristate structure. The organelle in this ancestor likely already had a highly reduced genome with most components being encoded by nuclear genes, these having been transferred over time from the organelle to the nuclear genome. Although we are not aware of any electron microscopy studies on *Endolimax*, *Entamoeba* species (especially *Ent. histolytica*) have been well examined by this method. No organelles reminiscent of mitochondria have been observed (Martínez-Palomo, 1993).

Not all studies agree on the phylogenetic position of *Entamoeba*. Analyses of certain protein molecules place *Entamoeba* on an early branch while others are more consistent with the rDNA results (Baldauf and Palmer, 1993; Shirakura *et al.*, 1994; Edlind *et al.*, 1996). One thing that is clear is that *Entamoeba* gene sequences are evolving fast (as seen by the length of branches leading to the organism) and this alone may be responsible for some of the discrepancies. One prediction of the rDNA placement of *Entamoeba* is that it once had genes encoding mitochondrial proteins in its nuclear genome. Based on the assumption that some of these may have been retained when the organelle was lost, a search for mitochondrial genes was undertaken in *Ent. histolytica*. Two genes were isolated and sequenced, one of which was suitable for phylogenetic analysis: the gene encoding chaperonin cpn60. This protein is only present in organelles in eukaryotes and is homologous to the GroEL family of bacterial chaperonins. Phylogenetic analysis of the *Ent. histolytica* cpn60 sequence confirmed its specific similarity to homologues of mitochondrial origin to the exclusion of all bacterial sequences (Clark and Roger, 1995) This result indicates that *Entamoeba* species are indeed descended from an ancestor that had mitochondria. Both of the genes isolated that are apparently of mitochondrial origin seem to encode signal peptides at their amino termini (Clark and Roger, 1995; Roger *et al.*, 1998). This would indicate that a subcellular compartment is present in the cell and that it may be a remnant of the mitochondrial organelle.

Dientamoeba may also be descended from a mitochondrion-bearing ancestor. Trichomonads, despite being on an earlier diverging branch of the eukaryotic tree, have hydrogenosomes – a double membrane bound organelle with some biochemical similarities to mitochondria but lacking cristae, a genome and an electron transport chain. The origins of the hydrogenosome have been much debated. One possibility is that it shares a common ancestor with or is derived from the mitochondrion. Recent studies have revealed the presence of a mitochondrial-type cpn60 homologue in another trichomonad, *Trichomonas vaginalis* (Roger, Clark and Doolittle, 1996; see also Chapters 6 and 7). This gene is likely to be present in *D. fragilis* and indicates that trichomonads also are descended from mitochondrion bearing ancestors. Thus the presence of these genes in *Entamoeba* does not confirm its late branch point as seen in the rDNA sequence based tree but only indicates descent from mitochondriate ancestors.

9.4 CONCLUSIONS

Given the paucity of characters available for use in species descriptions, the congruence of classically described species of *Entamoeba* and the ribosomal DNA sequence variants is remarkable. While additional diversity was detected by riboprinting, with one exception all species descriptions defined monophyletic molecular groups. What was unexpected is the depth of genetic divergence within the genus. The species in the three main lineages of *Entamoeba* defined by cyst nuclei number are more different from each other than are any two 'crown' organisms. Even allowing for faster rates of evolution in the *Entamoeba* lineage this is still a remarkable degree of divergence within a single genus.

Amoeboid protists have proven to be polyphyletic based on the limited data available so far (see Figure 9.3). That the Family Entamoebidae as originally described has also proven to be an artificial grouping is not unexpected. If anything, the possibility that *Entamoeba* and *Endolimax* may indeed be specifically related is more of a surprise. It is tempting to view the shared characteristics of these two genera as being present in their common ancestor. However, we cannot be sure at this time that they are in fact specifically related and, if they are, that their shared morphological and other characteristics are the result of common ancestry rather than convergence due to a shared niche. The amoeboid cell form appears to be the 'default' state for eukaryotic organisms and has been independently arrived at a number of times in unrelated lineages. Secondary loss of the cytoskeleton and secondary loss of organelles both appear to be quite frequent events in eukaryotic evolution. The absence of either character should not be taken as evidence for a primitive state.

NOTE ADDED IN PROOF

The organism referred to as *Entamoeba* sp. strain EC has been named *Entamoeba ecuadoriensis* (Clark and Diamond, 1997). In addition to Roger, Clark and Doolittle (1996), the cpn60 gene of *Trichomonas vaginalis* was independently isolated by two other groups (Horner *et al.*, 1996; Bui, Bradley and Johnson, 1996); the encoded protein was shown to be located in the hydrogenosome and to have a cleaved signal peptide (Bui, Bradley and Johnson, 1996) reinforcing the link between mitochondria and trichomonad hydrogenosomes.

9.5 REFERENCES

Bakker-Grunwald, T. and Wöstmann, C. (1993) *Entamoeba histolytica* as a model for the primitive eukaryotic cell. *Parasitology Today*, **9**, 27–31.

Baldauf, S.L. and Palmer, J.D. (1993) Animals and fungi are each other's closest relatives: congruent evidence from multiple proteins. *Proceedings of the National Academy of Sciences USA*, **90**, 11558–62.

Bui, E.T., Bradley, P.J. and Johnson, P.J. (1996) A common evolutionary origin for mitochondria and hydrogenosomes. *Proceedings of the National Academy of Sciences USA*, **93**, 9651–6.

Camp, R.R., Mattern, C.F.T. and Honigberg, B.M. (1974) Study of *Dientamoeba fragilis* Jepps & Dobell I. Electron microscopic observations of the binucleate stages II. Taxonomic position and revision of the genus. *Journal of Protozoology*, **21**, 69–82.

Clark, C.G. and Diamond, L.S. (1991) The Laredo strain and other *Entamoeba histolytica*-like amoebae are *Entamoeba moshkovskii*. *Molecular and Biochemical Parasitology*, **46**, 11–18.

Clark, C.G. and Diamond, L.S. (1997) Intraspecific variation and phylogenetic relationships in the genus *Entamoeba* as revealed by riboprinting. *Journal of Eukaryotic Microbiology*, **44**, 142–54.

Clark, C.G. and Pung, O.J. (1994) Host-specificity of ribosomal DNA variation in sylvatic *Trypanosoma cruzi* from North America. *Molecular and Biochemical Parasitology*, **66**, 175–9.

Clark, C.G. and Roger, A.J. (1995) Direct evidence for secondary loss of mitochondria in *Entamoeba histolytica*. *Proceedings of the National Academy of Sciences USA*, **92**, 6518–21.

Edlind, T.D., Li, J., Visvesvara, G.S. *et al.* (1996) Phylogenetic analysis of β-tubulin sequences from amitochondrial protozoa. *Molecular Phylogenetics and Evolution*, **5**, 359–67.

Fitch, W.M. and Margoliash, E. (1967) Construction of phylogenetic trees. *Science*, **155**, 279–84.

Grendon, J.H., DiGiacomo, R.F. and Frost, F.J. (1995) Descriptive features of *Dientamoeba fragilis* infections. *Journal of Tropical Medicine and Hygiene*, **98**, 309–15.

Horner, D.S., Hirt, R.P., Kilvington, S. *et al.* (1996) Molecular data suggest an early acquisition of the mitochondrion endosymbiont. *Proceedings of the Royal Society of London, Series B. Biological Sciences*, **263**, 1053–9

Martínez-Palomo, A. (1993) Parasitic amoebas of the intestinal tract, in *Parasitic Protozoa* (eds J.P. Kreier and J.R. Baker), Academic Press, San Diego, CA, pp. 65–141.

Nei, M. (1987) *Molecular Evolutionary Genetics*, Columbia University Press, New York, NY.

Nei, M. and Li, W.-H. (1979) Mathematical model for studying genetic variation in terms of restriction endonucleases. *Proceedings of the National Academy of Sciences USA*, **76**, 5269–73.

Ratcliffe, H.L. and Geiman, Q.M. (1934) Amebiasis in reptiles. *Science*, **79**, 324–5.

Roger, A.J., Clark, C.G. and Doolittle, W.F. (1996) A possible mitochondrial gene in the early-branching amitochondriate protist *Trichomonas vaginalis*. *Proceedings of the National Academy of Sciences USA*, **93**, 14618–22.

Roger, A.J., Svärd, S.G., Tovar, J. *et al.* (1998) A mitochondrial-like chaperonin 60 gene in *Giardia lamblia*: Evidence that diplomonads once harbored an endosymbiont related to the progenitor of mitochondria. *Proceedings of the National Academy of Sciences USA*, **95**, 229–34.

Shirakura, T., Hashimoto, T., Nakamura, Y. *et al.* (1994) Phylogenetic place of a mitochondria-lacking protozoan, *Entamoeba histolytica*, inferred from amino acid sequences of elongation factor II. *Japanese Journal of Genetics*, **69**, 119–35.

Silberman, J.D., Clark, C.G. and Sogin, M.L. (1996) *Dientamoeba fragilis* shares a recent common evolutionary history with the trichomonads. *Molecular and Biochemical Parasitology*, **76**, 311–4.

Walsh, J.A. (1986) Problems in recognition and diagnosis of amebiasis: Estimation of the global magnitude of morbidity and mortality. *Reviews of Infectious Diseases*, **8**, 228–38.

10

Relationships between amoeboflagellates

Johan F. De Jonckheere

Protozoology Laboratory, Biosafety and Biotechnology, Department of Microbiology, Institute of Hygiene and Epidemiology, B-1050 Brussels, Belgium.
E-mail: jdjonckh@ben.vub.ac.be

ABSTRACT

Members of the phylum Percolozoa may be the most ancient true Protozoa. The Heterolobosea and the Lyromomadea are the two classes in the Percolozoa that include amoeboflagellates. The family Vahlkampfiidae belongs to the class Heterolobosea. Analyses of ribosomal DNA sequences entail reassessment of intergenic relationships and species evaluation in the Vahlkampfiidae. *Naegleria* spp. are vahlkampfiids whose flagellate stage has two flagella and does not divide. *Willaertia magna* has been described as a member of the family Vahlkampfiidae. The flagellates of *W. magna* have four flagella and are able to divide, producing daughter cells with four flagella. These daughter cells can divide further. Recently, *W. minor* was described as the only other species of the genus *Willaertia*. It shares with *W. magna* the capacity to divide in the flagellate stage, but is distinguished morphologically from *W. magna* by the fact that the daughter cells have only two flagella. The daughter flagellates of *W. minor* not only fail to replace the two lost flagella, they also appear to divide only once. The SSUrDNA sequence of *W. magna* confirms that this amoeba is quite distinct from the genus *Naegleria*, whilst the sequencing results have placed *W. minor* within the genus *Naegleria*. It is speculated that the flagellate stages in *N. minor* might act as gametes, giving support to the suggestion arising from other methods of the presence of sex in *Naegleria*. Although the original description of *Adelphamoeba galeacystis* did not mention a cytostome and the ability to divide in the flagellate stage, these characters are present and thus similar to what is found in *Didascalus thorntoni*.

Evolutionary Relationships Among Protozoa. Edited by G.H. Coombs, K. Vickerman, M.A. Sleigh and A. Warren. Published in 1998 by Chapman & Hall, London. ISBN 0 412 79800 X

The SSUrDNA sequence of *D. thorntoni* is identical to that of *A. galeacystis*. Therefore, the latter becomes a junior synonym of *D. thorntoni*. Division and a possible cytostome are described in the biflagellate stage of *Paratetramitus jugosus*. Only a 1% difference in nucleotide sequences exists between the SSUrDNA of *D. thorntoni* and that of *P. jugosus*, so they probably belong to different species of the same genus, or may even belong to the same species. *Tetramitus rostratus* also has division and a cytostome in the flagellate stage, which is, however, persistent and carrying four flagella. On the other hand, in the genus *Vahlkampfia* a flagellate stage seems to be absent. The genetic distances of the SSUrDNA within the cluster *V. lobospinosa*, *T. rostratus* and *P. jugosus* are similar to the genetic distances between different species of *Naegleria*. Therefore, the different genera in the first cluster could be considered different species of one genus or, alternatively, the *Naegleria* spp. might be upgraded to distinct genera. The sequences of the SSUrDNA in the genus *Naegleria* has demonstrated already that all the described subspecies should be upgraded to species level. The SSUrDNA sequence differences between the different *Vahlkampfia* spp. suggest some might be different genera instead of belonging to different species of one genus. In the Percolozoa, the class Lyromonadea has been created for anaerobes such as the amoeboflagellate *Psalteriomonas* and the flagellate *Lyromonas*. The SSUrDNA of the amoeboflagellate *P. lanterna* has recently been sequenced and phylogenetic analysis confirms that it does not cluster with the Vahlkampfiidae. A group I intron is detected in the ribosomal DNA of the genus *Naegleria* but not in that of the other genera of amoeboflagellates. The group I intron in the SSUrDNA of *Naegleria* is approximately 1.3 kb long and has an open reading frame (ORF). In one *Naegleria* lineage the group I intron has lost the ORF. The group I intron seems to be of ancestral origin and has been lost from the SSUrDNA in the majority of *Naegleria* lineages. A group I intron has now also been found in the LSUrDNA of two undescribed *Naegleria* species.

Amoeboflagellates appear to have multiple origins as indicated by their scattered presence throughout phylogenetic trees. In this chapter only the amoeboflagellates belonging to the phylum Percolozoa will be considered. This is because members of the Percolozoa may be the most ancient true Protozoa (Cavalier-Smith, 1993). The Heterolobosea and the Lyromonadea are the two classes in the Percolozoa that include amoeboflagellates. The results from phylogenetic analyses based on small subunit ribosomal DNA (SSUrDNA) will be contrasted with the morphological differences between the different genera, especially of their flagellate forms. It will become obvious that morphological observations are unreliable and prone to personal interpretation.

10.1 THE CLASS HETEROLOBOSEA

The class Heterolobosea includes two orders, the Schizopyrenida and the Acrasida. The Schizopyrenida comprises two families, the Vahlkampfiidae and the Gruberellidae. Analyses of ribosomal DNA sequences have engendered reassessment of intergeneric relationships and species evaluation within the Vahlkampfiidae. No rDNA sequence data are yet available for the Gruberellidae.

10.1.1 The genus *Naegleria*

Naegleria spp. are vahlkampfiids whose flagellate stage has two flagella and does not divide (Table 10.1). The genus has attracted a lot of attention since 1970, when *Naegleria fowleri* was described as the causative agent of primary amoebic meningoencephalitis in man. This led to the use of modern techniques to differentiate between species and to the isolation of many different *Naegleria* strains, some of which have been described subsequently as new species or subspecies. Clark and Cross (1988) were the first to publish the SSUrDNA sequence of a *Naegleria* strain. Comparison of this sequence with the SSUrDNA sequences of other eukaryotes resulted in a phylogenetic tree that supports the polyphyletic origin of amoebae and suggests a flagellate ancestry for *Naegleria*. Analysis of partial sequences of the SSUrDNA in the genus *Naegleria* confirms the validity of species described by non-morphological methods, and demonstrates that all the described subspecies should be upgraded to species level (De Jonckheere, 1994a). To date, 11 well described *Naegleria* spp. are accepted (Table 10.2), whilst *N. gruberi* remains a species complex. The recently described *N. pussardi* (Pernin and De Jonckheere, 1996) seems to be the earliest branching *Naegleria* species (Figure 10.1). Hinkle and Sogin (1993) have sequenced the SSUrDNA of three other vahlkampfiid genera. Their results indicate that the vahlkampfiids are a monophyletic group and that *Vahlkampfia*, *Tetramitus* and *Paratetramitus* are closer relatives of each other than any of these three are to *Naegleria*. They estimated that the divergence of *Naegleria* from the other vahlkampfiid genera occurred about one billion years ago. In one *Naegleria* sp., formerly described as *Willaertia minor*, but on the basis of the SSUrDNA sequence renamed *N. minor* (Figure 10.1) (see 10.1.2), the flagellates can divide. This led to the speculation that the flagellate stages in *N. minor* might act as gametes giving support to the suggestion, arising from other studies, of the existence of sex in *Naegleria* (Cariou and Pernin, 1987).

Table 10.1 Characteristics of the flagellate stages in the genera belonging to the Vahlkampfiidae, of which the SSUrDNA sequenced

Naegleria spp.	*N. minor*	*Willaertia*	*Tetramitus*	*Paratetramitus*	*Didascalus*	*Vahlkampfia*
Two flagella No cytostome No division	Four flagella No cytostome Division Daughter flagellates: Two flagella No division	Four flagella No cystosome Division Daughter flagellates: Four flagella Division	Four flagella Cytostome Division Daughter flagellates: Four flagella Division permanent	Two flagella Cytostome? Division Daughter flagellates: ?	Two flagella Cytostome Division Daughter flagellates: ?	None

Table 10.2 List of *Naegleria* spp. of which the species status is supported by SSUrDNA sequences

Species	*Former designation*	*Geographic distribution*
N. fowleri	*N. fowleri*	Worldwide
N. lovaniensis	*N. lovaniensis*	Worldwide
N. jadini	*N. jadini*	Europe
N. australiensis	*N. australiensis australiensis*	Worldwide
N. italica	*N. australiensis italica*	Europe
N. andersoni	*N. andersoni andersoni*	Worldwide
N. jamiesoni	*N. andersoni jamiesoni*	Unknown
N. gruberi several clusters	*N. gruberi* several clusters	
N. clarki	*italica* related	Unknown
N. galeacystis	*Adelphamoeba galeacystis*	Unknown
N. minor	*Willaertia minor*	Unknown
N. pussardi		Unknown

10.1.2 The genus *Willaertia*

Willaertia magna has been described as a member of the family Vahlkampfiidae (De Jonckheere *et al.*, 1984). The flagellates of *W. magna* have four flagella and are able to divide, producing daughter cells with four flagella. These daughter cells can divide further (Robinson, Christy and De Jonckheere, 1989) (Table 10.1). Unlike the much bigger *W. magna*, amoebae of *W. minor* (Dobson *et al.*, 1993) have the size of *Naegleria* amoebae. *Willaertia minor* shares with *W. magna* the capacity to divide in the flagellate stage, but is distinguished from the latter by the fact that the daughter flagellates have only two flagella (Table 10.1). The dividing flagellates of *W. minor* not only fail to replace the two lost flagella, they also appear to divide only once. The SSUrDNA sequence of *W. magna* confirms that this amoeba is quite distinct from the genus *Naegleria*, whilst SSUrDNA analyses place *W. minor* within the genus *Naegleria* (De Jonckheere, 1996). *Willaertia magna* seems to be the last known amoeboflagellate genus to have diverged from the *Naegleria* lineage (Figure 10.1).

10.1.3 The genera *Adelphamoeba* and *Didascalus*

In the original description of *Adelphamoeba galeacystis* (Napolitano, Wall and Ganz, 1970) a cytostome was described but not division of the fla-

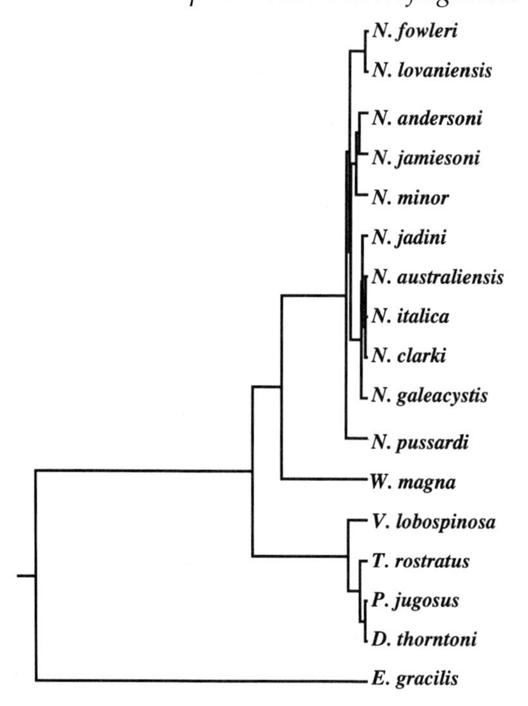

Figure 10.1 Phylogenetic tree (distance matrix) based on partial SSUrDNA sequences of *Naegleria* spp., *W. magna* (X93221, X93222 and X93223), and *D. thorntoni* (X93085) obtained in the author's laboratory and those of *P. tetramitus* (M98050), *T. jugosus* (M98051) and *V. lobospinosa* (M98052) available at EMBL. The accession number of the *N. minor* SSUrDNA sequence is X93224. *Euglena gracilis* (M12677) was included as an outgroup.

gellate. However, division in the flagellate stage has been demonstrated in a new isolate of this organism (De Jonckheere, Brown and Robinson, 1997). The supposed *A. glaeacystis* strain present in culture collections is a *Naegleria* strain and has been renamed *N. galeacystis*. Although the original description of *Didascalus thorntoni* (Singh, 1952) did not mention a cytostome and the ability to divide in the flagellate stage, these characters are present (De Jonckheere, Brown and Robinson, 1997) (Table 10.1). The SSUrDNA sequence of *D. thorntoni* is identical to that of the new isolate of *A. galeacystis* (De Jonckheere, Brown and Robinson, 1997). Therefore, the latter becomes a junior synonym of *D. thorntoni*. The SSUrDNA of *D. thorntoni* differs from that of *Paratetramitus jugosus* by only 1%.

10.1.4 The genus *Paratetramitus*

Division, and a possible cytostome, are described in the biflagellate stage of *P. jugosus* (Darbyshire, Page and Goodfellow, 1976) (Table 10.1). As indicated above (see 10.1.3) *D. thorntoni* and *P. jugosus* are closely related and probably belong to different species of the same genus, or may even belong to the same species. Analysis of partial SSUrDNA sequence (620 bp) of *V. aberdonica* revealed total homology with that of *P. jugosus* (see 10.1.6.). However, additional sequence information from this molecule differentiates between the two, indicating that *V. aberdonica* might belong to the same genus as *P. jugosus* (Brown and De Jonckheere, in preparation).

10.1.5 The genus *Tetramitus*

Tetramitus rostratus is an amoeboflagellate in which the flagellate, with four flagella, is relatively permanent (Table 10.1). Examination of the ultrastructure of the flagellate suggests *Tetramitus* is a more evolved *Naegleria*, while *Paratetramitus* is intermediate between the two (Balamuth, Bradbury and Schuster, 1983). Phylogenetic analysis based on SSUrDNA does not support this conclusion totally, as *Tetramitus* and *Paratetramitus* are more closely related to *Vahlkampfia* than any of these three are to *Naegleria* (Hinkle and Sogin, 1993). The genetic distances between the SSUrDNA within the cluster *Vahlkampfia lobospinosa, T. rostratus* and *P. jugosus* are similar to the genetic distances between some species of *Naegleria* (Figure 10.1). Therefore, the different genera in the former cluster could be considered different species of one genus or, alternatively, some *Naegleria* spp. could be upgraded to distinct genera.

10.1.6 The genus *Vahlkampfia*

The majority of vahlkampfiid amoebae that do not transform into flagellates have been assigned to the genus *Vahlkampfia* (Table 10.1). As such they should not be included in this chapter on amoeboflagellates. However, their close relationships with genera that are real amoeboflagellates (Hinkle and Sogin, 1993; Brown and De Jonckheere, in preparation) raises the suspicion that they might form flagellates under ideal conditions. Amongst *Naegleria* spp. there is no variation in the length of the SSUrDNA except when an intron is present (see 10.3). When the SSUrDNA of different *Vahlkampfia* spp. is amplified by PCR, significant length variation is observed, from 1.9 kb to 2.15 kb (Brown and De Jonckheere, 1994), and these length differences are not due to the presence of introns. This was the first indication of a high degree of sequence variation within the genus *Vahlkampfia*. Comparisons of complete SSUrDNA sequences confirmed that the *Vahlkampfia* spp. do not form a discrete group (Brown and De Jonckheere, in preparation). *Vahlkampfia enterica* clusters with *T. rostratus* but these two species are also grouped with *V. lobospinosa* and *P. jugosus* and, therefore, with *D. thorntoni* as well. *Vahlkampfia ustiana* forms a separate branch. Phylogenetic analysis with only part of the SSUrDNA sequences generated the same branching order. Confident that partial sequence information can be used for tree building, a phylogenetic analysis was performed using the first 620 bp of the SSUrDNA obtained from the seven *Vahlkampfia* reference strains (Figure 10.2). The conclusion from this work is that the grouping of these non-flagellate forming vahlkampfiid amoebae into a single genus should not be considered valid. The SSUrDNA sequence analysis indicates that the described *Vahlkampfia* spp. may represent at least three different genera, some of which may include species assigned to other genera that form flagellates. *Vahlkampfia aberdonica* seems to be closely related to *P. jugosus* (see 10.1.4), which forms flagellates. It is interesting to remember here that *P. jugosus* was initially described as a species of *Vahlkampfia* because no flagellates were

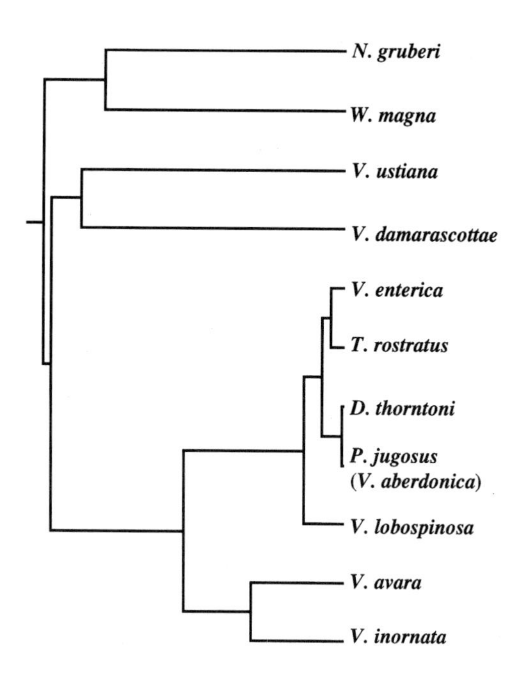

Figure 10.2 Phylogenetic tree (distance matrix) based on partial SSUrDNA sequences of seven *Vahlkampfia* spp. compared to other genera (see Figure 10.1) within the Vahlkampfiidae.

obtained originally from the reference strain. The genus *Paratetramitus* was created later when a flagellate stage was detected in this species (Darbyshire, Page and Goodfellow, 1976). As shown before (see 10.1.3), *D. thorntoni* is also closely related to *T. rostratus*. The three genera might, therefore, be species of the same genus. The close genetic relationship between flagellate-forming and non flagellate-forming vahlkampfiids suggests that the ability to transform to flagellates should be reassessed in the different *Vahlkampfia* spp. (Brown, personal communication).

10.1.7 Other Heterolobosea

The SSUrDNAs of other Vahlkampfiidae, such as the amoeboflagellate *Heteramoeba clara* (Droop, 1962) and the amoeba *Pernina chaumonti* (El Kadiri, Joyon and Pussard, 1992) (both marine species) and the amoeba *Singhamoeba horticola* (Sawyer, Nerad and Munson, 1992) need to be sequenced and compared. Morphological comparisons place *Pernina chaumonti* between the Vahlkampfiidae and the Gruberellidae. To date, no rDNA sequence information has been obtained from any member of the Gruberellidae. Also, members of the Acrasida should be included in this comparison as a flagellate stage is present in some of the different species belonging to this order, as is the case in the family Vahlkampfiidae. The Acrasida is the order in the class Heterolobosea that produces fruiting bodies, in contrast with the Schizopyrenida (Page, 1988). The order Acrasida also comprises two families, the Acrasidae and the Guttulinopsidae (Page, 1988).

10.2 THE CLASS LYROMONADEA

In the Percolozoa, the class Lyromonadea (Cavalier-Smith, 1993) has been created for anaerobes such as the amoeboflagellate *Psalteriomonas lanterna* (Broers *et al.*, 1990) and the flagellate *P. vulgaris* (Broers *et al.*, 1993). The latter is now placed in a separate genus, *Lyromonas*, created by Cavalier-Smith (1993), on the basis of morphological differences. The SSUrDNA of the amoeboflagellate *P. lanterna* has been sequenced recently and phylogenetic analysis confirms that it does not cluster with the Vahlkampfiidae (Weekers, Kleyn and Vogels, 1996). Also, the GC % is very low (33.4%) compared to members of the Vahlkampfiidae. Due to the lack of sequence data from possible phylogenetically closely related organisms it is still not possible to conclude whether *P. lanterna* should be included in the Heterolobosea, or even in the Percolozoa (Figure 10.3). The sequence of the SSUrDNA of *L. vulgaris* has been obtained even more recently (Weekers, Broers and De Jonckheere, 1996). The genus *Lyromonas* seems, indeed, to branch near the genus *Psalter-*

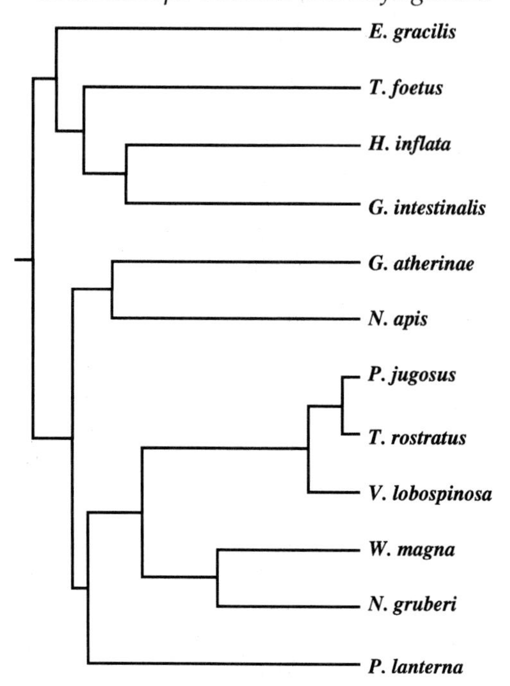

Figure 10.3 Phylogenetic tree (distance matrix) based on SSUrDNA sequences of *P. lanterna,* five genera of vahlkampfiids (see Figure 10.1), and six organisms presumed to be at the base of the eukaryotic tree (*Euglena gracilis,* M12677; *Tritrichomonas foetus,* M81842, *Hexamita inflata* L07836; *Giardia intestinalis,* X52949; *Glugea atherinae* U15987 and *Nosema apis,* X73894).

iomonas, as suggested by morphological data, but the sequence differences between the two are too big to infer that they belong to the same genus.

10.3 GROUP I INTRONS IN THE rDNA OF THE VAHLKAMPFIIDAE

Group I introns are self-splicing inserts with a characteristic secondary structure due to base pairing of short conserved sequence elements. Group I introns are detected in the ribosomal DNA of the genus *Naegleria* but not in that of the other genera of the Percolozoa which have been investigated by PCR amplification of the rDNA; *Vahlkampfia, Tetramitus, Paratetramitus, Didascalus, Willaertia, Psalteriomonas* and *Lyromonas.* However, except for the genus *Vahlkampfia,* the absence of group I

Table 10.3 The presence of group I introns in the rDNA of *Naegleria* spp.

Species	*SSUrDNA*	*LSUrDNA*
N. fowleri	–	–
N. lovaniensis	–	–
Naegleria sp. (NG872)	+ (+ ORF)	+ (– ORF)
Naegleria sp. (NG236)	–	+ (– ORF)
N. jadini	–	–
N. australiensis	–	–
N. italica	+ (+ ORF)	–
N. andersoni	+ (+ ORF)	–
N. jamiesoni	+ (+ ORF)	–
N. gruberi	{ + (+ ORF)	–
(species complex)	–	
N. clarki	+ (+ ORF)	–
Naegleria sp. (NG434, NG650, NG597)	+ (– ORF)	–
N. galeacystis	–	–
N. minor	–	–
N. pussardi	–	–

introns could be due to the fact that too few strains and species have been investigated. The group I intron, present in the SSUrDNA of only a few *Naegleria* spp., is approximately 1.3 kb long and has an open reading frame (ORF) (De Jonckheere, 1994b). The group I intron appears to be of ancestral origin and has been lost from the SSUrDNA of the majority of *Naegleria* linages. In one *Naegleria* lineage, the group I intron was not lost, but the ORF in the intron was not retained (De Jonckheere and Brown, 1994). In the genus *Naegleria*, a group I intron has been detected recently in the large subunit rDNA (LSUrDNA) but it is less common than in the SSUrDNA (De Jonckheere and Brown, in preparation). A summary of the presence of group I introns, with or without ORF, in the rDNA of the genus *Naegleria* is presented in Table 10.3.

10.4 CONCLUSIONS

Except for one species of *Naegleria*, it seems that *Naegleria* is the only genus within the Vahlkampfiidae in which the flagellate stage, if generated, does not divide. Analysis of the SSUrDNA of different *Naegleria* spp. has not changed fundamentally the species concept within this genus because most species were described on the basis of techniques other than morphological comparison. In contrast, the different *Vahlkampfia* spp. were described only on the basis of morphological and ultrastructural characters. The SSUrDNA sequence analyses demonstrate that morphology is unreliable and, consequently, both species and genus re-classifications are pertinent. Comparing SSUrDNA sequences differentiates between the genera, although some genera (for example, *Didascalus*, *Paratetramitus* and *Tetramitus*) might actually be considered species of one genus, whilst other genera (for example, *Vahlkampfia*) should be divided into different genera. *Willaertia* seems to have diverged as a separate genus from the *Naegleria* lineage after the latter diverged from the other vahlkampfiid genera. However, when complete sequences of the SSUrDNA of different so-called *Vahlkampfia* spp. become available, this picture might change, because the genus *Vahlkampfia* appears to comprise at least three distinct genera. Sequencing the rDNA of other genera belonging to the Vahlkampfiidae, and of other Heterolobosea, will further elucidate the phylogenetic relationships within this class.

The SSUrDNA sequences of *Psalteriomonas* and *Lyromonas* confirm that they are, indeed, related as suggested originally, but that they do not belong to the same genus. However, it is unclear at present whether they should be included in the Heterolobosea or not. Also, their inclusion in the Percolozoa is not certain. The SSUrDNA of more organisms that are considered to be at the base of the eukaryotic tree should be sequenced to shed light on this.

Acknowledgement

I wish to thank P.H.H. Weekers for sharing the data analysis and SSUrDNA sequence (EMBL accession number X94430) of *P. lanterna* prior to publication and Susan Brown for discussing the data on the genus *Vahlkampfia* from her PhD thesis.

10.5 REFERENCES

Balamuth, W., Bradbury, P.C. and Schuster, F.L. (1983) Ultrastructure of the amoeboflagellate *Tetramitus rostratus.*, *Journal of Protozoology*, **30**, 445–55.

Broers, C.A.M., Stumm, C.K., Vogels, G.D. and Brugerolle, G. (1990) *Psalteriomonas lanterna* gen. nov., sp. nov., a free-living amoeboflagellate isolated from freshwater anaerobic sediments. *European Journal of Protistology*, **25**, 369–80.

Broers, C.A.M., Meijers, H.H.M., Symens, J.C. *et al.* (1993) Symbiotic association of *Psalteriomonas vulgaris* n. spec. with *Methanobacterium formicicum. European Journal of Protistology*, **29**, 98–105.

Brown, S. and De Jonckheere, J.F. (1994) Identification and phylogenetic relationships of *Vahlkampfia* spp. (free-living amoebae) by riboprinting. *FEMS Microbiology Letters*, **115**, 241–6.

Cariou, M.L. and Pernin, P. (1987) First evidence for diploidy and genetic recombination in free-living amoebae of the genus *Naegleria* on the basis of electrophoretic variation. *Genetics*, **115**, 265–70.

Cavalier-Smith, T. (1993) Kingdom Protozoa and its 18 Phyla. *Microbiological Reviews*, 57, 953–94.

Clark, C.G. and Cross, G.A.M. (1988) Small-subunit ribosomal RNA sequence from *Naegleria gruberi* supports the polyphyletic origin of amoebas. *Molecular Biology and Evolution*, **5**, 512–8.

Darbyshire, J.F., Page, F.C. and Goodfellow, L.P. (1976) *Paratetramitus jugosus*, an amoeboflagellate of soil and freshwater, type-species of *Paratetramitus* nov. gen. *Protistologica*, **12**, 375–87.

De Jonckheere, J.F. (1994a) Comparison of partial SSUrDNA sequences suggests revisions of species names in the genus *Naegleria*. *European Journal of Protistology*, **30**, 333–41.

De Jonckheere, J.F. (1994b). Evidence for the ancestral origin of group I introns in the SSUrDNA of *Naegleria* spp. *Journal of Eukaryotic Microbiology*, **41**, 457–63.

De Jonckheere, J.F. (1997) The phylogenetic position of the amoeboflagellate *Willaertia* deduced from SSUrDNA sequences. *European Journal of Protistology*, **33**, 72–6.

De Jonckheere, J.F. and Brown, S. (1994) Loss of ORF in the SSUrDNA group I intron of one *Naegleria* lineage. *Nucleic Acids Research*, **22**, 3925–7.

De Jonckheere, J.F., Brown, S. and Robinson, B.S. (1996) On the identity of the amoeboflagellates *Didascalus thorntoni* and *Adelphamoeba galeacystis*. *Journal of Eukaryotic Microbiology*, **44**, 52–4.

De Jonckheere, J.F., Pussard, M., Dive, D.G. and Vickerman, K. (1984) *Willaertia magna* gen. nov., sp. nov. (Vahlkampfiidae), a thermophilic amoeba found in different habitats. *Protistologica*, **20**, 5–13.

Dobson, P.J., Robinson, B.S., Christy, P. and Hayes, S.J. (1993) Low genetic diversity in *Willaertia magna* from wide geographical sources, and characterisation of *Willaertia minor* n.sp. (Heterolobosea, Vahlkampfiidae). *Journal of Eukaryotic Microbiology*, **40**, 298–304.

Droop, M.R. (1962) *Heteramoeba clara* n. gen., n. sp., a sexual biphasic amoeba. *Archiv für Microbiologie*, **42**, 254–66.

El Kadiri, G., Joyon, L. and Pussard, M. (1992) *Pernina chaumonti*, n.g., n.sp., a new marine amoeba (Rhizopoda, Heterolobosea). *European Journal of Protistology*, **28**, 43–50.

Hinckle, G. and Sogin, M.L. (1993) The evolution of the vahlkampfiidae as deduced from 16S-like ribosomal RNA analysis. *Journal of Eukaryotic Microbiology*, **40**, 599–603.

Napolitano, J.J., Wall, M.E. and Ganz, C.S. (1970) *Adelphamoeba galeacystis* n.g., n. sp., an amoeboflagellate isolated from soil. *Journal of Protozoology*, **17**, 158–61.

Page, F.C. (1988) *A New Key to Freshwater and Soil Gymnamoebae*. Freshwater Biological Association, Ambleside.

Pernin, P. and De Jonckheere, J.F. (1996) *Naegleria pussardi*, a new *Naegleria* species phylogenetically related to the high temperature tolerant species at the molecular level. *European Journal of Protistology*, **32**, 403–11.

Robinson, B.S., Christy, P.E. and De Jonckheere, J.F. (1989) A temporary flagellate (mastigote) stage in the vahlkampfiid amoeba *Willaertia magna* and its possible evolutionary significance. *BioSystems*, **23**, 75–86.

Sawyer, T.K., Nerad, T.A. and Munson, D.A. (1992) *Singhamoeba horticola* (Singh & Hanumaiah, 1979) n. comb., type species of *Singhamoeba* n.g. *Journal of Protozoology*, **39**, 107–9.

Singh, B.N. (1952) Nuclear division in nine species of small free-living amoebae and its bearing on the classification of the order Amoebida. *Philosophical Transactions of the Royal Society of London* (B), **236**, 405–61.

Weekers, P.H.H., Broers, C.A.M. and De Jonckheere, J.F. (1996) *Phylogeny of the Lyromonadea based on SSU rDNA sequences* (abst.). Evolutionary relationships among protozoa. 10–11 September, London.

Weekers, P.H.H., Kleyn, J. and Vogels, G.D. (1997) Phylogenetic position of *Psalteriomonas lanterna* deduced from the SSU rDNA sequence. *Journal of Eukaryotic Microbiology*, **44**, 467–70.

11

Molecular phylogeny of kinetoplastids

Hervé Philippe

Laboratoire de Biologie Cellulaire (URA CNRS 2227), Bâtiment 444, Université Paris-Sud, 91405 Orsay Cedex. France. E-mail: hp@bio4.bc4.u-psud.fr

ABSTRACT

The Kinetoplastida are a well-characterized group of flagellates, among which are important human and animal pathogens. They have been intensively studied not only for their medical interest but also for some unusual processes occurring in their cellular biology, such as mRNA editing. They have been subdivided in two major subgroups that differ in morphological characters and life cycles: the trypanosomatids and the bodonids. Knowledge of their phylogenetic relationships is of primary interest in studying the evolution of parasitism and of mRNA editing. Numerous phylogenies have been published, yielding quite different results, especially concerning the genus *Trypanosoma*. In this chapter, I review briefly these phylogenies and discuss possible explanations for the observed discrepancies, in particular the problems of species sampling and of evolutionary rates.

11.1 INTRODUCTION

Kinetoplastids are a diversified group of small colourless flagellates united by several shared characters (for reviews see Lee and Hutner, 1985; Vickerman, 1989). In particular, they possess a single mitochondrion which contains a significant amount of DNA that forms a stainable structure (the kinetoplast) and that usually lies close to the flagellum base. Kinetoplastids were first grouped together with euglenids on the basis of ultrastructural characters (Kivic and Walne, 1984). This conclusion was first challenged by molecular phylogenies based

Evolutionary Relationships Among Protozoa. Edited by G.H. Coombs, K. Vickerman, M.A. Sleigh and A. Warren. Published in 1998 by Chapman & Hall, London. ISBN 0 412 79800 X

on SSU ribosomal RNA (Sogin, Elwood and Gunderson, 1986), but was rapidly confirmed when several additional protist sequences were included in the analysis (Sogin, 1991). According to the SSUrRNA sequences they represent the earliest branch of mitochondriate eukaryotes (Sogin, 1991), but this conclusion is strongly challenged by phylogenies based on protein sequences (Hashimoto *et al.*, 1995) (see Philippe and Adoutte, Chapter 2).

Kinetoplastids have been traditionally subdivided into two subgroups, bodonids and trypanosomatids, which differ in several features especially in the number of flagella (Lee and Hutner, 1985; Vickerman, 1989). Bodonids are characterized by the presence of two heterodynamic flagella (one directed anteriorly and locomotory, the other directed posteriorly and recurrent or trailing). They are mostly free-living phagotrophic organisms, abundant in organically-rich environments (e.g. *Bodo* and *Dimastigella* species), but some are parasites of aquatic organisms (e.g. *Cryptobia* and *Trypanoplasma* species). Trypanosomatids are characterized by the presence of a single locomotory flagellum, free or attached to the body as an undulating membrane. Their kinetoplasts always lie close to the flagellum base. All trypanosomatids are parasitic and some cause serious diseases of medical, veterinary, or agricultural importance. For example, in humans, they are responsible for Chagas' disease (*Trypanosoma cruzi*), for sleeping sickness (*Trypanosoma brucei*) and for leishmaniases (*Leishmania* spp.). In South America, the genus *Phytomonas* cause diseases of major economic importance in coconut, oil palm and coffee trees. Trypanosomatid species undergo complex life cycles with one or two successive hosts, the monogenetic and digenetic species, respectively. These hosts are frequently a vertebrate and/or an insect. During the life cycle, they display several major changes in their morphology (amastigote, promastigote, opisthomastigote, choanomastigote, epimastigote and trypomastigote forms according to the cell shape and relative position of kinetoplast–basal body complex).

Since free-living species are frequent among bodonids, it has generally been considered that trypanosomatids were derived from bodonids. If so, this latter group would be paraphyletic. The part played by different organisms as hosts has led to two hypotheses concerning the origin of parasitism. According to the first (Léger, 1904), parasitism arose in ancient invertebrates and was followed by a long coevolution with their hosts. Colonisation of vertebrates and plants arose when haematophagous invertebrates (insects and leeches) appeared. According to the second hypothesis (Minchin, 1908), parasitism occurred first in the gut of early vertebrates. The next step was the colonisation of the blood, which then allowed the introduction of the parasite into leeches and haematophagous insects (dipterans and hemipterans), leading to digenetic trypanosomatids. These species could then colonize new vertebrates and plants, or lose their primary vertebrate hosts, leading to monogenetic trypanosomatids. In this hypothesis the digenetic trypanosomatids are of basal emergence in kinetoplastid phylogeny whereas in the first one the monogenetic trypanosomatids are of basal emergence.

It has been assumed that molecular phylogeny would discriminate between these two hypotheses. In this chapter, I shall briefly review the most important results of molecular phylogenetic studies. A comparison of all available sequence data will enable me to suggest explanations for the observed contradictions. In particular, phylogenies based on few species are often misleading (Philippe and Douzery, 1994) and in general do not permit us to argue in favour of any given evolutionary scenario. Most importantly, strong differences in evolutionary rates within the kinetoplastids produce erroneous phylogenetic inferences (such as the paraphyly of the genus *Trypanosoma* observed in several SSUrRNA trees) and hinder the reconstruction of the correct kinetoplastid phylogeny.

11.2 FIRST MOLECULAR PHYLOGENIES

The first molecular phylogeny of kinetoplastids to be published (Lake *et al.*, 1988) was based on the mitochondrial small and large subunit ribosomal RNA (9S and 12S rRNA). The inferred topology ((*Crithidia*, *Leptomonas*), *Leishmania*, (*Trypanosoma brucei*, *Trypanosoma cruzi*)) was arbitrarily rooted on *Crithidia*, assuming that the monogenetic species are ancestral to the digenetic ones. Such a rooting implied a strong acceleration of evolutionary rates in the genus *Trypanosoma* (threefold). However, if the root is assumed to be located at the base of the genus *Trypanosoma*, the difference in evolutionary rates can be greatly reduced. In consequence, this first molecular phylogeny was questionable because of the lack of a convincing outgroup.

Because mitochondrial rRNAs are evolving very fast in kinetoplastids, it is not possible to use sequences of other eukaryotes as an outgroup. In contrast, the nuclear SSUrRNA is much more conserved, thus it makes alignments between kinetoplastids and other eukaryotes, such as *Euglena*, Metazoa or Fungi, possible. Based on SSUrRNA, rooted with *Euglena* sequences, Gomez *et al.* (1991) and Briones *et al.* (1992) obtained the phylogeny: (*Trypanosoma brucei*, (*Trypanosoma cruzi*, (*Crithidia*, (*Leishmania donovani*, *Leishmania tarentolae*)))). A striking result was the paraphyly of the genus *Trypanosoma*, although this is supported also by several morphological characters and host ranges.

This phylogeny seems to strongly contradict that based on mitochondrial rRNA, but as previously suggested, the two inferred unrooted topologies are in fact identical. The difference is simply due to an arbitrary rooting on either *Crithidia* (Lake *et al.*, 1988) or *Euglena* (Gomez *et al.*, 1991; Briones *et al.*, 1992). However, the distance between *Euglena* and the kinetoplastids is much greater than that between *Trypanosoma* and *Leishmania*, 58% and 13% respectively (Briones *et al.*, 1992). It is well known that a distant outgroup can generate an erroneous rooting of the tree (Swofford and Olsen, 1990). In summary, these first phylogenetic studies are of little reliability because of the use of too few species (Philippe and Douzery, 1994) and of distant outgroups.

11.3 THE SSUrRNA PHYLOGENY

Four more recent papers published between December 1993 and August 1994 avoided these two pitfalls, and provided new SSUrRNA sequences for the following kinetoplastids: *Bodo caudatus, Endotrypanum monterogei, Leptomonas sp., Leishmania major* and *Leishmania donovani* (Fernandes, Nelson and Beverley, 1993), four *Herpetomonas* species (Landweber and Gilbert, 1994), *Trypanoplasma borreli, Trypanosoma carassii, Blastocrithidia culicis* and *Herpetomonas muscarum* (Maslov *et al.*, 1994) and *Crithidia oncopelti* (Du, Maslov and Chang, 1994). Phylogenies presented in these publications confirmed the monophyly of the trypanosomatids with respect to the bodonids and supported the old idea that trypanosomatids are derived from bodonids. With these data available, outgroups could be used to root the trypanosomatid tree. However, the topology of the inferred tree was insignificantly modified by these new sequences, because the genus *Trypanosoma* remained paraphyletic in all the analyses (bootstrap values of 75%, 77%, 83% and 91%). These publications also confirmed the small genetic distances between *Leishmania, Leptomonas, Endotrypanum* and *Crithidia fasciculata* which contrasted with the 10 times larger distances within the genus *Trypanosoma*. However, *Crithidia oncopelti* did not cluster with *Crithidia fasciculata* and even branched between *T. cruzi* and the *Leishmania* group. *Blastocrithidia culicis* and *Herpetomonas muscarum* branched also in this part of the tree but the branching order of these four groups was not strongly supported. Interestingly, phylogenies obtained with different methods of tree reconstruction and with different species sampling gave almost identical results. The only difference observed concerned the paraphyly of *Trypanosoma*, when using slightly different alignments (Fernandes, Nelson and Beverley, 1993).

These congruent results, plus an uncritical faith in rRNA phylogenies, have led people to accept this tree as a close approximation of the real phylogeny. Since all the first emerging species, especially *T. brucei* and *T. cruzi*, are digenetic and display a high level of mRNA editing in their mitochondria, it has been suggested that (1) the ancestral trypanosomatid is a digenetic species and that the loss of vertebrate host occurred several times leading to monogenetic species, (2) editing, and especially pan-editing, is 'a more primitive mechanism than previously thought' (Maslov *et al.*, 1994).

These latter conclusions can be drawn because the genus *Trypanosoma* is seen as paraphyletic and because mRNA editing is known only from trypanosomatids (mainly for the genes NADH7, cox3 and ATPase6). The assumption that pan-editing is an ancestral phenomenon requires one evolutionary step, from pan-editing to limited editing while assuming that it is a derived feature requires two independent acquisitions. However, data on the editing in mitochondrial genome of the bodonid *Trypanoplasma borreli* indicate that only the 5′ and 3′ termini of the cytb and cox1 genes are cryptic (Lukěs *et al.*, 1994). This suggests that editing in bodonids is limited. If this result is confirmed for NADH7, cox3 and ATPase6, then the assumption of an ancestral or a derived nature of pan-editing will require the same number of steps, i.e. two. But, as discussed below, if one considers that the genus *Trypanosoma* is monophyletic, pan-editing of at least ATPase6 and NADH7 can be assumed to be a derived, recent character. It can be seen from this example that drawing an evolutionary scenario from just a few species is very hazardous and that the inferred ancestral state can significantly change by the addition of a few new species. This is particularly troublesome when we know that phylogenies based on few species are unreliable (Philippe and Douzery, 1994).

11.4 CHALLENGING THE rRNA PHYLOGENY

Shortly after the above work, a more disturbing issue arose. Berchtold *et al.* (1994) sequenced two isolates of the free-living bodonid, *Dimastigella trypaniformis*. The monophyly of the trypanosomatids was confirmed (bootstrap value 100%), but the relative order of emergence of the three bodonids, *Bodo*, *Trypanoplasma* and *Dimastigella* was impossible to decipher. An unrooted topology ((*Bodo*, *Trypanoplasma*), *Dimastigella*, trypanosomatids) was found with a bootstrap value of 90%, but, when rooting was attempted, the use of different outgroup species (for example *Stylonychia* versus *Saccharomyces*) in a species sample composed of 18 sequences, resulted in different bodonids emerging first (*Bodo* versus *Dimastigella* with a bootstrap support of 39% and 71% respectively). The results suggest that it is impossible to solve this question with rRNA sequences because the very long branch at the base of the kinetoplastids cannot be broken by the addition of another euglenozoan species. This long branch is probably an artefact of the rRNA gene, because for many protein genes this branch is much shorter (see Chapter 2). This non-resolution is currently due to the properties of the rRNA and perhaps not to the true phylogenetic pattern (too closely related speciations) and could thus be resolved by using other phylogenetic markers.

But the phylogeny obtained by Berchtold *et al.* (1994) differs from the previously published ones also on the question of the monophyly of the genus *Trypanosoma*. They found a strong signal for the monophyly (bootstrap values from 81 to 89%) whereas almost all the other studies have found a similar support for paraphyly. This discrepancy was confirmed by Marché *et al.* (1995) and by Maslov *et al.* (1996).

In their phylogenetic analysis, Marché *et al.* (1995) added two *Phytomonas* sequences and used all the available kinetoplastid sequences (25). In their reconstruction, on the one hand, *Phytomonas* was a sister-group of *Herpetomonas*, and thus sister-group to the *Leishmania/Crithidia* lineage, and on the other hand, the genus *Trypanosoma* was monophyletic, with a bootstrap support of only 60%. These authors argued that the observed paraphyly could be an artefact due to long branch attraction (Felsenstein, 1978) because, using the relative rate test, *T. brucei* was shown to evolve faster than *T. cruzi*, in that the distance between *Bodo* and *T. cruzi* was equal to 13.6% and that between *Bodo* and *T. brucei* was equal to 15%. The use of a large number of species allowed breakage of the long branches (Hendy and Penny, 1989) and the monophyly observed in their study could be more reliable.

Maslov *et al.* (1996) added four sequences of non-mammalian trypanosomes but used a total of only 17 sequences in all. *Trypanosoma* was paraphyletic, with a strong bootstrap support (90%), but the section Stercoraria and non-mammalian trypanosomes, represented by six species, were monophyletic. The important result of this analysis was a lack of co-evolution between the trypanosomes and their hosts, since for example an amphibian parasite was more closely related to those of fish rather than to those of amniotes. This suggested that trypanosomatid evolution was accompanied by secondary acquisition of hosts. Once again, the observed paraphyly of *Trypanosoma* seemed to be robust because it was found with different tree reconstruction methods and also with partial LSUrRNA sequences.

However, Maslov *et al.* (1996) did not use the *Dimastigella* sequences, which are useful in breaking the long branch of the outgroup, and thus in reducing the long branch attraction phenomenon. Therefore I carried out an analysis using all the kinetoplastid species presently available (38). One of the most parsimonious trees obtained (Figure 11.1) provides a clear summary of all the previous studies. It shows: (1) the monophyly of trypanosomatids; (2) the monophyly of all the trypanosomatids except *Trypanosoma*; (3) a group of very closely related species containing *Leishmania*, *Crithidia*, *Endotrypanum* and *Leptomonas*, in which the phylogeny is unstable; (4) *Phytomonas* as sister-group of *Herpetomonas*, which is itself a sister-group of the previous group; (5) a group

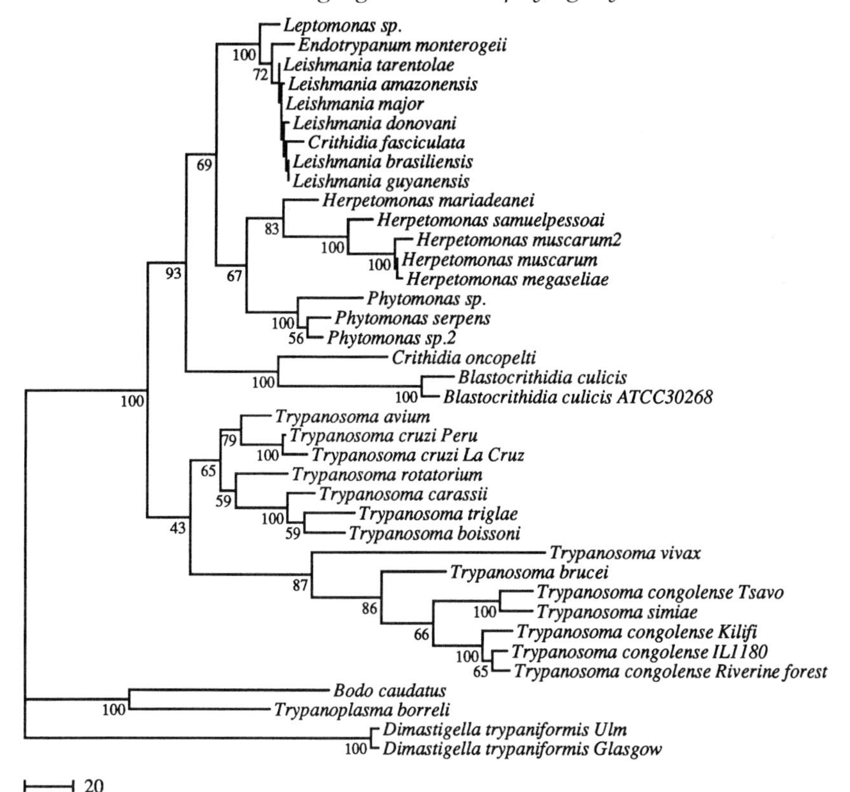

Figure 11.1 Phylogenetic tree based on SSUrRNA sequences. All the sequences were retrieved from Genbank and aligned by eye with the help of the MUST program (Philippe, 1993). On the 2546 aligned positions, 1744 sites can be considered as unambigously aligned. Seventy-two most parsimonious trees (number of steps = 1285) were obtained using the PAUP version 3.0 program (Swofford, 1993). On the other most parsimonious trees, *Trypanosoma* is sometimes paraphyletic and the relationships within the *Leishmania* group are very variable. Bootstrap values were computed with the PHYLIP package (Felsenstein, 1993) with 100 replicates. Branch lengths are proportional to the number (see scale bar) of inferred substitutions (ACCTRAN option).

composed of *Crithidia oncopelti* and *Blastocrithidia*; (6) uncertainty concerning the monophyly of the genus *Trypanosoma*.

The most parsimonious trees differed mainly on the last point, the one shown (Figure 11.1) with monophyly of *Trypanosoma* (bootstrap value 43%) and the other with their paraphyly (bootstrap value 55%). This new result, however, is perfectly compatible with the suggestion of

Marché *et al.* (1995). First of all, if only one bodonid species (*Trypanoplasma borreli*) is used as an outgroup, the bootstrap value for the paraphyly increases to 74%, whereas the monophyly is only supported by a value of 24%. Thus breaking the long branch of the outgroup by a second species weakens the support for paraphyly. Second, the new salivarian trypanosome sequences (*T. congolense*, *T. simae* and *T. vivax*, Urakawa and Majiwa, unpublished) are evolving faster than the previously used *T. brucei* (see the branch lengths on Figure 11.1), which itself evolves faster than *T. cruzi*. Thus, a weakening of the support for the monophyly of *Trypanosoma* observed in Figure 11.1 as compared with the tree of Marché *et al.* (1995) is expected because of the added salivarian sequences with a higher evolutionary rate. In conclusion, several lines of evidence suggest that the paraphyly of *Trypanosoma* seen in many trees is an artefact of long branch attraction between *T. brucei* and the outgroup.

11.5 PROTEIN PHYLOGENIES

In several phylogenies based on protein sequences (Drouin, Moniz de Sa and Zucker, 1995; Henze *et al.*, 1995; Wiemer *et al.*, 1995), the genus *Trypanosoma* has been found to be monophyletic but this interesting result has not been discussed. Only recently, Alvarez, Cortinas and Musto (1996) have pointed out that protein coding genes suggest monophyly of *Trypanosoma*. They used 14 different genes for which both *T. cruzi* and *T. brucei* data were present. However, in many cases, only one other trypanosomatid was available and there were rarely outgroup sequences. As a result, their evidence for the monophyly of *Trypanosoma* is mainly based on the higher level of similarity between *T. cruzi* and *T. brucei* than between any *Trypanosoma* and any other trypanosomatid, except for GAPDH and α-tubulin genes for which the phylogenetic tree can be rooted. The analysis of Alvarez, Cortinas and Musto (1996) is thus strongly dependent on the molecular clock hypothesis, because two taxa with the highest level of similarity can be related only distantly if the other taxa evolve faster. However, the existence of a molecular clock is highly controversial (Li, 1993; Li *et al.*, 1996) (see Chapter 2).

I extended the study of Alvarez, Cortinas and Musto (1996) including new species, especially outgroup ones, and analysing three additional protein encoding genes. As closely-related as possible outgroup species were chosen, first bodonids, second *Euglena* and third vertebrates. Indeed, almost all eukaryotic species are equidistant to Euglenozoa (see chapter 2) and the vertebrates have been well sampled for all the genes studied. One can thus compare the genetic diversity among the hosts and among the parasites. Three proteins used by Alvarez, Cortinas and

Table 11.1 Monophyly of *Trypanosoma*

Gene	*Outgroup*	*Monophyly Trypanosoma*	*d(T,L)*	*d(Tb,Tc)*
Actin	Vertebrates	+	12%	7%
Alpha tubulin	*Euglena*	+	6%	1%
COX II	*Trypanoplasma*	–	20%	16%
HGPRT[a]	Vertebrates	–	50%	45%
HSP70 (cytoplasmic)	Vertebrates	+	13%	7%
HSP70 (mitochondrial)	Vertebrates	+	16%	8%
HSP90	Vertebrates	+	16%	8.5%
GAPDH	*Trypanoplasma*	+	17%	9%
Topoisomerase II	Vertebrates	+	29%	19%
Trypanothione reductase	Vertebrates	+	31%	17%
Beta tubulin	*Euglena*	+	6%	2%
PGK[b]	Vertebrates	+	22%	15%
Thymidylate synthase	Vertebrates	+	24%	20%
nuclear rRNA (1750 bp)	*Trypanoplasma*	–/+	5–8%	7%
nuclear rRNA (1000 pb)	*Trypanoplasma*	+	1.5%	1%

[a] HGPRT, Hypoxanthine-Guanine ribosyltransferase; [b] PGK, Phosphoglycerate kinase. The distances calculated between *Trypanosoma* and *Leishmania* and between *T. brucei* and *T. cruzi* are equal to the per cent of observed differences. For the nuclear rRNA, two estimates were obtained by using the positions unambigously aligned within kinetoplastids and within all eukaryotes, respectively.

Musto (1996) were excluded due to lack of suitable outgroup or to insufficient number of informative sites.

The results of this analysis are summarized in Table 11.1. Except in two cases out of 13, *Trypanosoma* forms a monophyletic group. In addition, as previously observed, the frequency of amino acid differences between *T. cruzi* and *T. brucei* is always smaller than that between *Trypanosoma* and *Leishmania*. This new analysis is thus in full agreement with those of Alvarez, Cortinas and Musto (1996), and together with the analysis of the nuclear SSUrRNA discussed above, strongly supports monophyly of the genus *Trypanosoma*.

11.6 VARIATION OF EVOLUTIONARY RATES

Some of the protein phylogenies for which an appropriate sampling is available are shown: β-tubulin (Figure 11.2), HSP90 (Figure 11.3) and thymidylate synthase (Figure 11.4). On the β-tubulin tree, in which the monophyly of *Trypanosoma* and *Leishmania* is highly supported, another interesting feature appears when branch lengths are considered. The

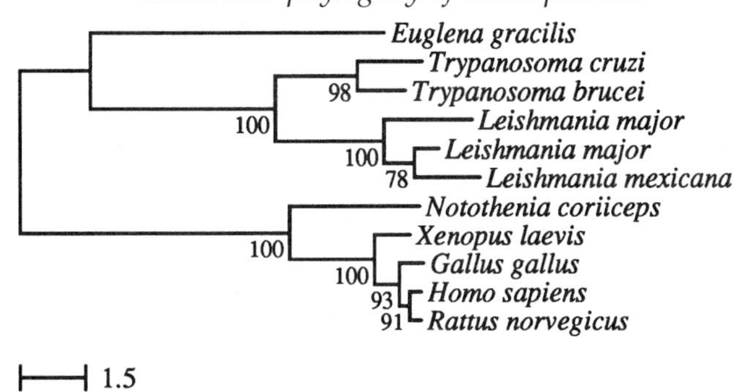

Figure 11.2 Phylogenetic tree based on β-tubulin sequences. The tree was constructed with the Neighbour-Joining method and the bootstrap values were obtained with 1000 replicates. The distances between taxa are the number of observed differences. All the computations have been carried out with the MUST package (Philippe, 1993).

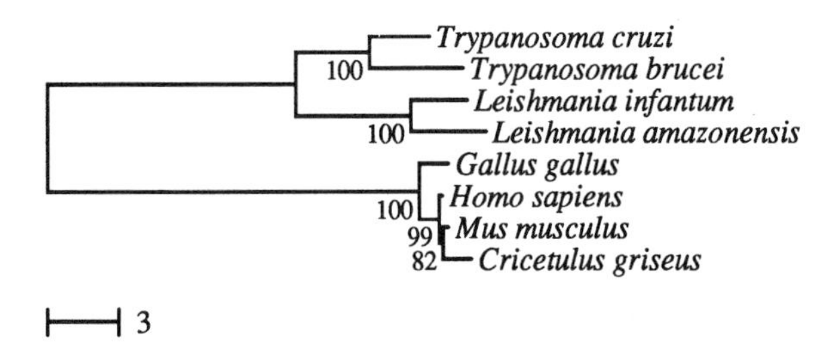

Figure 11.3 Phylogenetic tree based on HSP90 sequences. For method, see Figure 11.2.

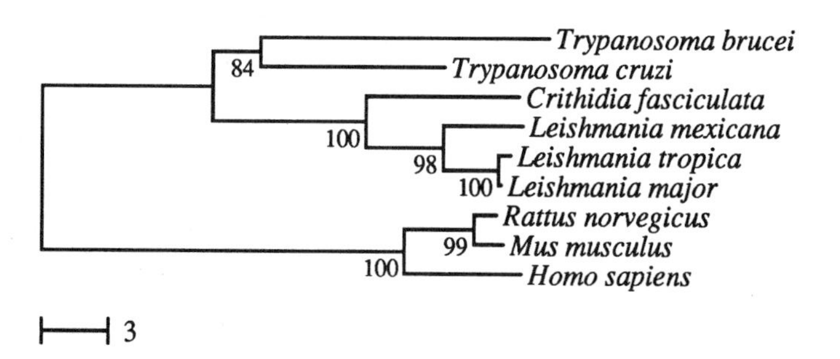

Figure 11.4 Phylogenetic tree based on thymidylate synthase sequences. For method, see Figure 11.2.

four following branches, from the base of the trypanosomatids to the base of *Trypanosoma* group (L_{bT}) and to the base of *Leishmania* group (L_{bL}), from the base of *Trypanosoma* group to extant *Trypanosoma* (L_{wT}) and from the base of *Leishmania* group to extant *Leishmania* (L_{wL}), are of similar length. This is quite different from the rRNA tree (Figure 11.1) in which L_{wT} is much greater than L_{bL}, itself much greater than L_{bT}, itself greater than L_{wL} ($L_{wT} \gg L_{bL} \gg L_{bT} > L_{wL}$). In the case of HSP90 (Figure 11.3), the branch lengths are similar to these observed for β-tubulin. However, in the case of thymidylate synthase (Figure 11.4), the relative branch lengths are again different from the previous cases ($L_{wT} = L_{bL} > L_{bT} = L_{wL}$).

When comparing the 16 phylogenetic markers, this heterogeneity of the four previous branch lengths is a general phenomenon. This suggests that there are strong differences in evolutionary rates within kinetoplastids and that this phenomenon is mainly dependent on the gene rather than on the taxon. In general, these strong differences in evolutionary rates are not reflected on the trees in different distances between the root of the tree and extant species. This is due to a general tendency of tree reconstruction methods, especially the distance method, to produce a clock-like behaviour, even if there is no molecular clock (Philippe and Adoutte, 1996) (chapter 2). This phenomenon is more and more apparent with increasing mutational saturation.

A general trend of this comparison, however, seems to be that the nuclear rRNA pattern ($L_{wT} \gg L_{wL}$) is not found with other genes. Since two *Trypanosoma*, two *Leishmania* and *Crithidia* are often available, a quantitative comparison can be carried out. As discussed in chapter 2, the method consists in computing the distance within a group by comparing two species of this group: in this case, one computed d(*T. brucei, T. cruzi*), d(*Leishmania* sp.1, *Leishmania* sp.2) and d(*Leishmania, Crithidia*) (Table 11.2). For *Leishmania*, the two most distantly available species were compared, implying an underestimate of the distance based on protein *versus* that based on nuclear rRNA, since numerous diverse sequences are available for it. However, this phenomenon could be compensated by the existence of unseen paralogous comparison (see an obvious example on Figure 11.2), since recent duplications of genes encoding proteins are common.

The heterogeneity of evolutionary rates described above can now be quantitated by computing the ratio α_{LL} = d(*Leishmania* sp. 1, *Leishmania* sp. 2/d(*T. brucei, T. cruzi*) and α_{LL} = d(*Leishmania, Crithidia*)/d(*T. brucei, T. cruzi*) for the different genes (Table 11.2). The first ratio varies from 0.03 to 1.5, the second from 0.07 to 1.29. There is more than an order of magnitude difference when rRNA and proteins are compared, but between proteins these ratios are highly variable. This confirms the lack of molecular clock behaviour for all these genes in kinetoplastids.

Table 11.2 Relative evolutionary rates within trypanosomatids

Gene	*d(Tb,Tc)*	*d(L,L)*	*d(C,L)*	$\frac{d(L, L)}{d(Tb, Tc)}$	$\frac{d(C, L)}{d(Tb, Tc)}$	$\frac{d_{prot}(L, L)/d_{prot}(Tb, Tc)}{d_{rRNA}(L, L)/d_{rRNA}(Tb, Tc)}$	$\frac{d_{prot}(C, L)/d_{prot}(Tb, Tc)}{d_{rRNA}(C, L)/d_{rRNA}(Tb, Tc)}$
ATPase (subunit 6)	14%		8%		0.57		8.00
COX II	16%		10%		0.63		8.75
HGPRT	45%		17%		0.38		5.29
HSP70 (cytoplasmic)	7%	3%		0.43		15.00	
HSP70 (mitochondrial)	8%		10%		1.25		17.50
HSP90	8.5%	6%		0.71		24.71	
Trypanothione reductase	17%		22%		1.29		18.12
Beta tubulin	2%	3%		1.50		52.50	
PGK	15%	5%	11%	0.33	0.73	11.67	10.27
Thymidylate synthase	20%	6%	12%	0.30	0.60	10.50	8.40
mitochondrial rRNAs	16%		10.5%		0.66		9.40
nuclear rRNA (1750 bp)	7%	0.2%	0.5%	0.03	0.07	1.00	1.00
nuclear rRNA (1000 pb)	1%	0.1%	0.2%	0.10	0.20	3.50	2.80

HGPRT, Hypoxanthine-Guanine ribosyltransferase; PGK, Phosphoglycerate kinase. The distances calculated between *T. brucei* and *T. cruzi*, between the most distantly related available species of *Leishmania* and between *Crithidia* and *Leishmania* are equal to the per cent of observed differences. The first distance is much greater than the two others when using the nuclear rRNA but is similar when using other genes. This suggests an acceleration of evolutionary rate of the nuclear rRNA in the genus *Trypanosoma*.

In order to compare the nuclear rRNA pattern ($L_{wT} \gg L_{wL}$) with the other patterns, two new ratios are computed, α_{LL}(protein)/α_{LL}(rRNA) and α_{LC}(protein)/α_{LC}(rRNA), which measure the relative evolutionary rates between the genus *Trypanosoma* and the genus *Leishmania* by comparing protein and rRNA. For example, for rRNA, the evolutionary rate is high for *Trypanosoma* (7%) and low for *Leishmania* (0.2%), yielding to α_{LL}(rRNA) = 0.03; however, for β-tubulin, the evolutionary rate is low for *Trypanosoma* (2%) and high for *Leishmania* (3%), yielding to α_{LL}(β-tubulin) = 1.5; the new ratio is thus 1.5/0.03 = 52.5. This ratio indicates that, in *Leishmania*, rRNA evolves much more slowly than β-tubulin, as compared with *Trypanosoma*. All the ratios (Table 11.2) indicate that the general trend observed below is real: the mean of the ratio for all the proteins is 21.1 for *Leishmania* and 10.8 for *Leishmania/Crithidia*, indicating that there is a marked difference in evolutionary rates for rRNA in kinetoplastids. The ratio of 21.1 can be interpreted as a 21-fold acceleration in *Trypanosoma*, as a 21-fold deceleration in *Leishmania* or as any intermediate combination of these. The fact that the ratio is only 10.8 for *Leishmania/Crithidia* strongly suggests that the evolutionary rates of rRNA are, indeed, highly variable with time, and that numerous accelerations and decelerations occurred in the past. It could be the same for proteins, but there are too few data to obtain a robust estimate.

Interestingly, if only the most conserved parts of the rRNA are used, i.e. those alignable for all eukaryotes, the previously observed differences in evolutionary rate are significantly reduced, since the mean ratios are 6 and 3.8 instead of 21.1 and 10.8 (Table 11.2). This indicates that the more variable parts of a molecule are the more sensitive to variation in evolutionary rates. Indeed, this becomes apparent when carrying out the alignment: all the *Leishmania* species are readily aligned for the full length of the rRNA whereas large parts are unalignable among *Trypanosoma* species, even between two *T. cruzi* strains. As a result, to reduce the possible bias due to heterogeneity, it could be fruitful to use the more conserved parts of rRNA, even if the quantity of informative sites decreases. For example, when using the 1000 highly conserved positions, *Trypanosoma* is almost always monophyletic, irrespective of what is the species sampling or the tree reconstruction method (data not shown).

11.7 CONCLUSION

In summary, Table 11.2 shows that the evolutionary rates of different genes are highly variable, i.e. that instead of a molecular clock-like behaviour, the phylogenetic markers in kinetoplastids display a

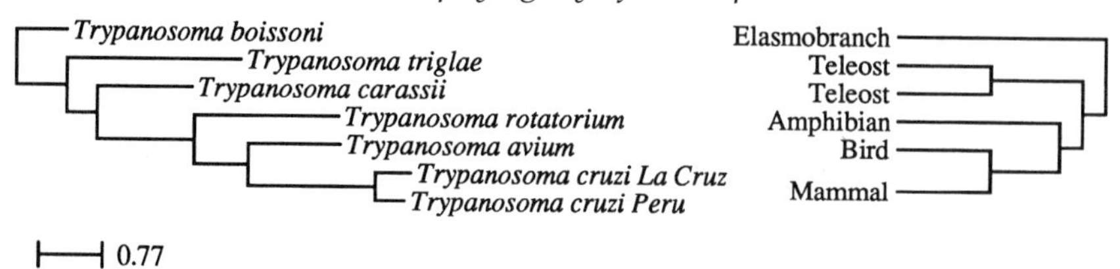

Figure 11.5 Phylogenetic tree of the non-salivarian trypanosomes based on SSUrRNA sequences. The tree was arbitrarily rooted on *T. boissoni*, a parasite of ray (elasmobranch). On the right of the figure is indicated the commonly accepted phylogeny of the hosts. For method, see Figure 11.2.

chaotic behaviour. This could be a general phenomenon for all genes and all taxa (Philippe, unpublished and see chapter 2). Such chaotic behaviour is known easily to produce erroneous branching patterns (Felsenstein, 1978). It is thus important to compare the phylogenies based on several independent genes, that display different evolutionary constraints and thus have different variation in evolutionary rates, and, before assuming that a robust phylogeny of a group is obtained, to construct a consensus between all these phylogenies (the best method to carry out this analysis is still a matter of debate, see Miyamoto and Fitch, 1995). In the case of the general phylogeny of eukaryotes, such an approach is described in chapter 2.

Knowing that the evolutionary rate of rRNA is highly variable in kinetoplastids, it is possible to question the inferred phylogeny. In the case of the non-salivarian trypanosomes, instead of rooting the tree with other trypanosomatids (Figure 11.1), I rooted the tree on *T. boissoni*, a parasite of elasmobranchs (Figure 11.5). Such a rooting is dictated by the host tree and allows avoidance of the long branch attraction artefact that could potentially occur with a distant outgroup. A striking point of this new analysis is that the phylogeny of the parasite is almost identical to the phylogeny of the host, except concerning the monophyly of teleosts. In contradiction to Maslov *et al.* (1996), this tree suggests a coevolution between hosts and parasites for the trypanosomes, but displays great differences in the branch length, that would not be unexpected from previous results (Table 11.2). It is thus necessary to obtain data from other genes in order to be able to choose between (i) lack of coevolution and a rough molecular clock, and (ii) coevolution and a difference of evolutionary rates of at least fivefold.

What about the general phylogeny of kinetoplastids? Only rRNA has been sequenced for a sufficient number of species and can provide a

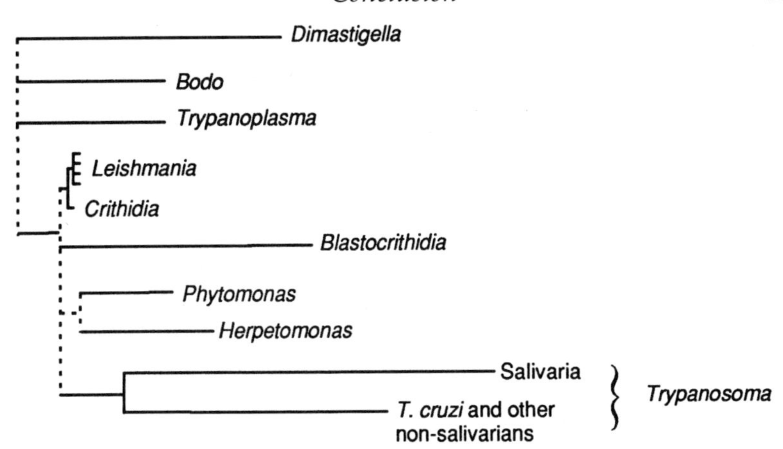

Figure 11.6 Schematic phylogenetic tree of the kinetoplastids, mainly inferred from SSUrRNA sequences. The differences in evolutionary rates are taken from the branch lengths of Figure 11.1 and from Table 11.2. The dotted lines indicate non-resolution.

broad view. Given the observed differences in evolutionary rates, I propose as a summary a schematic tree which states what is resolved (Figure 11.6).

Only a few branching patterns are strongly supported; these are monophyly of: (i) trypanosomatids, (ii) *Trypanosoma*, and (iii) the group containing *Leishmania*, *Crithidia*, *Endotrypanum* and *Leptomonas*. The sister-group of *Phytomonas* and *Herpetomonas*, although not strongly supported by bootstrap analysis, could be correct because it was found in the majority of the analyses despite a higher evolutionary rate of *Herpetomonas*. All the other branching patterns should be considered as unresolved. This lack of resolution is due to the failure of tree reconstruction methods to handle correctly the strong differences in evolutionary rates rather than to a real evolutionary radiation. A significant improvement of the kinetoplastid phylogeny requires the analysis of several protein genes, such as EF-1α, GAPDH or HSP70, with a large species sampling.

Acknowledgements

I am grateful to the editors for giving me the opportunity to present this work at the meeting on evolutionary relationships among Protozoa. I acknowledge Miklós Müller's critical reading of the manuscript.

11.8 REFERENCES

Alvarez, F., Cortinas, M.N. and Musto, H. (1996) The analysis of protein coding genes suggest monophyly of *Trypanosoma. Molecular Phylogenetics and Evolution*, **5**, 333–43.

Berchtold, M., Philippe, H., Breunig, A. *et al.* (1994) The phylogenetic position of *Dimastigella trypaniformis* within the parasitic kinetoplastids. *Parasitology Research*, **80**, 672–9.

Briones, M.R.S., Nelson, K., Beverley, S.M. *et al.* (1992) *Leishmania tarentolae* taxonomic relatedness inferred from phylogenetic analysis of the small subunit ribosomal RNA gene. *Molecular and Biochemical Parasitology*, **53**, 121–8.

Drouin, G., Moniz de Sá, M. and Zucker, M. (1995) The *Giardia lamblia* actin gene and the phylogeny of eukaryotes. *Journal of Molecular Evolution*, **41**, 841–9.

Du, Y., Maslov, D.A. and Chang, K.-P. (1994) Monophyletic origin of β-division proteobacterial endosymbionts and their coevolution with insect trypanosomatid protozoa *Blastocrithidia culicis* and *Crithidia* spp. *Proceedings of the National Academy of Sciences USA*, **91**, 8437–41.

Felsenstein, J. (1978) Cases in which parsimony or compatibility methods will be positively misleading. *Systematic Zoology*, **27**, 401–10.

Felsenstein, J. (1993) PHYLIP Manual Version 3.5. University Herbarium, University of California, Berkeley, California.

Fernandes, A.P., Nelson, K. and Beverley, S.M. (1993) Evolution of nuclear ribosomal RNAs in kinetoplastid protozoa: perspectives on the age and origins of parasitism. *Proceedings of the National Academy of Sciences USA*, **90**, 11608–12.

Gomez, E., Valdés, A.M., Piñero, D. and Hernández, R. (1991) What is a genus in the Trypanosomatidae family? Phylogenetic analysis of two small rRNA sequences. *Molecular Biology and Evolution*, **8**, 254–9.

Hashimoto, T., Nakamura, Y., Kamaishi, T. *et al.* (1995) Phylogenetic place of kinetoplastid protozoa inferred from a protein phylogeny of elongation factor 1α. *Molecular and Biochemical Parasitology*, **70**, 181–5.

Hendy, M.D. and Penny, D. (1989). A framework for the quantitative study of evolutionary trees. *Systematic Zoology*, **38**, 297–309.

Henze, K., Badr, A., Wettern, M. *et al.* (1995) A nuclear gene of eubacterial origin in *Euglena gracilis* reflects cryptic endosymbioses during protist evolution. *Proceedings of the National Academy of Sciences USA*, **92**, 9122–6.

Kivic, P.A. and Walne, P.L. (1984) An evaluation of a possible phylogenetic relationship between the Euglenophyta and Kinetoplastida. *Origins of Life*, **13**, 269–88.

Lake, J.A., De La Cruz, V.F., Ferreira, P.C.G. *et al.* (1988) Evolution of parasitism: kinetoplastid protozoan history reconstructed from mitochondrial rRNA gene sequences. *Proceedings of the National Academy of Sciences USA*, **85**, 4779–83.

Landweber, L.F. and Gilbert, W. (1994) Phylogenetic analysis of RNA editing: a primitive genetic phenomenon. *Proceedings of the National Academy of Sciences USA*, **91**, 918–21.

Lee, J.J. and Hutner, S.H. (1985) Kinetoplastida Honigberg, 1963 emend Vickerman, 1976, in *Illustrated Guide to the Protozoa* (eds J.J. Lee, S.H. Hutner and E.C. Bovee), pp.141–55.

Léger, L. (1904) Sur les affinités de l'*Herpetomonas subulata* et la phylogénie des trypanosomes. *Comptes Rendus des Séances de la Société Française de Biologie et de ses Filiales*, **56**, 615–17.

Li, W.-H. (1993) So, what about the molecular clock hypothesis? *Current Opinion in Genetics and Development*, **3**, 896–901.

Li, W.-H., Ellsworth, D.L., Krushkal, J. *et al.* (1996) Rates of nucleotide substitution in primates and rodents and the generation-time effect hypothesis. *Molecular Phylogenetics and Evolution*, **5**, 182–7.

Lukěs, J., Arts, G.J., Van den Burg, J. *et al.* (1994) Novel pattern of editing regions in mitochondrial transcripts of the cryptobiid *Trypanoplasma borreli*. *EMBO Journal*, **13**, 5086–98.

Marché, S., Roth, C., Philippe, H. *et al.* (1995) Characterization and detection of plant trypanosomatids by sequence analysis of the small subunit ribosomal RNA gene. *Molecular and Biochemical Parasitology*, **71**, 15–26.

Maslov D.A., Avila, H.A., Lakěs, J.A. and Simpson, L. (1994) Evolution of RNA editing in kinetoplastid protozoa. *Nature*, **368**, 345–8.

Maslov D.A., Lukěs, J., Jirku, M. and Simpson, L. (1996) Phylogeny of trypanosomes as inferred from the small and large subunit rRNAs: implications for the evolution of parasitism in the trypanosomatid protozoa. *Molecular and Biochemical Parasitology*, **75**, 197–205.

Minchin, E.A. (1908) Investigation on the development of trypanosomes in tsetse-flies and other Diptera. *Quarterly Journal of Microscopical Science*, **52**, 159–260.

Miyamoto, M.M. and Fitch, W.M. (1995) Testing species phylogenies and phylogenetic methods with congruence. *Systematic Biology*, **44**, 64–76.

Philippe, H. (1993) MUST: a computer package of Management Utilities for Sequences and Trees. *Nucleic Acids Research*, **21**, 5264–72.

Philippe, H. and Adoutte, A. (1996) What can phylogenetic patterns tell us about the evolutionary processes generating biodiversity? in *Aspects of the Genesis and Maintenance of Biological Diversity* (eds M. Hochberg, J. Clobert and R. Barbault), Oxford University Press, Oxford, pp. 41–59.

Philippe, H. and Douzery, E. (1994) The pitfalls of molecular phylogeny based on four species, as illustrated by the Cetacea/Artiodactyla relationships. *Journal of Mammalian Evolution*, **2**, 133–52.

Sogin, M.L. (1991) The phylogenetic significance of sequence diversity and length variations in eukaryotic small subunit ribosomal RNA coding regions, in *New Perspectives on Evolution* (eds L. Warren and H. Koprowski), Wiley-Liss, New York, pp. 175–88.

Sogin, M.L., H.J. Elwood, H.J. and Gunderson, J.H. (1986) Evolutionary diversity of eukaryotic small-subunit rRNA genes. *Proceedings of the National Academy of Sciences USA*, **83**, 1383–7.

Swofford, D.L. (1993). PAUP: Phylogenetic analysis using parsimony, Version 3.0. Illinois Natural History Survey, Champaign.

Swofford, D.L. and Olsen, G.J. (1990) Phylogeny reconstruction. In *Molecular systematics*. Eds. D.M. Hillis and C. Moritz, pp. 411-501.

Vickerman K. (1989) Phylum Zoomastigina. Class Kinetoplastida, in *Handbook of Protoctista* (eds L. Margulis, J.O. Corliss, M. Melkonian and D.J. Chapman), Jones and Bartlett, Boston, pp. 215–38.

Wiemer, E.A.C., Hannaert, V., van den Ijssel, P.R.L.A. *et al.* (1995) Molecular analysis of glyceraldehyde-3-phosphate dehydrogenase in *Trypanoplasma borreli*: an evolutionary scenario of subcellular compartmentation in Kinetoplastida. *Journal of Molecular Evolution*, **40**, 443–54.

12

Evolutionary relationships among the African trypanosomes: implications for the epidemiology and generation of human sleeping sickness epidemics

Geoff Hide

Wellcome Unit of Molecular Parasitology, Glasgow University, Anderson College, 56 Dumbarton Road, Glasgow, G11 6NU, UK. E-mail: gvwa12@udcf.gla.ac.uk

ABSTRACT

By nature of their importance as major pathogens of man and animals, the classical taxonomy of the African sleeping sickness trypanosomes has been based largely on medical criteria rather than evolutionary principles. As clinical diagnosis is a primary objective for anyone investigating the diversity of the African trypanosomes, characters such as human infectivity, disease pathology, host range and geographical location form the basis of the taxonomic criteria. An increasing number of exceptions and the requirement for a more detailed knowledge of strain structure for epidemiological interpretation, has stimulated a need for a greater understanding of the classification of this group. A system is described where restriction fragment length polymorphism analysis, of repetitive DNA sequences, is used to determine the evolutionary relationships among the African trypanosomes. This review highlights the importance of the need for an evolutionary approach to the identification of trypanosome strains to facilitate an understanding of the epidemiology, population genetics and origins of human sleeping sickness.

Evolutionary Relationships Among Protozoa. Edited by G.H. Coombs, K. Vickerman, M.A. Sleigh and A. Warren. Published in 1998 by Chapman & Hall, London. ISBN 0 412 79800 X

12.1 INTRODUCTION

By nature of their importance as major pathogens of man and animals, the classical taxonomy of the African sleeping sickness trypanosomes has been based largely on medical criteria rather than evolutionary principles. As clinical diagnosis is a primary objective for anyone investigating the diversity of the African trypanosomes, characters such as human infectivity, disease pathology, host range and geographical location form the basis of the taxonomic criteria. An increasing number of exceptions and the requirement for a more detailed knowledge of strain structure for epidemiological interpretation, has stimulated a need for a greater understanding of the classification of this group.

In this review, I will discuss the impact of recent molecular approaches on our understanding of the evolutionary relationships among trypanosomes of the *T. brucei* group and the epidemiology of human sleeping sickness in Africa.

12.1.1 *Trypanosoma brucei* and human disease

In Africa, human sleeping sickness is characterized by short epidemics interspersed with long periods of endemicity where disease prevalence is low or absent. It is confined to the areas of Africa infested with its vector, the tsetse fly, and associated with specific foci of disease. The disease itself is extremely debilitating and proceeds through two distinct phases: early and late. Following a tsetse fly bite, trypanosomes proliferate in the host bloodstream and undergo antigenic variation to evade the immune system. Symptoms of the early phase, nausea, fever and lethargy, are non-specific and easily confused with those of other diseases (e.g. malaria, influenza). In late phase, trypanosomes cross the blood–brain barrier and can be found in neural tissue and cerebrospinal fluid (CSF). Subsequent neurological damage causes the classical symptoms of sleeping sickness: disruption of biorhythms; inappropriate and irregular sleep patterns; loss of concentration and coordination. Unless treated, death will ensue. The progress of the disease varies from acute (death, 6–12 months) to chronic (death, 5–20 years). Treatment is effective in the early phase of the disease but in late phase infections few effective drugs, capable of passing the blood–brain barrier, are available. Those which do are toxic and drug treatment results in fatal encephalopathy in 5–10% of patients treated.

12.1.2 The classical taxonomy of the *Trypanosoma brucei* complex

In the currently accepted taxonomy of *T. brucei*, three subspecies of *T. brucei* are described (Hoare, 1972). *T. b. rhodesiense* causes acute human sleeping sickness in foci in east and south east Africa. *T. b. gambiense* causes chronic human sleeping sickness in west and central Africa. *T. b. brucei* is found throughout Africa, infects a wide range of animal hosts, but does not infect humans. This classification system is not without its problems. For example, it has been known for some time that human infective trypanosomes (e.g. *T. b. rhodesiense*) have animal reservoirs (Heisch, McMahon and Manson-Bahr, 1958). By implication an isolate taken from animals is not necessarily *T. b. brucei*. Secondly, the Ugandan and Zambian foci differ in their degree of disease virulence (high *vs* low) despite both being attributed to *T. b. rhodesiense* (Hide *et al.*, 1991).

12.1.3 Important epidemiological questions

To understand the epidemiology of human sleeping sickness, a wide variety of factors need to be considered. One of the most important of these is a correct identification of the species, subspecies and strains of parasite responsible for disease. A number of important questions need to be addressed in relation to the trypanosomes and disease epidemiology: (1) Does the diversity of trypanosome strains observed, using molecular markers, reflect the classical taxonomy? (2) Does human infectivity have a monophyletic origin? (3) What is the composition of strains circulating among man and animals during an epidemic of human sleeping sickness? (4) How stable are parasite strains in a human sleeping sickness focus over time? (5) What is the contribution of genetic exchange to the degree of diversity in parasite populations during an epidemic? (6) What is the contribution of animal reservoirs to the maintenance of epidemics? (7) What is the composition of strains circulating in an individual animal? (8) What is the composition of strains circulating in an endemic area of human sleeping sickness? (9) What are the origins of human sleeping sickness epidemics?

12.2 MOLECULAR AND BIOCHEMICAL APPROACHES TO INVESTIGATING EVOLUTIONARY RELATIONSHIPS

Two basic approaches have been used to analyse the evolutionary relationships among the African trypanosomes: isoenzyme analysis and molecular analysis of diversity. Isoenzyme electrophoresis has been extensively used for identification of trypanosome stocks (Gibson, Marshall and Godfrey, 1980; Tait, Babiker and Le Ray, 1984; Godfrey *et al.*, 1990). More recently, a number of different types of DNA analyses have been employed to elucidate the relationships of the African trypanosomes (reviewed in Hide and Tait, 1991; Hide 1996a). Briefly, the main approaches have been centred on the detection of restriction fragment length polymorphisms (RFLPs) in antigen genes (Paindavoine *et al.*, 1986) or repetitive DNA (Hide *et al.*, 1990, Hide 1996b) and the analysis of randomly amplified polymorphic DNAs (RAPDs) (Mathieu-Daude *et al.*, 1995; Stevens and Tibayrenc, 1995). In this review, I will concentrate on the impact of results obtained by RFLP analysis of repetitive DNA.

12.2.1 RFLP analysis of repetitive DNA

RFLP analysis of repetitive DNA in trypanosomes has been described in detail elsewhere (Hide *et al.*, 1990; Hide, 1996b) but briefly the procedure consists of six steps: (1) isolation and cloning of trypanosomes; (2) extraction of DNA; (3) gel electrophoresis of restriction enzyme digested DNA; (4) Southern blotting and hybridization with three repetitive DNA probes (a ribosomal RNA gene, a ribosomal RNA gene spacer and a dispersed repetitive sequence); (5) autoradiography and recording of banding patterns; (6) construction of a similarity matrix and a dendrogram based on similarity of banding patterns.

12.3 EVOLUTIONARY RELATIONSHIPS AND EPIDEMIOLOGY

12.3.1 Genetic diversity of *T. brucei* strains

RFLP analysis of repetitive DNA was used to examine the relationships between representatives of the three subspecies of *T. brucei* to investigate the robustness of the classical classification (Hide *et al.*, 1990). Five basic groups of *T. brucei* stocks were included in this analysis: (1) from West African sleeping sickness patients; (2) *T. b. rhodesiense* – Kenya focus; (3) *T. b. rhodesiense* – Zambian focus; (4) *T. b. brucei* – W. Africa; (5) *T. b. brucei* – East Africa. The results are summarized as a dendrogram (Figure 12.1). *T. b. brucei* populations from East and West Africa were clearly distinct as might be expected of two populations geographically separated. Trypanosomes from West African human infections fell into three groups.

Group 1 may be designated 'true' *T. b. gambiense*. The second group contained stocks previously designated as 'non-gambiense' (Tait, Babiker and Le Ray, 1984, Paindavoine *et al.*, 1986) or Type II *T. b. gambiense* (Gibson 1986). RFLP analysis of repetitive DNA showed that these stocks were completely unrelated to the true *T. b. gambiense* stocks but were more closely related to a collection of *T. b. brucei* stocks isolated from cattle in Nigeria (Figure 12.1). Despite this relationship with the Nigerian stocks, both types of gambiense stocks co-exist temporally and spatially. For example, a 'true' *T. b. gambiense* stock (NIPA, Figure 12.1) and the non-gambiense stocks (OUSOU, KOBIR and LIGO,

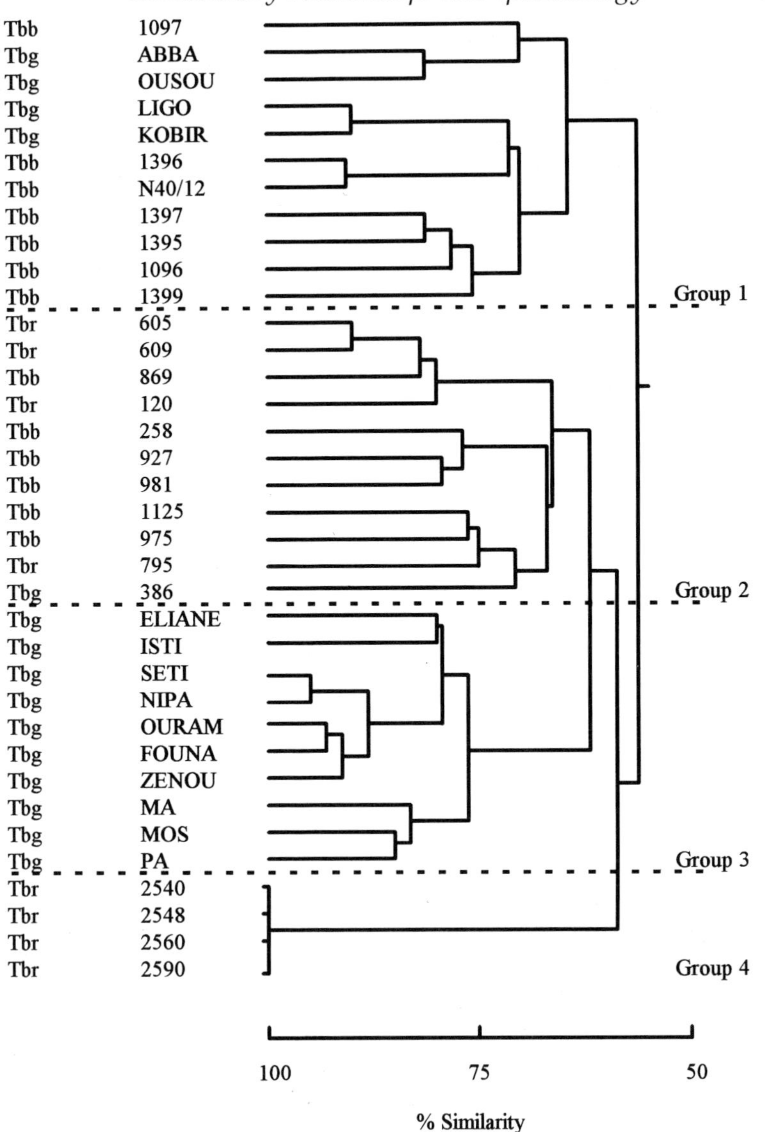

Figure 12.1 A dendrogram, showing the relationships between African trypanosomes, was constructed from a similarity matrix derived from banding patterns produced by restriction fragment length polymorphism (RFLP) analysis of repetitive DNA probes (Redrawn from Hide *et al.*, 1990.) Trypanosome stocks are indicated by stock name (1097, ABBA etc.) and are classified according to Hoare (1972) as *Trypanosoma brucei brucei* (Tbb), *T. b. gambiense* (Tbg) and *T. b. rhodesiense* (Tbr). Details of stocks can be found elsewhere (Hide *et al.*, 1990). The stocks grouped into four groups: (1) West African, (2) East African, (3) *T. b. gambiense* and (4) *T. b. rhodesiense* from the Zambian focus.

Figure 12.1) were isolated from patients who lived in neighbouring streets in Daloa, Ivory Coast (Hide, 1988).

By analogy to the close relationships observed between *T. b. rhodesiense* and *T. b. brucei* stocks in East Africa (see later), Hide *et al.*, (1990) proposed that the non-gambiense stocks were West African equivalents of the agent of acute sleeping sickness in East Africa, *T. b. rhodesiense.* To test this hypothesis, Hide (1988) examined the case histories of six patients from Daloa (corresponding hosts to three 'true' *T. b. gambiense* and three non-gambiense isolates) to determine whether an association could be made between acute or chronic disease and trypanosome strain. One patient had an acute disease profile (a non-gambiense infection) while the others were described as sub-acute. Clear interpretation of disease profiles, by the clinicians, was hampered by the patients not presenting themselves for treatment until the disease had progressed to advanced stages. This left the issue unresolved but it did highlight a further difficulty in using disease profile as a taxonomic character. It is, of course, an important issue to resolve whether these 'non-gambiense' strains are *T. b. rhodesiense* as the most recently identified trypanocidal drug, Difluoromethylornithine (DFMO), is effective against *T. b. gambiense* but not *T. b. rhodesiense.*

The third group of West African human stocks consisted of a single stock, STIB 386, which grouped alongside East African *T. b. rhodesiense* stocks. This, taken together with its ability to be crossed with East African stocks (Jenni *et al.*, 1986), led Hide *et al.* (1990) to propose that it was of East African origin and that it had been brought to West Africa relatively recently.

The situation with *T. b. rhodesiense* was also found to be complex. RFLP analysis showed that *T. b. rhodesiense* isolates from the Kenyan focus were unrelated to those from the Zambian focus (Figure 12.1) but were more closely related to *T. b. brucei* stocks isolated from Kenya.

Clearly, this study highlights the limitations of the classical taxonomy and suggests that the groups proposed by Hoare (1972) may be rather more genetically diverse than he supposed.

12.3.2 Does human infectivity have a monophyletic origin?

RFLP analysis of repetitive DNA sequences from *T. b. gambiense* stocks (Hide *et al.*, 1990) clearly showed the presence of at least two distinct types of unrelated human infective trypanosomes in West Africa. This, and the relationship of the non-gambiense group with *T. b. brucei* stocks isolated from cattle in Nigeria, suggest that the *T. b. gambiense* group has more than one origin and that the origins of the non-gambiense stocks may lie within *T. b. brucei* in West Africa.

Similarly, a clear distinction existed between the *T. b. rhodesiense* stocks from the Kenyan focus and those from the Zambian focus. As with the non-gambiense stocks, the Kenyan *T. b. rhodesiense* were more closely related to the *T. b. brucei* stocks isolated in Kenya than other *T. b. rhodesiense* stocks from Zambia. To investigate this further, a larger study was carried out with stocks from each area (Hide *et al.*, 1991). In this study, four collections of *T. b. rhodesiense* stocks were included: (1) stocks isolated in two separate villages in the Luangwa Valley focus (Zambia); (2) stocks isolated during the 1960 epidemic in Central Nyanza (Kenya); (3) stocks from the 1980 epidemic in the Busoga region of Uganda; (4) stocks isolated in the Tororo District of Uganda during the 1990 epidemic. The results showed that *T. b. rhodesiense* stocks from Zambia were unrelated to stocks from Kenya or Uganda. Furthermore, the two groups were as different from each other as they each were from *T. b. gambiense*. The stocks isolated from Uganda and Kenya fell into a single, homogeneous, group. These studies clearly showed that both *T. b. gambiense* and *T. b. rhodesiense* are each composed of at least two groups of strains with separate origins. Thus human infectivity has arisen more than once and is not a monophyletic trait. In terms of human disease epidemiology, this has important implications because it raises challenging questions about the acquisition of human infectivity: (1) How frequently does it arise? (2) From which strains or populations of trypanosomes does it arise? (3) What selective agents maintain it in the population? (4) How stable is the trait in the population? (5) Is it responsible for the generation of epidemics and the existence of foci of disease?

12.3.3 What is the composition of strains circulating during an epidemic of human sleeping sickness?

A recent human sleeping sickness epidemic in the Tororo District of southeast Uganda (1988–1992) provided a wealth of material from a series of villages within a 15 km radius. This collection, taken from humans, cattle, pigs and tsetse, opened up considerable possibilities for investigating the diversity and population genetics of trypanosome strains at the height of an epidemic. Hide *et al.* (1994), using RFLP analysis of repetitive DNA, analysed 88 stocks from this collection. A summary of the relationships is shown in the dendrogram in Figure 12.2.

Two basic groups of stocks were identified. One was a tightly clustered group of stocks which consisted of stocks from man, cattle and tsetse. All of these stocks were found to be resistant to human serum (i.e. human infective) and possessed isoenzyme profiles which had previously been associated with human infective trypanosomes. This group represents *T. b. rhodesiense*. Twenty-three per cent of the stocks isolated

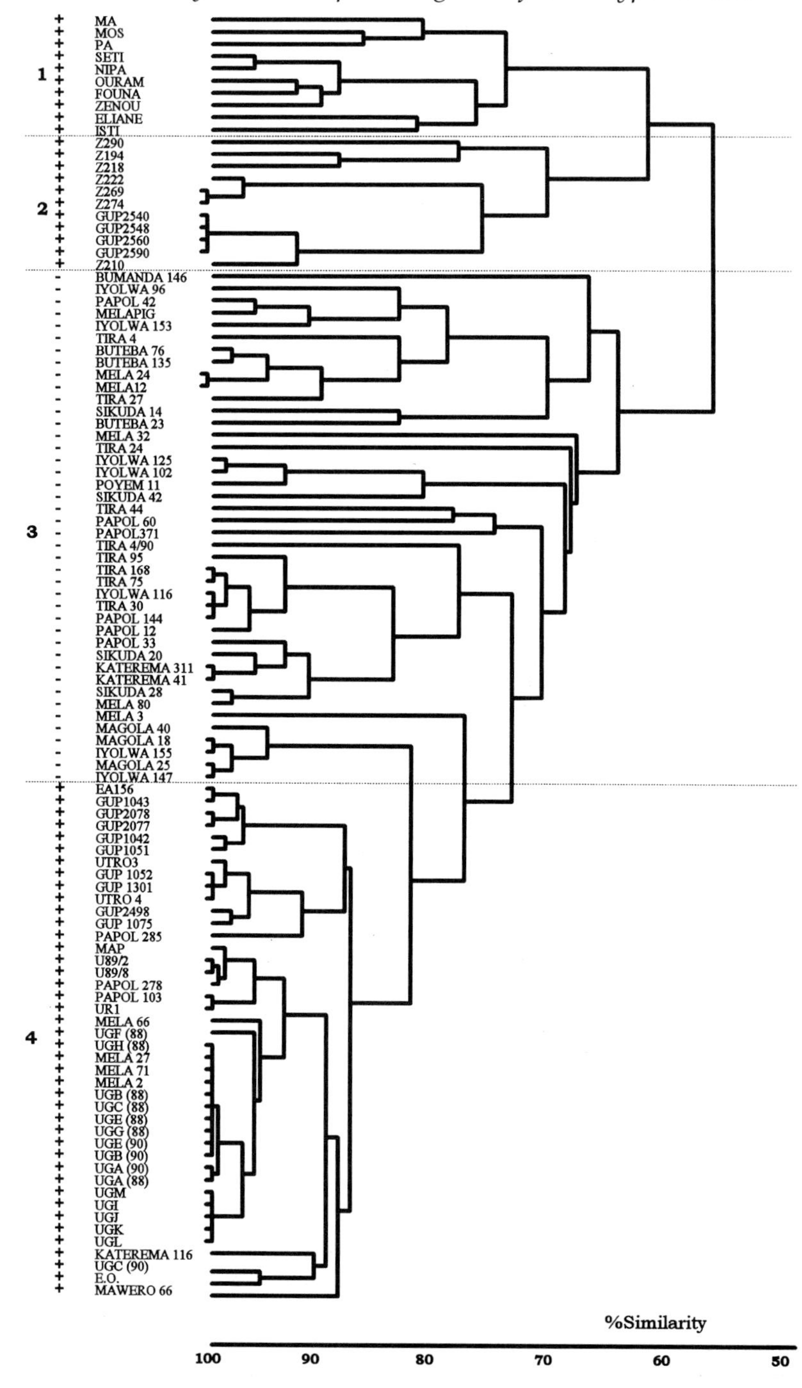

1
MA
MOS
PA
SETI
NIPA
OURAM
FOUNA
ZENOU
ELIANE
ISTI
2
Z290
Z194
Z218
Z222
Z269
Z274
GUP2540
GUP2548
GUP2560
GUP2590
Z210
3
BUMANDA 146
IYOLWA 96
PAPOL 42
MELAPIG
IYOLWA 153
TIRA 4
BUTEBA 76
BUTEBA 135
MELA 24
MELA12
TIRA 27
SIKUDA 14
BUTEBA 23
MELA 32
TIRA 24
IYOLWA 125
IYOLWA 102
POYEM 11
SIKUDA 42
TIRA 44
PAPOL 60
PAPOL371
TIRA 4/90
TIRA 95
TIRA 168
TIRA 75
IYOLWA 116
TIRA 30
PAPOL 144
PAPOL 12
PAPOL 33
SIKUDA 20
KATEREMA 311
KATEREMA 41
SIKUDA 28
MELA 80
MELA 3
MAGOLA 40
MAGOLA 18
IYOLWA 155
MAGOLA 25
IYOLWA 147
4
EA156
GUP1043
GUP2078
GUP2077
GUP1042
GUP1051
UTRO3
GUP 1052
GUP 1301
UTRO 4
GUP2498
GUP 1075
PAPOL 285
MAP
U89/2
U89/8
PAPOL 278
PAPOL 103
UR1
MELA 66
UGF (88)
UGH (88)
MELA 27
MELA 71
MELA 2
UGB (88)
UGC (88)
UGE (88)
UGG (88)
UGE (90)
UGB (90)
UGA (90)
UGA (88)
UGM
UGI
UGJ
UGK
UGL
KATEREMA 116
UGC (90)
E.O.
MAWERO 66
%Similarity
100
90
80
70
60
50

from cattle fell into this group showing that the domestic cattle in these Tororo villages were probably a significant reservoir host for *T. b. rhodesiense*.

The second group included a heterogeneous set of stocks which were all isolated from cattle, pigs and tsetse. None of them were human serum resistant and they all showed isoenzyme profiles previously associated with non-human infective trypanosomes. This group represents *T. b. brucei*. Previous analyses had failed to unambiguously distinguish *T. b. rhodesiense* from *T. b. brucei* (e.g. Tait *et al.*, 1985; Godfrey *et al.*, 1990) and had led to the notion that the *T. b. rhodesiense* was a host range variant of the latter. This study indicated that they were clearly quite distinct populations. Further evidence for this view is provided by the observation that the *T. b. rhodesiense* stocks isolated from the Tororo epidemic were closely related to stocks isolated from the 1960 epidemic in Central Nyanza (Kenya) and the 1980 epidemic in Busoga (Uganda) (Hide *et al.*, 1994). Thus stocks isolated 30 years previously from another part of the focus were more closely related to the Tororo human stocks than were the Tororo cattle stocks isolated from the same villages at the same time. The genetic distance (D) (Nei, 1972), calculated from isoenzyme data, showed that the *T. b. brucei* and *T. b. rhodesiense* populations from Tororo were more different from each other (D = 0.077; Hide *et al.*, 1994) than *T. b. brucei* populations from east and west Africa (*T. b. brucei* Uganda and Nigeria, D = 0.049; Tait *et al.*, 1985).

12.3.4 Stability of trypanosome strains over time

Hide *et al.* (1994) showed that the same *T. b. rhodesiense* genotype has been present in the 'Busoga' focus since the early 1960s. Thus the epi-

---

Figure 12.2 A dendrogram showing the relationships between African trypanosome stocks isolated during the height of a human sleeping sickness epidemic in Tororo, Uganda, in 1988–1990 (redrawn from Hide *et al.*, 1996). Trypanosome stocks are indicated by stock name (e.g. TIRA 24, where TIRA refers to the village of isolation) and full details of each stock can be found elsewhere (Hide *et al.*, 1990, 1991, 1994). Symbols (+) and (–) refer to human serum resistance or sensitivity, respectively. Groups 1 and 2 contain *T. b. gambiense* and *T. b. rhodesiense* (from the Zambian focus), respectively, for comparison purposes. Group 3 contains stocks isolated from domestic cattle and pigs during the epidemic. Group 4 contains human infective trypanosomes (*T. b. rhodesiense*) isolated from humans, cattle and tsetse during the epidemic and some older isolates from previous epidemics in Busoga, Uganda, in the 1980s and Central Nyanza, Kenya, in 1960.

demics in Central Nyanza (Kenya, 1960), Busoga (Uganda, 1980s) and Tororo (Uganda, 1988–1992) are almost certainly explained by the spread of the same parasite strain. This long term stability may explain the fact that human sleeping sickness is associated with disease foci which remain stable over many years. The question of when the Busoga focus *T. b. rhodesiense* strain arose is unclear but a recent re-evaluation of historical literature (Koerner, de Raadt and Maudlin, 1995) suggests that the same strain may have been responsible for the great epidemic in Busoga in 1900. This epidemic, which devastated Uganda with some 500 000 reported cases, was previously thought to be caused by *T. b. gambiense*. If the proposal of Koerner, de Raadt and Maudlin (1995) is correct then it would suggest a considerable degree of strain stability in natural populations.

12.3.5 The contribution of genetic exchange to diversity in trypanosome populations

In recent years much debate has surrounded the question of clonality and genetic exchange in *T. brucei*. On the one hand, genetic exchange has been predicted in natural populations (Tait, 1980) and demonstrated in the laboratory (Jenni *et al.*, 1986; Gibson, 1989; Tait and Turner, 1990) while on the other hand population genetic studies do not support the concept of random mating in natural populations (Tibayrenc, Kjellberg and Ayala, 1990; Stevens and Welburn, 1993; Mathieu-Daude *et al.*, 1995). However, other studies (Hide *et al.*, 1994) have demonstrated that genetic exchange is detectable in natural populations. The source of this controversy lies with four possible explanations: (1) Different populations of *T. brucei* show different degrees of clonality (Hide *et al.*,1994, 1996; Tibayrenc, 1995; Stevens and Tibayrenc, 1996). (2) Selection and/or restrictions in transmission cycles lead to an epidemic population structure (Maynard-Smith *et al.*, 1993; Hide *et al.*, 1994) dominated by the over-representation of a small number of strains. (3) Sampling bias: *T. brucei* population geneticists are in the habit of pooling isolates from different populations to maximise sample sizes. If such pools contain isolates which never have the opportunity of mating then linkage disequilibrium will occur. (4) Complex genetic mechanisms, perhaps involving selfing (Tait, *et al.*, 1996), may result in linkage disequilibrium. In conclusion, results show that *T. brucei* exhibits both genetic exchange and clonality depending on the population under study (Hide *et al.*, 1994, 1996; Tibayrenc, 1995; Stevens and Tibayrenc, 1996).

The existence of genetic exchange in *T. brucei* raises some important epidemiological and evolutionary issues. Firstly, does genetic exchange occur regularly between *T. b. brucei* and *T. b. rhodesiense*? If it did, a mechanism for the maintenance and spread of the human infectivity trait would be implied. Although successful laboratory crosses have been carried out between *T. b. rhodesiense* and *T. b. brucei* (Gibson, 1989), Hide *et al.* (1994) failed to find any evidence of recombination between *T. b. brucei* and *T. b. rhodesiense* in the field. Furthermore, the stability of the *T. b. rhodesiense* group of stocks with time and the homogeneity of RFLP patterns suggest that this does not occur. Secondly, does genetic exchange occur rarely between *T. b. brucei* and *T. b. rhodesiense*? Thirdly, does genetic exchange occur between strains of *T. b. brucei* resulting in the occasional generation of human serum resistant trypanosomes? These last two possibilities suggest mechanisms for the infrequent generation of new human infective strains. As yet these questions remain unanswered but offer hypotheses which can be tested.

12.3.6 The contribution of animal reservoirs to the maintenance of epidemics

In the Tororo District epidemic, 23% of domestic cattle were found to harbour human infective trypanosomes (Hide *et al.*, 1996). The importance of the cattle reservoir in this epidemic was investigated by comparing the probabilities of human–human and cattle–human transmission (Hide *et al.*, 1996). Figure 12.3 summarizes these probabilities and shows that humans are five times more likely to become infected with trypanosomes from cattle than from other humans. Thus the animal reservoir is a very significant factor in generating and maintaining epidemics.

12.3.7 The genotypic composition of strains circulating within an individual animal

The important involvement of the domestic animal reservoir in the overall diversity of circulating trypanosome strains raises the issue of the diversity of strains within a single animal. To date this has not been investigated. However, using isolates taken from cattle and pigs, it has been shown that identical genotypes can be recovered from a single animal throughout a three month time period (Hide *et al.*, 1998, Angus and Welburn, unpublished) suggesting that one predominant strain is found in a single animal.

12.3.8 The diversity of trypanosome strains in an endemic area when compared with an epidemic area

Little has been done to examine the diversity of strains circulating in endemic areas where the prevalence of human sleeping sickness is low. The Busia region of Kenya offers opportunities to investigate this

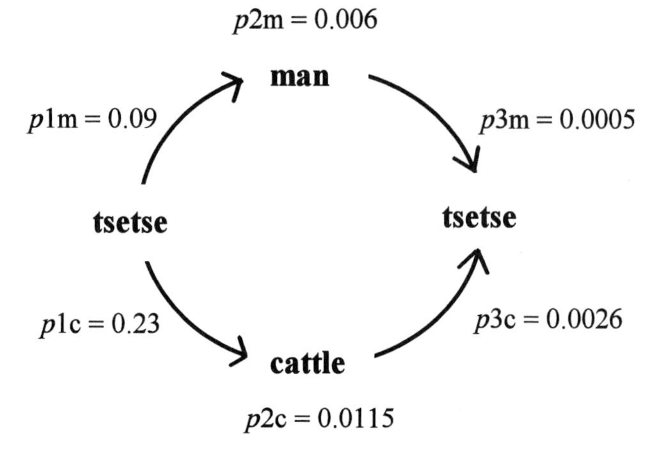

Figure 12.3 The probability of transmission of human infective trypanosomes by tsetse feeding on cattle (*p*3c) is five times that of tsetse feeding on man (*p*3m). Using data obtained during the Tororo epidemic (Hide *et al.*, 1996), the probability of tsetse feeding on man or cattle was estimated by bloodmeal analysis (*p*1m and *p*1c, respectively) and the frequency of infected humans or cattle in the population was measured (*p*2m or *p*2c*, respectively); The frequency of cattle infected with human infective trypanosomes (23%) can be calculated (*p*2c = *p*2c* × 23%). From these data, the probability of a tsetse carrying a *T. b. rhodesiense* infection derived from cattle, *p*3c (= *p*1c × *p*2c), as opposed to being derived from man, *p*3m (= *p*1m × *p*2m), is five times greater.

question further. Located only 30 km from Tororo, it had only seven cases of human sleeping sickness reported in 1991 compared with 350 in neighbouring Tororo. Analysis suggests that *T. b. rhodesiense* strains are circulating in the domestic cattle population there (Hide *et al.*, 1998) and, therefore, that low level endemicity is not caused by an absence of appropriate strains in the animal reservoir.

12.4 A MODEL FOR THE ORIGINS AND MAINTENANCE OF HUMAN SLEEPING SICKNESS FOCI AND EPIDEMICS

A model for the origins and maintenance of human sleeping sickness foci can be proposed. (1) Infrequently, throughout Africa, human infective trypanosomes arise from the background population of *T. b. brucei*. These mutants could arise in domestic animal, wild animal or tsetse fly populations. Such mutants can be readily selected in the laboratory (Hadjuk, Hager and Esko, 1992). (2) The genotype for human infectivity is propagated at low frequencies through the animal and tsetse populations by either genetic exchange or genetic drift. (3) Once a transmission cycle involving humans is established, selection favours retention of the human infectivity trait in the population. (4) Further selection, results in the generation of a localized human infective strain which becomes increasingly less able to recombine with non-human infective strains due to the exclusion of human serum sensitive strains from transmission cycles involving the human host. Thus, a speciation event begins. (5) Under certain socio-ecological conditions transmission of this strain increases. The locality is hit by a human sleeping sickness epidemic which is maintained by the domestic animal reservoir. Gene frequencies show considerable linkage disequilibrium due to the over-representation of certain genotypes, caused by a rapid population expansion, and resulting in an epidemic population structure. This was probably the situation occurring in Tororo. (6) The epidemic population becomes clonal due to selection through human transmission cycles during the epidemic. (7) In the long-term co-evolution occurs between the strain and the human host population such that the parasite appears to become less virulent. This may reflect the situation seen with Zambian *T. b. rhodesiense* and to a greater extent with *T. b. gambiense.*

This model provides a framework which could explain the origins of human sleeping sickness but many questions remain unanswered. DNA sequencing may provide the means to investigate the validity of this model as it affords the unique opportunity of using phylogenetic approaches to directly trace the origins of human strain groups, measure rates of evolution, infer times of divergence and track the spread of specific parasite strain groups. Furthermore, such an approach will allow us to develop a taxonomic structure, based on evolutionary principles, which will be of use to clinicians, taxonomists and epidemiologists alike.

Acknowledgements

I thank the Wellcome Trust for financial support and colleagues who have contributed much to the work reported here: Sue Welburn, Stephen Angus, Ian Maudlin and Andy Tait.

12.5 REFERENCES

Gibson, W.C. (1986) Will the real *Trypanosoma b. gambiense* please stand up. *Parasitology Today*, **2**, 255–7.

Gibson, W.C. (1989) Analysis of a genetic cross between *Trypanosoma brucei rhodesiense* and *T. b. brucei. Parasitology*, **99**, 391–402.

Gibson, W.C., Marshall, T.F. de C. and Godfrey, D.G. (1980) Numerical analysis of enzyme polymorphism. A new approach to the epidemiology and taxonomy of trypanosomes of the genus *Trypanozoon*. *Advances in Parasitology*, **18**, 175–245.

Godfrey, D.G., Baker, R.D., Rickman, L.R. and Mehlitz, D. (1990) The distribution, relationships and identification of enzymic variants within the subgenus *Trypanozoon*. *Advances in Parasitology*, **29**, 1–74.

Hadjuk, S.L., Hager, K. and Esko, J.D. (1992) High-density lipoprotein-mediated lysis of trypanosomes. *Parasitology Today*, **8**, 95–8.

Heisch, R.B., McMahon, J.P. and Manson-Bahr, P.E.C. (1958) The isolation of *Trypanosoma rhodesiense* from bushbuck. *British Medical Journal*, **2**, 1203.

Hide, G. (1988) *Variation in repetitive DNA in African trypanosomes*. PhD thesis, University of Edinburgh.

Hide, G., Cattand, P., Le Ray, D. *et al.* (1990) The identification of *T. brucei* subspecies using repetitive DNA sequences. *Molecular and Biochemical Parasitology*, **39**, 213–26.

Hide, G. and Tait, A. (1991) The molecular epidemiology of parasites. *Experientia*, **47**, 128–42.

Hide, G., Buchanan, N., Welburn, S.C. *et al.* (1991) *Trypanosoma brucei rhodesiense*: Characterisation of stocks from Zambia, Kenya and Uganda using repetitive DNA probes. *Experimental Parasitology*, **72**, 430–9.

Hide, G., Tait, A., Maudlin, I. and Welburn, S.C. (1994) Epidemiological relationships of *Trypanosoma brucei* stocks from South East Uganda: Evidence for different population structures in human and non-human trypanosomes. *Parasitology*, **109**, 95–111.

Hide, G., Tait, A., Maudlin, I. and Welburn, S.C. (1996) The origins, dynamics and generation of *Trypanosoma brucei rhodesiense* epidemics in East Africa. *Parasitology Today*, **12**, 50–5.

Hide, G. (1996a) The molecular epidemiology of trypanosomatids, in *Trypanosomiasis and Leishmaniasis: Biology and Control* (eds G. Hide, J.C. Mottram, G.H. Coombs and P.H. Holmes), British Society for Parasitology / CAB International, Oxford, 289–303.

Hide, G. (1996b) The molecular identification of trypanosomes. *Methods in Molecular Biology*, **50**, 243–63.

Hide, G., Angus, S.B., Holmes, P.H. *et al.* (1998) *Trypanosoma brucei*: comparison of circulating strains in an endemic and an epidemic area of a sleeping sickness focus. *Experimental Parasitology* (in press).

Hoare, C.A. (1972). *The Trypanosomes of Mammals*. Blackwell Scientific Publications, Oxford.

Jenni, L., Marti, S., Schweizer, J. *et al.* (1986) Hybrid formation between African trypanosomes during cyclical transmission. *Nature*, **322**, 173–5.

Koerner, T., de Raadt, P. and Maudlin, I. (1995) The 1901 Uganda sleeping sickness epidemic revisited – a case of mistaken identity? *Parasitology Today*, **11**, 303–6.

Mathieu-Daude, F., Stevens, J., Welsh, J. *et al.* (1995) Genetic diversity and population structure of *Trypanosoma brucei*: clonality versus sexuality. *Molecular and Biochemical Parasitology*,. **72**, 89–101.

Maynard-Smith, J., Smith, N.H., O'Rourke, M. and Spratt, B.G. (1993) How clonal are bacteria? *Proceedings of the National Academy of Sciences USA*, **90**, 4384–8.

Nei, M. (1972) Genetic distance between populations. *American Naturalist*, **106**, 283–92.

Paindavoine, P., Pays, E., Laurent, M. *et al.* (1986) The use of DNA hybridization and numerical taxonomy in determining relationships between *T. brucei* stocks and subspecies. *Parasitology*,. **92**, 31–50.

Stevens, J.R. and Welburn, S.C. (1993) Genetic processes within an epidemic of sleeping sickness. *Parasitology Research*, **79**, 421–7.

Stevens, J.R. and Tibayrenc, M. (1995) Detection of linkage disequilibrium in *Trypanosoma brucei* isolated from tsetse flies and characterised by RAPD analysis and isoenzymes. *Parasitology*, **110**, 181–6.

Stevens, J.R. and Tibayrenc, M. (1996) *Trypanosoma brucei* s.l.: evolution, linkage and the clonality debate. *Parasitology*, **112**, 481–8.

Tait, A. (1980) Evidence for diploidy and mating in trypanosomes. *Nature*, **287**, 536–8.

Tait, A., Babiker, E.A. and Le Ray, D. (1984) Enzyme variation in *Trypanosoma brucei* spp. I. Evidence for the subspeciation of *Trypanosoma brucei gambiense*. *Parasitology*, **89**, 311–26.

Tait, A., Barry, J.D., Wink, R. *et al.* (1985) Enzyme variation in *Trypanosoma brucei* ssp II. Evidence for *T. b. rhodesiense* being a subset of variants of *T. b. brucei*. *Parasitology*, **90**, 89–100.

Tait, A. and Turner, C.M.R. (1990) Genetic exchange in *Trypanosoma brucei*. *Parasitology Today*, **6**, 70–5.

Tait, A., Buchanan, N., Hide, G. and Turner, C.M.R. (1996) Evidence for self-fertilisation in *T. brucei*. *Molecular and Biochemical Parasitology*, **76**, 31–42.

Tibayrenc, M., Kjellberg, F. and Ayala, F.J. (1990) A clonal theory of parasitic protozoa: the population structures of *Entamoeba, Giardia, Leishmania, Naegleria, Plasmodium, Trichomonas* and *Trypanosoma* and their medical and taxonomic consequences. *Proceedings of the National Academy of Sciences USA*, **87**, 2414–18.

Tibayrenc, M. (1995) Population genetics of parasitic protozoa and other microorganisms. *Advances in Parasitology*, **36**, 47–115.

13

Organelle and enzyme evolution in trypanosomatids

Fred R. Opperdoes, Paul A.M. Michels, Christiane Adje and Véronique Hannaert

ICP–TROP 74.39, Avenue Hippocrate 74–75, B–1200 Brussels. E-mail: opperdoes@trop.ucl.ac.be

ABSTRACT

The Euglenozoa, comprising the Euglenida and Kinetoplastida (with the trypanosomatids and the bodonids) represent the first organisms that adapted to the appearance of atmospheric oxygen by the acquisition of mitochondria and catalase-containing microbodies (peroxisomes and glycosomes). It is generally accepted that the mitochondrion as well as the chloroplast (which is present in the euglenids) are the remnants of, respectively, an alpha-proteobacterial and a cyanobacterial endosymbiont that invaded the primitive ancestral eukaryotic cell. However, the evolutionary origin of the other subcellular organelle, the microbody, so far has remained obscure. Nevertheless, the fact that microbodies (peroxisome, glyoxysome or glycosome) were acquired at the same time as mitochondria and plastids may be indicative of an endosymbiotic origin of this cell organelle as well. Most of the genes coding for the glycolytic enzymes that are present in the glycosomes of the Trypanosomatidae have now been cloned and sequenced. Contrary to expectation, all but one of the enzymes have typical eukaryotic features and branch off from their respective trees at positions in agreement with the evolutionary position of the Kinetoplastida. Moreover, a study on the evolution of the glyceraldehyde-3-phosphate dehydrogenase genes in the Kinetoplastida suggests that in the course of its evolution some glycolytic enzymes may have been transferred from the cytosol to the glycosome. This observation will certainly complicate any future interpretation of glycosome evolution. Another complicating factor is the high tendency of the trypa-

Evolutionary Relationships Among Protozoa. Edited by G.H. Coombs, K. Vickerman, M.A. Sleigh and A. Warren. Published in 1998 by Chapman & Hall, London. ISBN 0 412 79800 X

nosomatid genome for undergoing gene conversion by homologous recombination. In the case of phosphoglycerate kinase (PGK), two genes now coding for two glycosomal PGKs have been created by the duplication of a gene that coded for a cytosolic isoenzyme. So far, only the glycosomal glycerol-3-phosphate dehydrogenase carries typical prokaryotic features and phylogeny has revealed that the enzyme clusters with its prokaryotic rather than with its eukaryotic homologues. The gene coding for the microbody enzyme thiolase has undergone an early gene duplication leading to the formation of acetoacetylCoA and a 3-ketoacylCoA thiolases. This has allowed us to unambiguously root the thiolase tree. The 3-ketoacylCoA thiolase branch of the tree provides evidence for two events of horizontal gene transfer. A prokaryotic gene that entered the ancestral eukaryotic cell gave rise to a mitochondrial thiolase; a second event of horizontal gene transfer coming from another prokaryotic organism led to the incorporation of a separate thiolase isoenzyme into peroxisomes. Finally, in the case of the enzyme catalase, the glyoxysomal catalase of plants on the one hand and the peroxisomal catalase of animals and fungi, on the other hand, are not related by a conventional descent. At least one of these groups of catalases must have entered the eukaryotic cell through an event of horizontal gene transfer. The intriguing question as to whether all these observations are indicative of an endosymbiotic origin of the microbody is discussed.

13.1 INTRODUCTION

A great variety of protists appear to lack mitochondria, peroxisomes and plastids. Some major taxa such as the metamonads, the microsporidians and the archamoebae have been grouped in the superkingdom Archezoa, because the absence of these organelles is interpreted as a primitive trait (Cavalier-Smith, 1987). They are considered the most primitive eukaryotic cells known to date and they are essentially anaerobes without facilities to respire oxygen. In all other eukaryotic taxa mitochondria and peroxisomes can be found, although in some lineages such organelles may have been lost secondarily. The distribution of mitochondria and peroxisomes over the various taxonomic groups suggests that these organelles have originated after the separation of the Archezoa from the main line of eukaryotic evolution, but before the development of the lineage leading to the Euglenozoa. The latter group includes the photosynthetic and heterotrophic euglenoids and the heterotrophic kinetoplastids (comprising the families Trypanosomatidae and Bodonidae). The monophyletic assemblage of the Euglenozoa (Patterson and Sogin, 1993), therefore, most likely represents the first lineage that has adapted to the appearance of atmospheric oxygen by the acquisition of both mitochondria and peroxisomes. The euglenoids, but not the kinetoplastids, probably acquired plastids at a later stage in evolution from an endosymbiotic event involving a primitive chloroplast-bearing eukaryote (Gillman, 1994).

It is generally accepted that both mitochondrion and chloroplast are remnants of respectively an α-proteobacterial and a cyanobacterial endosymbiont that have invaded the primitive ancestral eukaryotic cell. The evolutionary origin of the peroxisome, the third major cell organelle of eukaryotes has, so far, remained obscure. However, the fact that peroxisomes, like plastids and mitochondria, multiply by growth and binary fission and were acquired at approximately the same time as the mitochondrion, may be indicative of an endosymbiotic origin of peroxisomes as well.

Glycosomes are the peroxisomes typical of the order Kinetoplastida (Opperdoes, 1987). These organelles are found both in the Bodonidae and the Trypanosomatidae, but not in organisms belonging to any other lineage. The most prominent feature of glycosomes is that they contain the majority of the enzymes of the glycolytic pathway (Opperdoes, 1987, 1988; Opperdoes and Michels, 1993; Hannaert and Michels, 1994), hence their name. Glycosomes are surrounded by a single unit membrane, lack any trace of DNA, RNA and ribosomes, vary from 0.2–1.0 μm in diameter and may, or may not, contain catalase.

Despite the diversity in enzyme content of the various members of the peroxisome family, we consider a monophyletic origin for these organelles most plausible (Opperdoes and Michels, 1993; Michels and Hannaert, 1994). This conclusion is based on the fact that some enzymes are shared by all peroxisomes, on their similar morphology, on the route of biogenesis and, most importantly, on the observation that all peroxisomes (including glyoxysomes and glycosomes) have their proteins imported with the help of several similar import signals. These are peroxisomal targeting signal-1 (PTS-1), i.e. a C-terminal tripeptide comprising the amino acids -SKL or permutations thereof (Gould *et al.*, 1989; Swinkels, Gould and Subramani, 1992, and PTS-2, i.e. a conserved motif within a nonapeptide located near the N-terminus of the proteins (Swinkels *et al.*, 1991). Moreover, the proteins involved in the import machinery in peroxisomes of widely different organisms appear to be homologous (Subramani, 1996).

Several scenarios can be put forward to explain the origin of peroxisomes. The organelle may have originated by budding from a membrane within the ancient unicellular eukaryote, and engulfing part of the cell content. The newly formed vesicle would subsequently have evolved into a functional organelle. This would not explain why peroxisomes multiply independently and by binary fission. Second, peroxisomes usually contain large parts of functional pathways, or even entire pathways (Borst, 1989). It is difficult to imagine that a complete set of enzymes would be present and functional at once in a newly formed vesicle (Borst and Swinkels, 1989; Michels and Opperdoes, 1991). Successive transfer of individual enzymes of a pathway to the newly formed organelle is equally unlikely. Intermediate steps would not be an advantage, but rather a burden to the cell. We, therefore, prefer the alternative hypothesis (de Duve, 1982; Borst, 1989; Cavalier-Smith, 1990) that peroxisomes are derived from an endosymbiont. This would imply that the pathways would have been present from the beginning and that the present peroxisome diversity is the result of loss of enzymes and complete pathways rather than their acquisition. In addition to the loss of individual enzymes the entire genome would have been lost from the organelle, while some useful genes would have been transferred to the host nucleus. Of course such a scenario would not exclude the possibility that certain enzymes may have been acquired by the organelle later in evolution.

In the case of the latter scenario it might be expected that some of the enzymes that were originally present in the endosymbiont and that were not lost from the resulting organelle during evolution must have retained some of their prokaryotic characteristics. Amino-acid sequence analysis of peroxisomal proteins might thus reveal such an endosymbiotic ancestry. By now many of such sequences have become available, but in most cases there is no cytosolic counterpart available for comparison. In the case of the glycolytic enzymes present in the glycosome such an analysis can be made however. Both prokaryotes and eukaryotes have glycolytic enzymes and in eukaryotes the latter are usually cytosolic.

13.2 GLYCOLYTIC ENZYMES OF THE KINETOPLASTIDA

Glycolytic enzymes are present in all types of cell, both prokaryotic and eukaryotic, where they are usually present in the cytosol, while in the trypanosomatids the enzymes are sequestered in the glycosomes (Opperdoes and Borst, 1977). Moreover, trypanosomatids have for some of their glycosomal enzymes a cytosolic counterpart as well. Thus it was reasoned that if the peroxisome would have originated from a prokaryotic endosymbiont a clear difference in amino-acid sequence between the glycosomal enzymes and their cytosolic counterparts from either trypanosomatids or eukaryotes would become apparent. For this purpose, we have cloned and sequenced most of the genes coding for the glycosomal enzymes together with some of their cytosolic counterparts from both trypanosomatids and a related bodonid organism *Trypanoplasma borreli*.

Most of the glycosomal enzymes have been relatively well conserved throughout evolution and have identities that range between 45 and

55% when compared with homologous sequences from either prokaryotic or eukaryotic organisms. Where the distances of the trypanosomatid sequences from both eukaryotes and prokaryotes are more or less equal, such analyses do not always allow clear predictions as to the evolutionary origin (prokaryotic endosymbiont or eukaryotic invention) of glycosomes in these organisms (Opperdoes and Michels, 1993; Michels and Hannaert, 1994). However, the use of phylogenetic inference programs and a detailed inspection of the sequences themselves has revealed that some of the glycolytic enzymes must have had a clear eukaryotic origin (Figure 13.1). Typical eukaryotic signatures are the S-loop domain in glyceraldehyde-3-phosphate dehydrogenase (GAPDH) (Hannaert *et al.*, 1992) (Figure 13.2) and the C-terminus of triosephosphate isomerase (TIM) (Swinkels *et al.*, 1986). Other peroxisomal enzymes, however, branch with their prokaryotic homologues and carry some typical prokaryotic features instead (see below).

13.2.1 Phosphoglucose isomerase, aldolase and triosephosphate isomerase

The genes coding for phosphoglucose isomerase (PGI), aldolase and TIM have been cloned and sequenced (Marchand *et al.*, 1988, 1989; Swinkels *et al.*, 1986). In *Trypanosoma brucei* evidence for the presence of only single genes for PGI and TIM was found, while the gene coding for aldolase is present in two identical copies in the form of a tandem repeat. For the PGI (Nyame *et al.*, 1994) and TIM (Kohl *et al.*, 1994) genes in *Leishmania mexicana*, a situation similar to that in *T. brucei* has been observed. The respective genes for PGI and TIM each encode a glycolytic enzyme of typical eukaryotic nature, as follows from the presence of typical sequence motifs (Borst, 1989) and the branching observed for the trypanosomatid lineage in the respective phylogenetic trees for PGI (not shown) and TIM (Figure 13.1). For the class-I aldolase of *T. brucei* there is no prokaryotic homologue available for comparison.

In *T. brucei* bloodstream forms the enzymes PGI, aldolase and TIM are predominantly found in the glycosome. PGI carries at its C-terminus a typical peroxisomal targeting signal (PTS-1), responsible for its import into the glycosome, while aldolase carries a PTS-2 type signal (Blattner, Dörsam and Clayton, 1995). The import signal of TIM remains to be identified and this is at the moment under investigation. In procyclic insect stages less than 50% of the activities of both PGI and TIM are found associated with the glycosomes and the remainder is found in the cytosol (Opperdoes, Markos and Steiger, 1981). In *L. mexicana* less than 10% of the PGI and about 50% of the TIM activity is associated with the glycosomes whereas the remainder is found in the

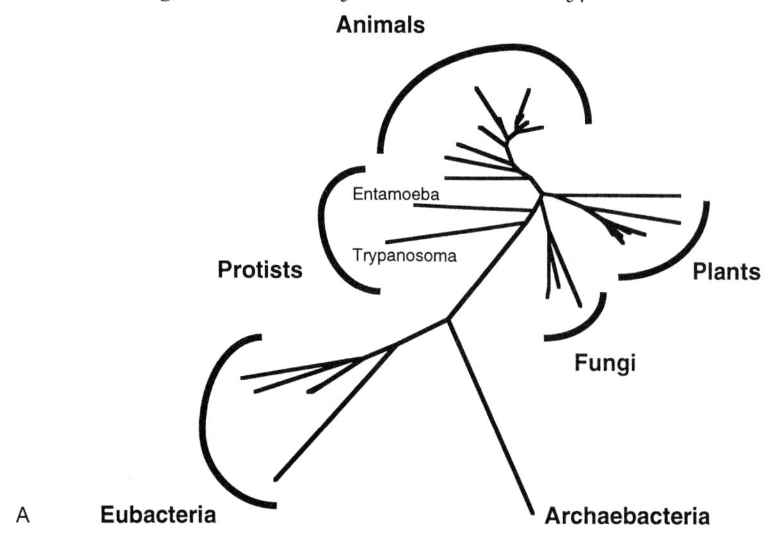

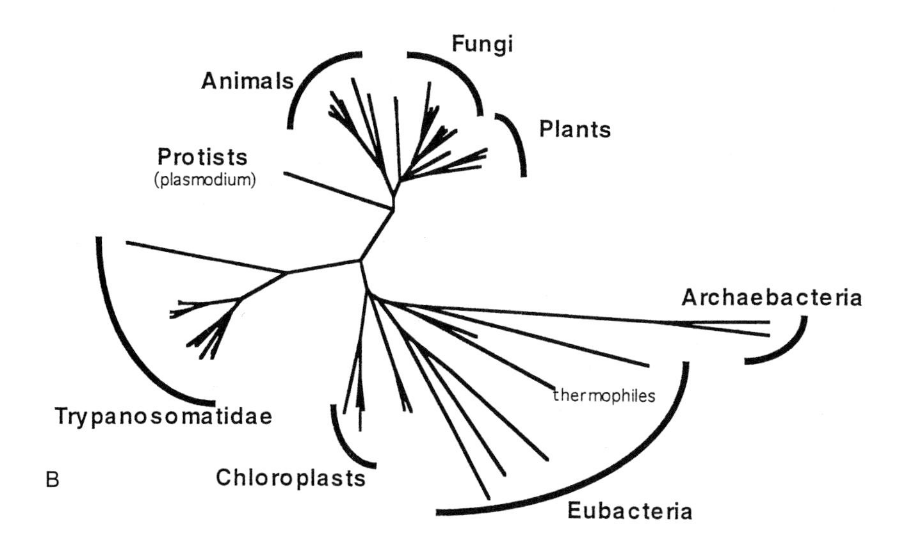

Figure 13.1 Comparison of unrooted phylogenetic trees based on the amino-acid sequences of three glycolytic enzymes: enolase (a cytosolic enzyme in *Trypanosoma brucei*), phosphoglycerate kinase and triosephosphate isomerase (both glycosomal in *T. brucei*). In all three cases the trypanosomatids take branching positions close to the other protists and that are intermediate between the other eukaryotes, such as the animals, fungi and plants on one side and the prokaryotes on the other side. (A) Tree based on partial enolase sequences. The position of the two available protists *T. brucei* and *Entamoeba histolytica* is indicated.

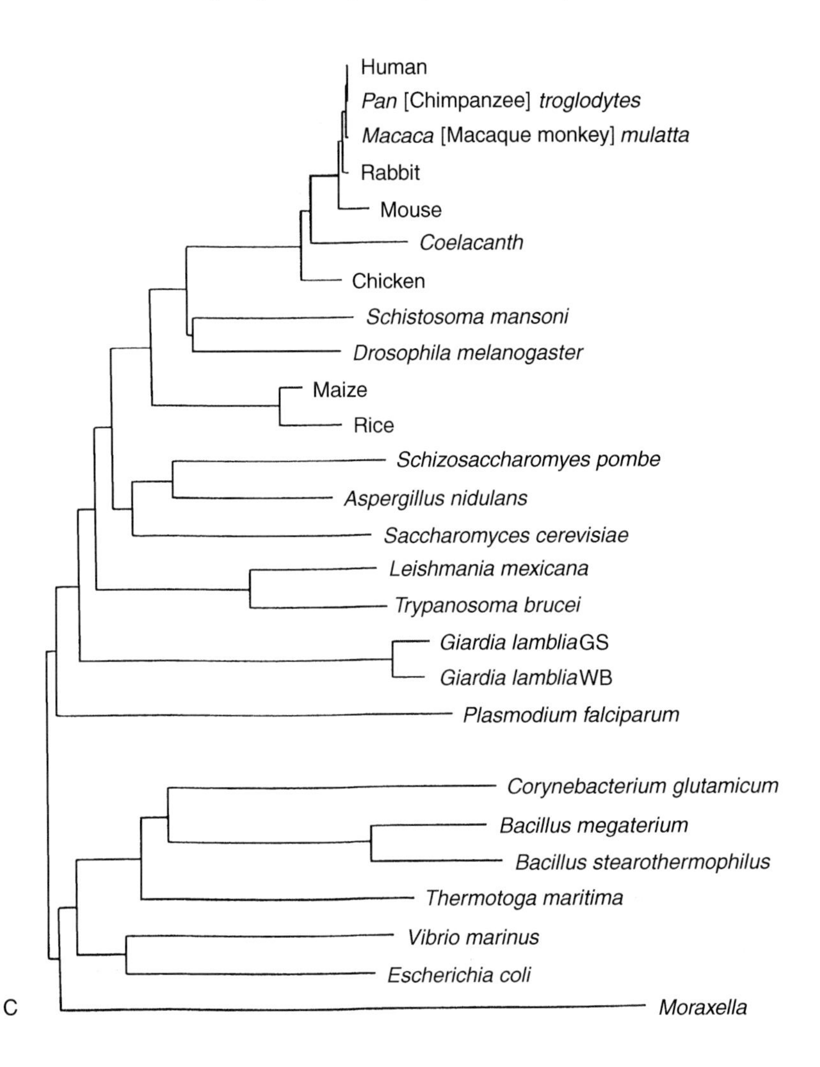

(B) Tree based on all available prokaryotic, protist, fungal, animal and plant sequences of phosphoglycerate kinase. (C) Tree based on a selected number of prokaryotic and eukaryotic triosephosphate isomerase sequences. Sequences were taken from the SwissProt database, version 32 and aligned. Trees were made using the Neighbour-joining method. For enolase only partial sequences were used. Insertions or deletions were not included in the analysis.

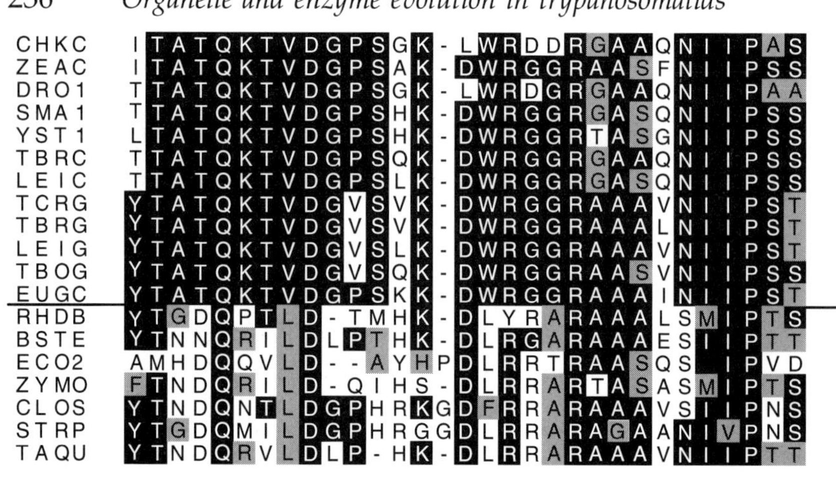

Figure 13.2 Alignment of the S-loop domain of a number of eukaryotic and prokaryotic glyceraldehyde-3-phosphate dehydrogenases. Residues that are identical in more than 50% of cases have been shaded. The horizontal line represents the separation between eukaryotes (above) and prokaryotes (below). According to this analysis the glycosomal sequences have a typical eukaryotic signature. TBRC, LEIC, EUGC are the cytosolic isoenzymes and TBRG, LEIG, TBOG the glycosomal isoenzymes of *T. brucei, L.mexicana, T. borreli* and *E.gracilis*, respectively.

soluble fraction (Nyame *et al.*, 1994; Kohl *et al.*, 1994). It is not clear in this case how a single enzyme would distribute between two cell compartments. Both *Leishmania* and *Trypanosoma* PGI have C-terminal tripeptides that have been shown to be very effective in targeting reporter proteins to glycosomes of *T. brucei* (Blattner *et al.*, 1992; Sommer *et al.*, 1992). The fact that the extent of differential targeting varies with the life-cycle stage, suggests that we are dealing here with a physiologically important mechanism (Nyame *et al.*, 1994; Michels and Hannaert, 1994).

13.2.2 Glyceraldehyde-3-phosphate dehydrogenase

A study on the molecular evolution of the GAPDH genes in the Kinetoplastida suggested that in the course of evolution the enzyme GAPDH may have been transferred from the trypanosomatid's cytosol to its glycosome (Wiemer *et al.*, 1995). The euglenoid *Euglena gracilis* has one typical euglenozoan GAPDH gene, that encodes a cytosolic isoenzyme (Henze *et al.*, 1995). *Euglena* does not have any GAPDH (or other glycolytic enzyme) associated with its peroxisomes (Opperdoes *et al.*, 1988). In the bodonid *T. borreli* a similar situation is encountered: only a single GAPDH gene can be detected. However, in this organism GAPDH activity is present in both the cytosol and in the glycosome. These activities could be attributed to a single enzyme that is present in the two compartments (Wiemer *et al.*, 1995). In addition this GAPDH has now acquired at its C-terminal end a typical peroxisomal targeting signal (-ARL). This situation is reminiscent of that observed for PGI in the trypanosomatids (see above).

In three members of the trypanosomatid family, i.e. *T. brucei, Trypanosoma cruzi* and *L. mexicana*, the corresponding GAPDH also carries the same, or a very similar, C-terminal targeting signal, but is exclusively glycosomal (Kendall *et al.*, 1990; Michels *et al.*, 1991, Hannaert *et al.*, 1992). Interestingly, in trypanosomatids a second gene, which is only distantly related to the glycosomal GAPDH and which codes for a cytosolic isoenzyme, is present (Michels *et al.*, 1991; Hannaert *et al.*, 1992). This gene was most likely acquired by an event of horizontal gene transfer. This has enabled the Trypanosomatidae to develop isoenzymes specialized for functioning in each of the two compartments. The original enzyme became devoted to glycolysis in the glycosome, the newly acquired enzyme evolved to perform other functions in the cytosol (Wiemer *et al.*, 1995; Michels and Hannaert, 1994).

These observations do not provide us with information as to the origin of the glycosome and its early evolution. However, our analyses of the kinetoplastid GAPDH do allow the conclusion that an originally cytosolic enzyme can become fully compartmentalized during evolution.

13.2.3 The evolution of the phosphoglycerate kinase genes of Trypanosomatidae

Another complicating factor in our search for the origin of the glycosome is the high tendency of trypanosomatid genomes to undergo gene conversion (Michels, 1987; Le Blancq *et al.*, 1988). This is clearly demonstrated in the case of phosphoglycerate kinase (PGK). At least two members of the genus *Trypanosoma* (i.e. *T. brucei brucei*, Osinga *et al.*, 1985 and *T. congolense*, Parker *et al.*, 1995), as well as another trypanosomatid, *Crithidia fasciculata* (Swinkels, Evers and Borst, 1988), all have three PGK genes: pgkA, pgkB and pgkC. The genomic organization of the three PGK genes in *T. brucei* is shown in Figure 13.3. These genes encode respectively a glycosomal PGK isoenzyme of 56.000 molecular mass (PGK56), a cytosolic isoenzyme (PGK45) and a second glycosomal isoenzyme (PGK47). Genes B and C are highly similar (72–100%), whereas gene A is less similar (53–84%) to B and C. In *Leishmania* the situation is somewhat different; no evidence for the presence of a pgkA

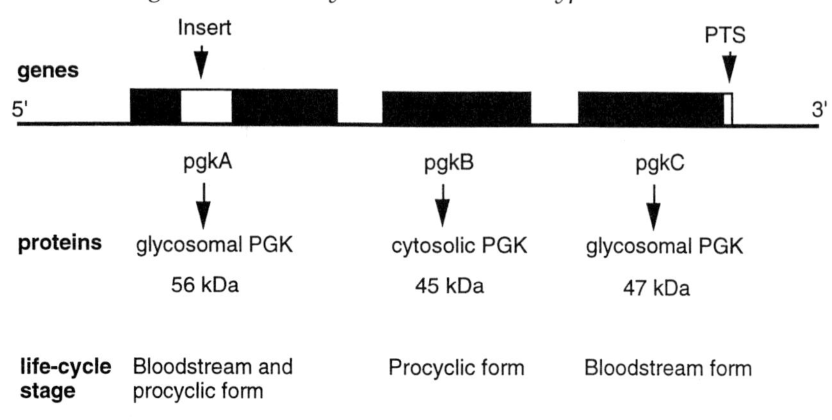

Figure 13.3 Tandem array of PGK genes in *T. brucei.*

gene was found in any of the members of this genus (i.e. *L. mexicana* and *L. donovani*, this study, and *L. major* (D. Hart, personal communication)).

Several evolutionary scenarios can be put forward to explain this important difference between two, so closely related, members of the same trypanosomatid family, as are *Crithidia* and *Leishmania*. The most likely scenario would be the one where in a trypanosomatid ancestor with a single gene for PGK, gene duplications have resulted in the formation of three tandemly linked genes: respectively pgkA, pgkB and pgkC. When the various trypanosomatid genera separated from each other, all three genes were propagated to the genera *Trypanosoma* and *Crithidia*, while in the genus *Leishmania* the pgkA gene was lost again. Any other scenario is less parsimonious in that it would require many parallel gene duplications leading to the independent formation of the genes A, B and C.

To discriminate between the different scenarios, we have attempted to retrace the evolution of the pgk genes by using the sequence information for the corresponding proteins. Firstly, all the aligned full-length PGK sequences were used for the creation of an unrooted phylogenetic tree using the neighbour-joining method. Figure 13.1B shows that the trypanosome PGKs constitute a monophyletic group that clusters nicely between the bacterial and chloroplast PGKs on one side and the eukaryotic PGKs on the other side. This branching pattern is similar to that observed for the cytosolic enzyme enolase (Figure 13.1A). A subset of sequences containing the trypanosomatid PGKs, together with an appropriate outgroup, was then used for the construction of a bootstrap consensus tree (Figure 13.4A). Because this tree is rooted by the inclusion of the *Plasmodium falciparum* PGK, it could be inferred that an event of gene duplication, leading to the formation, in *Crithidia*, of a separate pgkA gene, must have preceded the separation of the various trypanosomatid species. In addition the tree suggests a number of separate gene-duplication events leading to the formation of B and C genes in *Crithidia* and *Leishmania* and of A, B and C genes in the genus *Trypanosoma*. However, the nodes leading to the various *Trypanosoma* PGKs are not well supported by bootstrap analysis. This might be taken as indication that different PGK domains in the protein in this part of the tree have been subject to different rates of evolution.

For a closer analysis of this possibility, the N-terminal domain of the PGKs in front of the 80 amino-acid insertion in the protein encoded by the pgkA gene, as well as the domain C-terminal of the same insertion, were used to create two new sets of aligned N- and C-terminal domain sequences. The C-terminal extensions, only present in the glycosomal proteins encoded by the pgkC genes, were excluded from the analyses. Neighbour-joining and bootstrap analysis carried out on these two sets of sequences gave strikingly different topologies and support for robustness of the trees. The tree obtained for the C-terminal domain (Figure 13.4C) indicates that within a single species the proteins encoded by the A, B and C genes are more closely related to each other than to their autologous counterparts present in the other species. On the contrary, the tree obtained for the N-terminal PGK domains (Figure 13.4B) suggests an early event of gene duplication resulting in the formation of a pgkA gene prior to the separation of the different trypanosomatid genera.

Our interpretation for these two completely different tree topologies is as follows. Because of the rooting of the tree it can be concluded that there must have occurred two gene duplications in trypanosomatid ancestors, resulting in the creation of separate pgkA, pgkB and pgkC genes. The pgkA gene then followed a normal evolution from ancestral to, respectively, crithidial and trypanosomal pgkA, while the B and C genes seem to have been created by gene duplications in each (sub)species independently. Since this is highly unparsimonious and thus less likely to be true, we tend to interpret these results in a different way: an early event of gene duplication in an ancestral trypanosomatid has led to the formation of the A, B and C genes for PGK arranged in the form of a tandem repeat. Owing to the high degree of homology between these three genes, the known high capacity for homologous recombination in the Trypanosomatidae (Cruz and Beverley, 1990; Ten Asbroek *et al.*, 1990) and the fact that the three genes remained together as tandem repeats, until today, we conclude that frequent events of gene conversion must have occurred. Moreover,

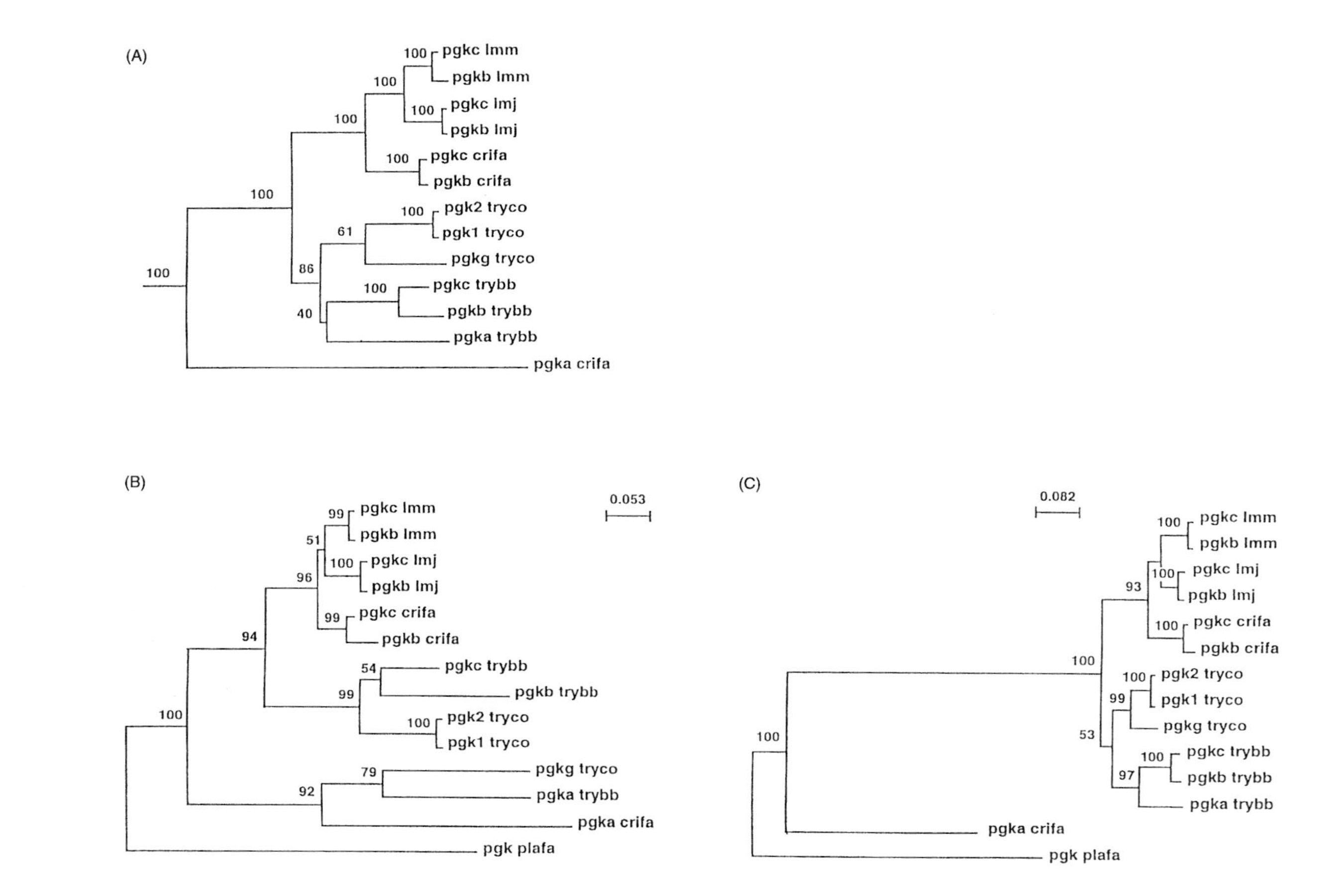

(A)
pgkc lmm
pgkb lmm
pgkc lmj
pgkb lmj
pgkc crifa
pgkb crifa
pgk2 tryco
pgk1 tryco
pgkg tryco
pgkc trybb
pgkb trybb
pgka trybb
pgka crifa
(B)
0.053
pgkc lmm
pgkb lmm
pgkc lmj
pgkb lmj
pgkc crifa
pgkb crifa
pgkc trybb
pgkb trybb
pgk2 tryco
pgk1 tryco
pgkg tryco
pgka trybb
pgka crifa
pgk plafa
(C)
0.082
pgkc lmm
pgkb lmm
pgkc lmj
pgkb lmj
pgkc crifa
pgkb crifa
pgk2 tryco
pgk1 tryco
pgkg tryco
pgkc trybb
pgkb trybb
pgka trybb
pgka crifa
pgk plafa

Figure 13.4 Rooted consensus trees constructed from trypanosomatid PGK protein sequences. The trees were obtained by the neighbour-joining method on 100 random resamplings of the sequences. The numbers at the nodes represent the number of times this node was present in the bootstrap analysis. (A) Tree rooted by all other non-trypanosomatid PGK sequences. (B) Consensus tree for the N-terminal domain. As outgroup the *Plasmodium falciparum* sequence was included in the analysis. (C) Consensus tree for the C-terminal domain. crifa, *Crithidia fasciculata*; lmm, *Leishmania mexicana*; lmj, *Leishmania major*; tryco, *Trypanosoma congolense*; trybb, *Trypanosoma brucei*; plafa, *Plasmodium falciparum*. pgk1 tryco and pgk2 tryco represent the *T. congolense* pgkB and C genes, which are identical, because the C gene has lost its C-terminal targeting signal by the creation of a premature stop codon. pgkg tryco represents the pgkA gene (Parker *et al*., 1995)

such events must have continued to occur, not only after the separation of the different trypanosomatids into genera, but even after their separation into species. This would explain the high degree of identity between the B and C genes in each of the species analysed. However, for the pgkA gene, due to the presence of its 250 nucleotide-long insertion lacking any homology with the other pgk genes, its 5′-part had less possibility to undergo homologous recombination with the other members of this gene family. As a consequence the first 480 nucleotides of the pgkA gene evolved independently and undisturbed by recombinational events. However, the mutations incorporated in the 3′-half of the same gene, as well as those in the pgkB and pgkC gene, were regularly erased again (either completely or partly) by repeated events of homologous recombination (c.f. Le Blancq *et al.*, 1988).

Apparently in some genera the pgkA gene was lost again. It would be interesting to find out when precisely in evolution this loss took place and whether it has affected as well other genera closely related to *Leishmania*, such as *Phytomonas*, *Leptomonas* and *Herpetomonas*.

13.2.4 Glycerol-3-phosphate dehydrogenase

The situation with the glycosomal NAD-dependent glycerol-3-phosphate dehydrogenase (GPD) appears to be fundamentally different from what has been observed for the other glycosomal enzymes. In trypanosomes GPD is present exclusively in glycosomes and is targeted by a C-terminal tripeptide, while in the related genera *Leishmania* and *Crithidia* the enzyme is present both in the mitochondrion and in the glycosome. It has a short N-terminal extension reminiscent of a mitochondrial presequence in addition to its C-terminal glycosomal targeting signal (Kohl *et al.*, 1996).

GPD is an enzyme that has evolved much faster than the majority of the glycolytic enzymes. The identity between the trypanosomatid sequence and that of all other organisms ranges from 25–36%. Furthermore, the trypanosomatid GPD shares the highest percentage of identical residues with bacterial GPDs, such as those of *Escherichia*, *Haemophilus*, *Bacillus* and *Pseudomonas* (32–36%). With eukaryotic GPD it shares only 25–30% identical residues. Finally, the two trypanosomatid GPDs contain a number of typical prokaryotic features that are not found in the eukaryotic sequences (Figure 13.5). At this stage it is not clear how to interpret these observations. One possibility might be that the trypanosomatid GPD is the true glycosomal enzyme in the sense that it is directly derived from a prokaryotic endosymbiont that developed into a glycosome. Another possibility would be that, because in *Leishmania* this enzyme seems to carry a mitochondrial targeting signal as well, it is of pre-mitochondrial origin, and only in a later stage it

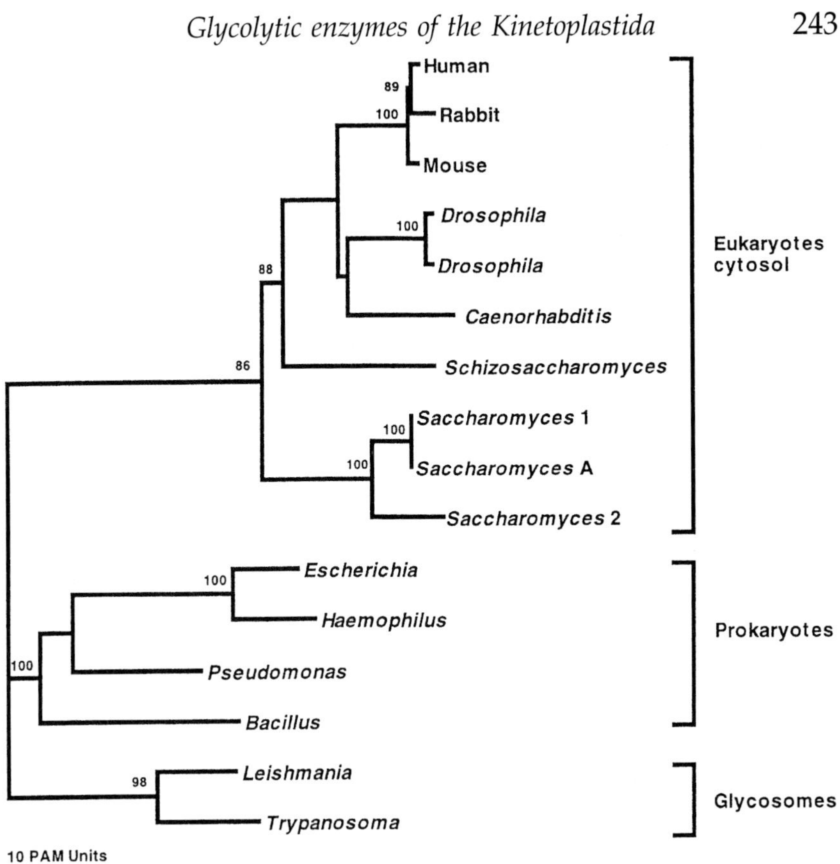

Figure 13.5 Unrooted consensus trees constructed from the protein sequences of NAD-dependent glycerol-3-phosphate dehydrogenase. The trees were obtained by the neighbour-joining method on 100 random resamplings of the sequences. The numbers at the nodes represent the number of times this node was present in the bootstrap analysis.

acquired a C-terminal glycosomal targeting signal in addition to its already present mitochondrial routing signal. Finally, it cannot be excluded that the GPD gene has entered the trypanosomatid ancestor by an event of horizontal gene transfer unrelated to one leading to the formation of an organelle.

13.2.5 ATP-dependent phosphofructokinase

The glycosomal phosphofructokinase (PFK) of *Trypanosoma brucei* is, like other ATP-dependent PFKs, a tetrameric protein. However, it is rendered unique by its 50 kDa subunits and the fact that it is neither regulated by fructose 2,6-bisphosphate nor by any of the other typical effectors of eukaryotic PFKs (Nwagwu and Opperdoes, 1982; Cronin and Tipton, 1985). In bacteria PFK is a tetrameric protein of identical 36 kDa subunits, while in mammals it is made up of four identical or similar 80 kDa subunits. In the latter case, each 80 kDa subunit consists of two homologous domains which relate to the 36 kDa subunit of the bacterial PFKs. In addition, plants, some protists, and bacteria, most of them capable of growing under anaerobic conditions, have a PFK which is dependent on inorganic pyrophosphate (PP_i) as the phosphoryl donor, rather than on ATP. ATP- and PP_i-dependent PFKs represent rather different, but yet homologous, groups of enzymes: they can be aligned with each other and share some conserved sequence motifs (Fothergill-Gilmore and Michels, 1993).

Recently we succeeded in cloning and sequencing the *T. brucei* PFK gene. In accordance with its subcellular location, it carries at its C-terminus a peroxisomal targeting signal (AKL). The *T. brucei* enzyme is only distantly related to other eukaryotic and bacterial ATP-dependent PFKs (20–25% identity), while, surprisingly, it is more closely related to the family of the PP_i-dependent PFKs. It shares 39% identical residues with the *Entamoeba histolytica* PP_i-dependent PFK. A phylogenetic tree comprising both bacterial and eukaryotic ATP-PFKs, as well as the PP_i-PFKs, is shown in Figure 13.6. Although this is an unrooted tree, it illustrates that the two families of PP_i- and ATP-dependent PFKs had a common ancestor that must have been located somewhere between the branching points for the actinomycete *Amycolatopsis* PP_i-PFK and the eubacterial *Thermus* ATP-PFK. The tree also suggests that with the separation of the mammalian and bacterial PFKs, an event of gene doubling/fusion must have occurred that has led to the formation of separate N- and C-terminal domains in the yeast and mammalian enzymes. In line with the result of the sequence analysis described above, the ATP-PFK of *Trypanosoma* clusters with the family of PP_i-PFKs, rather than with the ATP-PFK family. Yet, despite its close relationship with the PP_i-PFKs, the *T. brucei* enzyme has in its sequence some signatures that are reminiscent of an ATP-dependent enzyme, and the purified enzyme is highly specific for ATP and displays no activity with inorganic pyrophosphate.

In addition to the free-living *Naegleria fowleri*, the parasitic anaerobic protists *Giardia lamblia* and *Entamoeba histolytica,* the hydrogenosome-containing parasite of the mammalian urogenital tract *Trichomonas vaginalis*, and *Isotricha prostoma*, a rumen ciliate without mitochondria, all have PP_i-PFKs, whereas the mitochondria-containing ciliates *Tetrahymena pyriformis* and *Paramecium caudatum* have an ATP-PFK

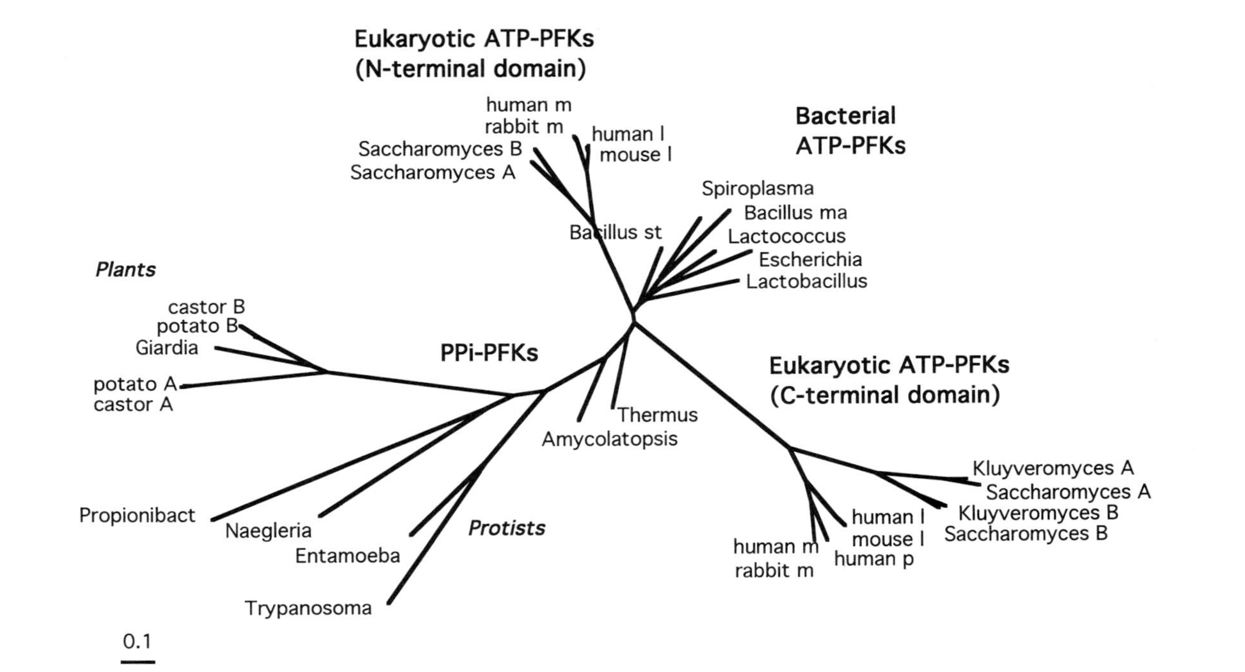

Figure 13.6 Unrooted neighbour-joining tree constructed from the PP_i-dependent phosphofructokinases and the N- and C-terminal domains of the ATP-phosphofructokinases. l, liver; m, muscle.

(Mertens, 1991). By contrast the plastid-containing *Euglena gracilis* has both an ATP- and a PPi-dependent enzyme.

A possible explanation for the fact that the above protists have a glycolytic pathway with a PP_i-dependent PFK, while eukaryotes that have adapted to more aerobic conditions all have an ATP-PFK, could be that the former enzyme provides a significant energetic advantage under anaerobic conditions. PP_i-PFK uses as high-energy phosphate donor a by-product of biosynthetic reactions (PP_i), which would otherwise be hydrolysed to inorganic phosphate. This advantage (three, rather than two, molecules of ATP produced per glucose) would be important for organisms that lack the capacity for oxidative phosphorylation (Mertens, 1991).

We hypothesize that originally the ancestral eukaryotic cell which lacked mitochondria synthesized its ATP by substrate-linked phosphorylation in a PP_i-dependent glycolysis. If the acquisition of mitochondria by an ancestral eukaryote has been a monophyletic event, resulting from endosymbiosis with a proteobacterium of the α-group, this eukaryote probably acquired an ATP–PFK from the endosymbiont, as well. This ATP–PFK was retained in those organisms that adapted to the presence of molecular oxygen, while the original PP_i–PFK became obsolete and was lost. Some protists may neither have acquired mitochondria nor an ATP–PFK and remained true anaerobes. Others retained their mitochondrion but lost again the ATP–PFK. This is probably what happened in the ancestral euglenozoan that evolved to the extant members of the order Kinetoplastida. An adaptation of this organism to a more aerobic environment, eventually resulted in a change from a PP_i-dependent to an ATP-dependent energy metabolism, by which it had to modify its PP_i–PFK in such a way that it could accommodate ATP, rather than PP_i, as the phosphoryl donor. Because PFK is usually a cytosolic enzyme in eukaryotes, except in the Kinetoplastida where it is glycosomal, relocation of the enzyme must have occurred by the acquisition of a peroxisomal targeting signal. Finally plants as well as *E. gracilis*, which all have a PP_i–PFK that is allosterically regulated by fructose 2,6-bisphosphate, may have acquired the enzyme secondarily from a cyanobacterial endosymbiont that eventually gave rise to the chloroplast.

This hypothesis predicts that anaerobic protists lacking mitochondria will harbour the typical PP_i–PFK originally present in the ancestral eukaryote. ATP–PFKs may either have evolved from this ancestral enzyme by changing their substrate specificity as occurred in the Kinetoplastida, or a true ATP–PFK may have been acquired from a proteobacterial endosymbiont. It would be interesting to test this hypothesis by sequencing the PP_i- and ATP-dependent PFKs from both aerobic and anaerobic eukaryotic microorganisms.

13.3 CATALASE

All Archezoa lack catalase, while many representatives of the Euglenozoa (including some trypanosomatids) have the enzyme. Thus the acquisition of peroxisomes by the eukaryotic cell and the appearance of this typical peroxisomal marker enzyme seem to coincide in evolution. Therefore, it is not unlikely that catalase may have entered the eukaryotic cell together with an endosymbiont that may have been reduced to a peroxisome. Unfortunately the sequence of a trypanosomatid catalase is not yet available, but a phylogenetic analysis carried out on all available catalase A protein sequences in the SwissProt database (Figure 13.7) suggests that at least one event of horizontal gene transfer must have occurred during evolution. Peroxisomal/glyoxysomal catalases of plants cluster together, but are well separated from the cluster of the other peroxisomal catalases by a number of catalases of prokaryotic origin. Although this may seem to point to a polyphyletic origin

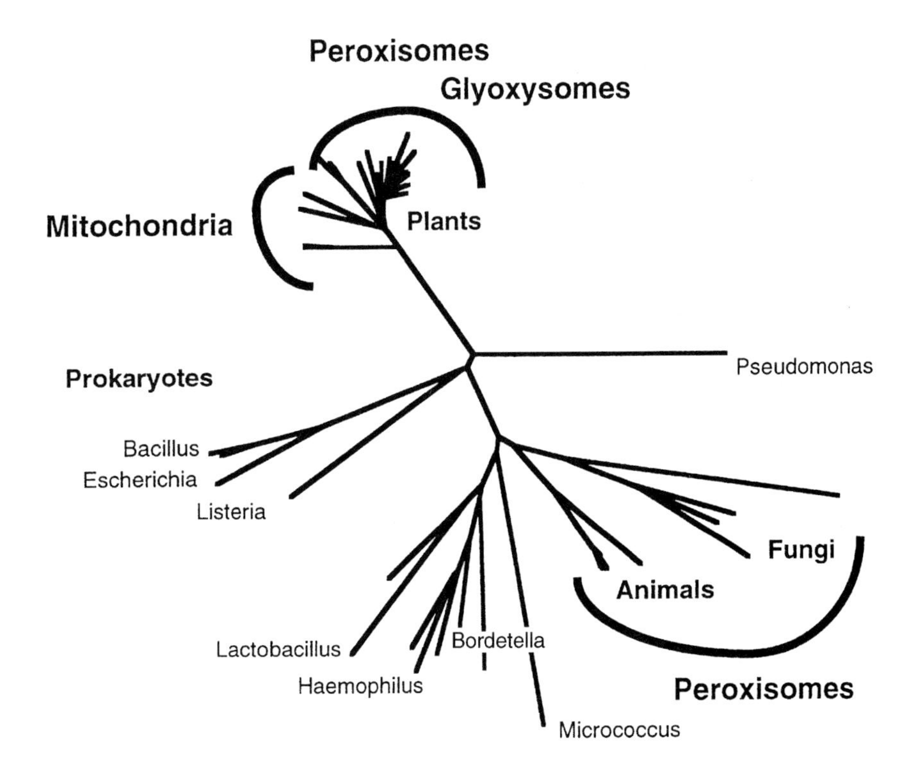

Figure 13.7 Unrooted neighbour-joining tree constructed from the amino-acid sequences of a number of prokaryotic, plant, fungal and animal catalases.

of peroxisomes, it may also indicate that catalases might have entered the organelle through mechanisms not related to its formation. For instance in plants, an already present mitochondrial catalase may have acquired a peroxisomal targeting signal.

13.4 THIOLASE AS A POSSIBLE MARKER OF PEROXISOMAL EVOLUTION

Another enzyme that is shared by most peroxisomes (including the trypanosomatid glycosome) is thiolase. Although the trypanosome thiolase sequence is not yet available, a phylogenetic analysis carried out on all thiolase protein sequences revealed interesting information.

Present-day prokaryotes and eukaryotes contain two thiolase enzymes that are homologous. This can be inferred from the fact that (1) all 25 thiolases can easily be aligned, (2) they are all similar in molecular mass (40–45 kDa) and (3) all thiolases display pair-wise identities of 40% or more. Figure 13.8 shows that apparently an early duplication of a thiolase gene in a prokaryotic organism must have led to the appearance of two separate thiolase enzymes. One evolved to acetoacetyl CoA thiolase (EC 2.3.1.9), an enzyme with a preference for acyl coenzyme A of very short chain length, that in prokaryotes is involved in butyric acid fermentation and in eukaryotes in ketone-body formation. The other thiolase evolved to 3-keto-acyl CoA thiolase (EC 2.3.1.16), with a preference for fatty acid CoAs of (very) long chain lengths and active in the β-oxidation pathway in both bacteria and in the mitochondria and peroxisomes of eukaryotes.

This event of gene duplication has allowed us to unambiguously root the two thiolase trees. The eukaryotic acetoacetyl CoA thiolase is a mitochondrial enzyme and probably entered the eukaryotic cell by horizontal gene transfer when a eukaryotic ancestor acquired the mitochondrion. In the case of the 3-keto-acyl CoA thiolase, there probably must have occurred two events of horizontal gene transfer. The prokaryotic gene first entered the ancestral eukaryotic cell, probably together with the endosymbiont that gave rise to the formation of the mitochondrion. Later, a second event of horizontal gene transfer, from another prokaryotic organism to the eukaryotic cell, may have occurred that has led to the acquisition of a separate thiolase isoenzyme by peroxisomes, so conferring β-oxidation capacity on these organelles. Later events of gene duplication within the eukaryotic line of descent then have led to the formation of additional peroxisomal isoenzymes, each with its different chain length specificity. The intriguing question as to whether this second event of horizontal gene transfer in the case of 3-keto-acylCoA thiolase represented the acquisition of the peroxisome as a new type of organelle by the primitive eukaryotic cell via an endo-

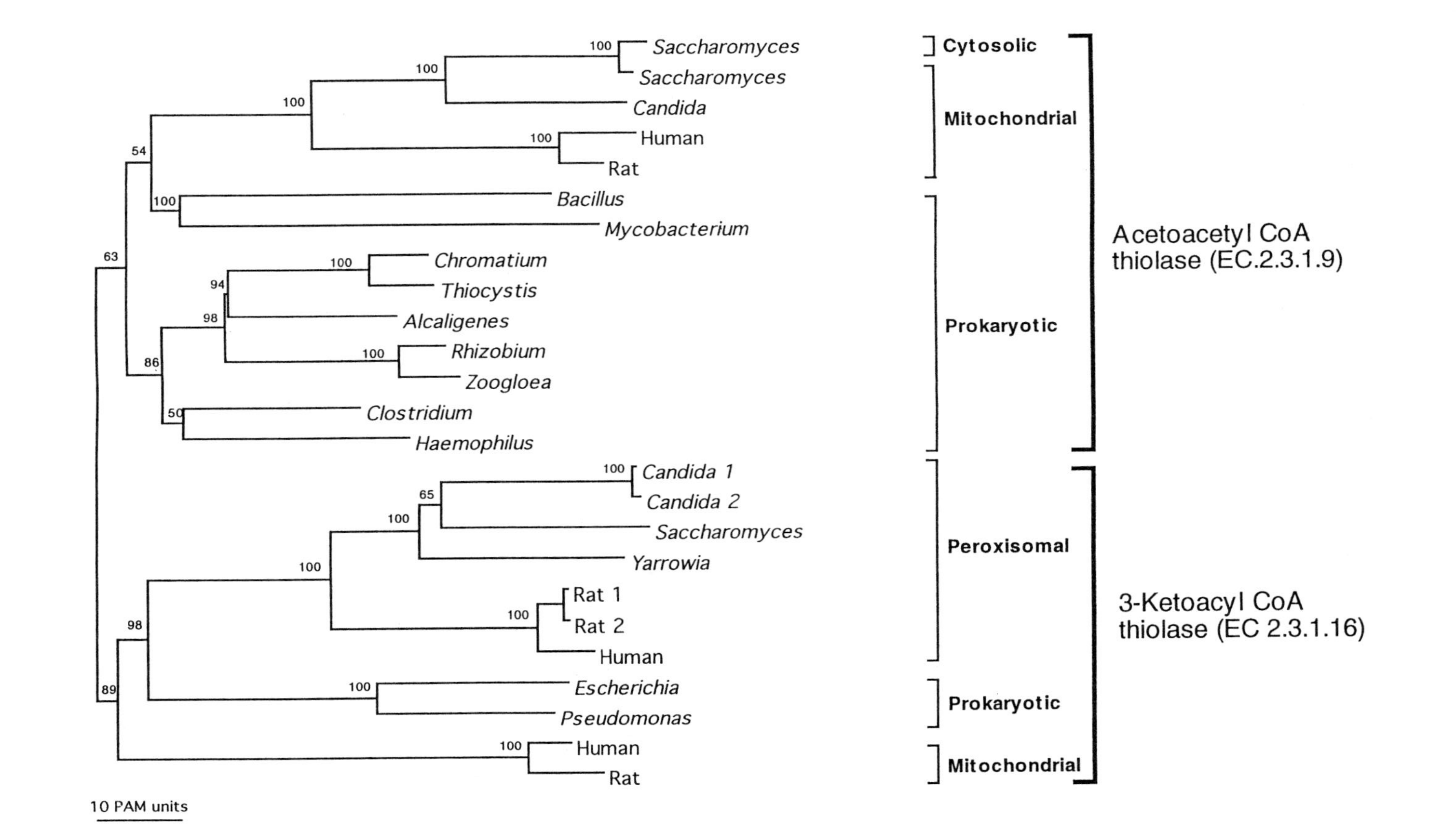

Figure 13.8 Unrooted neighbour-joining tree constructed from the amino-acid sequences of thiolases.

symbiotic event, or the acquisition of a single gene, of which the product became sequestered inside peroxisomes, can not be answered at present.

13.5 CONCLUSION

At present it seems difficult, if not impossible, to retrace the evolutionary origin of the peroxisome as an organelle. It turns out that peroxisomes are chimaeric organelles that contain enzymes of typical eukaryotic as well as prokaryotic signature. Sequence information on enzymes is still scarce and our analyses are not yet based on a sufficient number of polypeptide sequences. Moreover, our analyses have shown that in a number of events traces that may have led us to the origin of the peroxisome as an organelle have been erased. Some of those events are: (1) the relatively easy relocation of enzymes from either cytosol or mitochondrion to peroxisomes by the acquisition of a C-terminal tripeptide, a specific motif in the N-terminal sequence, or any other, as yet not well defined, alternative peroxisomal targeting signal; (2) The frequent events of horizontal gene transfer that seem to have occurred; and (3) gene conversions which, particularly in protists such as the Trypanosomatidae, occur quite frequently.

13.6 REFERENCES

Blattner, J., Swinkels, B., Dörsam, H. *et al.* (1992) Glycosome assembly in trypanosomes: variations in the acceptable degeneracy of a COOH-terminal microbody targeting signal. *Journal of Cell Biology*, **119**, 1129–36.

Blattner, J., Dörsam, H. and Clayton, C.E. (1995) Function of N-terminal import signals in trypanosome microbodies. *FEBS Letters*, **360**, 310–14.

Borst, P. (1989) Peroxisome biogenesis revisited. *Biochimica Biophysica Acta*, **1008**, 1–13.

Borst, P. and Swinkels, B.W. (1989) The evolutionary origin of glycosomes: how glycolysis moved from cytosol to organelle in evolution, in *Evolutionary Tinkering in Gene Expression*, (eds M. Grunberg-Manago, B.F.C. Clark and H.G. Zachau), Plenum Publishing, New York, pp. 163–74.

Cavalier-Smith, T. (1987) The origin of cells: a symbiosis between genes, catalysts, and membranes. *Cold Spring Harbor Symposium on Quantitative Biology*, **52**, 805–24.

Cavalier-Smith, T. (1990) Symbiotic origin of peroxisomes, in *Endocytobiology IV* (eds P. Nardon, V. Gianinazzi-Pearson, A. Grenier, L. Margulis and D.C. Smith), Villeurbanne, pp. 515–19.

Cronin, C.N. and Tipton, K.F. (1985) Purification and regulatory properties of phosphofructokinase from *Trypanosoma (Trypanozoon) brucei brucei*. *Biochemistry Journal*, **227**, 113–24.

Cruz, A. and Beverley, S.M. (1990) Gene replacement in parasitic protozoa. *Nature*, **348**, 171–3.

de Duve, Ch. (1982) Peroxisomes and related particles in historical perspective. *Annals of the New York Academy of Science*, **386**, 1–4.

Fothergill-Gilmore, L.A. and Michels, P.A.M. (1993) Evolution of glycolysis. *Progress in Biophysical and Molecular Biology*, **59**, 105–235.

Gillman, N.W. (1994) *Organelle Genes and Genomes*. Oxford University Press, Oxford.

Gould, S.J., Keller, G.A., Hosken, N., *et al.* (1989) A conserved tripeptide sorts proteins to peroxisomes. *Journal of Cell Biology*, **108**, 1657–64.

Hannaert, V., Blaauw, M., Kohl, L. *et al.* (1992) Molecular analysis of the cytosolic and glycosomal glyceraldehyde-3-phosphate dehydrogenase in *Leishmania mexicana*. *Molecular and Biochemical Parasitology*, **55**, 115–26.

Hannaert, V. and Michels, P.A.M. (1994) Structure, function and biogenesis of glycosomes in Kinetoplastida. *Journal of Bioenergetics and Biomembranes*, **20**, 205–12.

Henze, K., Badr, A., Wettern, M. *et al.* (1995) A nuclear gene of eubacterial origin in *Euglena gracilis* reflects cryptic endosymbioses during protist evolution. *Proceedings of the National Academy of Sciences USA*, **92**, 9122–6.

Kendall, G., Wilderspin, A.W.F., Ashall, F. *et al.* (1990) *Trypanosoma cruzi* glycosomal glyceraldehyde-3-phosphate dehydrogenase does not conform to the 'hotspot' model of topogenesis. *EMBO Journal*, **9**, 2751–8.

Kohl, L., Callens, M., Wierenga, R.K. *et al.* (1994) Triose-phosphate isomerase of *Leishmania mexicana*. Cloning and characterization of the gene, overexpression in *Escherichia coli* and analysis of the protein. *European Journal of Biochemistry*, **220**, 331–8.

Kohl, L., Drmota, T., Do Thi, C.-D. *et al.* (1996) Cloning and characterization of the NAD-linked glycerol-3-phosphate dehydrogenase of *Trypanosoma brucei* and *Leishmania mexicana* and the expression of the trypanosome enzyme in *Escherichia coli*. *Molecular and Biochemical Parasitology*, **76**, 159–73.

Le Blancq, S.M., Swinkels, B.W., Gibson, W.C. *et al.* (1988) Evidence for gene conversion between the phosphoglycerate kinase genes of *Trypanosoma brucei*. *Journal of Molecular Biology*, **200**, 439–47.

Marchand, M., Poliszczak, A., Gibson, W.C. *et al.* (1988) Characterization of the genes for fructose-bisphosphate aldolase in *Trypanosoma brucei*. *Molecular and Biochemical Parasitology*, **29**, 65–76.

Marchand, M., Kooystra, U., Wierenga, R.K. *et al.* (1989) Glucosephosphate isomerase from *Trypanosoma brucei*. Cloning and characterization of the gene and analysis of the enzyme. *European Journal of Biochemistry*, **184**, 455–64.

Mertens, E. (1991) Pyrophosphate-dependent phosphofructokinase, an anaerobic glycolytic enzyme? *FEBS Letters*, **285**, 1–5.

Michels, P.A.M. (1987) Genomic organization and gene structure in African trypanosomes, in *Gene Structure in Eukaryotic Microbes*, SGM Special Publication 22, (ed. J.R. Kinghorn), IRL Press, Oxford, pp. 243–62.

Michels, P.A.M., Marchand, M., Kohl, L. *et al.* (1991) The cytosolic and glycosomal isoenzymes of glyceraldehyde-3-phosphate dehydrogenase in *Trypanosoma brucei* have a distant evolutionary relationship. *European Journal of Biochemistry*, **198**, 421–8.

Michels, P.A.M. and Opperdoes, F.R. (1991) The evolutionary origin of glycosomes. *Parasitology Today*, **7**, 105–9.

Michels, P.A.M. and Hannaert, V. (1994) The evolution of kinetoplastid glycosomes. *Journal of Bioenergetics and Biomembranes*, **20**, 213–19.

Nyame, K., Do-Thi, C.D., Opperdoes, F.R. *et al.* (1994) Subcellular distribution and characterization of glucosephosphate isomerase in *Leishmania mexicana mexicana*. *Molecular and Biochemical Parasitology*, **67**, 269–79.

Nwagwu, M. and Opperdoes, F.R. (1982) Regulation of glycolysis in *Trypanosoma brucei*. *Acta Tropica*, **36**, 61–72.

Opperdoes, F.R., Markos, A. and Steiger, R.F. (1981) Localization of malate dehydrogenase, adenylate kinase and glycolytic enzymes in glycosomes and the threonine pathway in the mitochondrion of cultured procyclics of *Trypanosoma brucei*. *Molecular and Biochemical Parasitology*, **4**, 291–309.

Opperdoes, F.R. (1987) Compartmentation of carbohydrate metabolism in trypanosomes. *Annual Review of Microbiology*, **41**, 127–51.

Opperdoes, F.R. (1988) Glycosomes may provide clues to the import of microbody proteins. *Trends in Biochemical Sciences*, **13**, 255–60.

Opperdoes, F.R., Nohynkova, E., Van Schaftingen, E. *et al.* (1988) Demonstration of glycosomes (microbodies) in the Bodonid flagellate *Trypanoplasma borreli* (Protozoa, Kinetoplastida). *Molecular and Biochemical Parasitology*, **30**, 155–64.

Opperdoes, F.R. and Borst, P. (1977) Localization of nine glycolytic enzymes in a microbody-like organelle in *Trypanosoma brucei*: the glycosome. *FEBS Letters*, **80**, 360–4.

Opperdoes, F.R. and Michels, P.A.M. (1989) Biogenesis and evolutionary origin of peroxisomes, in *Organelles of Eukaryotic Cells: Molecular Structure and Interaction* (eds J.M. Tager, F. Guerrieri, S. Azzi, and S. Papa), Plenum Publishing, New York, pp. 187–95.

Opperdoes, F.R. and Michels, P.A.M. (1993) The glycosomes of the Kinetoplastida. *Biochimie*, **75**, 231–4.

Osinga, K.A., Swinkels, B.W., Gibson, W.C. *et al.* (1985) Topogenesis of microbody enzymes: a sequence comparison of the genes for the glycosomal (microbody) and cytosolic phosphoglycerate kinases of *Trypanosoma brucei*. *EMBO Journal*, **4**, 3811–17.

Parker, H.L., Hill, T., Alexander, K. *et al.* (1995) Three genes and two isozymes: gene conversion and the compartmentalization and expression of the phosphoglycerate kinases of *Trypanosoma* (Nannomonas) *congolense*. *Molecular and Biochemical Parasitology*, **69**, 269–79.

Patterson, D.J. and Sogin, M.L. (1993) Eukaryotic origins and protistic diversity, in *The Origin and Evolution of the Cell* (eds H. Hartman and K. Matsuno), World Scientific Publishing, River Edge, N.J., pp 13–47.

Sommer, J.M., Cheng, Q.-L., Keller, G.A. *et al.* (1992) In vivo import of firefly luciferase into the glycosomes of *Trypanosoma brucei* and mutational analysis of the C-terminal targeting signal. *Molecular Biology of the Cell*, **3**, 749–59.

Subramani, S. (1996) Convergence of model systems for peroxisome biogenesis. *Current Opinion in Cell Biology*, **8**, 513–18.

Swinkels, B.W., Gibson, W.C., Osinga, K.A. *et al.* (1986) Characterization of the gene for the microbody (glycosomal) triosephosphate isomerase of *Trypanosoma brucei*. *EMBO Journal*, **5**, 1291–8.

Swinkels, B.W., Evers, R. and Borst, P. (1988) The topogenic signal of the glycosomal (microbody) phosphoglycerate kinase of *Crithidia fasciculata* resides in a carboxy terminal extension. *EMBO Journal*, **7**, 1159–65.

Swinkels, B.W., Gould, S.J., Bodnar, A.G. *et al.* (1991) A novel, cleavable peroxisomal targeting signal at the amino-terminal of the rat 3-ketoacyl-CoA thiolase. *EMBO Journal*, **10**, 3255–62.

Swinkels, B.W., Gould, S.J. and Subramani, S. (1992) Targeting efficiencies of various permutations of the consensus C-terminal tripeptide peroxisomal targeting signal. *FEBS Letters*, **305**, 133–6.

Ten Asbroek, A.L.M.A., Ouellette, M. and Borst, P. (1990) Targeted insertion of the neomycin phosphotransferase gene into the tubulin gene cluster of *Trypanosoma brucei*. *Nature*, **348**, 174–5.

Vickerman, K. and Preston, T. (1976) Comparative cell biology of the Kinetoplastid flagellates, in *Biology of the Kinetoplastida*, vol. 1 (eds W.H.R. Lumsden and D.A. Evans) Academic Press, New York, pp 35–130.

Wiemer, E.A.C., Hannaert, V., Van den Ijssel, P.R.L.A. *et al.* (1995) Molecular analysis of glyceraldehyde-3-phosphate dehydrogenase in *Trypanoplasma borreli*. *Journal of Molecular Evolution*, **40**, 443–54.

14

The phylum Apicomplexa: an update on the molecular phylogeny

J.T. Ellis, D.A. Morrison and A.C. Jeffries

Molecular Parasitology Unit, Department of Cell and Molecular Biology, Faculty of Science, University of Technology, Sydney, Westbourne Street, Gore Hill, NSW 2065, Australia. E-mail: j.ellis@uts.edu.au

ABSTRACT

The sporozoans (Phylum Apicomplexa) form a group of diverse parasitic protozoa whose motile invasive stages are characterized by the presence of an apical complex consisting of a polar ring connected with cortical microtubules, two apical rings associated with a conoid apparatus and stalked secretory organelles called rhoptries. A plastid remnant with its own circular DNA may also be common to members of the phylum and suggests a photosynthetic ancestor. Phylogenetic relationships within the Apicomplexa are still the subject of considerable debate and of several competing hypotheses. Phylogenetic trees based on small subunit rDNA sequences have been generated for 46 species belonging to the apicomplexan classes Coccidea and Haematozoea (which includes the piroplasms and malaria parasites), to the supposed flagellated apicomplexan *Perkinsus* and to the related Dinozoa (dinoflagellates). Conclusions that can be drawn from the resulting phylogenetic analyses are: (1) the phylum Apicomplexa is monophyletic only if *Perkinsus* is excluded, with the analyses placing it in the Dinozoa; (2) the Coccidea are monophyletic if *Cryptosporidium* is excluded, with most analyses placing *Cryptosporidium* as sister to the Coccidea plus the Haematozoea; (3) the coccidean *Eimeria* is sister to the clade that contains the coccidean genera *Sarcocystis, Toxoplasma* and *Neospora*; (4) the Haematozoea and the Piroplasmida are monophyletic; (5) the genus *Babesia* is paraphyletic. The difference in G + C content of the

Evolutionary Relationships Among Protozoa. Edited by G.H. Coombs, K. Vickerman, M.A. Sleigh and A. Warren. Published in 1998 by Chapman & Hall, London. ISBN 0 412 79800 X

genomes of coccidians and piroplasms is consistent with the effects of directional mutation pressure. The classification schemes of Vivier and Desportes (1990) and Hausmann and Hülsmann (1991), based on phenotypic characters, most closely reflects the conclusions of molecular phylogeny. With the removal of *Perkinsus* from the Apicomplexa the older name Sporozoa can be justifiably retained for the phylum.

14.1 INTRODUCTION

The phylum Apicomplexa Levine 1970 contains over 4600 named species of parasitic protozoa many of which are pathogens in man and livestock. Of the approximately 180 genera (Levine, 1988), eight (*Babesia, Eimeria, Gregarina, Haemogregarina, Haemoproteus, Isospora, Plasmodium* and *Sarcocystis*) contain more than one half of the named species. The taxon Apicomplexa is distinctly broader than the class Sporozoa Leuckart, 1879 (in that the latter as originally defined encompassed only the gregarines and coccidians) but for many parasitologists today the names Apicomplexa and Sporozoa have become synonymous (Corliss, 1994; Cox, 1991, 1994).

The Apicomplexa are unicellular parasites characterized by the presence of an apical complex in their infective life cycle stages (zoites). The apical complex is made up of polar rings and associated cortical microtubules, rhoptries, micronemes, and usually a conoid. The basic life cycle consists of three successive phases: gamogony, the sexual phase with production of gametes and fertilisation; sporogony, the asexual production of numerous sporozoites from the zygote within a cyst; and schizogony (merogony), a phase of asexual multiplication (growth) usually by multiple fission. Extensive variations on the theme exist, however (Current, Upton and Long, 1990; Vivier and Desportes, 1990).

The Apicomplexa is usually considered to contain four clearly-defined groups: the coccidians, the gregarines, the haemosporidians, and the piroplasms. However, the appropriate taxonomic arrangement of these groups is under considerable dispute: in recent times at least seven different taxonomic schemes have been proposed (Table 14.1). These schemes all agree to a certain extent, but there are few areas where they are all in agreement. They are based largely on biological characteristics of the parasites, such as host, tissue and vector specificities; and are designed to be utilitarian rather than to reflect evolutionary history.

Cladistic analysis is an approach to phylogeny reconstruction that groups taxa in such a way that those with historically more recent ancestors form groups nested within groups of taxa with more distant ancestors. This nested set of taxa can be represented as a branching diagram or tree (a cladogram), which is an hypothesis of the evolutionary history of the taxa. The analysis is performed by searching for nested groups of shared derived character states. Phylogenies based on morphological (phenotypic) and/or molecular (genotypic) data can therefore contribute to the development of a more stable taxonomy. They can, for example, indicate whether the currently-recognized taxonomic groups are monophyletic (all members descended from a common ancestor) or not, and they reveal the evolutionary relationships within and between the groups.

Early initiatives aimed at elucidating the molecular phylogeny of apicomplexans were based on sequencing 18S ribosomal RNA (rRNA) (see review by Johnson and Baverstock, 1989). Much of this work has now been extensively criticized because of the poor quality of the sequence data used (see for example the criticisms of Cai, McDonald and Thompson, 1992) and because the sequence alignments did not rigorously establish or fulfil the criterion of homology necessary for phylogenetic analysis (see Barta, Jenkins and Danforth, 1991). Nevertheless, this work was important in stimulating further research in the discipline.

More recently, the polymerase chain reaction has been used as a means of amplifying 18S rRNA genes (rDNA) prior to sequencing (Schlegel, 1991; Ellis *et al.*, 1994a) and sequence data generated in this way appear to be much more reliable. However, given the low fidelity of the thermostable *Taq* polymerase (which has been frequently used in these studies) and its capacity to introduce errors, some concern has been expressed in interpreting analyses where there is only a small amount of nucleotide variation between sequences (Ellis *et al.*, 1995). Despite this drawback, intraspecies variation in 18S rDNA has been rigorously demonstrated in *Cryptosporidium parvum*, for example, leading to the hypothesis that this species is composed of at least two subgroups (Carraway, Tzipori and Widmer, 1996).

During the last 5 years or so there has been a tremendous increase in the amount of 18S rDNA sequence data available from the Apicomplexa (see Morrison and Ellis, 1997), and the main aim of the discussion presented here is to review this recent work on the molecular phylogeny (predominantly from 18S rDNA sequence comparisons) of this very important group of parasites. Since the methods for phylogenetic analysis have been reviewed elsewhere (Morrison, 1996) they will not be addressed here. The classification scheme of Corliss (1994; see Table 14.1) will be adopted as a basis for this discussion.

14.2 MOLECULAR PHYLOGENY OF THE COCCIDEA AND HAEMATOZOEA

At the time of writing, no information is available on 18S rDNA sequences of the gregarines. Most research on the phylogeny of the

Table 14.1 The changing face of Apicomplexan taxonomy. Seven taxonomic schemes for the Apicomplexa produced during the last decade. The rank at which each of the groups of organisms is recognized is indicated by the column in which its name occurs

Phylum	*Subphylum*	*Infraphylum*	*Superclass*	*Class*	*Subclass*	*Superorder*	*Order*	*Suborder*	*Family*
(Levine, 1985, 1988)									
Apicomplexa				Perkinsasida			Perkinsorida		
				Sporozoasida	Gregarinasina		Archigregarinorida		
							Eugregarinorida	Blastogregarinorina	
								Aseptatorina	
								Septatorina	
							Neogregarinorida		
					Coccidiasina		Agamococcidiorida		
							Protococcidiorida		
							Eucoccidiorida	Adeleorina	
								Eimeriorina	
								Haemospororina	
					Piroplasmasina		Piroplasmorida		
(Mehlhorn & Walldorf, 1988)									
Sporozoa (Apicomplexa)				Perkinsea					
				Sporozoea	Gregarinia		Archigregarinida		
							Eugregarinida		
							Neogregarinida		
					Coccidia	Agamococcidea			
						Protococcidea			
						Eucoccidea	Eucoccidida	Adeleina	
								Eimeriina	
							Haemosporida	Conoidina	Haemogregarinidae
								Aconoidina	Haemosporidae
									Piroplasmidae
(Vivier & Desportes 1990)									
Apicomplexa				Gregarina			Blastogregarinida		
							Archigregarinida		
							Eugregarinida		
							Neogregarinida		
				Coccidia			Coelotrophiida		
							Adeleida		
							Eimeriida		
				Haematozoa			Haemosporida		
							Piroplasmida		
(Cox 1991, 1994)									
Sporozoa				Gregarinea			Archigregarinida		

					Eugregarinida
					Neogregarinida
				Coccidea	Agamococcidiida
					Protococcida
					Adelaida
					Eimeriida
				Haemosporidea	Haemosporidida
				Piroplasmea	Piroplasmida
					Dactylosomida
(Cavalier-Smith 1993)					
Apicomplexa	Apicomonada			Apicomonadea	Perkinsida
					Colpodellida
	Gamontozoa	Sporozoa	Gregarinia	Eogregarinea	Blastogregarinida
					Archigregarinida
					Eugregarinida
				Neogregarinea	Neogregarinida
			Coccidia	Coelotrophea	Coelotrophiida
				Eucoccidea	Adeleida
					Eimeriida
		Haematozoa		Haemosporea	Haemosporida
				Piroplasmea	Anthemosomida
					Ixoplasmida
(Corliss 1994)					
Apicomplexa				Perkinsidea	Perkinsida
					Colpodellida
				Gregarinidea	
				Coccidea	Coelotrophiida
					Adeleida
					Eimeriida
				Haematozoea	Haemosporida
					Piroplasmida
(Hausmann & Hülsmann 1996)					
Apicomplexa				Gregarinea	Blastogregarinida
					Archigregarinida
					Eugregarinida
					Neogregarinida
				Coccidea	Coelotrophiida
					Adelaide
					Eimeriida
				Haematozoea	Haemosporida
					Piroplasmida

Apicomplexa has been undertaken on the classes Coccidea and Haematozoea because they contain the major pathogens of man and livestock. For a recent review of the cyst-forming coccidians (*Sarcocystis, Toxoplasma, Neospora*) see Tenter and Johnson (1997)

14.2.1 Class Coccidea

14.2.1.1 Order Eimeriida

Many of the coccidia are of medical or veterinary importance and so have come under close scrutiny. Amongst the cyst-forming coccidia, the phylogenetic relationships of *Toxoplasma, Neospora, Isospora* and *Sarcocystis* has been extensively investigated (Tenter, Baverstock and Johnson, 1992; Ellis *et al.*, 1994b; Holmdahl *et al.*, 1994; Ellis *et al.*, 1995; Morrison and Ellis, 1997). An extensive analysis of partial 18S rRNA sequences which included *T. gondii* and six species of *Sarcocystis* showed support for two monophyletic groups of *Sarcocystis*: one group contained two species which have felids as definitive hosts and a second group contained four species that have canids as the definitive host. These two clades together did not form a monophyletic group unless *T. gondii* was included (Tenter, Baverstock and Johnson, 1992). When the paraphyly of *Sarcocystis* was subsequently re-assessed by a phylogenetic analysis of 18S rDNA sequences, however, the results were inconclusive and monophyly could not be excluded (Ellis *et al.*, 1995). Exhaustive, large scale analyses of the 18S rDNA sequences have subsequently provided evidence for the monophyly of *Sarcocystis* (Ellis and Morrison, 1995; Morrison and Ellis, 1997). The argument for monophyly is based on the observation that the phylogenetically informative sites in the 18S rDNA of these taxa are predominantly located in areas corresponding to the helical regions of the rRNA. Exclusion of much of the noise associated with the data found in the single-stranded regions of the rRNA greatly improves the phylogenetic signal leading to much clearer pictures of phylogenetic history.

Neospora caninum, Isospora felis and *T. gondii* have been found to represent a monophyletic group (Ellis *et al.*, 1994b; Holmdahl *et al.*, 1994; Carreno *et al.*, 1997; Holmdahl *et al.*, 1997). *Neospora caninum* is now recognized as a major cause of paralysis in dogs, and foetal mortality in the livestock industry (Dubey and Lindsay, 1996). The complete life cycle of *Neospora* is unknown, and because of its similarity in morphology to *T. gondii*, it has frequently been misidentified. Sequence comparisons of 18S rDNA of *Neospora* derived from a cow when compared with those derived from dogs did not yield any informative evidence on the speciation of bovine and canine isolates (Marsh *et al.*, 1995).

Isospora was previously believed to be the sister group to *Eimeria*. *I. felis* has a monoxenous life cycle involving the cat, although experimental and natural infections have shown that this parasite is capable of infecting a variety of mammals (Lindsay, 1990). Given (1) that the life cycle of *I. felis* is intermediate between the strict monoxenous life cycle shown by *Eimeria* species and the facultatively-heteroxenous life cycle of *T. gondii*; (2) the monophyly of *I. felis* and *Isospora suis* with *N. caninum* and *T. gondii* as shown by 18S rDNA sequence comparisons (Carreno *et al.*, 1997; see Fig. 14.1); and (3) the oocysts of *Toxoplasma, Sarcocystis* and *Isospora* all contain two sporocysts, each containing four sporozoites; it is probably appropriate to follow the recommendations of Cox (1994) and place *Isospora* in the family Sarcocystidae.

Analyses of 18S rDNA consistently support the monophyly of *Eimeria* with *Sarcocystis* and *Toxoplasma* (Barta *et al.*, 1991; Ellis *et al.*, 1995; Ellis and Morrison, 1995; Morrison and Ellis, 1996; Barta *et al.*, 1997). Molecular phylogeny within the genus *Eimeria* has become a focus of attention because of the surprising phylogenetic association of the recently-described human pathogen *Cyclospora cayetanensis* with members of this group (Relman *et al.*, 1996). This species has been linked with human diarrhoeal disease (Bendall *et al.*, 1993; Ortega *et al.*, 1993) and its complete life cycle is not yet known. Morphological distinction between mature oocysts of the coccidians relies on the number of sporocysts present and the number of sporozoites in each sporocyst. Oocysts of *Cyclospora* contain only two sporocysts (compared with four in *Eimeria*) but the number of sporozoites present is the same as in *Eimeria* (two per sporocyst). Trees based on 18S rDNA comparisons show that *Eimeria* spp. are monophyletic only if *Cyclospora* is included (Relman *et al.*, 1996).

Barta and colleagues (1997) concluded from a phylogenetic analysis of small subunit rDNA of *Eimeria* spp. that the avian parasites formed four clades which were consistent with oocyst size and shape as well as with the site of intestinal development in the definitive host. *Eimeria bovis* was the sister taxon to the avian *Eimeria* species. Unfortunately the 18S rDNA sequence of *Cyclospora* was not included in this study.

14.2.2 Class Haematozoea

14.2.2.1 Order Piroplasmida

The piroplasms are largely of veterinary importance, though occasionally those parasitizing cattle or rodents infect man. They are transmitted as sporozoites in the saliva of ticks to a vertebrate, where binary fission occurs either directly in erythrocytes (*Babesia*) or in lymphocytes (*Theileria*). Several taxonomically problematic species of *Babesia* are known, however, for example *B. equi*, which undergoes an initial round of schizogony in the lymphocytes of the horse before invading red blood cells. This behaviour has led to suggestions that it should be assigned to a separate genus (*Nuttallia* or *Nicollia*, see Pierce, 1975) or reclassified as a species of *Theileria* (Zapf and Schein, 1994).

Comparisons of 18S rDNA sequences have shown that the Piroplasmida are monophyletic. The analyses show support for the splitting of the *Babesia* into at least two groups (one containing the true babesias of cattle (e.g. *B. bovis*, *B. bigemina*) and dogs (*B. canis*), the other containing the rodent parasites *B. microti* and *B. rodhaini*); the renaming of *B. equi* as a *Theileria* species is also supported (Ellis *et al.*, 1992; Allsopp *et al.*, 1994; Mackenstedt *et al.*, 1994).

14.2.2.2 Order Haemosporida

The molecular phylogeny of malaria parasites, *Plasmodium* spp., has been debated in recent years (Waters, Higgins and McCutchan, 1991, 1993; Escalante and Ayala, 1995). From 18S rDNA comparisons, Waters *et al.* (1991) concluded that the parasite of human malignant tertain malaria, *P. falciparum*, belongs to a monophyletic group with the avian malaria parasites – an observation that led to the suggestion that this species arose 'by an event akin to lateral transfer'; although it is not clear what events were involved (Brooks and McLennan, 1992; Siddall and Barta, 1992). Subsequently, Escalante and Ayala (1994) showed *P. reichenowi* (a parasite of the chimpanzee) to be the sister taxon to *P. falciparum* and that this clade was only remotely related to other *Plasmodium* species. This study found no support for monophyly of *P. falciparum* with avian malarial parasites although it confirmed the observation that the *Plasmodium* species which infect humans (*P. falciparum*, *P. vivax* and *P. malariae*) are only distantly related to each other. These observations have been confirmed by Qari and colleagues (1996), who also assigned *Plasmodium* spp. to two clades – one containing species that infect birds and reptiles, the other containing the rest.

The evolutionary relationships between *Plasmodium* and other haemosporidian genera have not yet been the subject of report, although analyses of 18S rDNA of *Leucocytozoon* and others are in progress (Kissinger, Li and McCutchan, unpublished).

14.3 MOLECULAR PHYLOGENY OF THE APICOMPLEXA – AN OVERALL PERSPECTIVE

14.3.1 Phylogeny from rDNA

Phylogenetic trees based on both morphological and molecular data have been produced for only a small subset of the known species of

Apicomplexa. The morphological data analysed (Figure 14.1a) are mainly ultrastructural and developmental features, which are the most commonly-studied characters of the Protozoa and are taken from Barta (1989). The molecular data (Figure 14.1b) are the nucleotide sequences of the gene coding for the 18S rRNA and amended from Morrison and Ellis (1997). These two trees allow comments on several of the current taxonomic conflicts.

First, the genus *Perkinsus* (and sometimes also the genus *Colpodella*) is often included within the Apicomplexa, as it has several (but not all) of the features of the apical complex (Goggin and Barker, 1993). However, the sequence data indicate that the Apicomplexa is a monophyletic phylum only if the genus *Perkinsus* is excluded and included within the phylum Dinozoa (the dinoflagellates). What is more, when data sets from 18S-like rDNA and actin sequences are combined in simultaneous analysis, the resulting cladogram strongly corroborates the placement of *Perkinsus* species among the dinoflagellates (Siddall *et al.*, 1997).

Second, the Gregarinidea are usually considered to represent the most distinct group within the Apicomplexa; and the morphological phylogeny indicates that they represent the monophyletic sister group to the rest of the phylum. Three or four subgroups within the gregarines are usually recognized, depending on whether the blastogregarines are included as part of the eugregarines or not. There are currently no sequence data for the gregarines that could be used to address these issues.

Third, both the morphological and molecular data support the monophyly of the piroplasms as a group. However, the molecular data indicate that some of the recognized genera, notably *Babesia*, are not monophyletic, and should either be subdivided or united with other genera (as described in 14.2.2.1). Data are needed for more taxa before this issue can be resolved.

Fourth, both the morphological and molecular data support the monophyly of the Haemosporida as a group. However, the taxonomic placement of the haemosporidians in relation to other groups has been somewhat labile; they have usually been bracketed with the piroplasms as the Haematozoea, but sometimes as an order within the Coccidea, or left as an isolated group. The molecular phylogeny strongly indicates a relationship with the piroplasms while the morphological phylogeny slightly favours a relationship with part of the Coccidea (Barta, 1989). The genus *Plasmodium*, in particular, has morphological features in common with both the piroplasms and coccidians; a large unique insertion in its 18S rDNA sequence makes it a distinctive taxon that would repay further investigation.

Finally, neither the morphological nor the molecular data support the monophyly of the coccidians as a group. Four subgroups within the

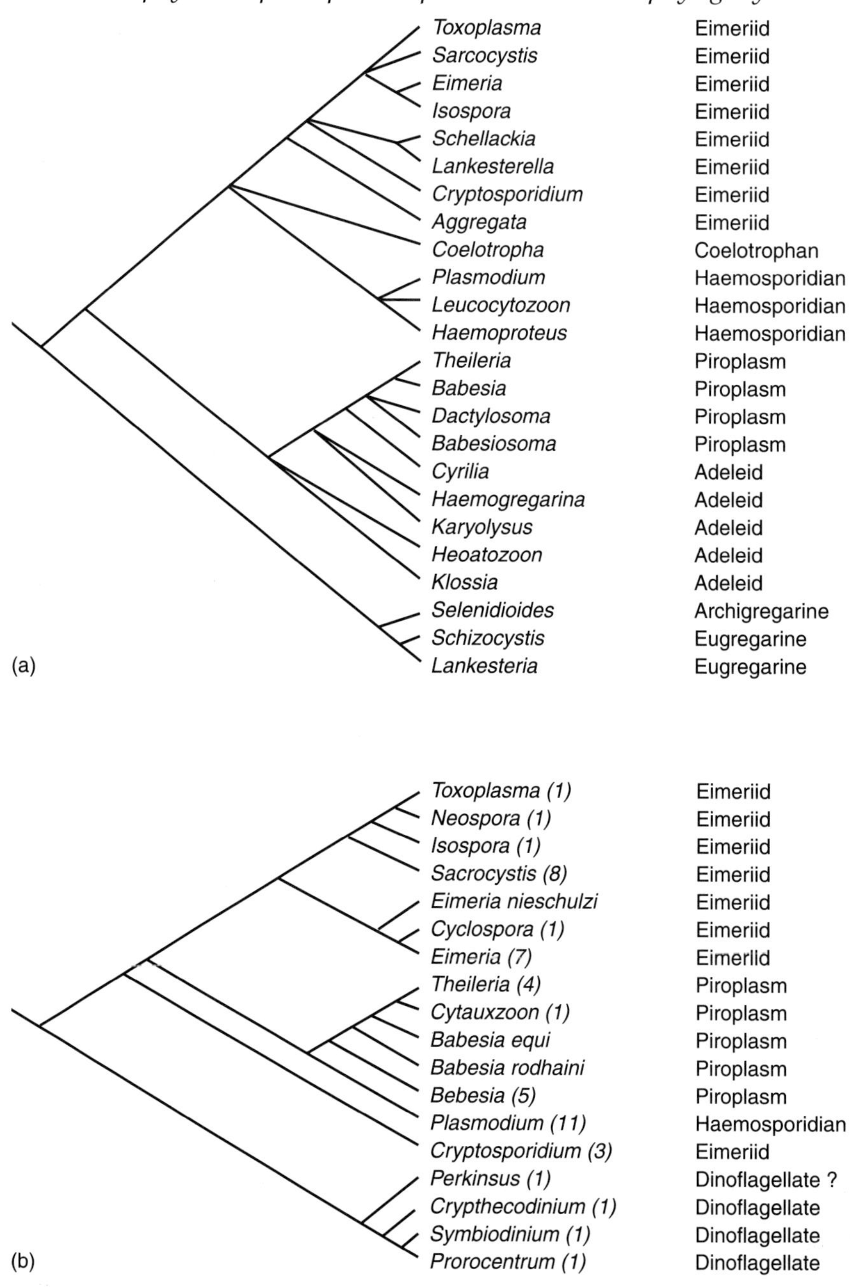

Figure 14.1 Cladograms showing the phylogenetic relationships among some members of the Apicomplexa (and related organisms) as inferred from (a) morphological and (b) molecular data. The group to which each of the genera or

coccidians are often recognized: the coelotrophans (or protococcidians), the agamococcidians (which are sometimes included with the coelotrophans), the adeleids, and the eimeriids. No molecular data exist yet for taxa within the first two small groups, although the morphological data suggest that the coelotrophans are related to the eimeriids rather than to the adeleids. The taxonomy of the latter two large groups is under dispute, with, for example, the genus *Cryptosporidium* being alternatively included in either group, and the haemogregarines being either included with the adeleids or with the haematozoans.

The morphological phylogeny strongly indicates a relationship of *Cryptosporidium* with the eimeriids, which are a monophyletic group. However, the molecular phylogeny suggests that *Cryptosporidium* is the sister to the Coccidea *plus* the Haematozoea, and that the eimeriids are monophyletic only if it is excluded. This part of the molecular data therefore conflicts with all of the taxonomic schemes, and this genus therefore deserves further molecular studies. Furthermore, the molecular data indicate that some of the recognized genera, especially *Eimeria*, are not monophyletic (see Section 14.2.1.1), and should be either subdivided or united with other genera. Data are needed for more taxa before this issue can be resolved.

The morphological phylogeny also indicates that the adeleids are not a monophyletic group, nor are the haemogregarines. However, this phylogeny does suggest a relationship between the haemogregarines and the piroplasms. There are currently no sequence data for the haemogregarines or other adeleids that could resolve these issues.

Thus, the classification scheme of Vivier and Desportes (1990) and that given in the textbook by Hausmann and Hülsmann (1996) are the ones that most closely reflect the current estimates of the phylogeny of the Apicomplexa, as they exclude *Perkinsus* from the group and unite the piroplasms and haemosporidians.

14.3.2 Phylogeny from plastid-like DNA

During investigations into mitochondrial function in malaria parasites, two maternally-inherited, extrachromosomal DNA molecules were isolated (reviewed in Feagin, 1994; Wilson, Williamson and Preiser, 1994; Jeffries and Johnson, 1996). One of these was a 6-kbp long, linear

---

species belongs is indicated. The numbers in brackets indicate how many species of each genus have had sequence data collected. The morphological phylogeny is based on Barta (1989) and the molecular data on Morrison and Ellis (1997).

DNA molecule that has unequivocally been shown to be the mitochondrial genome (Vaidya, Akella and Suplick, 1989). The other was a 35-kbp long, circular DNA molecule which has features suggesting that it has evolved from a chloroplast genome and is therefore referred to as plastid-like DNA (plDNA) (see chapters 8 and 16 for further discussion on apicomplexan plastid DNA).

The 35-kbp molecule has been identified by hybridization experiments in a small number of apicomplexans, mainly *Plasmodium* species. The complete sequences of the *Plasmodium falciparum* (Wilson *et al.*, 1996) and *T. gondii* (D. Roos, unpublished data) plDNAs have been obtained while the plDNA SSUrRNA genes from *B. bovis* (Gozar and Bagnara, 1995), *T. gondii* (Egea and Lang-Unnasch, 1995) and a number of *Sarcocystis* species (A. Jeffries, unpublished data) have also been sequenced. The plDNA is probably located in a multimembraned organelle referred to as the *hohlzylinder* (Siddall, 1992; McFadden *et al.*, 1996; Wilson *et al.*, 1996). Such organelles have been reported from a wide range of apicomplexans, including coelotrophiid and adeleid coccidians, piroplasms and a gregarine (Siddall, 1992) and are generally believed, though not yet proven, to be present in all apicomplexans.

The most obvious phylogenetic questions about the plDNAs and their organelles relate to their origin – whether they resulted from a primary or a secondary endosymbiotic event, and when this endosymbiosis occurred in relation to the origin of the phylum. Is it possible that the loss of photosynthetic capacity by this organelle was coincident with the origin of parasitism in the Apicomplexa? Can the phylogeny of the plDNA tell us anything about the phylogeny of the Apicomplexa?

Unfortunately, the available data do not allow us to answer with certainty any of these questions. The widespread presence of the plDNA/*hohlzylinder* strongly suggests that apicomplexans have a common origin. From sequence and morphological data the best evolutionary relationships so far inferred suggest an organelle ancestry in common with the chloroplasts of either rhodophytes (Williamson *et al.*, 1994) or euglenoids (Egea & Lang-Unnasch, 1995). How far back in evolutionary history the endosymbiotic event occurred, however, is uncertain. It has been noted (Palmer, 1992; Cavalier-Smith, 1993) that there are a number of possible timings for the origin of the organelle. The phyla Apicomplexa, Dinozoa and Ciliophora form a monophyletic group known as the Alveolata Cavalier-Smith, 1993 (Wolters, 1991; Gajadhar *et al.*, 1991, Cavalier-Smith, 1993). Many members of the Dinozoa have functional chloroplasts while none of the Ciliophora do. Therefore, the plDNA/organelle may have originated in the early alveolates with photosynthesis being secondarily lost in the ciliates, some of the dinoflagellates and the apicomplexans. Alternatively, the plDNA/organelle may have evolved in one of the phyla of the Alveolata group. If specific to the Apicomplexa, a separate origin for the plastids of dinoflagellates will have to be proposed. With no data available from dinoflagellate plastids and no knowledge of a relict plastid in the ciliates, we are unable to decide between these alternative evolutionary scenarios.

The origin of plastid-like organelles via a secondary endosymbiotic event is indicated by the presence of more than two membranes surrounding the organelle. The number is not yet known with certainty for the apicomplexan *hohlzylinder,* but three or four seems the most likely, suggesting that a secondary endosymbiotic event involving a photosynthetic eukaryote has occurred, but whether the symbiont was from a euglenoid, chromophyte or rhodophyte lineage cannot be determined. An algal origin for the plDNA/organelle is further supported by the presence and location of specific plastid genes such as the ribosomal subunit genes together with algal-specific sequence features in a number of these and other genes (Wilson *et al.*, 1996). Preliminary phylogenetic inferences from sequences of: the *P. falciparum* *rpo*C_1 (Howe, 1992) and *rpo*B (Gardner *et al.*, 1994) genes, the 16S-like rDNA (including *T. gondii* and *B. bovis* sequences) (Gardner *et al.*, 1993; Egea and Lang-Unnasch, 1995) and the putative nuclear-encoded but plastid DNA-derived *psb*A gene from *Sarcocystis muris* (Hackstein *et al.*, 1995), have suggested an evolutionary relationship with euglenoid plastids. This conclusion may be in doubt on account of the high A+T bias of the sequences, though the LogDet transformation method was used by Egea and Lang-Unnasch (1995) to allow for this bias. More importantly, perhaps, the low taxon density in these trees close to the apicomplexans does not allow an assessment of the evolutionary relationship with the photosynthetic dinoflagellates, as noted above. Confounding the proposed euglenoid relationship is the presence in the plDNAs of an open reading frame, ORF470 (Williamson *et al.*, 1994), which has significant sequence similarity with ORFs from red algae, suggesting a rhodophyte ancestry. Sequence data from a larger number of taxa, both apicomplexan and dinoflagellate, than is available at present will be required before the pathway of descent of the apicomplexan plastid can be resolved.

14.4 MOLECULAR EVOLUTION OF THE APICOMPLEXA

Since the available evidence from molecular analyses shows that the Apicomplexa are a monophyletic group, it would appear appropriate (and an interesting exercise at that) to try and identify conserved molecular traits that correlate with the predicted evolutionary relationships that exist among its constituent taxa. Unfortunately, comparative data are not available for many of the apicomplexan groups, and therefore no firm conclusions may be drawn. However, it is possible to speculate on a number of issues given the data that are available.

Genome G+C content is recognized as a strong characteristic of an organism, and it is believed to be closely related to phylogeny (Osawa *et al.*, 1992). The phenomenon of directional mutation pressure (DMT) has been invoked to explain variation in genome G+C content across taxa (Sueoka, 1988). DMT is dependent on the types of mutations that are incorporated into the DNA during replication, which in turn lead to the variations observed in genome G+C content. Although there are few data available to make reliable conclusions about the Apicomplexa (Table 14.2), the genome G+C content does appear to vary among them in a fashion that appears consistent with DMT. In support of this hypothesis are the results of studies on codon usage among Apicomplexa (Ellis *et al.*, 1993; Ellis *et al.*, 1994c; Morrison, Ellis and Johnson, 1994; Char *et al.*, 1996). Metric multidimensional scaling of codon usage data was used to investigate overall similarities in the pattern of codon usage between the taxa studied and these analyses revealed relationships among the data which were in general agreement with the phylogeny of these organisms.

Two additional lines of evidence indicate that the coccidia and the piroplasms differ in their molecular organization. Like all eukaryotes, the genomes of apicomplexan parasites vary considerably in size (Table 14.2). The available data indicate that *Toxoplasma* and *Eimeria* have similar numbers of chromosomes (14, 11) which are reflected in their similar genome size (40Mb), while *Babesia* and *Theileria* both have only four chromosomes and a genome size of 8–10Mb.

Table 14.2 Variation in genome size, GC content and chromosome number among the Apicomplexa

Species	*Number of chromosomes*	*Genome size/Mb*	*Genome %GC*[g]
Plasmodium falciparum[a]	14	30	18
Cryptosporidium parvum[b]	5	6–8	?
Theileria parva[c]	4	9–10	38
Babesia bovis[d]	4	8–9	?
Eimeria tenella[e]	14	40	?
Toxoplasma gondii[f]	11	40	50

[a]Lanzer *et al.* (1995) *Mol. Biochem. Parasitol.* 70, 1; [b]Kim *et al.* (1992) *Mol. Biochem. Parasitol.* 50, 105; [c]Morzaria *et al.* (1990) *Mol. Biochem. Parasitol.* 40, 203; [d]Jones *et al.* (1997) in preparation; [e]Shirley (1994) *Parasitol. Res.* 80, 366; [f]Sibley and Boothroyd (1992) *Mol. Biochem. Parasitol.* 51, 291; [g]cited in Ellis *et al.* (1994c) *Parasitology* 109, 265.

The genomic organization of 18S rDNA also appears to vary in a consistent way between the Coccidea and Haematozoea. In *T. gondii* for example, there are approximately 110 copies of 18S rDNA repeat per genome (Guay *et al.*, 1992), and in *Eimeria* spp. this rDNA is also highly repeated (Ellis and Bumstead, 1990). The 18S rDNA of *Plasmodium*, *Babesia* and *Theileria*, however, exists as single non-tandemly repeated units that are dispersed in the genome (Reddy *et al.*, 1991; Kibe *et al.*, 1994).

In contrast, the organization of the 5S rDNA varies considerably among the different Apicomplexa. *T. gondii* is the only Apicomplexan known to contain a 5S rRNA gene linked to the 18S and 28S rDNA (Guay *et al.*, 1992); the 5S gene is highly repeated in *Eimeria tenella*, but here these tandem repeats are unlinked to the rest of the rDNA (Stucki, Braun, and Roditi, 1993). *C. parvum* and *P. falciparum* both contain only three copies of the 5S rDNA and these are also unlinked to the rest of the rDNA (Shippen-Lentz and Vezza, 1988; Taghi-Kilani, Remacha-Moreno and Wenman, 1994).

The evidence described above testifies to the considerable genetic diversity that is known to exist among the Apicomplexa. Nevertheless, given the monophyletic nature of the phylum, it is likely that a sizeable number of molecular characteristics will be found that are conserved among these organisms. One particularly interesting line of research is the random DNA sequencing of expressed sequence tags (Wan, Blackwell and Ajioka, 1996). This approach will provide a wide range of genetic markers which ultimately will be used to examine the apicomplexan genome.

14.5 REFERENCES

Allsopp, M.T.E.P., Cavalier-Smith, T., De Waal, D.T. and Allsopp, B.A. (1994) Phylogeny and evolution of the piroplasms. *Parasitology*, **108**, 147–52.

Barta, J.R. (1989) Phylogenetic analysis of the class Sporozoea (phylum Apicomplexa Levine, 1970): evidence for the independent evolution of heteroxenous life cycles. *Journal of Parasitology*, **75**, 195–206.

Barta, J.R., Jenkins, M.C. and Danforth, H.D. (1991) Evolutionary relationships of avian *Eimeria* species among other apicomplexan protozoa: monophyly of the Apicomplexa is supported. *Molecular Biology and Evolution*, **8**, 345–55.

Barta, J.R., Martin, D.S., Liberator, P.A. *et al.* (1997) Phylogenetic relationships among eight *Eimeria* species infecting domestic fowl inferred using complete small subunit ribosomal DNA sequences. *Journal of Parasitology*, **83**, 262–71.

Bendall, R.P., Lucas, S., Moody, A. *et al.* (1993) Diarrhoea associated with cyanobacterium-like bodies: a new coccidian enteritis of man. *Lancet*, **341**, 590–2.

Brooks, D.R. and McLennan, D.A. (1992) The evolutionary origin of *Plasmodium falciparum*. *Journal of Parasitology*, **78**, 564–6.

Cai, J., McDonald, V. and Thompson, D.E. (1992) PCR cloning and nucleotide sequence determination of the 18S rRNA gene and internal transcribed spacer 1 of the protozoan parasites *Cryptosporidium parvum* and *Cryptosporidium muris*. *Biochimica Biophysica Acta*, **1131**, 317–20.

Carraway, M., Tzipori, S. and Widmer, G. (1996) Identification of genetic heterogeneity in the *Cryptosporidium parvum* ribosomal repeat. *Applied and Environmental Microbiology*, **62**, 712–16.

Carreno, R.A., Schnitzler, B.E., Jeffries, A.C. *et al.* (1997) Phylogenetic analysis of coccidia based on 18S rDNA sequence comparison indicates that *Isospora* is most closely related to *Toxoplasma* and *Neospora*. Submitted.

Cavalier-Smith, T. (1993) Kingdom Protozoa and its 18 phyla. *Microbiological Reviews*, **57**, 953–94.

Char, S., Kelly, P., Naeem, A. and Farthing, M.J.G. (1996) Codon useage in *Cryptosporidium parvum* differs from that in other Eimeriorina. *Parasitology*, **112**, 357–62.

Corliss, J.O. (1994) An interim utilitarian ('user-friendly') hierarchical classification and characterisation of the protists. *Acta Protozoologica*, **33**, 1–51.

Cox, F.E.G. (1991) Systematics of parasitic protozoa, in *Parasitic Protozoa*, 2nd edn (eds J.P. Kreier and J.R. Baker), Academic Press, San Diego, pp. 55–80.

Cox, F.E.G. (1994) The evolutionary expansion of the Sporozoa. *International Journal for Parasitology*, **24**, 1301–16.

Current, W.L., Upton, S.J. and Long, P.L. (1990) Taxonomy and life cycles, in *Coccidiosis of Man and Domestic Animals* (ed. P.L. Long), CRC Press, Boca Raton, pp. 1–16.

Dubey, J.P. and Lindsay, D.S. (1996) *Neospora caninum* and neosporosis. *Veterinary Parasitology*, **67**, 1–59.

Egea, N. and Lang-Unnasch, N. (1995) Phylogeny of the large extrachromosomal DNA of organisms in the phylum Apicomplexa. *Journal of Eukaryotic Microbiology*, **42**, 679–84.

Ellis, J. and Bumstead, J. (1990) *Eimeria* species: studies using rRNA and rDNA probes. *Parasitology*, **101**, 1–6.

Ellis, J., Hefford, C., Baverstock, P.R. *et al.* (1992) Ribosomal DNA sequence comparison of *Babesia* and *Theileria*. *Molecular and Biochemical Parasitology*, **54**, 87–96.

Ellis, J., Griffin, H., Morrison, D. and Johnson, A.M. (1993) Analysis of dinucleotide frequency and codon usage in the phylum Apicomplexa. *Gene*, **126**, 163–70.

Ellis, J., Luton, K., Baverstock, P.R. *et al.* (1994a) The phylogeny of *Neospora caninum*. *Molecular and Biochemical Parasitology*, **64**, 303–11; erratum (1994), *Molecular and Biochemical Parasitology*, **67**, 341–2.

Ellis, J., Morrison, D.A. and Johnson, A.M. (1994b) Molecular phylogeny of sporozoan parasites. *Today's Life Science*, **6**, 30–4.

Ellis, J.T., Morrison, D.A., Avery, D. and Johnson, A.M. (1994c) Codon usage and bias among individual genes of the coccidia and piroplasms. *Parasitology*, **109**, 265–72.

Ellis, J. and Morrison, D.A. (1995) Effects of sequence alignment on the phylogeny of *Sarcocystis* deduced from 18S rDNA sequences. *Parasitology Research*, **81**, 696–9.

Ellis, J., Luton, K., Baverstock, P.R. *et al.* (1995) Phylogenetic relationships between *Toxoplasma* and *Sarcocystis* deduced from a comparison of 18S rDNA sequences. *Parasitology*, **110**, 521–8.

Escalante, A.A. and Ayala, F.J. (1995) Evolutionary origin of *Plasmodium* and other Apicomplexa based upon rRNA genes. *Proceedings of the National Academy Sciences USA*, **92**, 5793–7.

Feagin, J.E. (1994) The extrachromosomal DNAs of apicomplexan parasites. *Annual Review of Microbiology*, **48**, 81–104.

Gajadhar, A.A., Marquardt, W.C., Hall, R. *et al.* (1991) Ribosomal RNA sequences of *Sarcocystis muris, Theileria annulata* and *Crypthecodinium cohnii* reveal evolutionary relationships among apicomplexans, dinoflagellates, and ciliates. *Molecular and Biochemical Parasitology*, **45**, 147–54.

Gardner, M.J., Feagin, J.E., Moore, D.J. *et al.* (1993) Sequence and organization of large subunit rRNA genes from the extrachromosomal 35 kb circular DNA of the malaria parasite *Plasmodium falciparum. Nucleic Acids Research*, **21**, 1067–71.

Gardner, M.J., Goldman, N., Barnett, P. *et al.* (1994) Phylogenetic analysis of the *rpo*B gene from the plastid-like DNA of *Plasmodium falciparum. Molecular and Biochemical Parasitology*, **66**, 221–31.

Goggin, C.L. and Barker, S.C. (1993) Phylogenetic position of the genus *Perkinsus* (Protista, Apicomplexa) based on small subunit ribosomal RNA. *Molecular and Biochemical Parasitology*, **60**, 65–70.

Gozar, M.M.G. and Bagnara, A.S. (1995) An organelle-like small subunit ribosomal RNA gene from *Babesia bovis*: Nucleotide sequence, secondary structure of the transcript and preliminary phylogenetic analysis. *International Journal for Parasitology*, **25**, 929–38.

Guay, J-M., Huot, A., Gagnon, S. *et al.* (1992) Physical and genetic mapping of cloned ribosomal DNA from *Toxoplasma gondii*: primary and secondary structure of the 5S gene. *Gene*, **114**, 165–71.

Hackstein, J.H.P., Mackenstedt, U., Mehlhorn, H. *et al.* (1995) Parasitic apicomplexans harbor a chlorophyll a-D1 complex, the potential target for therapeutic triazines. *Parasitology Research*, **81**, 207–16.

Hausmann, K. and Hülsmann, N. (1996) *Protozoology*, Thieme Medical Publishers, New York, pp. 87–98.

Holmdahl, O.J.M., Mattson, J.G., Uggla, A. and Johnasson, K-E. (1994) The phylogeny of *Neospora caninum* and *Toxoplasma gondii* based on ribosomal RNA sequences. *FEMS Microbiology Letters*, **119**, 187–92.

Holmdahl, O.J.M., Ellis, J., Harper, P.A.W. and Morrison, D.A. (1997) Phylogenetic relationships, based on small subunit rRNA sequences, of two isolates of *Besnoitia besnoiti* and other cyst-forming coccidian parasites, in *International Coccidiosis Conference* (eds M.W. Shirley and F.M. Tomley), pp. 73–4.

Howe, C.J. (1992) Plastid origin of an extrachromosomal DNA molecule from *Plasmodium*, the causative agent of malaria. *Journal of Theoretical Biology*, **158**, 199–205.

Jeffries, A.C. and Johnson, A.M. (1996) The growing importance of the plastid-like DNAs of the Apicomplexa. *International Journal for Parasitology*, **26**, 1139–50.

Johnson, A.M. and Baverstock, P.R. (1989) Rapid ribosomal RNA sequencing and the phylogenetic analysis of protists. *Parasitology Today*, **5**, 102–5.

Jones, S.H., Lew, A.E., Jorgensen, W.K. and Barker, S.C. (1997) *Babesia bovis*:

genome size, number of chromosomes and telomeric probe hybridisation. *International Journal for Parasitology*. **27**, 1569–73.

Kibe, M.K., Ole-MoiYoi, O.K., Nene, V. *et al.* (1994) Evidence for two single copy units in *Theileria parva* ribosomal RNA genes. *Molecular and Biochemical Parasitology*, **66**, 249–59.

Kim, K., Gooze, L., Petersen, C. *et al.* (1992) Isolation, sequence and molecular karyotype analysis of the actin gene of *Cryptosporidium parvum*. *Molecular and Biochemical Parasitology*, **50**, 105–13.

Lanzer, M. Fischer, K. and Le Blancq, S.M. (1995) Parasitism and chromosome dynamics in protozoan parasites: is there a connection? *Molecular and Biochemical Parasitology*, **70**, 1–8.

Levine, N.D. (1985) Phylum II. Apicomplexa Levine 1970, in *An Illustrated Guide to the Protozoa* (eds J.J. Lee, S.H. Hunter and E.C. Bovee), Society of Protozoologists, Lawrence, pp. 322–74.

Levine, N.D. (1988) *The Protozoan Phylum Apicomplexa*. CRC Press, Boca Raton.

Lindsay, D.S. (1990) *Isospora*: infections of intestine: Biology, in *Coccidiosis of Man and Domestic Animals* (ed. P.L. Long), CRC Press, Boca Raton, pp. 77–89.

Mackenstedt, U., Luton, K., Baverstock, P.R. and Johnson, A.M. (1994) Phylogenetic relationships of *Babesia divergens* as determined from comparison of small subunit ribosomal RNA gene sequences. *Molecular and Biochemical Parasitology*, **68**, 161–5.

Marsh, A.E., Barr, B.C., Sverlow, K. *et al.* (1995) Sequence analysis and comparison of ribosomal DNA from bovine *Neospora* to similar coccidial parasites. *Journal of Parasitology*, **81**, 530–5.

McFadden, G.I., Reith, M.E., Munholland, J. and Lang-Unnasch, N. (1996) Plastid in human parasites. *Nature*, **381**, 482.

Mehlhorn, H. and Walldorf, V. (1988) Life cycles, in *Parasitology in Focus, Facts and Trends* (ed. H. Mehlhorn), Springer-Verlag, Berlin, pp. 18–44.

Morrison, D.A., Ellis, J.T. and Johnson, A.M. (1994) An empirical comparison of distance matrix techniques for estimating codon usage divergence. *Journal of Molecular Evolution*, **39**, 533–6.

Morrison, D.A. (1996) Phylogenetic tree-building. *International Journal for Parasitology*, **26**, 589–617.

Morrison, D.A. and Ellis, J.T. (1997) Some effects of nucleotide sequence alignment on phylogeny estimation. *Molecular Biology and Evolution*, **14**, 428–41.

Morzaria, S.P., Spooner, P.R., Bishop, R.P. *et al.* (1990) Sfil and Notl polymorphisms in *Theileria* stocks detected by pulsed field gel electrophoresis. *Molecular and Biochemical Parasitology*, **40**, 203–11.

Ortega, Y.R., Sterling, C.R., Gilman, R.H. *et al.* (1993) *Cyclospora* species – a new protozoan pathogen of humans. *New England Journal of Medicine*, **328**, 1308–12.

Osawa, S., Jukes, T.H., Watanabe, K. and Muto, A. (1992) Recent evidence for evolution of the genetic code. *Microbiological Reviews*, **56**, 229–64.

Palmer, J.D. (1992) Green ancestry of malarial parasites? *Current Biology*, **2**, 318–20.

Pierce, M.A. (1975) *Nuttalia* França, 1909 (Babesiidae) preoccupied by *Nuttalia* Dall, 1898 (Psammobiidae): a re-appraisal of the taxonomic position of the avian piroplasms. *International Journal for Parasitology*, **5**, 285–7.

Qari, S.H., Shi, Y.P., Pieniazek, N.J. *et al.* (1996) Phylogenetic relationship among the malaria parasites based on small subunit rRNA gene sequences: monophyletic nature of the human malaria parasite, *Plasmodium falciparum*. *Molecular Phylogenetics and Evolution*, **6**, 157–65.

Reddy, G.R., Chakrabati, G., Yowell, C.A. and Dame, J.B. (1991) Sequence microheterogeneity of three small subunit ribsomal RNA genes of *Babesia bigemina*: Expression in erythrocyte culture. *Nucleic Acids Research*, **19**, 3641–5.

Relman, D.A., Schmidt, T.M., Gajadhar, A. *et al.* (1996) Molecular phylogenetic analysis of *Cyclospora*, the human intestinal pathogen, suggests that it is closely related to the *Eimeria*. *Journal of Infectious Diseases*, **173**, 440–5.

Schlegel, M. (1991) Protist evolution and phylogeny as discerned from small subunit ribosomal RNA sequence comparisons. *European Journal of Protistology*, **27**, 207–19.

Shippen-Lentz, D.E. and Vezza, A.C. (1988) The three 5S rRNA genes from the human malaria parasite *Plasmodium falciparum* are linked. *Molecular and Biochemical Parasitology*, **27**, 263–74.

Shirley, M.W. (1994) The genome of *Eimeria tenella*: further studies on its molecular organisation. *Parasitology Research*, **80**, 366–73.

Sibley, L.D and Boothroyd, J.C. (1992) Construction of a molecular karyotype for *Toxoplasma gondii*. *Molecular and Biochemical Parasitology*, **51**, 291–300.

Siddall, M.E. (1992) Hohlzylinders. *Parasitology Today*, **8**, 90–1.

Siddall, M.E. and Barta, J.R. (1992) Phylogeny of *Plasmodium* species: estimation and inference. *Journal of Parasitology*, **78**, 567–8

Siddall, M.E., Reece, K. S., Graves, J. E. and Burreson, E M. (1997) 'Total evidence' rejects the inclusion of *Perkinsus* species in the phylum Apicomplexa. *Parasitology*, **115**, 165–76.

Stucki, U., Braun, R.E. and Roditi, I. (1993) *Eimeria tenella*: Characterisation of a 5S ribomosal RNA repeat unit and its use as a species-specific probe. *Experimental Parasitology*, **76**, 68–75.

Sueoka, N. (1988) Directional mutation pressure and neutral molecular evolution. *Proceedings of the National Academy of Sciences USA*, **85**, 2653–7.

Taghi-Kilani, R., Remacha-Morneo, M. and Wenman, W.M. (1994) Three tandemly repeated 5S ribosomal RNA-encoding genes identified, cloned and characterised from *Cryptosporidium parvum*. *Gene*, **142**, 253–8.

Tenter, A.M., Baverstock, P.R. and Johnson, A.M. (1992) Phylogenetic relationships of *Sarcocystis* species from sheep, goats, cattle, and mice based on ribosomal RNA sequences. *International Journal for Parasitology*, **22**, 503–13.

Tenter, A.M. and Johnson, A.M. (1997) Phylogeny of the tissue cyst-forming Coccidia. *Advances in Parasitology*, **39**, 69–139.

Vaidya, A.B., Akella, R. and Suplick, K. (1989) Sequences similar to genes for two mitochondrial proteins and portions of ribosomal RNA in tandem arrayed 6-kilobase-pair DNA of a malarial parasite. *Molecular and Biochemical Parasitology*, **35**, 97–108.

Vivier, E. and Desportes, I. (1990) Phylum Apicomplexa, in *Handbook of Protoctista* (eds L. Margulis, J.O. Corliss, M. Melkonian and D.J. Chapman), Jones and Bartlett, Boston, pp. 549–73.

Wan, K-L., Blackwell, J.M. and Ajioka, J.W. (1996) *Toxoplasma gondii* expressed sequence tags: insight into tachyzoite gene expression. *Molecular and Biochemical Parasitology*, **75**, 179–186.

Waters, A.P., Higgins, D.G. and McCutchan, T.F. (1991) *Plasmodium falciparum* appears to have arisen as a result of lateral transfer between avian and human hosts. *Proceedings of the National Academy of Sciences USA*, **88**, 3140–4.

Waters, A.P., Higgins, D.G. and McCutchan, T.F. (1993) The phylogeny of malaria: a useful study. *Parasitology Today*, **9**, 246–50.

Williamson, D.H., Gardner, M.J., Preiser, P. *et al.* (1994) The evolutionary origin of the 35 kb circular DNA of *Plasmodium falciparum*: new evidence supports a possible rhodophyte ancestry. *Molecular and General Genetics*, **243**, 249–52.

Wilson, R.J.M., Williamson, D.H. and Preiser, P.R. (1994) Malaria and other apicomplexans: the "plant" connection. *Infectious Agents and Disease*, **3**, 29–37.

Wilson, R.J.M., Denny, P.W., Preiser, P.R. *et al.* (1996) Complete gene map of the plastid-like DNA of the malaria parasite *Plasmodium falciparum. Journal of Molecular Biology*, **261**, 155–72.

Wolters, J. (1991) The troublesome parasites – molecular and morphological evidence that Apicomplexa belong to the dinoflagellate–ciliate clade. *BioSystems*, **25**, 75–83.

Zapf, F. and Schein, E. (1994) The development of *Babesia* (*Theileria*) *equi* in the gut and the haemolymph of the vector tick, *Hyalomma* species. *Parasitology Research*, **80**, 297–302.

15

Origin of plastids

Christopher J. Howe, Adrian C. Barbrook and Peter J. Lockhart*

*Department of Biochemistry, University of Cambridge, Tennis Court Road, Cambridge, CB2 1QW, UK. E-mail: c.j.howe@bioc.com.ac.uk

ABSTRACT

Plastids are found in a wide range of organisms from higher plants to unicellular flagellate organisms such as *Cyanophora paradoxa* and the amoeboid alga *Chlorarachnion*. In addition, many organisms that are not now photosynthetic have been shown to retain a remnant plastid genome. Such organisms include non-photosynthetic strains of *Euglena*, and Apicomplexa such as *Plasmodium falciparum*. The function of the remnant plastid genome in such species is not yet clear. There is now widespread acceptance of the hypothesis that plastids are ultimately derived from endosymbiotic oxygenic photosynthetic bacteria. In some cases present-day plastids represent a secondary endosymbiosis, where a plastid-containing eukaryote has been engulfed by another host. However, there is still much dispute over whether all plastids originate (1) from a single endosymbiotic event followed by alteration in the light harvesting pigments used in different species (a 'monophyletic' origin) or (2) from multiple endosymbioses, perhaps involving endosymbionts with different light harvesting systems (a 'polyphyletic' origin). Sequence data have been widely used to attempt to discriminate between the hypotheses, and many analyses have favoured the monophyletic hypothesis. However, we have shown that such analyses can be misled by the AT-bias of the plastid genomes. Distortion of tree topologies is found even when third codon position data are excluded from the analysis or predicted amino acid sequences are used instead of DNA sequences. Application of the LogDet transformation may allow correct phylogenetic inferences to be made even in the presence of biased base composition. However, other factors may also lead to the occurrence of artefactual patterns giving spurious support to a hypothesis

Evolutionary Relationships Among Protozoa. Edited by G.H. Coombs, K. Vickerman, M.A. Sleigh and A. Warren. Published in 1998 by Chapman & Hall, London. ISBN 0 412 79800 X

of monophyly. One such factor may be variation among the taxa compared in the number and distribution of sites free to vary. Although we have considered these sources of potential error in the context of plastid origins, they should be kept in mind in any phylogenetic analysis.

15.1 INTRODUCTION

Plastids are found in a wide range of organisms, from higher plants through multicellular and unicellular algae, unicellular flagellate protists such as *Cyanophora paradoxa* and *Euglena gracilis* to amoeboid algae such as *Chlorarachnion*. This chapter will deal with the whole range of plastids, rather than artificially restricting itself to those in any particular group of organisms.

The most familiar function of plastids is photosynthesis, but there are many others too. Green plants, for example, contain a diverse collection of plastids in different tissue types with specific functions. These plastid types are to a large extent interconvertible and include proplastids in meristematic tissue, amyloplasts which function in starch storage, and chromoplasts which are the sites of pigment deposition in many fruits and flowers. In addition, organisms that were once photosynthetic, but are no longer, retain a remnant plastid. This has been shown for higher plants, such as the parasite *Epifagus virginiana* (beechdrops) and protists such as *Astasia longa*, a non-photosynthetic relative of *Euglena* (Wolfe, Morden and Palmer, 1992; Siemeister & Hachtel, 1989). More recently, the presence of a remnant plastid has also been inferred for the apicomplexan *Plasmodium*, the causative agent of malaria (Howe, 1992; Williamson *et al.*, 1994) (see chapter 16).

Plastids all contain a genome, which varies in size from one organism to another. In green plants, it is of the order of 120–140 kbp, and primarily encodes polypeptides of the complexes of the photosynthetic electron transfer chain, the ATP synthase complex, small and large subunits of the ribosome and an RNA polymerase as well as rRNA and tRNA. The plastid genomes of the red alga *Porphyra purpurea* and the diatom *Odontella sinensis* are ca. 190 and 120 kbp respectively (Reith and Munholland, 1995; Kowallik *et al.*, 1995). Those of *Euglena gracilis* and *Cyanophora paradoxa* are 143 kbp and 136 kbp respectively (Hallick *et al.*, 1993; Stirewalt *et al.*, 1995). A number of genes are present in algae or protists that are not present in higher plant plastids, including those for components of certain biosynthetic pathways, and the SecA and SecY components of the protein translocation machinery. Plastid polypeptides that are not plastid encoded are encoded in the nucleus and imported post-translationally. In non-photosynthetic organisms the plastid genome is much reduced, being 70 kbp in the case of *Epifagus*, 73 kbp in *Astasia* and 35 kbp in *Plasmodium*. Indeed, it is not clear why it needs to be retained at all in non-photosynthetic organisms, but likely explanations include the need to retain a plastid $tRNA^{Glu}$ gene which is required for activating glutamate for synthesis of tetrapyrroles for the rest of the plant, and a protease subunit gene (Howe and Smith, 1991; Wolfe, Morden and Palmer, 1992). Often the presence of the remnant plastid genome is the only convincing indicator of the presence of a remnant plastid.

15.2 ENDOSYMBIOTIC ORIGIN OF PLASTIDS

It is now widely accepted that plastids arose from endosymbiosis between an oxygenic photosynthetic organism and a non-photosynthetic host. The idea was proposed over 100 years ago by Schimper (1883) on the basis of microscopical observations, with the suggestion that if plastids divided by binary fission rather than being assembled *de novo*, then their relationship to the cell that contained them resembled a symbiosis. It has been restated and amplified, notably by Mereschkowsky (1905) and then by Margulis (1981). Of the many pieces of evidence in support of the idea, the most convincing was the demonstration of the presence of DNA in chloroplasts. The only reasonable alternative hypothesis of plastid origins, that they were derived by partitioning of existing photosynthetic machinery into a separate compartment of the cell (the so-called 'cluster-clone hypothesis'), does not satisfactorily account for this (Bogorad, 1975). Furthermore, many features of chloroplast gene organization, such as the structure of the plastid encoded RNA polymerase, and the organization of genes for components of the ATPsynthase complex and of the ribosome are very similar to those in prokaryotes, suggesting an ancestry of oxygenic photosynthetic prokaryotes. Indeed, the plastid of the flagellate *Cyanophora paradoxa* (in this organism called the cyanelle) retains a remnant peptidoglycan cell wall round the organelle (Pfanzagl *et al.*, 1996).

However, if one takes the endosymbiotic origin as accepted, there are still many important questions outstanding. These include how and why some, but not all, genetic information was transferred from the endosymbiont to the host nucleus, how the expression of genes within the endosymbiont and host is controlled and integrated, and, most importantly for the purposes of this review, whether all plastids ultimately owe their origin to a single endosymbiosis involving a photosynthetic prokaryote, or to several independent endosymbioses. This last question is raised by the fact that plastids in different organisms vary widely in their structure (such as the number of membranes round the plastid) and the nature of their light-harvesting pigments, as summarized in Table 15.1. In this review we will use the term 'chromophyte' to include the chlorophyll *a,c*-containing algae, and 'rhodophyte'

Table 15.1 Presence of chlorophylls and phycobiliprotein, and number of membranes surrounding the plastid for oxygenic photosynthetic prokaryotes and plastid-containing eukaryotes

Group	*Pigments*	*Membranes*
Cyanophytes	Chlorophyll *a* Phycobiliproteins	–
Prochlorophytes	Chlorophyll *a,b,c*	–
Chlorophytes	Chlorophyll *a,b*	2
Euglenophytes	Chlorophyll *a,b*	3
Chlorarachniophytes	Chlorophyll *a,b*	4
Glaucophytes	Chlorophyll *a* Phycobiliproteins	2 (+peptidoglycan)
Apicomplexa	?	?
Rhodophytes (Red algae)	Chlorophyll *a* Phycobiliproteins	2
Chromophytes		
Dinophytes	Chlorophyll a,c_2	3
Bacillariophytes	Chlorophylls $a,c_{1,2}$	4
Chrysophytes	Chlorophyll $a,c_{1,2}$	4
Crytophytes	Chlorophyll a,c_2 Phycobiliproteins	4
Eustigmatophytes	Chlorophyll *a*	4
Phaeophytes	Chlorophyll $a,c_{1,2}$	4
Prymnesiophytes	Chlorophyll $a,c_{1,2}$	4
Raphidophytes	Chlorophyll *a,c*	4
Xanthophytes	Chlorophyll *a,c*	4

to include the red algae, containing chlorophyll *a* and phycobiliprotein (but excluding species with cyanelles).

In a 'monophyletic' model of plastid origins, the origin of all plastids can be traced back to a single endosymbiotic prokaryote. This prokaryote may have had a full range of light-harvesting systems, which were subsequently reduced so that different lineages now have their own characteristic systems. Alternatively, the endosymbiotic prokaryote may have had a simple light-harvesting system, such as the chlorophyll *a*/phycobiliprotein system of cyanobacteria, with other systems being developed after endosymbiosis. In the 'polyphyletic' model, the origins of plastids can be traced back to a number of different endosymbiotic prokaryotes, which presumably had different light harvesting systems. These separate endosymbioses would then have given rise to different plastid lineages. In the rest of this chapter, we will look at the use of molecular biological data to distinguish between the two models for plastid origin. The examples we will look at deal only with genes that derive from the endosymbiont, although some of them may have subsequently been transferred to the host cell nucleus. It is important to be aware of two limitations to such considerations (Howe *et al.*, 1992). One is that a monophyletic origin of plastids, with a single primary endosymbiotic event, does not exclude the possibility of secondary endosymbioses, in which photosynthetic *eukaryotes* were engulfed by non-photosynthetic hosts (see Section 7). That this has occurred is widely accepted (Palmer and Delwiche, 1996; van de Peer *et al.*, 1996) but it is equally compatible with monophyletic and polyphyletic hypotheses. The second limitation is that even if phylogenetic analysis based on plastid sequences seemed to indicate a monophyletic origin of the endosymbiont, multiple primary endosymbioses might nevertheless have occurred, if members of the **same** taxon had become endosymbionts with different hosts on different occasions. Although this might seem unlikely, it is quite possible that some feature of a particular organism might predispose it to forming endosymbioses with a number of different hosts.

15.3 APPROACHES TO TESTING THE MONOPHYLETIC AND POLYPHYLETIC HYPOTHESES

A number of different molecular biological approaches have been used to try to distinguish between the monophyletic and polyphyletic models of plastid origins. A critical analysis of the data generated so far in these different approaches shows that the issue has not yet been successfully resolved.

15.3.1 Occurrence of proteins

The simplest approach is based on the presence or absence of particular proteins, and there has been particular interest in the light-harvesting chlorophyll-binding proteins. The diatom *Phaeodactylum* was shown some time ago to contain a polypeptide immunochemically related to the chlorophyll *a/b*-binding proteins of green plastids (Grossman, Manodori and Snyder, 1990), and another such protein was also demonstrated in the rhodophyte *Porphyridium cruentum* (Wolfe *et al.*, 1994) and others. The presence of the protein in *Porphyridium*, combined with a failure to detect it in oxygenic photosynthetic bacteria, was interpreted as indicating a monophyletic origin for plastids (Wolfe *et al.*,

1994). However, a potential homologue has subsequently been shown to exist in oxygenic photosynthetic bacteria (Dolganov, Bhaya and Grossman, 1995). This indicates that the possession of such a protein is an ancestral characteristic, rather than a derived one shared by the plastids, and is therefore not phylogenetically informative.

15.3.2 Gene organization data

The organization of certain groups of genes provides limited evidence in support of a monophyletic origin of plastids, although the collection of further gene organization data may help to resolve the issue. For example, the *rpoBC1C2*, *rps2*, and *atpIHGFDA* genes are clustered in the rhodophyte *Porphyra* and in green chloroplasts, but not in cyanobacteria. This has been interpreted as indicative of a shared derived gene reorganization occurring after endosymbiosis but before divergence (Reith and Munholland, 1993). However, it is difficult to know how much significance to attach to this kind of observation since there is no probabilistic model available for plastid genome rearrangement. It is clear that certain regions of the plastid genome are involved in rearrangements at particularly high frequencies, so that the possibility of similar rearrangements occurring in different lineages cannot be discounted (Howe *et al.*, 1988). The *rpoBC1C2* and *rps2* genes are in a different position from the *atpIHGFDA* genes in the plastid genome of the diatom *Odontella sinensis* (Kowallik *et al.*, 1995), which indicates that this region of the genome may indeed be subject to rearrangement. Furthermore, the assumed ancestral organization is based on a rather small number of cyanobacterial genomes, and differences have been observed among species in *atp* gene organization (van Walraven, Lutter and Walker, 1993). Information from more loci and more taxa will be vital in furthering this approach.

15.3.3 Sequence based comparisons

There have been very many studies using sequence data from genes for plastid proteins or plastid RNAs to try to discriminate between monophyletic and polyphyletic hypotheses. We will deal first with studies on the genes for the subunits of ribulose *bis*-phosphate carboxylase (*rbchL* and *rbcS*), as these have given particularly unexpected results. We will then consider analyses with other genes, where problems of a more general nature have become apparent.

Analyses based on the *rbc* genes for subunits of ribulose *bis*-phosphate carboxylase have given a number of unexpected results. Firstly, phylogenetic analyses based on sequence data suggested that while oxygenic photosynthetic bacteria, green plastids and cyanelles had originated among the gamma-group of anoxygenic purple photosynthetic bacteria, the chromophyte and rhodophyte plastids originated among the beta group of anoxygenic photosynthetic bacteria (Morden and Golden, 1991). This would indicate not only a polyphyletic origin of plastids, but multiple origins of *oxygenic* photosynthesis. That oxygenic photosynthesis might have developed independently more than once was unexpected, although it is by no means impossible, as oxygenic photosystems are very similar to anoxygenic ones. However, immunochemical analysis of the extrinsic 33 kDa polypeptide associated with the oxygen-evolving activity of Photosystem II of green plants and algae suggested that the same polypeptide was present in rhodophyte and chromophyte algae, but absent from ancestral species (Fairweather, Packer and Howe, 1994). It is highly unlikely that immunochemically related polypeptides (of the same size) would have arisen independently, and the 33 kDa polypeptide data therefore suggest that oxygenic photosynthesis indeed had a single origin. One possible explanation for the apparent conflict between the *rbc* genes and other data is that the former in rhodophytes and chromophytes were acquired by horizontal transfer (i.e. transfer of DNA between taxa) from anoxygenic photosynthetic bacteria. However, an even more surprising result came from the dinoflagellates. These were recently shown to possess a very different form of the enzyme, which has previously been reported only from anaerobic proteobacteria (Morse *et al.*, 1995). The occurrence of this form of the enzyme in dinoflagellates may again be the result of horizontal transfer. More detailed analysis of *rbc* sequences suggests that gene duplication and horizontal transfer have occurred many times in different lineages (Delwiche and Palmer, 1996).

Analyses with other sequences have given less obviously anomalous results, indicating a single origin of oxygenic photosynthesis, although not necessarily of plastids. Although many analyses have been consistent with a monophyletic origin, a significant number have supported a polyphyletic one instead, which suggests that there may be an underlying weakness with the use of primary sequence data (Howe, Barbrook and Lockhart, 1995). Indeed, two important problems can be highlighted. These are (1) variations in nucleotide composition among lineages, presumably arising from differences in nucleotide substitution patterns ('substitutional bias') in different lineages, and (2) variations in the distribution of sites free to vary. As discussed below, these have the potential to mislead phylogenetic analysis greatly.

15.4 SUBSTITUTIONAL BIAS

15.4.1 The problem

Plastid genomes are generally much more AT-rich than nuclear or cyanobacterial genomes. Why this should be so is not clear, but it presumably represents differences in the kinds of DNA damage occurring in the different compartments (whether through misincorporation by DNA polymerase or as a result of chemical influences) or differences in the repair machinery. However, the models used until recently for inferring phylogenetic trees have not allowed for this. A consequence of substitutional bias is that different lineages can have multiple convergent substitutions (i.e. the same mutation may occur independently in these lineages). The occurrence of convergent substitutions will cause lineages to be grouped more closely in sequence-based phylogenetic analysis. If the substitutional bias reflects phylogeny (i.e. organisms with similar bias are genuinely phylogenetically related) then these convergent substitutions will reinforce the 'historical' information in the sequences. If the bias does not reflect phylogeny (i.e. sequences with similar bias are not genuinely related) then it may cause sequences that are not specifically related (but have similar bias) to be artefactually grouped to the exclusion of others.

The importance of this problem when analysing plastid evolution was first identified in work on the origin of the cyanelle of *Cyanophora paradoxa* (Lockhart *et al.*, 1992). This analysis considered the inferred position of the cyanelle with respect to the green chloroplasts and cyanobacteria when three different gene sequences, from *atpB*, *tufA* and *atpD*, were used. First, second and third codon positions were considered separately. With the first two genes, the cyanelle sequence and the sequences for the green chloroplast proteins share a similar bias. With *atpD*, the gene for the chloroplast protein is in the nucleus and shows a different bias from the cyanelle gene. The implementations of parsimony and maximum likelihood used did not allow for substitutional bias. Strikingly, different preferred tree topologies were found when different codon positions and different genes were used. Where the pattern of bias was the same between cyanelles and chloroplasts, and codon positions showing the bias most clearly were included, the cyanelle was grouped with the chloroplast. Where the pattern differed, the cyanelle was grouped with the cyanobacteria. In other words, the grouping of cyanelle and chloroplast was influenced by the bias in the sequences.

15.4.2 Dealing with substitutional bias

A number of approaches have been adopted to deal with this problem, and not all of them have been justified by close examination of the sequences under study. It is important to be aware that substitutional bias may affect a phylogeny even if the sequences involved do not differ greatly in their nucleotide composition, as the fraction of sites influencing the tree topology may be rather low. Three main approaches have been taken.

15.4.2.1 *Use of amino acid sequences*

Some workers have chosen to construct trees using amino acid sequences, rather than nucleotide sequences, in the belief that the former will not generally be affected by bias. Regrettably, this belief is not justified, since, for example, the cyanelle analyses described above showed that all three codon positions (and therefore the amino acid sequence) can be affected by bias (Lockhart *et al.*, 1992).

15.4.2.2 *Corrected branch lengths and the identification of spurious groupings*

It is possible both to correct parsimony tree lengths for the effect of differences in base composition and also to test whether a particular grouping of taxa may be an artefact of base composition bias, although these approaches have not been used widely (Steel, Lockhart and Penny, 1993).

15.4.2.3 *Use of sequences with different biases*

It may be possible to test whether a tree topology has been distorted by substitutional bias by repeating the phylogenetic analysis with a different set of sequences which show different bias patterns and comparing the results. In the case we are considering, this could be done by comparing (i) a tree constructed with a gene which was located in the plastid in all the eukaryotic organisms under consideration with (ii) a tree constructed using a different gene which was located in the plastid in some organisms and in the nucleus (where genes in general are less AT rich than those in the plastid) in others. A difference in topology between trees (i) and (ii) would therefore indicate that they were being influenced by bias. The majority of plastid sequences used for phylogenetic inference correspond to category (i). The *secA* gene, encoding a component of the thylakoid protein translocation machinery, is an example of a sequence from category (ii). The *secA* gene is located in the plastid in rhodophyte and chromophyte algae, but in the nucleus in green plants and algae. Tree construction using *secA* sequences from oxygenic photosynthetic bacteria, green plants and rhodophyte and chromophyte algae indicated a polyphyletic origin of plastids, with rhodophyte and chromophyte algae forming a sister group to green plastids and photosynthetic bacteria, as shown in Figure 15.1 (Howe *et al.*, 1995). This suggests that some of the evidence from category (i) sequences for a monophyletic origin is indeed due to substitutional bias.

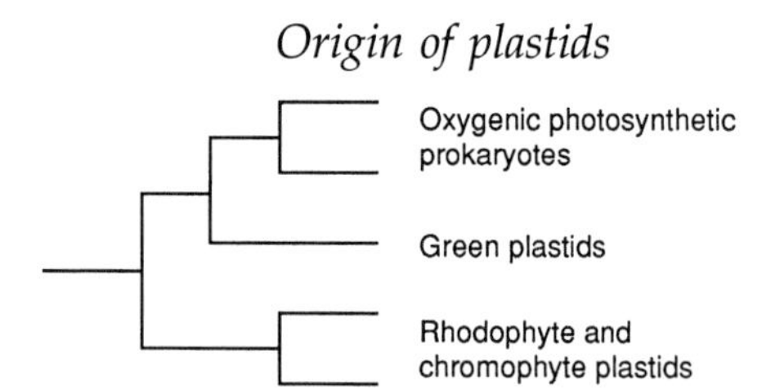

Figure 15.1 Schematic representation of trees constructed from *secA* sequences. Modified from Howe, Barbrook and Lockhart (1995).

However, just as using sequences with similar bias may group them artificially closely, using sequences with different biases may lead to their being placed artificially far apart, so one cannot say *a priori* that category (ii) trees will necessarily be the correct ones.

15.4.2.4 Transformation of data

Methods have now been developed that allow the correct recovery of trees when different lineages are evolving under different substitutional biases. They are based on a relatively simple transformation of the data, generally termed the LogDet transformation (Lockhart *et al.*, 1994). This involves the generation of a 'divergence matrix' for each pair of taxa being considered. This would be a 4×4 matrix when DNA sequences are being considered, or a 20×20 matrix for protein sequences. The ijth entry in such a matrix is the proportion of sites at which the taxa being considered have character states i and j (such as A in one taxon and G in the other). The next stage is to calculate a 'dissimilarity value' for each pair of taxa. This is the negative natural log of the determinant of the divergence matrix. These dissimilarity values can then be used under different tree selection criteria, such as split decomposition or neighbour-joining, to determine the correct tree, although lengths of 'branches' within the tree cannot be determined directly except under restricted conditions. A current limitation of the method is the assumption that different sequence positions evolve at the same rate. Nevertheless, some rate heterogeneity can be accommodated by holding a proportion of constant sites invariable in the data.

Application of the LogDet transformation to a collection of *atpB* sequences generated a tree of the form shown in Figure 15.2 (Howe, Barbrook and Lockhart, 1995). Importantly, this has an internal edge (i.e. division between taxa) separating photosynthetic bacteria from plastids. The existence of this edge indicates that the plastids group to the exclusion of the photosynthetic bacteria, implying a monophyletic plastid origin. Similar analyses (Lockhart *et al.*, in preparation) indicate

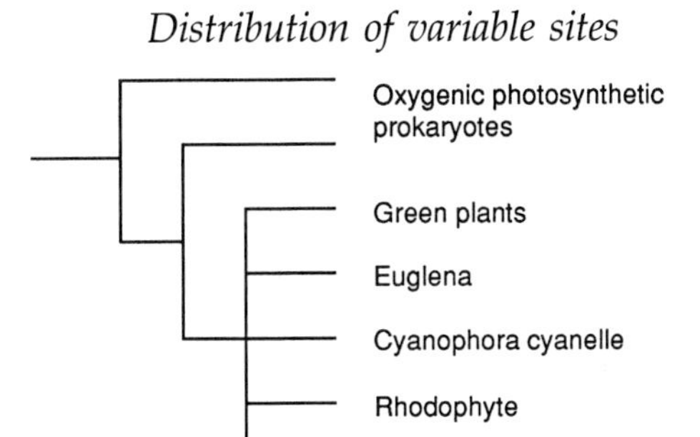

Figure 15.2 Schematic representation of tree generated with LogDet transformed *atpB* sequences using Split Decomposition. Modified from Howe, Barbrook and Lockhart (1995).

a similar result with 16S rDNA, but no separation of the bacteria from plastids when *tufA* sequences are used. It is of interest why these results differ. It seems that the apparent support for plastid monophyly in other studies with *tufA* sequences is an artefact of nucleotide composition differences (since the edge separating plastids from bacteria is not supported when the LogDet transformation is used). However, this is not a sufficient explanation for the support for monophyly with 16S rDNA and *atpB*. With these sequences, there must be some other determinant of the phylogenetic structure that leads to the internal edge separating plastids from bacteria. However, before this edge is definitively interpreted as evidence for a monophyletic origin, a second possible artefact has to be considered, which is the effect of differences in the distribution of variable sites.

15.5 DISTRIBUTION OF VARIABLE SITES

15.5.1 The nature of the problem

Another assumption underlying the models conventionally used for phylogenetic inference is that the same set of sites is free to vary (i.e. undergo mutational change) in all the sequences considered. If this assumption were violated, then artefactual patterns of substitution could arise linking together taxa that shared variable sites to the exclusion of other taxa regardless whether that linkage represented genuine historical relatedness (Lockhart *et al.*, 1996). An example is shown in Figure 15.3. In this, positions 7–10 are fixed in oxygenic photosynthetic bacteria, but free to vary in plastids (and in this example randomized). It is important to note that this shared pattern

	1	AGCTCCTATC
PLASTIDS	2	AGCTCCGGCG
	3	AGCTCCTGCG
	4	AGCTCCCTCG
	1	AGCTCCTAAT
OXYGENIC	2	AGCTCCTAAT
PHOTOSYNTHETIC	3	AGCTCCTAAT
PROKARYOTES	4	AGCTCCTAAT

Figure 15.3 Hypothetical example of differing distributions of variable sites in plastids and oxygenic photosynthetic prokaryotes. For explanation see text.

of variable sites need not necessarily reflect evolutionary relatedness, but the substitutions at positions 7–10 will nevertheless give rise to an internal edge in a tree separating the bacteria from the plastids. If the variable sites in the plastid are also affected by substitutional bias, then the patterns grouping plastids to the exclusion of the others will be even more pronounced. But in the examples under consideration, are there indeed differences in the number of sites free to vary in the taxa used? In the example with *atpB* discussed above, it was found that 33% of codons were free to vary within the oxygenic prokaryotes, 43% were free to vary within the plastids, and 48% were free to vary within oxygenic prokaryotes and plastids combined. In other words, there must be sites that are free to vary within the oxygenic prokaryotes that are not free to vary within the plastids, and similarly there must be sites free to vary within the plastids that are not free to vary within the oxygenic prokaryotes. (It is not clear why there are sites that are variable in one group but not in another. It might, for example, be due to differences between the two groups in the proteins that the molecule under study has to interact with.) There will therefore be a tendency for these data to generate an edge separating plastids collectively from the photosynthetic prokaryotes (and therefore indicating a monophyletic origin of plastids), which does not necessarily reflect evolutionary history.

15.5.2 Quantifying the effect

How can the effect of differences in the number of variable sites among taxa be estimated? One approach is to construct a family of separate datasets (by sampling with replacement) with differing numbers of variable sites and examine how the support given to the internal edge separating prokaryotes from plastids varies with the number of variant sites included. This was done (Lockhart *et al*, in preparation) in the case

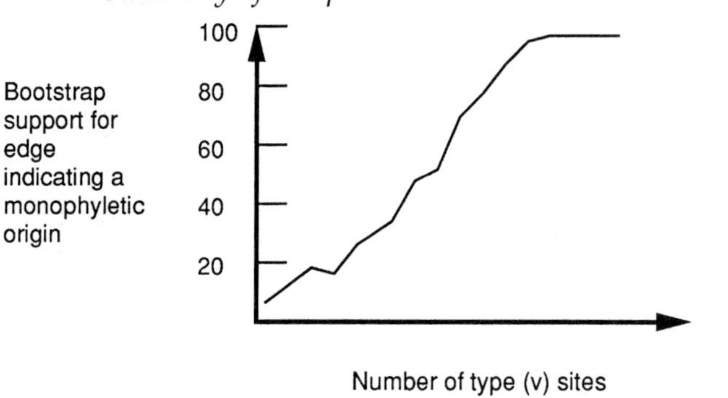

Figure 15.4 Effects of adding increasing numbers of type (v) sites on support for a monophyletic origin of plastids. For explanation see text.

of the *atpB* and 16S rDNA sequences by constructing datasets that contained:

1. the positions where there is variation both in plastids and in prokaryotes
2. the positions which do not vary in the plastids but do vary in the prokaryotes
3. the positions where the plastids all have the same nucleotide and where the prokaryotes all have the same nucleotide, which is different from the one in the plastids
4. varying numbers of positions which are invariant across all taxa
5. varying numbers of positions where there is no variation in prokaryotes but there is variation in plastids.

Our results with *atpB* and 16S rDNA are shown schematically in Figure 15.4. They suggest that only when positions that are invariable in prokaryotes and variable in plastids are included in the tree building do we see support for the edge separating plastids and prokaryotes. These findings support the idea that it is the uneven distribution of invariable sites between prokaryotes and plastids which gives rise to the apparent structure in trees reconstructed from these sequences, and thus the support for the monophyletic hypothesis.

15.6 SUMMARY OF THE PROBLEMS ENCOUNTERED

It is therefore clear that the outcome of attempts to use sequence data to differentiate between hypotheses of monophyletic or polyphyletic origin of plastids can be influenced very significantly by two kinds of bias. The first is a bias in the kinds of nucleotide substitution that take place in different lineages, with an apparent tendency for substitutions that increase the AT content of plastid genomes. The use of the LogDet correction allows artefacts caused by this to be avoided. The second problem is a bias in those sites that are free to vary. The effects of this can be estimated to some extent by the construction of artificial datasets from the original. It is important to recognize that the effects caused by these biases may not necessarily be misleading; they may be reinforcing genuine historical information, rather than giving an illusion of historical information when none actually exists. However, it will be important to understand more about how these phenomena arise in order to assess realistically the reliability of trees generated using sequence data.

15.7 SECONDARY ENDOSYMBIOSES

Work on the cryptophytes and chlorarachniophytes suggests that they have arisen by a process of serial endosymbiosis. As indicated in Table 15.1, these organisms have a plastid surrounded by a total of four membranes. They have an extra DNA-containing body, the nucleomorph, which is sandwiched between two of the membranes. These organisms are interpreted as having arisen in two stages, with a prokaryote being engulfed to form a eukaryote, with nuclear and plastid genomes. This was followed by a second endosymbiosis in which the photosynthetic eukaryote was taken up by a non-photosynthetic host, with the nucleus of the intermediate eukaryote remaining as the nucleomorph, lodged between membranes representing a vacuolar membrane of the second host and the plasma membrane of the first. Sequence-based phylogenetic trees suggest that the intermediate eukaryote for the cryptophytes may have been a red alga, and that for the chlorarachniophytes a green one (Douglas *et al.*, 1991; McFadden *et al.*, 1994; McFadden, Gilson and Waller, 1995; van de Peer *et al.*, 1996). Other plastids bounded by three or four membranes may have been derived in a similar way with complete loss of the nucleomorph.

15.8 GENERAL REMARKS

When one takes account of spurious patterns imposed on sequence data arising from differences in base composition or the number and distribution of sites free to vary, it is at present difficult to tell using sequence data alone whether plastids ultimately had a monophyletic or a polyphyletic origin. Nevertheless, even if plastids were ultimately derived from a single endosymbiosis, the process must subsequently have happened again several times to generate the cryptophytes and the chlorarachniophytes. In addition, another endosymbiosis (and possibly several) must be postulated to account for the existence of mitochondria. We believe the significance of other biochemical and morphological data should not be undervalued in comparison with sequence data when studying these ancient divergences. It is also important to be aware, too, that the problems we have highlighted in recovering plastid phylogeny need by no means be restricted to plastids. They should be kept in mind when using sequence data to study any ancient divergences.

15.9 REFERENCES

Bogorad, L. (1975) Evolution of organelles and eukaryotic genomes. *Science*, **188**, 891–8.

Delwiche, C.F. and Palmer, J.D. (1996) Rampant horizontal transfer and duplication of rubisco genes in eubacteria and plastids. *Molecular Biology and Evolution*, **13**, 873–82.

Dolganov, N.A.M., Bhaya, D. and Grossman, A.R. (1995) Cyanobacterial protein with similarity to the chlorophyll a/b-binding proteins of higher plants – evolution and regulation. *Proceedings of the National Academy of Sciences USA*, **92**, 636–40.

Douglas, S.E., Murphy, C.A., Spencer, D.F. and Gray, M.W. (1991) Cryptomonad algae are evolutionary chimaeras of two phylogenetically distinct unicellular eukaryotes. *Nature*, **350**, 148–51.

Fairweather, M.S., Packer, J.C.L. and Howe, C.J. (1994) The extrinsic proteins of Photosystem II in photosynthetic organisms: distribution, properties and evolutionary implications. *Biochemical and Biophysical Research Communications*, **205**, 1497–1502.

Grossman, A., Manodori, A. and Snyder, D. (1990) Light-harvesting proteins of diatoms: their relationship to the chlorophyll *a/b* binding proteins of higher plants and their mode of transport into plastids. *Molecular and General Genetics*, **224**, 91–100.

Hallick, R.B., Hong, L., Drager, R.G., *et al.* (1993) Complete sequence of *Euglena gracilis* chloroplast DNA. *Nucleic Acids Research*, **21**, 3537–44.

Howe, C.J. (1992) Plastid origin of an extrachromosomal DNA molecule from *Plasmodium*, the causative agent of malaria. *Journal of Theoretical Biology*, **158**, 199–205.

Howe, C.J., Barbrook, A.C. and Lockhart, P.J. (1995) Plastid origins, in *Photosynthesis: from Light to Biosphere*, vol. 1 (ed. P. Mathis), Kluwer Academic, the Netherlands, pp. 939–44.

Howe, C.J., Barker, R.F., Bowman, C.M. and Dyer, T.A. (1988) Common features of three inversions in wheat chloroplast DNA. *Current Genetics*, **13**, 343–9.

Howe, C.J., Beanland, T.J., Larkum, A.W.D. and Lockhart, P.J. (1992) Plastid origins. *Trends in Ecology and Evolution*, **7**, 378–83.

Howe, C.J. and Smith, A.G. (1991) Plants without chlorophyll. *Nature*, **349**, 109.

Kowallik, K.V., Stoebe, B., Schaffran, I. *et al.* (1995) The chloroplast genome of a chlorophyll a+c-containing alga, *Odontella sinensis. Plant Molecular Biology Reporter*, **13**, 336–42.

Lockhart, P.J., Howe, C.J., Bryant, D.A. *et al.* (1992) Substitutional bias confounds inference of cyanelle origins from sequence data. *Journal of Molecular Evolution*, **34**, 153–62.

Lockhart, P.J., Larkum, A.W.D., Steel, M.A. *et al.* (1996) Evolution of chlorophyll and bacteriochorophyll: the problem of invariant sites in sequence analysis. *Proceedings of the National Academy of Sciences USA*, **93**, 1930–4.

Lockhart, P.J., Steel, M.A., Hendy, M.D. and Penny, D. (1994) Recovering evolutionary trees under a more realistic model of sequence evolution. *Molecular Biological and Evolution*, **11**, 605–12.

Margulis, L. (1981) *Symbiosis in Cell Evolution*. W.H. Freeman, San Francisco.

McFadden, G.I., Gilson, P.R., Hofmann, C.J.B. *et al.* (1994) Evidence that an amoeba acquired a chloroplast by retaining part of an engulfed eukaryotic alga. *Proceedings of the National Academy of Sciences USA*, **91**, 3690–4.

McFadden, G.I., Gilson, P.R. and Waller, R.F. (1995) Molecular phylogeny of chlorarachniophytes based on plastid ribosomal RNA and rbcL sequences. *Archiv fur Protistenkunde*, **145**, 231–9.

Mereschkowsky, C. (1905) Über Natur und Ursprung der Chromatophoren im Pflanzenreiche. *Biologisches Centralblatt*, **25**, 593–604.

Morden, C.W. and Golden, S.S. (1991) Sequence analysis and phylogenetic reconstruction of the genes encoding the large and small subunits of ribulose-1,5-bisphosphate carboxylase/oxygenase from the chlorophyll *b*-containing prokaryote *Prochlorothrix hollandica. Journal of Molecular Evolution*, **32**, 379–95.

Morse, D., Salois, P., Markovic, P. and Hastings, J.W. (1995) A nuclear-encoded form II RuBisCo in dinoflagellates. *Science*, **268**, 1622–4.

Palmer, J.D. and Delwiche, C.F. (1996) Second-hand chloroplasts and the case of the disappearing nucleus. *Proceedings of the National Academy of Sciences USA*, **93**, 7432–5.

Pfanzagl, B., Zenker, A., Pittenauer, E. *et al.* (1996) Primary structure of cyanelle peptidoglycan of *Cyanophora paradoxa*: a prokaryotic cell wall as part of an organelle envelope. *Journal of Bacteriology*, **178**, 332–9.

Reith, M. and Munholland, J. (1993) A high-resolution map of the chloroplast genome of the red alga *Porphyra purpurea. Plant Cell*, **5**, 465–75.

Reith, M. and Munholland, J. (1995) Complete nucleotide sequence of the *Porphyra purpurea* chloroplast genome. *Plant Molecular Biology Reporter*, **13**, 333–5.

Schimper, A.F.W. (1883) Uber die Entwickelung der Chlorophyllkörner und Farbkörper. *Botanische Zeitung.*, 41, 105–112.

Siemeister, G. and Hachtel, W. (1989) A circular 73kb DNA from the colourless flagellate *Astasia longa* that resembles the chloroplast DNA of *Euglena*: restriction and gene map. *Current Genetics*, **15**, 435–41.

Steel, M.A., Lockhart, P.J. and Penny, D. (1993) Confidence in evolutionary trees from biological sequence data. *Nature*, **364**, 440–2.

Stirewalt, V.L., Michalowski, C.B., Löffelhardt, W.L. *et al.* (1995) Nucleotide sequence of the cyanelle genome from *Cyanophora paradoxa. Plant Molecular Biology Reporter*, **13**, 327–32.

van de Peer, Y., Rensing, S.A., Maier, U.-G. and de Wachter, R. (1996) Substitution rate calibration of small subunit ribosomal RNA identifies chlorarachniophyte endosymbionts as remnants of green algae. *Proceedings of the National Academy of Sciences USA*, **93**, 7732–6.

van Walraven, H.S., Lutter, R. and Walker, J.E. (1993) Organization and sequences of genes for the subunits of ATP synthase in the thermophilic cyanobacterium *Synechococcus 6716*. *Biochemical Journal*, **294**, 239–51.

Williamson, D.H., Gardner, M.J., Preiser, P. *et al.* (1994) The evolutionary origin of the 35kb circular DNA of *Plasmodium falciparum*: new evidence supports a possible rhodophyte ancestry. *Molecular and General Genetics*, **243**, 249–52.

Wolfe, G.R., Cunningham, F.X., Durnford, D. *et al.* (1994) Evidence for a common origin of chloroplasts with light-harvesting complexes of different pigmentation. *Nature*, **367**, 566–8.

Wolfe, K.H., Morden, C.W. and Palmer, J. (1992) Function and evolution of a minimal plastid genome from a nonphotosynthetic parasitic plant. *Proceedings of the National Academy of Sciences USA*, **89**, 10648–52.

16

Plastid-like DNA in apicomplexans

R. J. M. Wilson

National Institute for Medical Research, Mill Hill, London NW7 1AA. E-mail: r-wilson@nimr.mrc.ac.uk

ABSTRACT

The extrachromosomal DNAs of malaria parasites (*Plasmodium* spp.) and related apicomplexans show unexpected structural and molecular diversity. This article focuses on the 35 kb plastid-like, circular DNA (plDNA), of significance to evolutionary biology in general, and with possible implications for chemotherapeutics. The plDNA molecule of apicomplexans carries ~60 genes largely concerned with transcription, the genetic content bearing a superficial resemblance to the vestigial plastid DNA of non-photosynthetic parasitic plants. Unlike some of the latter, the malarial plDNA encodes a wide enough spectrum of tRNAs to accommodate a minimal translation system for its own protein-encoding genes. Of this subset of genes, only two have been identified that have functions outwith the machinery of genetic expression. One appears to be related to the ubiquitous family of Clp molecular chaperones whilst the other (ORF470) is a highly conserved protein of unknown function found in bacteria as well as the plastids of rhodophytic and chromophytic algae and a higher plant. The location of the plDNA, long predicted to be in the multi-membraned 'Golgi-adjunct' of toxoplasma ('spherical body' of *Plasmodium*) has been confirmed. The origin of this organelle, either in the form of a plastid or a photosynthetic bacterium, has still to be decided unequivocally but our working hypothesis is firmly based on the former possibility. We have proposed further that the organelle may be of secondary endosymbiotic origin, following an ancient interaction between a progenitor of the Apicomplexa and an alga. An important implication is that many algal genes vital for maintenance of plastid function lie in the host nucleus where they still await discovery. It is likely that these genes will provide clues explaining the maintenance of the vestigial

Evolutionary Relationships Among Protozoa. Edited by G.H. Coombs, K. Vickerman, M.A. Sleigh and A. Warren. Published in 1998 by Chapman & Hall, London. ISBN 0 412 79800 X

plastid over what has probably been an extensive evolutionary period. Further comparative analysis of apicomplexan plDNAs might help to elucidate the evolutionary, functional and chemotherapeutic possibilities of this organelle whose special functions remain obscure.

16.1 INTRODUCTION

Reports of extrachromosomal DNA in malaria parasites date from the early 1970s (reviewed in Feagin, 1994; Wilson and Williamson, 1997). At first, the findings were overshadowed by the potential problem of physical contamination with DNA from other sources, and the belief that the extrachromosomal DNA must be solely of mitochondrial origin – thought to be 'boring'! More interesting alternatives were considered only when DNA sequences were obtained that did not fit with these scenarios. Despite our early conversion to the belief that malaria parasites and related apicomplexans carry two separate sources of extrachromosmal DNA, one mitochondrial (mtDNA) and the other plastid (plDNA) (Wilson *et al.*, 1991), others maintained that the molecular criteria we used were not foolproof in distinguishing between two forms of mtDNA. This dilemma has continued up to the present day, as more primitive (bacteria-like) mt and plDNAs have been found with greater overlap in their genetic content than was known hitherto (Gray and Spencer, 1996). The nucleotide sequence of the apicomplexan plDNA itself did not help readily to resolve the issue because, at least in *Plasmodium,* it is extremely rich in A.T residues (86%), a property that tends to undermine standard phylogenetic comparisons with conventional DNAs.

The emphasis of this chapter is to explain why we believe one of the two extrachromosomal DNAs of apicomplexans is carried in a vestigial plastid. Moreover, reasons will be set out for suggesting that the plastid is likely to be of algal origin, acquired by an early progenitor of the Apicomplexa through secondary endosymbiosis. Besides the evolutionary implications of this work, an intriguing issue to emerge is that pathogenic apicomplexans carry a novel organelle derived from a plastid, which could conceivably be a focus for the discovery of new drug targets. But much yet remains to be done to identify such a target.

16.2 THE plDNA

A gene map of the plDNA of the human malaria parasite *Plasmodium falciparum* is shown in Figure 16.1. As detailed descriptions of this map have been given elsewhere (Wilson *et al.*, 1996), attention will be drawn

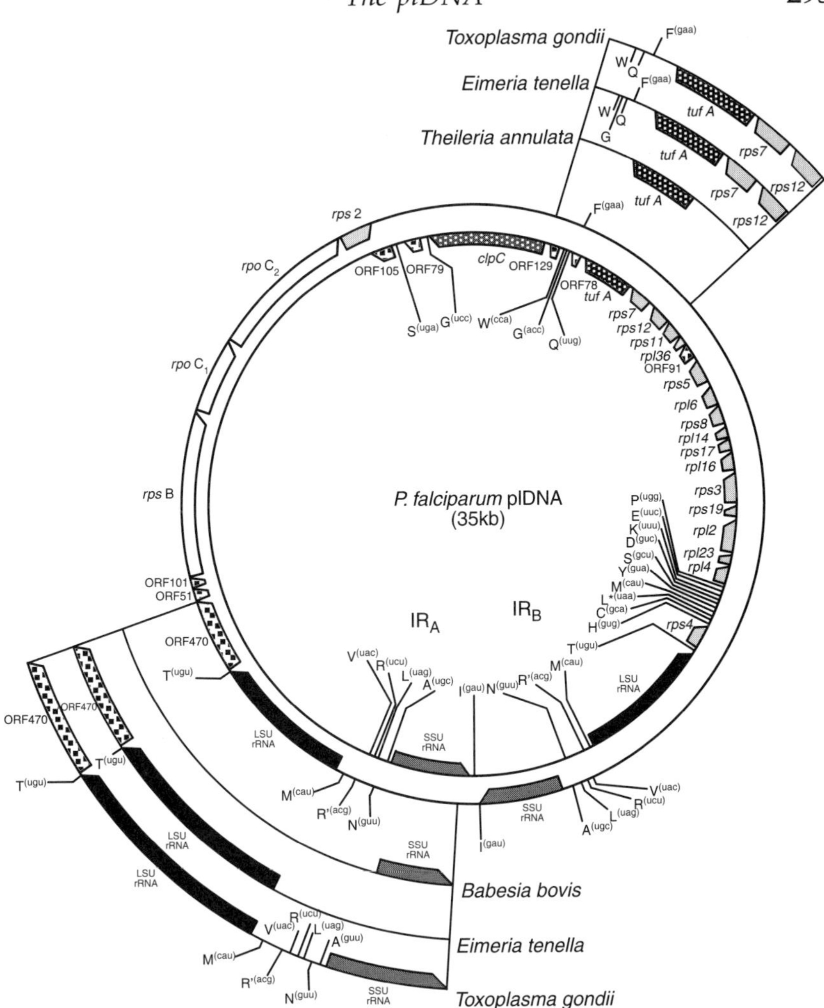

Figure 16.1 The gene map (Wilson *et al.*, 1996) of the 35 kb plDNA of *Plasmodium falciparum* (in the centre) is aligned with sequences from the plDNAs from *Toxoplasma gondii, Eimeria tenella* and *Theileria annulata* (Paul Denny, N.I.M.R. personal communication) and *Babesia bovis* (Gozar and Bagnara, 1995).

only to features relevant to a discussion of the molecule's origin, conservation, and function.

The inverted repeat (IR), containing duplicated copies of large and small subunit ribosomal RNA genes (rRNA) as well as nine duplicated tRNA genes, is diagnostic of apicomplexan plDNA as the gene arrangement is presently unique. It should be emphasized, however, that knowledge of apicomplexan plDNA is very limited, only the molecules from *Plasmodium* (Wilson *et al.*, 1996) and toxoplasma (David Roos, Philadelphia, personal communication) having been completely sequenced.

Downstream of the IR_B arm of the inverted repeat, a string of genes encoding ribosomal proteins (rps) bears a striking resemblance to the truncated forms of the *S10, spc, alpha* and *str* operons of the *Escherichia coli* genome found in plastids. Unlike the plDNAs of higher plants, but like those of algae, the series of ribosomal protein genes on the 35 kb circle is succeeded by a *tuf* gene, encoding the elongation factor Tu (EF-Tu). In contrast with the rps, which are often poorly conserved at the amino acid level, EF-Tu is one of the most highly conserved predicted protein sequences encoded by the circle (45% identity with *E.coli*).

Downstream of *tuf*, a member of the *clp* gene family has been identified tentatively. These ubiquitous, nucleotide-binding, high molecular weight heat shock proteins, known as Hsp100 proteins, are molecular chaperones with the ability to disaggregate proteins in the presence of ATP (Schirmer *et al.*, 1996). However, the *Plasmodium* gene is peculiar, being similar only to the 3′-half of members of the *clp*A/B/C family, and so the relationship of its product to conventional clp proteins still has to be verified.

Downstream of the IR_A arm of the inverted repeat, lies an open reading frame encoding 470 amino acids (ORF470). This ORF has a high level of similarity (50% amino acid identity) to a gene of unknown function (*ycf*24) found in various bacteria, as well as on the plDNAs of rhodophyte and chromophyte algae (Williamson *et al.*, 1994; Kowallick *et al.*, 1995).

Finally, mention should be made of the *rpo* genes lying downstream of ORF470. These encode three subunits (β, β′, β″) of a prokaryotic type of RNA polymerase typically found in plastids and distinct from that of mitochondria (Gardner, Williamson and Wilson, 1991).

16.3 CONSERVATION OF THE plDNA IN APICOMPLEXANS

The plDNAs of *Plasmodium* spp. and *Toxoplasma gondii* are physically similar in size and like the *Eimeria tenella* homologue can form a large cruciform structure within the IR (Borst *et al.*, 1984; Wilson *et al.*, 1993). We have recently compared gene organization on the plDNA of *Plasmodium* with segments of the corresponding DNAs of *Toxoplasma, Eimeria* and *Theileria* (Denny *et al.*, unpublished). Figure 16.1 shows the gene arrangement is well conserved, although we note that $tRNA^{gly(acc)}$ is absent from its anticipated position on the *T. gondii* plDNA and the intergenic distances vary between genera. Although none of the control regions on the plDNA are known, intergenic regions may be under less selective pressure than the coding regions, hence their variability. Because the juxtaposition (including strandedness) of some of the genes is characteristic, these results are compatible with the hypothesis that the apicomplexan plDNAs have a single origin. In this regard, it is interesting that there is a significant difference in tryptophan codon usage between *T. gondii* and *P. falciparum* (P. Denny, personal communication); in *T. gondii* TGA (usually a stop codon) is sometimes used in certain conserved tryptophan positions. Further comment on the extent of diversity that might yet be found in both organization and gene content of the apicomplexan 35 kb plDNAs is probably premature. Suffice it to say that the features described above for *T. gondii* have been confirmed independently by David Roos and his colleagues at the University of Pennsylvania, Philadelphia (personal communication).

16.4 PLASTID-LIKE FEATURES OF THE 35 KB CIRCLE

As would be anticipated from their different evolutionary origins, the codon usage of the mt and plDNAs of *Plasmodium* is distinct. Additionally, some organizational features of the 35 kb DNA are more consistent with a plastid origin rather than a bacterial one – see Table 16.1.

The split *rpo*C gene is a feature of cyanobacteria and plastids, but not *E. coli*. Moreover, the juxtaposition of *rpo*C_2 and *rps* 2 genes parallels the arrangement found in plastids (Reith and Munholland, 1993) and these two *Plasmodium* genes are co-transcribed (Figure 16.2). By contrast, *rps* 2 is located separately from the *rpo* genes in extant (sampled) cyanobacteria.

The intron on the anticodon stem of the malarial 35kb gene for $tRNA^{leu(uaa)}$ (Preiser, Williamson and Wilson, 1995), provides another link with cyanobacterial and plastid DNAs, because this tRNA gene is uninterrupted in eubacteria for which information is available (see references in Delwiche, Kuhsel and Palmer, 1995).

The presence of the *tuf* gene on the malarial circle also can be mentioned in the context of plastid origins because it occurs on the plastid genomes of most algae, whereas in higher plants and green algae on the

Table 16.1 Plastid-like gene arrangements of the 35 kb DNA

*rpo*C	Split into *rpo*C_1 and *rpo*C_2.
rpoC_2	Followed by *rps* 2.
$tRNA^{leu}$	Has an intron in the anticodon stem.
rps	Clustered like truncated bacterial operons.

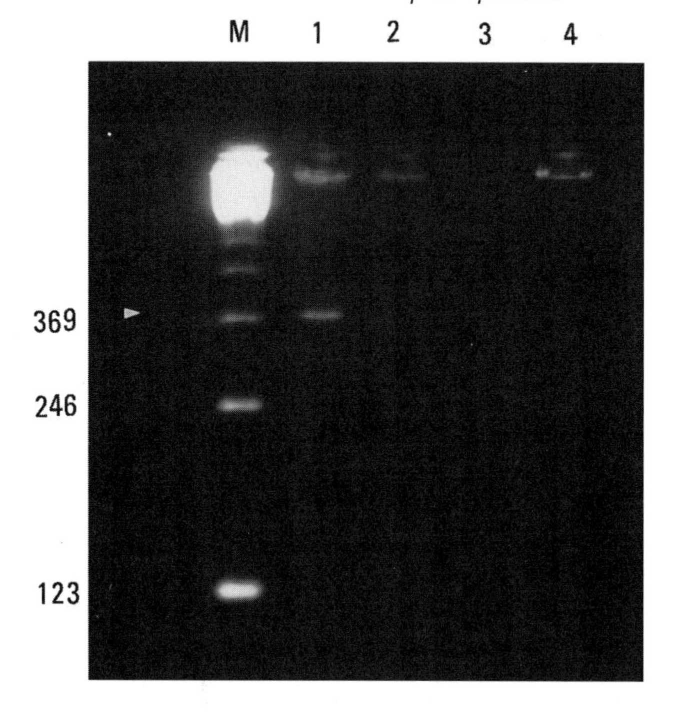

Figure 16.2 Evidence obtained by reverse transcription polymerase chain reaction (RTPCR) that *rpo*C_2 and *rps*2 are co-transcribed in *P. falciparum*. Lane 1: DNase-treated total RNA was amplified using random primers and Superscript II reverse transcriptase. The cDNA was then amplified using DNA polymerase (AmpliTaq) and two specific primers designed to bridge the adjacent genes on the 35 kb circle. The expected product size was 370 nt (arrow). M: 123 nt markers. The following controls were used: Lane 2: as in 1, but template pretreated with RNase; Lane 3: as in 1, but without RNA template; Lane 4: as in 1, but without both RNA and RT.

same lineage it has been translocated to the nucleus (Delwiche, Kuhsel and Palmer, 1995).

Phylogenetic analyses of *rpo*B (Gardner *et al.*, 1994), *rpo*C_1 (Howe, 1992), *tuf* (Kohler *et al.*, 1997) and rDNA (see Egea and Lang-Unnasch, 1995), uniformly place these 35 kb circle genes on a lineage distantly related to euglenoid plastids. Earlier phylogenies of *Euglena* itself (based on the rDNApl) placed its plastid on the rhodo-chromophyte lineage but this is now regarded as an artefact due to the unusually rich A.T base composition of the *Euglena* sequence; more recent analyses associate it with green plastids (Delwiche, Kuhsel and Palmer, 1995). Basic phylogenetic problems, similar to that just cited for *Euglena*, apply to the phylogenetic analysis of genes on the 35 kb plDNA – the sequences are invariably placed on long branches, and closely related sequences (e.g. from algae or from dinoflagellates which are notoriously phagotrophic and have plastids derived from different sources) are unavailable. It can be concluded that these analyses suffer from inadequate sampling in the relevant areas of divergence from the global tree.

Resolution of this problem might help to establish the origin of the apicomplexan plastid organelle. It is worth noting that the *Euglena* plastid gene for $tRNA^{leu(uaa)}$ does not have an intron, unlike its counterpart on many other plastid genomes, including that of the malaria parasite. This emphasizes the obvious point that success in seeking an apicomplexan plastid ancestor within the photosynthetic lineages by phylogenetic means will depend ultimately on what sequences are tested. Nonetheless, obtaining the complete gene map of the 35 kb circle has borne out the prediction (Howe, 1992) that it would contain genes, besides the *rpo* group, homologous with those of chloroplasts.

16.5 THE PLASTID ORGANELLE

It has been widely held for a number of years that the most likely location of plDNA in the apicomplexan cell is the small ovoid, or dumbbell-shaped, body known as the 'spherical body' in malaria and the 'Golgi-adjunct' in *Toxoplasma* (for a fuller list of names for this organelle in other apicomplexans, see Siddall, 1992). Only one of these bodies is evident in malaria and toxoplasma parasites but they may be more abundant in gregarines (see Siddall, 1992). High resolution *in situ* hybridization with DNA and SSUrRNA probes has confirmed the predicted localization of the plDNA in toxoplasma (McFadden *et al.*, 1996; Kohler *et al.*, 1997).

This finding leaves two interesting issues to be addressed, namely, how did the organelle arise and what does it do? With regard to the first point, McFadden and his colleagues maintain that the plastid is surrounded by at least two membranes (with the possibility of a third), but suggest that additional ones seen in some sections are likely to be merely folds of the host cell's endoplasmic reticulum. Others, ourselves included, have regarded the 'multi-membrane' aspect of the organelle in section as an integral feature (four membranes have been observed – see Dubremetz, 1995; Kohler *et al.*, 1997).

Conventional plastids (resulting from primary endosymbiosis with a cyanobacterium) have a double membrane whereas the presence of additional plastid membranes signifies the occurrence of secondary endosymbiotic events (Palmer and Delwiche, 1996). It is most unlikely that a progenitor of apicomplexans took up and maintained a 'free' plastid, because it would lack the essential genes that control plastid biogenesis carried in the nucleus of all other plastid-containing cells. The prospect that the progenitor of apicomplexans independently took into partnership a photosynthetic bacterium (by primary endosymbiosis) is against the monophyletic theory of plastids, as well as being counter to both the plastid-like organization of the organelle's DNA (as outlined above), and the various phylogenetic analyses.

To account for the multi-membraned envelope of the organelle, we proposed it is the remnant of an ancient secondary endosymbiotic event that took place between a progenitor of the Apicomplexa and an algal cell (Wilson, Williamson and Preiser, 1994), just as has been proposed for the plastids of euglenoids, dinoflagellates, and chromists. The first two of these other groups usually have plastids with three membranes, whereas the chromists, a diverse group of organisms with plastids derived from both red and green algae, have four membranes (Van de Peer *et al.*, 1996).

16.6 EVOLUTIONARY ORIGINS

Recognition of the plDNA's existence in apicomplexans points to a phototrophic alveolate ancestor of the phylum. This is consistent with its placement close to the dinoflagellate/ciliate lineages, as demonstrated by phylogenetic analysis of nuclear SSUrRNA genes (Van de Peer, Van de Auwera and De Wachter, 1996). All apicomplexans are parasitic and have diverged considerably during an extended evolutionary period. But because organellar genomes evolve at a faster rate than nuclear ones – as is clearly evident in this case from the extreme A.T-richness of the plDNA, it might be argued that further phylogenetic analysis of the apicomplexan plDNA is less likely to be as useful in determining ancestry as exploring nuclear genes used to maintain the plastid.

Curiously, apicomplexan nuclear genes with higher plant-like sequences have been found, such as the gene for enolase in *Plasmodium* (Read *et al.*, 1994), and the *hsp*30 gene in toxoplasma (Bohne, Gross and Heesemann, 1995). Similarities to higher plant genes were also hinted at by earlier phylogenies based on calmodulin and histones. Some of these genes appear to occur as single copies. But the argument that this makes them unlikely to represent genes translocated from other sources, for example from the algal symbiont we have proposed as the source of the apicomplexan plDNA, can now be challenged. This is because dinoflagellates themselves have recently been found to harbour replacement genes of unexpected origin in the nucleus, probably acquired by lateral transfer from a proteobacterial endosymbiont (reviewed by Palmer, 1995). Hence the so-called 'plant connection' with apicomplexans remains enigmatic and deserves further study. Endosymbiosis is increasingly seen as a means to explain the chimaeric nature of the eukaryotic genome (Martin, 1996).

16.7 EVIDENCE FOR PLASTID FUNCTION

Apart from photorespiratory functions, chloroplasts carry out many processes, including amino acid and fatty acid biosynthesis, and nitrogen metabolism (Wallsgrove, 1991). Plastid biogenesis necessitates chaperone and processing activities as many components now encoded in the cell's nucleus are targeted to the plastid and imported (Schnell, Kessler and Bobel, 1994). In addition, RNAs encoded on the plDNA are transcribed and used to synthesize intrinsic plastid protein components on the organelle's own 70S ribosomes.

By comparison with ordinary plDNAs, that of apicomplexans has undergone massive reduction in size and complexity. Nevertheless, its highly skewed genetic content suggests it is functional, albeit perhaps playing only a role in maintenance of the plastid compartment in which it resides. The highly diverged ribosomal protein sequences encoded by the plDNA suggest they have evolved under different constraints from the better conserved products of the ORF470, *tuf*, and *clp* genes. These last presumably interact with other conserved components specified by the nucleus and imported into the plastid by means of chaperones and membrane proteins recognizing transit peptides.

Transcription of many of the genes on the plDNA of *Plasmodium* has been verified (Preiser, Williamson and Wilson, 1995) and evidence for active protein synthesis is beginning to be gathered. The organelle appears to contain homogeneous, small, ribosome-like particles, internal membranous structures not being evident. The antibiotic clindamycin that blocks peptide bond formation and inhibits growth of *T. gondii* is believed to act at the level of the plastid compartment (Pfefferkorn, Nothnagel and Borotz, 1992). Evidence for a subset of plastid polysomes has been obtained using probes specific for rRNA encoded by the 35 kb circle (A. Roy *et al.*, N.I.M.R., unpublished). In these experiments, rapidly sedimenting rRNApl found in a polysome fraction obtained from *P. falciparum* was converted to a less rapidly sedimenting form by incubation with puromycin – incorporation of this analogue of amino acyl-tRNA into nascent peptide chains causing termination and dissociation of polysomes has been well documented (Gale *et al.*, 1981).

A different perspective has come from the claim that several apicomplexans carry a nuclear gene for the photosynthetic protein D1 of photosystem II (Hackstein *et al.*, 1995), the implication being that the product of this gene is (or was) targeted to the plastid compartment.

There is no evidence in *P. falciparum*, however, that the special plastid synthetic pathway for porphyrins utilizing $tRNA^{glu(uuc)}$ is active, although this tRNA is expressed from the plDNA (Preiser, Williamson and Wilson, 1995). Indeed, the conventional eukaryotic pathway, utilizing glycine as a precursor in the mitochondrion, has been demonstrated in *P. falciparum* (Surolia and Padmanaban, 1992).

In looking for a lead to a possible function of the apicomplexan plastid, it should perhaps be borne in mind that in specialized organisms such as malaria parasites the organelle may have been reduced to a vestige of its counterpart in the other apicomplexans yet to be examined. Almost nothing is known about stage-specific activity of the plastid in alternate phases of apicomplexan life cycles, but that differential activity might exist has been suggested by the altered levels of sensitivity to clindamycin in tachyzoite and bradyzoite forms of *T. gondii* (Tomavo and Boothroyd, 1995). A more extensive investigation of the plastid organelle and its genetic content in disparate genera of apicomplexans is becoming overdue and may contribute to our overall understanding of its function.

16.8 CONCLUSION

The origin and function of apicomplexan plDNA, as well as that of the plastid organelle of which it is an integral part, remain unknown. But the first steps have been taken to show that the plastid compartment maintains a separate transcription and translation system. Although the product of the highly conserved ORF470 gene (function unknown) might be of some direct importance, it could just as well have merely a housekeeping function required for maintenance of the organelle. It must be assumed then that the plastid's function depends on cytoplasmic factors encoded in the nucleus and imported into the organelle.

Acknowledgements

I thank my colleagues at N.I.M.R. for their expert help, as well as many others elsewhere who have collaborated with us over the years. This work was supported by the UNDP/World Bank/WHO Special Programme for Research in Tropical Diseases (TDR).

16.9 REFERENCES

Bohne, W., Gross, U., Ferguson, D.J.P. and Heesemann, J. (1995) Cloning and characterization of a bradyzoite-specifically expressed gene (*hsp*30/bag1) of *Toxoplasma gondii*, related to genes encoding small heat-shock proteins of plants. *Molecular Microbiology*, **16**, 1221–30.

Borst, P., Overdulve, J.P., Weijers, P.J. *et al.* (1984) DNA circles with cruciforms from *Isospora (Toxoplasma) gondii. Biochimica Biophysica Acta*, **781**, 100–11.

Delwiche, C.F., Kuhsel, M. and Palmer, J.D. (1995) Phylogenetic analysis of *tuf*A sequences indicates a cyanobacterial origin of all plastids. *Molecular and Phylogenetic Evolution*, **4**, 110–28.

Dubremetz, J.F. (1995) *Toxoplasma gondii*: Cell biology update, in *Molecular Approaches to Parasitology* (eds J.C. Boothroyd and R. Komuniecki), Wiley–Liss, New York, pp. 345–58.

Egea, N. and Lang-Unnasch, N. (1995) Phylogeny of the large extra-chromosomal DNA of organisms in the phylum Apicomplexa. *Journal of Eukaryotic Microbiology*, **42**, 679–84.

Feagin, J.E. (1994) The extrachromosomal DNAs of Apicomplexan parasites. *Annual Review of Microbiology*, **48**, 81–104.

Gale, E.F., Cundcliffe, E., Reynolds, P.E. *et al.* (1981) Antibiotic inhibitors of ribosome function, in *The Molecular Basis of Antibiotic Action* (eds E.F. Gale, E. Cundcliffe, P.E. Reynolds, M.H. Richmond and M.J. Waring), Wiley, Chichester, pp. 402–547.

Gardner, M.J., Goldman, N., Barnett, P. *et al.* (1994) Phylogenetic analysis of the *rpo*B gene from the plastid-like DNA of *Plasmodium falciparum. Molecular and Biochemical Parasitology*, **66**, 221–31.

Gardner, M.J., Williamson, D.H. and Wilson, R.J.M. (1991) A circular DNA in malaria parasites encodes an RNA polymerase like that of prokaryotes and chloroplasts. *Molecular and Biochemical Parasitology*, **44**, 115–23.

Gozar, M.M.G. and Bagnara, A.S. (1995) An organelle-like small subunit ribosomal RNA gene from *Babesia bovis*: Nucleotide sequence, secondary structure of the transcript and preliminary phylogenetic analysis. *International Journal of Parasitology*, **25**, 929–38.

Gray, M.W. and Spencer, D.F. (1996) Organellar evolution, in *Evolution of Microbial Life*, Society of General Microbiology, Symposium 54 (eds D. McL. Roberts, P.M. Sharp, C. Alderson and M. Collins), Cambridge University Press, Cambridge, pp.109–26.

Hackstein, J.H.P., Mackenstedt, U., Mehlhorn, H. *et al.* (1995) Parasitic apicomplexans harbor a chlorophyll *a* – D1 complex, the potential target for therapeutic triazines. *Parasitology Research*, **81**, 207–16.

Howe, C.J. (1992) Plastid origin of an extrachromosomal DNA molecule from *Plasmodium*, the causative agent of malaria. *Journal of Theoretical Biology*, **158**, 199–205.

Kohler, S., Delwiche, C.F., Denny, P. *et al.* (1997) A plastid of probable green algal origin in apicomplexan parasites. *Science*, **275**, 1485–9.

Kowallik, K. V., Stobe, B., Schaffran, I. *et al.* (1995) The chloroplast genome of a chlorophyll a+c- containing alga, *Odontella sinensis. Plant Molecular Biology Reporter*, **13**, 336–42.

Martin, W.F. (1996) Is something wrong with the tree of life? *Bioessays*, **18**, 523–7.

McFadden, G.I., Reith, M.E., Munholland, J. and Lang-Unnasch, N. (1996) Plastid in human parasites. *Nature*, **381**, 482.

Palmer, J.D. (1995) Rubisco rules fall; gene transfer triumphs. *Bioessays*, **17**, 1005–8.

Palmer, J.D. and Delwiche, C.F. (1996) Second-hand chloroplasts and the case of the disappearing nucleus. *Proceedings of the National Academy of Sciences USA*, **93**, 7432–5.

Pfefferkorn, E.R., Nothnagel, R.F. and Borotz, S.E. (1992) Parasiticidal effect of clindamycin on *Toxoplasma gondii* grown in cultured cells and selection of a drug-resistant mutant. *Antimicrobial Agents and Chemotherapy*, **36**, 1091–6.

Preiser, P., Williamson, D.H. and Wilson, R.J.M. (1995) tRNA genes transcribed from the plastid-like DNA of *Plasmodium falciparum. Nucleic Acids Research*, **23**, 4329–36.

Read, M., Hicks, K.E., Sims, P.F.G. and Hyde, J.E. (1994) Molecular characterisation of the enolase gene from the human malaria parasite *Plasmodium falciparum. European Journal of Biochemistry*, **220**, 513–20.

Reith, M. and Munholland, J. (1993) A high resolution gene map of the chloroplast genome of the red alga *Porphyra purpurea. The Plant Cell*, **5**, 465–75.

Schirmer, E.C., Glover, J.R., Singer, M.A. and Lindquist, S. (1996) Hsp100/Clp proteins: A common mechanism explains diverse functions. *Trends in Biochemical Sciences*, **21**, 289–96.

Schnell, D.J., Kessler, F. and Blobel, G. (1994) Isolation of components of the chloroplast protein import machinery. *Science*, **266**, 1007–12.

Siddall, M.E. (1992) Hohlzylinders. *Parasitology Today*, **8**, 90–1.

Surolia, N. and Padmanaban, G. (1992) *De novo* biosynthesis of heme offers a new chemotherapeutic target in the human malarial parasite. *Biochemical and Biophysical Research Communications*, **187**, 744–50.

Tomavo, S. and Boothroyd, J.C. (1995) Interconnection between organellar functions, development and drug resistance in the protozoan parasite, *Toxoplasma gondii. International Journal of Parasitology*, **25**, 1293–9.

Van de Peer, Y., Rensing, S.A., Maier, U.-G. and De Wachter, R. (1996) Substitution rate calibration of small subunit ribosomal RNA identifies chlorarachniophyte endosymbionts as remnants of green algae. *Proceedings of the National Academy of Sciences USA*, **93**, 7732–6.

Van de Peer, Y., Van de Auwera, G. and De Wachter, R. (1996) The evolution of stramenopiles and alveolates as derived by 'substitution rate calibration' of small ribosomal subunit RNA. *Journal of Molecular Evolution*, **42**, 201–10.

Wallsgrove, R. M. (1991) Plastid genes and parasitic plants. *Nature*, **350**, 664.

Williamson, D.H., Gardner, M.J., Preiser, P. *et al.* (1994) The evolutionary origin of the malaria parasite's 35 kb circular DNA; new evidence supports a possible rhodophyte ancestry. *Molecular and General Genetics*, **243**, 249–52.

Wilson, I., Gardner, M., Rangachari, K. and Williamson, D. (1993) Extrachromosomal DNA in the Apicomplexa, in *Toxoplasmosis*. NATO ASI Series, H78 (ed. J.E. Smith), Springer-Verlag, Heidelberg, pp. 51–60.

Wilson, R.J.M., Denny, P.W., Preiser, P.R. *et al.* (1996) Complete gene map of the plastid-like DNA of the malaria parasite *Plasmodium falciparum. Journal of Molecular Biology*, **261**, 155–72.

Wilson, R.J.M., Gardner, M.J., Feagin, J.E. and Williamson, D. H. (1991) Have malaria parasites three genomes? *Parasitology Today*, **7**, 134–6.

Wilson, R.J.M. and Williamson, D.H. (1997) Extrachromosomal DNA in the Apicomplexa. *Microbiology and Molecular Biology Reviews*, **61**, 1–16.

Wilson, R.J.M., Williamson, D.H. and Preiser, P. (1994) Malaria and other Apicomplexans: The 'plant' connection. *Infectious Agents and Disease*, **3**, 29–37.

17

The karyorelictids (Protozoa: Ciliophora), a unique and enigmatic assemblage of marine, interstitial ciliates: a review emphasizing ciliary patterns and evolution

Wilhelm Foissner

Universität Salzburg, Institut für Zoologie, Hellbrunnerstrasse 34, A-5020 Salzburg, Austria

ABSTRACT

This review updates morphology, ecomorphology, and evolution of karyorelictids, a small (~135 described species, but many more very likely exist) but unique assemblage of mainly marine, interstitial ciliates having paradiploid, non-dividing macronuclei originating from micronuclei. Thus, they have been widely considered to represent an ancestral state of the dimorphic ciliate nuclear apparatus. Most of the gross morphological peculiarities of the karyorelictids (e.g. filiform shape, high regeneration capacity) are apparent adaptations to the spatial structure and unstable conditions of their preferred biotope, coastal sands. Cladistic analysis, based on a re-investigation of most main groups of karyorelictids, produced two major branches, one containing geleiids and another with loxodids and trachelocercids. The geleiids are completely ciliated, like the supposed ancestors (heterotrichs) of the karyorelictids, and have unique monokinetidal oral structures very different from those of other karyorelictids. The loxodid/trachelocercid clade has a very strong synapomorphy, viz. a highly specialized ciliary row (bristle

Evolutionary Relationships Among Protozoa. Edited by G.H. Coombs, K. Vickerman, M.A. Sleigh and A. Warren. Published in 1998 by Chapman & Hall, London. ISBN 0 412 79800 X

kinety) surrounding a glabrous (unciliated) stripe on the left side of the cell. Ultrastructural and molecular data have suggested a close relationship between karyorelictids and heterotrichs s. str. (e.g. *Stentor*). Surprisingly, such a relationship is hardly recognizable in the somatic and oral ciliary pattern, which shows some (analogous?) characters (e.g. the trachelocercid oral apparatus) highly reminiscent of those found in haptorid gymnostomes and especially prostomatids (e.g. *Coleps*). Stomatogenesis of *Loxodes* is buccokinetal and thus links karyorelictids with oligohymenophorans rather than with heterotrichs. The karyorelictid infraciliature is rather complex and diverse. Thus, the nuclear peculiarities are very likely not ancestral but derived and probably evolved several times, as indicated by the quite different organization of geleiids and loxodids/trachelocercids.

17.1 INTRODUCTION

Ciliates are unicellular, heterokaryotic organisms having a macronucleus and a micronucleus of distinctly different size and function within the same cytoplasm (Raikov, 1982). The macronucleus, which is usually highly polyploid, divides amitotically during asexual reproduction and controls mainly somatic functions (e.g. RNA synthesis, morphogenesis, regeneration). The diploid micronucleus is active mainly during sexual reproduction (conjugation), although recent experiments indicate that it also plays an important role during asexual morphogenesis (Ng, 1990). However, the macronuclei of a restricted group of ciliates, the Karyorelictea, are diploid or nearly diploid (paradiploid) and cannot divide but differentiate from micronuclei during and after cell division (Raikov, 1958, 1982). These peculiarities were interpreted by Corliss (1974, 1979) as being ancestral (relict), and thus he named the whole group 'Karyorelictea'. However, recent molecular evidence (see chapter 18) and the morphological data summarized in this paper indicate that the special nuclear features of the karyorelictids could be derived (apomorph), i.e. evolved secondarily from polyploid ciliate nuclei.

Morphological analysis of the karyorelictids was limited for a long time by their extreme fragility, although many basic features were explored in the pioneering studies by Dragesco (1960), Dragesco and Dragesco-Kernéis (1986), Raikov, Gerassimova-Matvejeva and Puytorac (1975), and Wilbert (1986). Using a new, very 'strong' fixative and Wilbert's protargol technique, Foissner (1995, 1996a–c) and Foissner and Dragesco (1996a,b) obtained excellent preparations from all main groups of karyorelictids, showing a world of new details.

The present paper is a brief overview of the group, emphasizing recent morphological data, phylogeny, and possible relationships between karyorelictids and other ciliates. The very restricted space allowed does not permit much detail, but I hope to summarize the main points in a way attractive to both beginners and specialists. Although outdated in some respects, the last comprehensive reviews on karyorelictids by Corliss and Hartwig (1977) and Raikov (1982, 1994) are still useful and should be consulted, especially for details on their nuclear features and ecology.

17.2 COMPARATIVE MORPHOLOGY, CLADISTICS, AND CLASSIFICATION OF KARYORELICTIDS

The karyorelictids are a small group comprising about 135 species classified into two subclasses, three orders, six families and 11 genera (Table 17.1). I agree with the supraordinal classification suggested by Puytorac (1994) and Puytorac *et al.* (1987), however, with *Protocruzia* excluded because it has mitotically dividing macronuclei and recent molecular data group it with the hypotrichs (Hammerschmidt *et al.* 1996). It is thus doubtful whether this enigmatic genus (see Grolière *et al.* (1980) and Raikov (1982) for detailed accounts) can serve as a model for the origin of nuclear dimorphism in ciliates, as suggested by Bardele and Klindworth (1996); rather, it seems to be a dead and/or specialized route like that of the karyorelictids. *Stephanopogon*, another 'eociliate' classified by Corliss (1979), Corliss and Hartwig (1977) and others near the karyorelictids, has been proven to be a flagellate (Lipscomb and Corliss, 1982).

Our studies on the infraciliature provide a rather clear picture of the evolution within the karyorelictids, at least as concerns the main groups. Consideration of the evidence extends only to family level (Figure 17.1); see Foissner (1996a) and Foissner and Dragesco (1996b) for details on genera.

Using heterotrichs s. str. as outgroup, as suggested by the ultrastructural and molecular data, the karyorelictids can be founded as a monophyletic group by two unique characters (apomorphies), viz. the loss of adoral membranelles and of dividing macronuclei (see final section for detailed discussion). The cladogram then splits into two major branches, one containing loxodids and trachelocercids and the other with geleiids (Figure 17.1).

The geleiids are a very conspicuous component of the interstitial ciliate fauna because some attain a length of up to 5 mm and most have brown pigment granules in the cortex making them dark at low magnification (Figure 17.7). Comparatively little is known about the infraciliature of the geleiids, the most important studies being those of Dragesco and Dragesco-Kernéis (1986) and Nouzarede (1977). These investigations showed that geleiids, unlike all other karyorelictids, are completely ciliated and have oral monokinetids forming a right and left oral ciliary field (Figure 17.9). These 'paracytostomal' monokinetids are

Table 17.1 Classification of karyorelictids and number of species within taxa

Taxa[a]	*Number of species*[b]
Class Karyorelictea Corliss, 1974	135
Subclass Protoheterotrichia Puytorac *et al.*, 1987	20
Order Protoheterotrichida Nouzarede, 1977	20
Family Geleiidae Kahl, 1933	20
Genera *Avelia* Nouzarede, 1977	3
Geleia gen. nov.[c]	17
Subclass Trachelocercia Puytorac *et al.*, 1987	65
Order Loxodida Jankowski, 1978	45
Family Kentrophoridae Jankowski, 1980	14
Genus *Kentrophoros* Sauerbrey, 1928	14
Family Cryptopharyngidae Jankowski, 1980	7
Genera *Apocryptopharynx* Foissner, 1996	2
Cryptopharynx Kahl, 1928	5
Family Loxodidae Bütschli, 1889	24
Genera *Loxodes* Ehrenberg, 1830	6
Remanella Foissner, 1996	18
Order Trachelocercida Jankowski, 1978	65
Family Trachelocercidae Kent, 1881	62
Genera *Trachelocerca* Ehrenberg, 1840[d]	15
Trachelolophos Foissner & Dragesco, 1996	2
Tracheloraphis Dragesco, 1960	45
Family Prototrachelocercidae Foissner, 1996	3
Genus *Prototrachelocerca* Foissner, 1996	3
Incertae sedis: *Ciliofaurea* Dragesco, 1960 (4 species), and *Corlissia* Dragesco, 1960 (monotypic). Both possibly belong to the Protoheterotrichia or Loxodida.	

[a] Authorship and dating is controversial in some taxa. Most nomenclatural problems were discussed and solved by Foissner (1995, 1996a–c) and Foissner and Dragesco (1996b). Generic classification is also according to these papers. Most genera do not have taxonomic synonyms. Only *Trachelonema* Dragesco, 1960 has been synonymized with *Tracheloraphis* by Foissner and Dragesco (1996b).
[b] Mainly according to Carey (1992). Note that the actual number is very likely much higher in most genera (see ecomorphology section).
[c] Kahl (1933) founded *Geleia* with three new species, without, unfortunately, designating a type. The genus is thus invalid according to the ICZN. This was overlooked not only by Kahl (1935) but also by later workers. I thus declare *Geleia* Kahl, 1933 to be a nomen nudum, but reinstall *Geleia* as new genus to avoid an inflation of names. Furthermore, I fix *Geleia fossata* (Kahl, 1933) nov. comb. as type species of the new genus.
[d] Dating of this genus is uncertain and needs special investigation.

the most important autapomorphy of the geleiids (Figure 17.1). Further unique features are an enigmatic kinety in a groove near the anterior end and the distinct preoral suture caused by the subapical position of the oral apparatus (Figure 17.8).

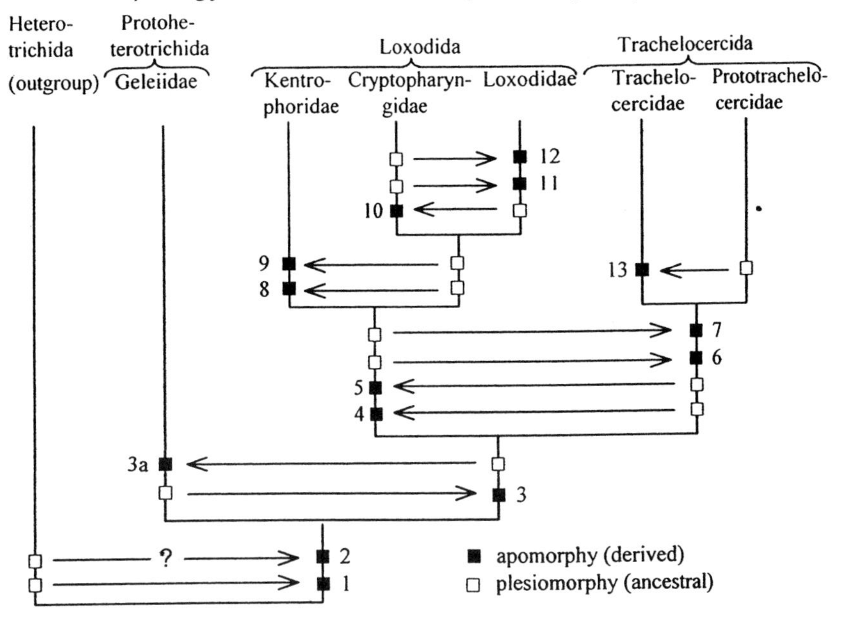

Figure 17.1 A phylogenetic (cladistic) system of karyorelictid ciliates. The analysis was restricted to classical morphological traits because ontogenetic data are lacking for most taxa. The heterotrichs were chosen as outgroup because ultrastructural data and molecular trees argue for a sistergroup relationship with the karyorelictids. Character states (apomorph/plesiomorph): 1, adoral membranelles highly modified or reduced/of typical structure; 2, macronucleus nondividing/dividing; 3, highly specialized bristle kinety framing glabrous stripe/without, i.e. completely and uniformly ciliated; 3a, paracytostomal monokinetids; 4, epipellicular scales or mucilage/without; 5, dorsolateral kinety/without; 6, brosse/without; 7, oral apparatus apical/ventrolateral; 8, epibiontic/symbiotic bacteria on glabrous stripe/without; 9, oral apparatus almost completely reduced/complete; 10, dorsolateral kinety elongated to ventral side/restricted to dorsal and posterior margin of cell; 11, Müller organelles/without; 12, buccal kineties interrupted at anterior buccal vertex/uninterrupted, Figure 17.25; 13, circumoral kinety (ciliature) simple/compound.

The loxodid/trachelocercid clade has a strong synapomorphy, viz. a highly specialized (bristle) kinety surrounding a more or less wide glabrous (nonciliated) stripe on the left side of the cell (Figures 17.1, 17.10, 17.13, 17.14, 17.16, 17.19, 17.20, 17.22, 17.26). The ontogenesis of the bristle kinety is not known. However, light and electron microscopical investigations showed that its fibrillar associates are distinctly different from those of the somatic kineties and that it very likely

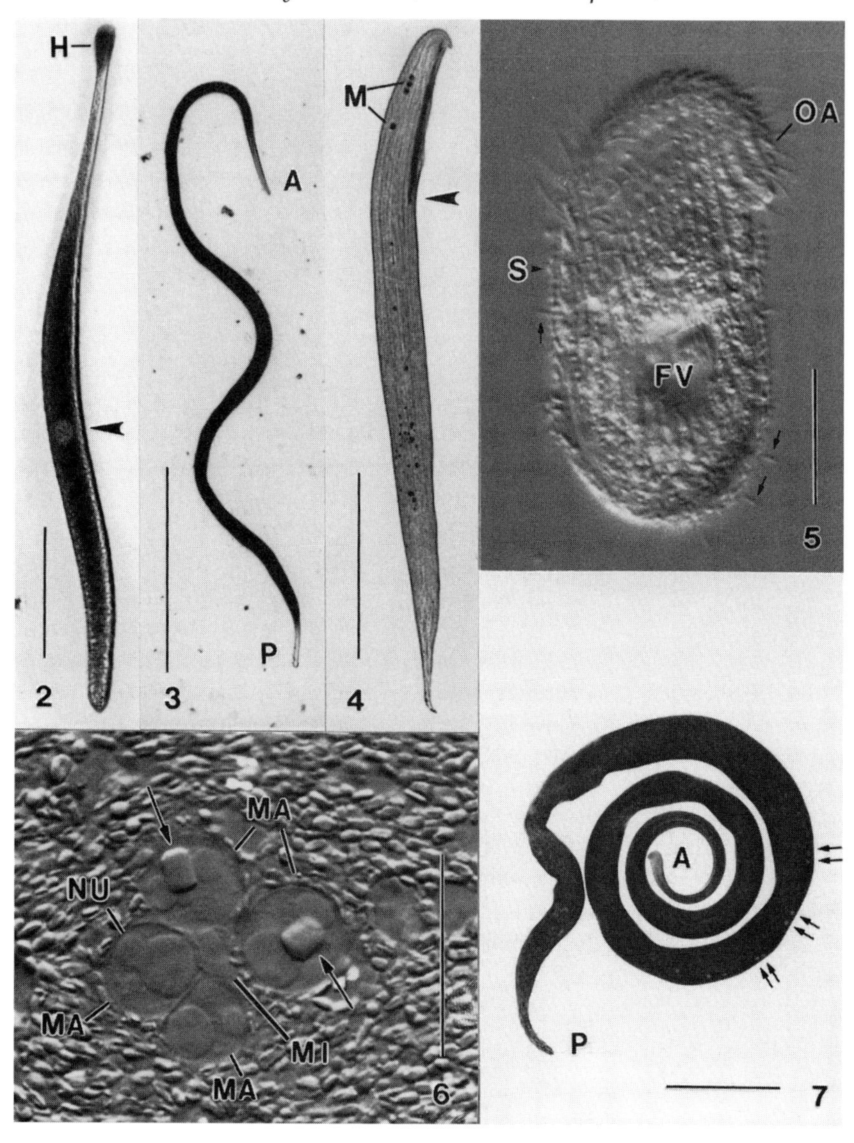

Figures 17.2–17.7 Living karyorelictids (from Foissner, 1995, 1996a,b; Foissner and Dragesco, 1996b). 17.2, 17.6: *Trachelocerca ditis*. Arrowhead marks nuclear capsule containing four macronuclei (Figure 17.6). Arrows mark crystalloid inclusions. 17.3: *Kentrophoros fistulosus* is black due to the symbiotic bacteria lawn growing on its left surface. 17.4: *Remanella multinucleata*. Arrowhead marks posterior end of oral apparatus. 17.5: *Apocryptopharynx hippocampoides*. Arrows mark bristles of bristle kinety. 17.7: A very long (3 mm), still undescribed geleiid having many nuclear groups (arrows). A, anterior end; FV, food vacuole; H,

consists of a small anterior and a large, U-shaped posterior fragment (Foissner and Dragesco, 1996b). The site where the fragments of the bristle kinety abut is clearly marked at the right margin of the glabrous stripe. Here, the dikinetids of the anterior segment have the anterior basal bodies ciliated, whereas those of the posterior segment have the posterior basal bodies ciliated (Figures 17.13, 17.14, 17.22, 17.26).

The loxodids are ventrostome, almost acontractile, usually leaf-like flattened ciliates preferring microaerobic habitats. Most *Loxodes, Remanella*, and *Kentrophoros* species are slender or filiform (Figures 17.3, 17.4), whereas cryptopharyngids are elliptical (Figures 17.5, 17.19). The left side is unciliated, except near the margin where the bristle kinety extends (Figures 17.15, 17.16). The oral apparatus commences at the anterior end and extends as a narrow slit posteriorly on the thin side of the cell. The oral ciliature is composed of several dikinetidal ciliary rows forming a complex pattern (Figure 17.15). Literature on loxodids is rather voluminous, the most important contributions being those by Bardele and Klindworth (1996), Fenchel and Finlay (1986), Foissner (1995, 1996a,b), Klindworth and Bardele (1996), and Raikov (1971, 1978).

The loxodid clade is defined by two comparatively inconspicuous synapomorphies, viz. epipellicular scales or mucilage and, more importantly, a unique dorsolateral kinety (Figure 17.1). Complex epipellicular scales occur in the Cryptopharyngidae (Figure 17.5), whereas a thick layer of mucous material is used by kentrophorids to attach the symbiotic 'kitchen garden' on the unciliated left side of the cell (see ecomorphology section). *Loxodes* and *Remanella* apparently lack scales and mucus. The dorsolateral kinety is present on the dorsolateral margin of the cell as a ciliary row which is more or less distinctly shortened anteriorly. The kinetids of this kinety are more closely spaced than those of the neighbouring somatic kineties and are associated with special fibres (Figure 17.17). *Kentrophoros* has such a kinety too, at least the fibres are clearly identifiable (Figure 17.18). Thus, the order Protostomatida Small and Lynn, 1985, uniting the Kentrophoridae and Trachelocercidae but excluding the Loxodidae, is very likely artificial. The Kentrophoridae are distinguished from the other loxodids by their highly reduced oral apparatus (Foissner, 1995) and the symbiotic kitchen garden (Raikov, 1971; Foissner, 1995). The Loxodidae have a unique apomorphy associated with the bristle kinety, viz. the Müller organelles (Figures 17.4, 17.10) used for gravity perception (Fenchel and Finlay, 1986). The Cryp-

---

head; M, Müller organelles; MA, macronuclei; MI, micronucleus; NU, nucleoli; OA, oral apparatus; P, posterior end; S, epipellicular scales. Scale bars 20 μm (Figures 17.5, 17.6), 200 μm (Figures 17.1, 17.4, 17.7), 500 μm (Figure 17.3).

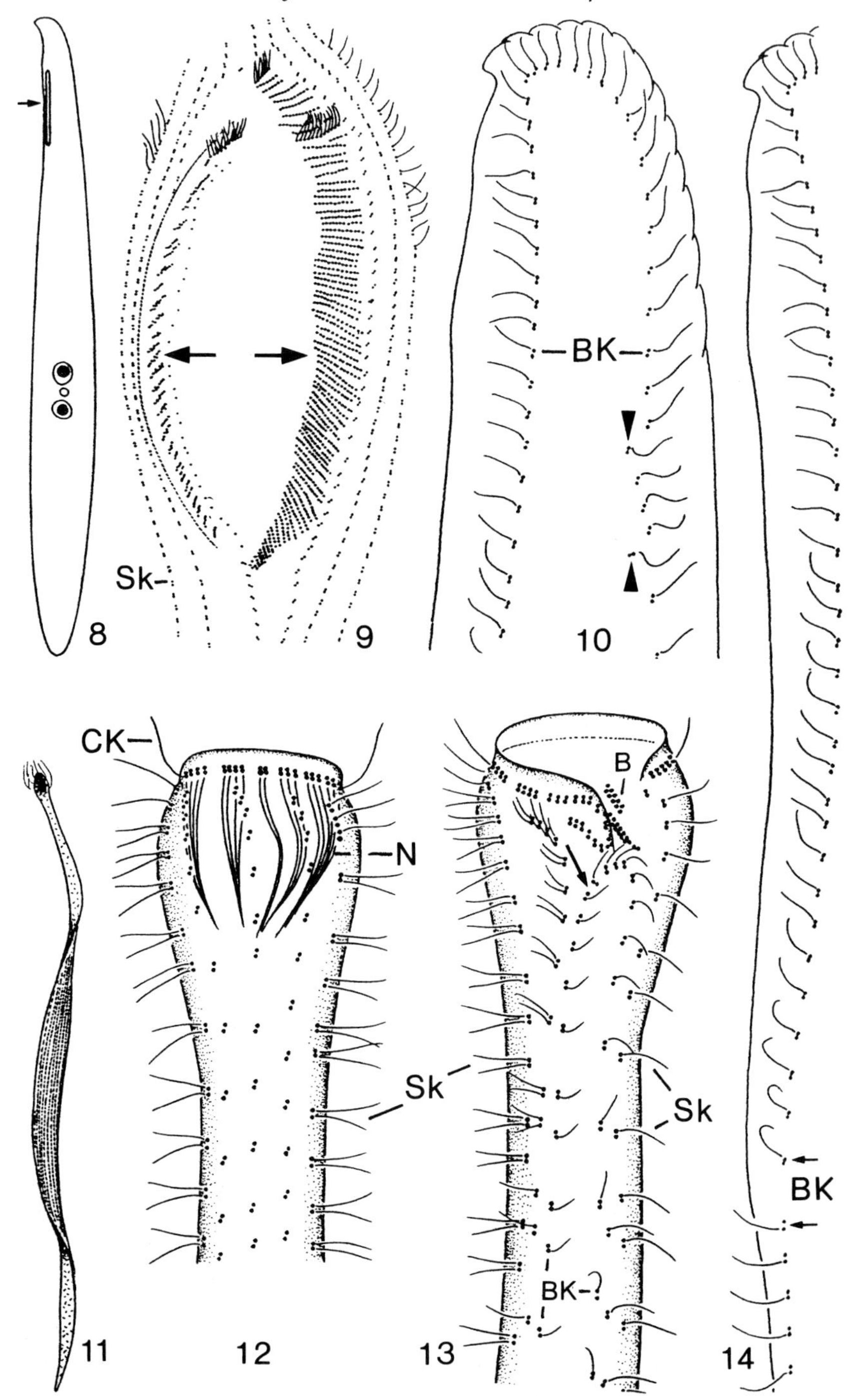
Sk
8
9
BK
10
CK
N
B
Sk
Sk
BK-?
BK
11
12
13
14

topharyngidae are very near the Loxodidae and provide an example of how the complex loxodid oral infraciliature may have evolved (Figures 17.15, 17.19, 17.25).

All trachelocercids are filiform, highly contractile organisms, typically composed of a globular head bearing the oral apparatus, a thin neck, and a more or less distinctly flattened, rounded or tapered trunk (Figures 17.2, 17.11). Trachelocercids are widespread in coastal sands and the most characteristic component of the interstitial ciliate fauna. Benchmark studies include papers by Dragesco (1960), Dragesco and Dragesco-Kernéis (1986), Foissner (1996c), Foissner and Dragesco (1996a, b), Raikov (1958, 1982), and Raikov, Gerassimova-Matvejeva and Puytorac (1975).

The trachelocercid clade is defined by two strong synapomorphies, viz. the apicalization of the oral apparatus and the brosse (Figures 17.1, 17.12, 17.13, 17.20–17.24, 17.26). Admittedly, the first character is rather speculative and partially based on Eisler's (1992) hypothesis that ancestral ciliates had ventrolaterally located oral structures as, for instance, found in *Remanella* (Figures 17.4, 17.15). However, there is also direct support for an apicalization of the trachelocercid oral apparatus, viz. the location of the site where the ends of the bristle kinety meet. In trachelocercids, this site is close beneath the circumoral kinety because the anterior arch of the bristle kinety is short or, as in *Trachelocerca*, even lacking (Figures 17.13, 17.20, 17.22, 17.26). In the sister group, the loxodids, the right anterior branch of the bristle kinety is much longer and extends along the oral slit and thus meets the other end only at the level of the posterior buccal vertex (Figure 17.14). It is easy to imagine

---

Figures 17.8-17.14 Somatic and oral infraciliature (ciliary pattern) of geleiid, loxodid, and trachelocercid karyorelictids after protargol impregnation. 17.8, 17.9: *Geleia decolor*, length 500 μm (from Dragesco and Dragesco-Kernéis, 1986). The oral ciliary fields (arrows) consist of rows of single basal bodies (monokinetids), unlike those in all other karyorelictids. 17.10, 17.14: *Remanella multinucleata*, left side infraciliature (from Foissner, 1996b). The left side is glabrous (unciliated), except for the bristle kinety whose fragments meet at the level of the proximal buccal vertex, as evident from the opposed ciliation of the dikinetids (Figure 17.14). Arrowheads mark bristle kinetids, slightly out of line, associated with the Müller organelles (cp. Figure 17.4). 17.11-17.13: *Tracheloraphis longicollis*, length about 800 μm (from Foissner and Dragesco, 1996b). Right and left side view of head, which bears distinct oral structures composed of a circumoral kinety (CK), nematodesmata (N) and brosse kineties (B). Arrow marks site where the ciliation of the bristle kinety is opposed (cp. Figure 17.14). B, brosse composed of two short, oblique kineties; BK, bristle kinety; CK, circumoral kinety; N, nematodesmata; Sk, somatic kineties.

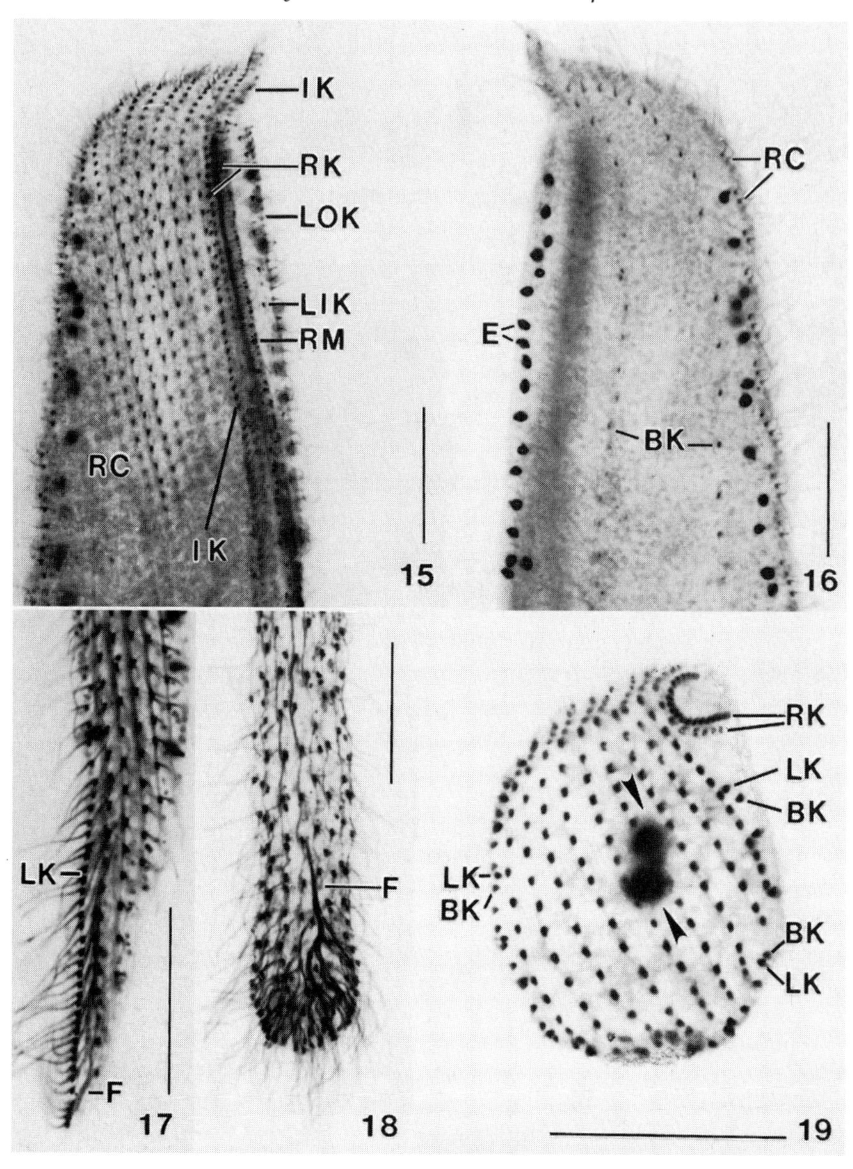

Figures 17.15-17.19 Infraciliature of loxodid karyorelictids after protargol impregnation (from Foissner, 1995, 1996a,b). 17.15, 17.16: *Remanella multinucleata*, right and left side view of anterior oral region. 17.17, 17.18: Posterior end of *R. multinucleata* and *Kentrophoros fistulosus* showing fibres (F) originating from the dorsolateral kinety. 17.19: *Cryptopharynx setigerus*, a very flat species, showing dorsolateral kinety (LK) extending around posterior half. Arrowheads mark macronuclei. BK, bristle kinety; E, extrusomes; F, fibres; IK, intrabuccal kinety; LK,

that a trachelocercid pattern arises if, for instance, the oral apparatus of *Remanella* is shifted anteriorly, i.e. apicalized (Foissner and Dragesco, 1996b). The second apomorphy, the brosse, is a highly distinctive character. Usually, the brosse consists of 1–3 short, oblique dikinetidal kineties located beneath the circumoral kinety in a more or less deep cavity (Figures 17.13, 17.22, 17.26). The brosse has been modified to a tuft of cilia in *Trachelolophos* (Figure 17.23) and, very likely, has been secondarily reduced in *Trachelocerca* (Figure 17.20).

Evolution within trachelocercids is difficult to follow for several reasons (Foissner and Dragesco, 1996b). *Prototrachelocerca* was separated at family level because of its unique circumoral ciliature, which consists of 2–3 closely spaced dikinetidal rows, somewhat reminiscent of the loxodid oral structures (Figure 17.24).

17.3 ECOMORPHOLOGY OF KARYORELICTIDS

Comprehensive reviews on the ecology of marine micrometazoa and protozoa, including karyorelictids, were published by Fenchel (1987), Patterson, Larsen and Corliss (1989), and Remane (1933). The reader is referred to these publications for details and specific literature. Here, I want to highlight only some peculiarities and problems relating to the karyorelictids.

Karyorelictid ciliates are marine, benthic organisms, except for some *Loxodes* species, which are widespread in microaerobic freshwater habitats. Most karyorelictids are obligate interstitial inhabitants, i.e. they live in sheltered microporal (grain size 120–400 μm) and mesoporal (400–1800 μm) coastal sands, preferring the tidal zone, particle sizes between 120 μm and 250 μm, and the upper, well-oxygenated, nutrient-rich 10 mm of the sediment. Here, they may reach high abundances of up to 5000 individuals per cm^2. Thus, in some sediments ciliates play an important role in the benthic energy transfer and are at least as important as metazoa as consumers (Fenchel, 1987).

The most conspicuous morphological features of the karyorelictids are the large size (some species attain a length of 5 mm; Figure 17.7), the vermiform shape often combined with leaf-like lateral flattening, the high contractility and regeneration capacity (geleiids, for instance, can regenerate the oral body half within 3–4 hours), and the ability to attach firmly to the sand grains (thigmotactism) by means of the cilia (Figures

---

dorsolateral kinety; LIK, LOK, left inner and outer buccal kinety; RC, ends of right lateral ciliary rows; RK, right buccal kineties; RM, right margin of buccal overture. Scale bars 20 μm.

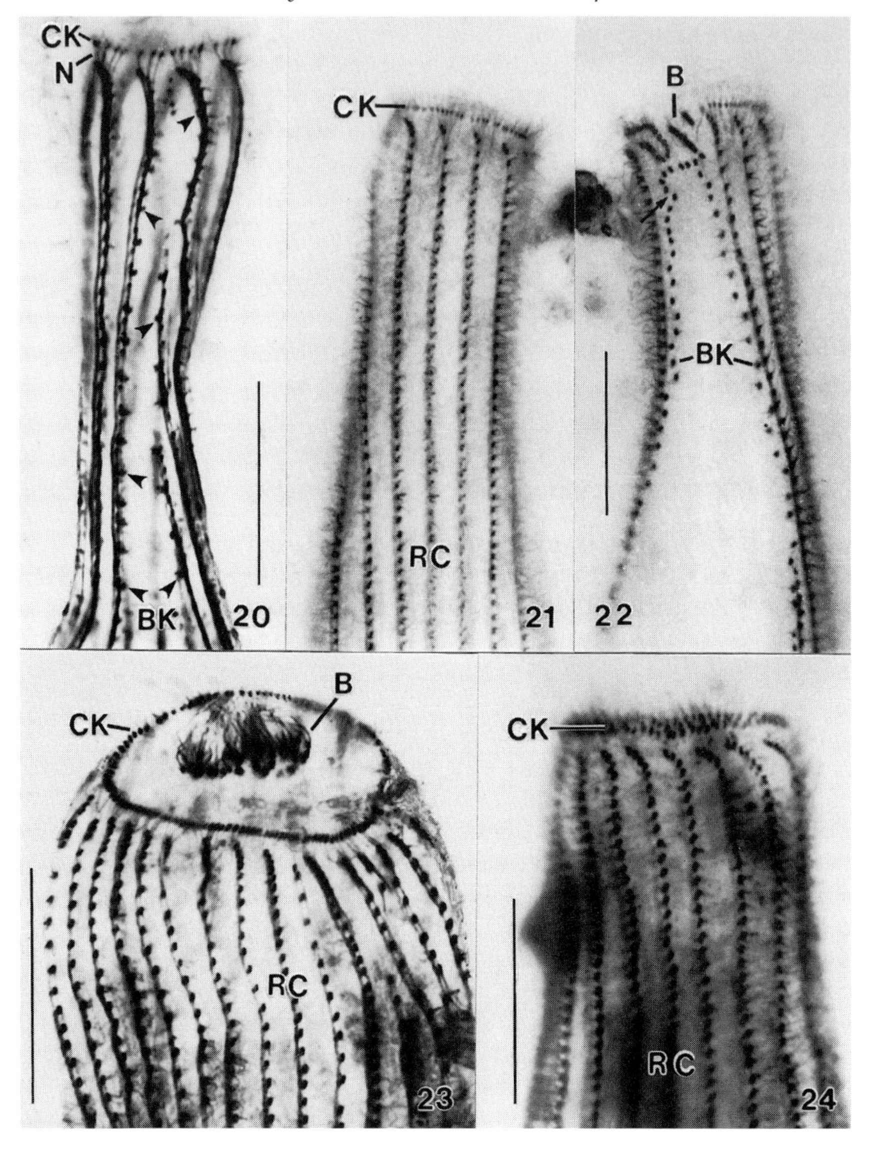

Figures 17.20-17.24 Infraciliature of trachelocercid karyorelictids after protargol impregnation (from Foissner, 1996c; Foissner and Dragesco, 1996a, b). All figures show the anterior (head) region with the oral apparatus. 17.20: *Trachelocerca sagitta*, left side view. Arrowheads mark bristle kinety close to somatic kineties. 17.21, 17.22: *Tracheloraphis longicollis*, right and left side view. Note distinct brosse. Arrow marks site where fragments of bristle kinety abut. 17.23: *Trachelolophos gigas* has the brosse in a pit near the centre of the oral field. 17.24: *Prototrachelocerca caudata* has a compound circumoral ciliature composed of closely

17.2–5, 17.7). Thus, the marine sand ciliate community is distinctly different, both morphologically and ecologically, from the ciliate communities found in freshwater and soil (Table 17.2). Very likely, most of these peculiarities evolved as adaptations to the porous structure and unstable conditions of the biotope. For instance, the risk of being washed out is minimized by thigmotactism and by long, flattened shapes entangling the organisms between the sand grains. Furthermore, a vermiform shape allows them to exploit the fine pores for food and other resources. Thus, this shape type is common also in mesopsammal micrometazoa (Remane, 1933) and among soil ciliates (Foissner, 1987). A high regeneration capacity is advantageous in a biotope which exerts strong mechanical forces by moving sand grains during tides and storms.

Freshly collected karyorelictids are usually packed with large food vacuoles containing a variety of algae, ciliates, and even micrometazoa such as rotifers and harpacticoid copepods. Surprisingly, the mode of food uptake is not known in most species; nobody has ever seen how a trachelocercid ingests these large food items. *Ad hoc*, one would expect that ingestion occurs via the oral apparatus because all karyorelictids, except for kentrophorids, have more or less elaborate oral structures comparable to those found in other ciliates (Figures 17.9, 17.12, 17.13, 17.15, 17.19–24). However, Lenk, Small and Gunderson (1984) and Lenk, Hollander and Small (1989) claimed that ingestion in at least some trachelocercids occurs via the glabrous stripe. In my opinion, this is not very likely because the glabrous stripe is rather narrow in many species and Lenk's micrographs do not show the process unequivocally. On the other hand, Lenk's observations cannot be denied entirely because *Kentrophoros*, which has the oral structures reduced to inconspicuous vestiges (Foissner, 1995), does indeed use the glabrous stripe for food uptake. *Kentrophoros* spp. are unique in having a symbiotic kitchen garden of sulphur bacteria on the left side (Raikov, 1971). The bacteria, which are embedded in a mucous substance, divide on the ciliate and are phagocytosed through the glabrous stripe (Raikov, 1971; Foissner, 1995).

Karyorelictids have a huge variety of extrusomes, some of which are highly reminiscent of hydrozoan cnidocysts (Foissner, 1996a,b; Raikov, 1978; Figure 17.16). Their function is not known. Possibly, they are used for prey capture and/or attachment to solid particles. *Loxodes* and *Remanella* have unique organelles, the Müller vesicles (Figures 17.4, 17.10) for gravity perception (Fenchel and Finlay, 1986).

---

spaced dikinetidal rows. B, brosse; BK, bristle kinety; CK, circumoral kinety; N, nematodesmata; RC, right side ciliary rows. Scale bars 20 μm.

Table 17.2 A comparative description of the ciliate communities in freshwater, soil, and marine sand[a]

Character	*Freshwater*	*Soil*	*Marine sand*[f]
Mean biomass (mg) of 10^6 individuals	1076 (n = 200)[b]	98 (n = 238)	872 (n = 200)
Mean body length (μm)	162 (n = 200)	110 (n = 238)	424 (n = 200)
Mean body width (μm)	56 (n = 422)	36 (n = 238)	54 (n = 200)
% Colpodea	5 (n = 422)	18 (n = 238)	0 (n = 292)
% Hypotrichida	11 (n = 422)	37 (n = 238)	20 (n = 292)
% Peritrichida	21 (n = 422)	3 (n = 238)	0.3 (n = 292)
% cyst forming species	⩽80	>95	<2
% species with reduced ciliature[c]	41 (n = 182)	53 (n = 229)	53 (n = 200)
% species with nodulated macronucleus[d]	8 (n = 200)	25 (n = 238)	43 (n = 200)
Ploidy of macronucleus	generally high	generally high	often low
Body shape	often cylindrical	often flattened, elongated, worm-like	often flattened, elongated, worm-like
Caudal prolongation	uncommon	common	very common
'Cephalization'	very rare	very rare	rare
Contractility	generally low	generally low	generally high
Fragility	generally low	generally low	generally high
Cytological peculiarities	cytoplasm seldom strongly vacuolated	cytoplasm seldom strongly vacuolated	cytoplasm often strongly vacuolated; skeletal rodlets
Movement	thigmotactic creepers common only in the Aufwuchs; sessile forms common	thigmotactic creepers common; sessile forms nearly absent	thigmotactic creepers common; sessile forms nearly absent
Nutrition	great majority of common species are bacterivorous, or macrophagous	great majority of autochthonous species are bacterivorous predaceous, or mycophagous	great majority of autochthonous species are macrophagous (predaceous)
Symbiotic bacteria on the body surface of species within the sulfide system	present	unknown	present
Number of species	about 4000	about 400	about 1000
Abundance[e] range m^{-2}	$5 \times 10^4 - 5 \times 10^6$	$0 - 4 \times 10^6$	$5 \times 10^6 - 3 \times 10^7$

[a] From Foissner (1987). See this publication for literature on data sources.
[b] n refers to the number of species considered.
[c] Peritrichs and suctorians were excluded because of their high degree of specialization. Calculated from the same data set as used for 'mean biomass'.
[d] Only species with more than two nodules have been considered as having a nodulated macronucleus. Two nodules are 'normal' in many groups (e.g., hypotrichs). Calculated from the same data set as used for 'mean biomass'.
[e] These are only a few, perhaps not representative, examples from mesosaprobic rivers, alpine grassland soils, and marine sands.
[f] Includes not only karyorelictids but also other ciliates inhabiting marine sands.

The diversity of karyorelictids is apparently small, i.e. only 135 species have been described (Table 17.1), most of which are assumed to be cosmopolitan. Some species are believed to have a restricted geographic range, for instance, *Geleia murmanica* to Europe, and *Avelia martinicense* to Martinique. However, I agree with Patterson, Larsen and Corliss (1989) that 'Statements relating to the distribution of species must be viewed with some (I would say, great) caution because of uncertainty over taxonomic practice and because of undersampling'. Although the diversity of karyorelictids is undoubtedly small as compared with the rest of the ciliates, many more species than are presently known very likely exist. New species are being described continuously and I expect that, when our new fixative and silver impregnation are more widely used, the number of known species will quickly and greatly increase, as happened with the soil ciliates (Foissner, 1987). Dragesco and I found at least two new genera (Foissner 1996a; Foissner and Dragesco, 1996a) and several new species during two weeks of work at Roscoff, a well-investigated site, although we did not particularly look for new taxa. And the geleiid depicted in Figure 17.7 is also a new species, having, unlike all described species, several groups of macronuclei. Furthermore, karyorelictids are very patchily distributed and many of them are, as usual, rare. Thus, I would not be surprised if forthcoming generations of scientists established that we knew only 10% of the species actually existing at the turn of the millenium.

17.4 RELATIONSHIPS OF KARYORELICTIDS WITH OTHER CILIATES

Structural similarities of the SSUrRNA gene sequences of several heterotrichs s. str. (e.g. *Stentor, Climacostomum*), heterotrichs s. l. (*Metopus* spp.), trachelocercids (*Tracheloraphis, Loxodes* spp.), and *Protocruzia* sp. were analysed with parsimony and distance algorithms (Baroin-Tourancheau *et al.*, 1992; Hirt *et al.*, 1995; Hammerschmidt *et al.*, 1996). The results showed, with strong bootstrap support, the karyorelictids as a sister group of the heterotrichs s. str., while *Protocruzia* clustered with the hypotrichs and *Metopus*, surprisingly, with the haptorid gymnostomes; however, bootstrap values were weak for *Protocruzia* and *Metopus* and their phylogenetic relationships are thus still open for discussion (Hirt *et al.*, 1995). A close relationship between heterotrichs s. str. and karyorelictids was proposed long ago also by Raikov, Gerassimova-Matvejeva and Puytorac (1975) because of distinct similarities in the somatic cortical ultrastructure. Surprisingly, such a close relationship is not evident from our studies of the somatic and oral infraciliature, and not even from the few reliable ontogenetic data available (Bardele and Klindworth, 1996).

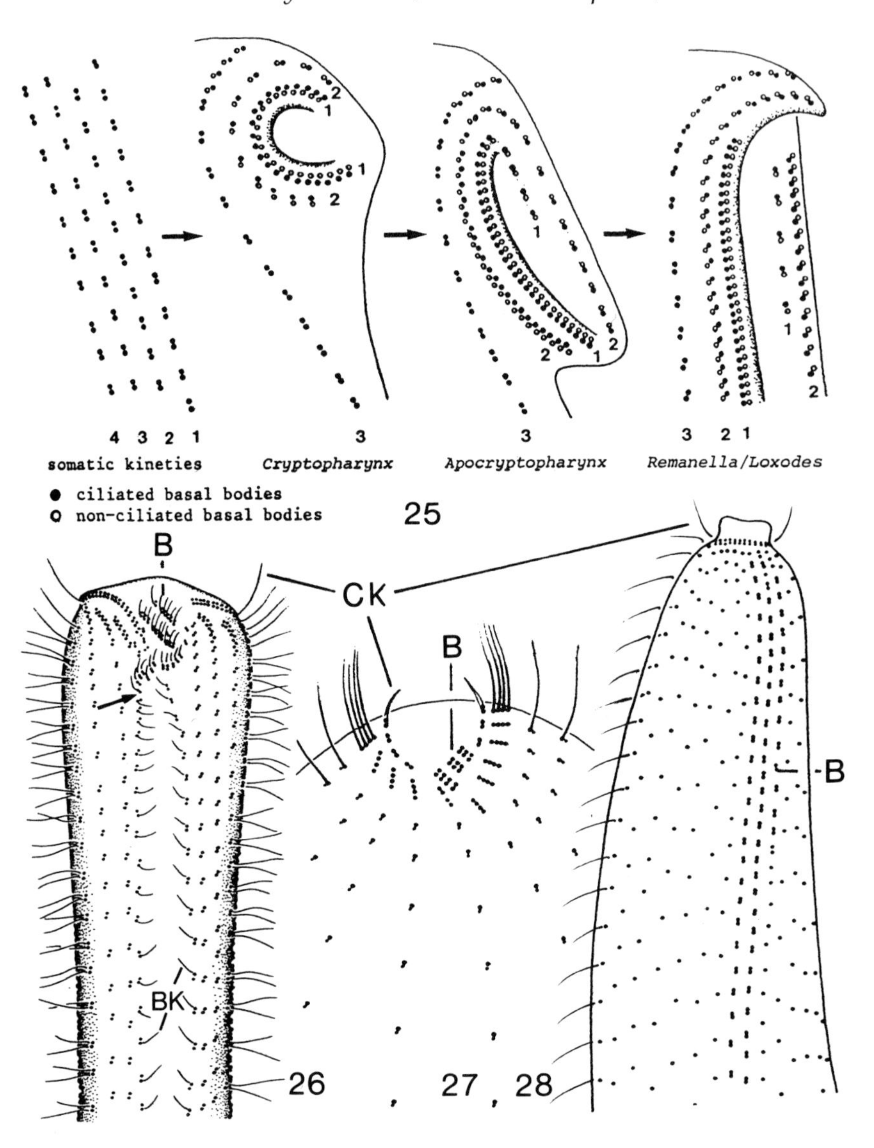

Figures 17.25–17.28 Relationships within karyorelictids and between karyorelictids and other ciliates. 17.25: Origin and evolution of the loxodid oral ciliature from somatic ciliary rows (cp. Figures 17.15, 17.19; from Foissner, 1996a). 17.26–17.28 *Tracheloraphis phoenicopterus* (26, Karyorelictea, from Foissner and Dragesco, 1996b), *Coleps hirtus* (27, Prostomatea, from Foissner, 1984), and *Enchelyodon terrenus* (28, Gymnostomatea, from Foissner, 1984) have a dikinetidal circumoral kinety and a brosse. Arrow marks site where fragments of bristle kinety abut. B, brosse; CK, circumoral kinety; BK, bristle kinety.

The somatic and oral infraciliature of the karyorelictids is distinctly different from that of the proposed sister group, the heterotrichs s. str., except for some general similarities, like somatic dikinetids and oral structures composed of elements which might be interpreted as paroral and/or adoral ciliature. Specifically, all karyorelictids lack classic adoral membranelles, as typical for heterotrichs, composed of several closely spaced ciliary rows; even the geleiids have the left oral ciliary field, which is densely ciliated and thus reminiscent of an adoral zone of membranelles, composed of single (monokinetidal) ciliary rows (Figure 17.9). On the other hand, the heterotrichs lack the glabrous stripe and the bristle kinety (Figures 17.10, 17.13, 17.16, 17.19, 17.20, 17.22), which are so conspicuous in all loxodids and trachelocercids, but absent in geleiids, indicating that they might be more ancestral than trachelocercids and loxodids.

The oral structures of the Geleiidae and Loxodida are unique, hardly bearing any resemblance to those of other ciliates, although there is some evidence that the loxodid oral kineties have somatic progenitors (Figure 17.25). This contrasts with the trachelocercid oral infraciliature (Figure 17.26), which strongly resembles that found in haptorid gymnostomes (e.g. *Enchelyodon*, Figure 17.28) and especially in prostomes (e.g. *Coleps*, Figure 17.27). All have a circumoral kinety composed of dikinetids, and a brush or brosse, i.e. some small kineties near the oral opening. In prostomatids, the brosse is interpreted as adoral ciliature (Huttenlauch and Bardele, 1987). Similarly, the trachelocercid brosse kineties could be remnants of heterotrich adoral membranelles. There is also another remarkable similarity between prostomatids and karyorelictids, viz. the rectangular or hexagonal cortical fibrillar system found in prostomatids and hymenostomes such as *Urotricha* (Foissner and Pfister, 1997) and *Paramecium* (Wichtermann, 1986) and karyorelictids like *Geleia swedmarki* (Dragesco, 1960) and *Trachelolophos gigas* (Foissner and Dragesco, 1996a).

Unfortunately, the molecular and morphological investigations can hardly be compared with ontogenetic data because these are almost completely lacking for karyorelictids. Only *Loxodes* has been recently studied in some detail. Bardele and Klindworth (1996) showed that stomatogenesis of *Loxodes* is buccokinetal, i.e. the daughter oral structures originate by direct participation of the parental oral apparatus. Thus, stomatogenesis of *Loxodes* at least is quite different from that of heterotrichs s. str. and of prostomatids, which form the daughter oral apparatus parakinetally or telokinetally, i.e. from parental somatic infraciliature (Foissner, 1996d). Typically, buccokinetal stomatogenesis is found in the Oligohymenophorea, a large group of ciliates comprising hymenostomatids (e.g. *Tetrahymena, Paramecium*), scuticociliates (e.g. *Uronema, Pleuronema*) and, possibly, also prostomatids (e.g. *Urotricha, Coleps*).

Thus, as with morphology, ontogenesis indicates some link between karyorelictids and oligohymenophorans, especially when the prostomatids are included in the latter. However, the molecular data do not indicate any relationship between karyorelictids and oligohymenophoreans/prostomatids (chapter 18).

Karyorelictids were frequently used as model organisms for the origin of nuclear dimorphism and oral structures in ciliates (Corliss, 1974, 1979; Orias, 1976; Small, 1984; Herrick, 1994). None of these hypotheses gained wide acceptance and most were based on incomplete data. Orias (1976), for instance, proposed *Kentrophoros* to be most primitive among all ciliates, partly because of its supposed mouthlessness. However, recent observations showed that *Kentrophoros* very likely has vestiges of an oral infraciliature, suggesting that it became secondarily mouthless due to its highly specialized mode of nutrition (Foissner, 1995). Likewise, Small (1984) based his hypothesis on the origin of the ciliate oral apparatus on the assumption that orally specialized kinetids and ciliation are 'totally absent in *Tracheloraphis*'. This has been clearly disproved by later investigations (Figures 17.12, 17.13, 17.20–17.24, 17.26). In the light of the highly specialized somatic and oral infraciliature, as well as the molecular evidence, I agree with Hirt *et al.* (1995) and (chapter 18) and Hammerschmidt *et al.* (1996), that the specific nuclear features of the karyorelictids are not ancestral but derived. I would not even be too surprised if non-dividing macronuclei had evolved twice, possibly due to ecological constraints (Fauré-Fremiet, 1961), considering the highly different somatic and oral infraciliatures of the Protoheterotrichida and Trachelocercia.

Acknowledgements

The technical assistance of Dr E. Herzog, B. Moser, and Mag. E. Strobl is greatly acknowledged. Financial support was provided by the Salzburg University, the Linnean Society and the Systematics Association. Special thanks to Dr Alan Warren (Natural History Museum, London) for inviting this review, and to Prof. Dr Jean Dragesco for stimulating my interest in karyorelictids.

17.5 REFERENCES

Bardele, C.F. and Klindworth, T. (1996) Stomatogenesis in the karyorelictean ciliate *Loxodes striatus*: a light and scanning microscopical study. *Acta Protozoologica*, **35**, 29–40.

Baroin-Tourancheau, A., Delgado, P., Perasso, R. and Adoutte, A. (1992) A broad molecular phylogeny of ciliates: identification of major evolutionary trends and radiations within the phylum. *Proceedings of the National Academy of Sciences USA*, **89**, 9764–8.

Carey, P.G. (1992) *Marine Interstitial Ciliates*. Chapman and Hall, London.

Corliss, J.O. (1974) Remarks on the composition of the large ciliate class Kinetofragminophora de Puytorac *et al*., 1974, and recognition of several new taxa therein, with emphasis on the primitive order Primociliatida n. ord. *Journal of Protozoology*, **21**, 207–20.

Corliss, J.O. (1979) *The Ciliated Protozoa. Characterization, Classification and Guide to the Literature*, 2nd edn, Pergamon Press, Oxford.

Corliss, J.O. and Hartwig, E. (1977) The 'primitive' interstitial ciliates: their ecology, nuclear uniquenesses, and postulated place in the evolution and systematics of the phylum Ciliophora. *Mikrofauna des Meeresbodens*, **61**, 65–88.

Dragesco, J. (1960) Ciliés mésopsammiques littoraux. Systématique, morphologie, écologie. *Des Travaux de la Station Biologique de Roscoff (Nouvelle série)*, **12**, 1–356.

Dragesco, J. and Dragesco-Kernéis, A. (1986) Ciliés libres de l'Afrique intertropicale. *Faune tropicale*, **26**, 1–559.

Eisler, K. (1992) Somatic kineties or paroral membrane: which came first in ciliate evolution? *BioSystems*, **26**, 239–54.

Fauré-Fremiet, E. (1961) Quelques considérations sur les ciliés mésopsammiques a propos d'un récent travail de J. Dragesco. *Cahiers de Biologie Marine*, **2**, 177–86.

Fenchel, T. (1987) *Ecology of Protozoa. The Biology of Free-Living Phagotrophic Protists*. Springer, Berlin.

Fenchel, T. and Finlay, B.J. (1986) The structure and function of Müller vesicles in loxodid ciliates. *Journal of Protozoology*, **33**, 69–76.

Foissner, W. (1984) Infraciliatur, Silberliniensystem und Biometrie einiger neuer und wenig bekannter terrestrischer, limnischer und mariner Ciliaten (Protozoa: Ciliophora) aus den Klassen Kinetofragminophora, Colpodea und Polyhymenophora. *Stapfia*, **12**, 1–165.

Foissner, W. (1987) Soil protozoa: fundamental problems, ecological significance, adaptations in ciliates and testaceans, bioindicators, and guide to the literature. *Progress in Protistology*, **2**, 69–212.

Foissner, W. (1995) *Kentrophoros* (Ciliophora, Karyorelictea) has oral vestiges: a reinvestigation of *K. fistulosus* (Fauré-Fremiet, 1950) using protargol impregnation. *Archiv für Protistenkunde*, **146**, 165–79.

Foissner, W. (1996a) The infraciliature of *Cryptopharynx setigerus* Kahl, 1928 and *Apocryptopharynx hippocampoides* nov. gen., nov. spec. (Ciliophora, Karyorelictea), with an account on evolution in loxodid ciliates. *Archiv für Protistenkunde*, **146**, 309–27.

Foissner, W. (1996b) A redescription of *Remanella multinucleata* (Kahl, 1933) nov. gen., nov. comb. (Ciliophora, Karyorelictea), emphasizing the infraciliature and extrusomes. *European Journal of Protistology*, **32**, 234–50.

Foissner, W. (1996c) Updating the trachelocercids (Ciliophora, Karyorelictea). II. *Prototrachelocerca* nov. gen. (Prototrachelocercidae nov. fam.), with a redescription of *P. fasciolata* (Sauerbrey, 1928) nov. comb. and *P. caudata* (Dragesco and Raikov, 1966) nov. comb. *European Journal of Protistology*, **32**, 336–55.

Foissner, W. (1996d) Ontogenesis in ciliated protozoa, with emphasis on stomatogenesis, in *Ciliates, Cells as Organisms* (eds K. Hausmann and P.C. Bradbury), Fischer, Stuttgart-Jena, pp. 95–177.

Foissner, W. and Dragesco, J. (1996a) Updating the trachelocercids (Ciliophora, Karyorelictea). I. A detailed description of the infraciliature of *Trachelolophos gigas* n. g., n. sp. and *T. filum* (Dragesco and Dragesco-Kernéis, 1986) n. comb. *Journal of Eukaryotic Microbiology*, **43**, 12–25.

Foissner, W. and Dragesco, J. (1996b) Updating the trachelocercids (Ciliophora, Karyorelictea). III. Redefinition of the genera *Trachelocerca* Ehrenberg and *Tracheloraphis* Dragesco, and evolution in trachelocercid ciliates. *Archiv für Protistenkunde*, **147**, 43–91.

Foissner, W. and Pfister, G. (1997) Taxonomic and ecologic revision of urotrichs (Ciliophora, Prostomatida) with three or more caudal cilia, including a user-friendly key. *Limnologica*, **27**, 311–47.

Grolière, C. A., Puytorac, P. de and Detcheva, R. (1980) A propos d'observations sur la stomatogenèse et l'ultrastructure du cilié *Protocruzia tuzeti* Villeneuve-Brachon, 1940. *Protistologica*, **16**, 453–66.

Hammerschmidt, B., Schlegel, M., Lynn, D.H. *et al.* (1996) Insights into the evolution of nuclear dualism in the ciliates revealed by phylogenetic analysis of rRNA sequences. *Journal of Eukaryotic Microbiology*, **43**, 225–30.

Herrick, G. (1994) Germline-soma relationships in ciliated protozoa: the inception and evolution of nuclear dimorphism in one-celled animals. *Developmental Biology*, **5**, 3–12.

Hirt, R.P., Dyal, P.L., Wilkinson, M. *et al.* (1995) Phylogenetic relationships among karyorelictids and heterotrichs inferred from small subunit rRNA sequences: resolution at the base of the ciliate tree. *Molecular Phylogenetics and Evolution*, **4**, 77–87.

Huttenlauch, I. and Bardele, C.F. (1987) Light and electron microscopical observations on the stomatogenesis of the ciliate *Coleps amphacanthus* Ehrenberg, 1833. *Journal of Protozoology*, **34**, 183–92.

Kahl, A. (1933) Ciliata libera et ectocommensalia. *Die Tierwelt der Nord- und Ostsee*, **23** (Teil II, c_3), 29–146.

Kahl, A. (1935) Urtiere oder Protozoa. I: Wimpertiere oder Ciliata (Infusoria) 4. Peritricha und Chonotricha. *Die Tierwelt Deutschlands und der angrenzenden Meeresteile*, **30**, 651–886.

Klindworth T. and Bardele, C.F. (1996) The ultrastructure of the somatic and oral cortex of the karyorelictean ciliate *Loxodes striatus*. *Acta Protozoologica*, **35**, 13–28.

Lenk, S.E., Small, E.B. and Gunderson, J. (1984) Preliminary observations of feeding in the psammobiotic ciliate *Tracheloraphis*. *Origins of Life*, **13**, 229–34.

Lenk, S.E., Hollander, B.A. and Small, E.B. (1989) Ultrastructure of feeding in the karyorelictean ciliate *Tracheloraphis* examined by scanning electron microscopy. *Tissue and Cell*, **21**, 189–94.

Lipscomb, D.L. and Corliss, J.O. (1982) *Stephanopogon*, a phylogenetically important 'ciliate', shown by ultrastructural studies to be a flagellate. *Science*, **215**, 303–4.

Ng, S.F. (1990) Embryological perspective of sexual somatic development in ciliated protozoa: implications on immortality, sexual reproduction and inheritance of acquired characters. *Philosophical Transactions of the Royal Society of London Series B*, **329**, 287–305.

Nouzarede, M. (1977) Cytologie fonctionnelle et morphologie experimentale de quelques protozoaires cilies mesopsammiques geants de la famille des Geleiidae (Kahl). *Bulletin de la Station Biologique d'Arcachon (Nouvelle série)*, **28** (Suppl.; year 1976), Vol. I, IX + 315 pp. and Vol. II (plates, without pagination).

Orias, E. (1976) Derivation of ciliate architecture from a simple flagellate: an evolutionary model. *Transactions of the American Microscopical Society*, **95**, 415–29.

Patterson, D.J., Larsen, J. and Corliss, J.O. (1989) The ecology of heterotrophic flagellates and ciliates living in marine sediments. *Progress in Protistology*, **3**, 185–277.

Puytorac, P. de (1994) Classe des Karyorelictea Corliss, 1974. *Traité de Zoologie*, **2**(2), 21–34.

Puytorac, P. de, Grain, J. and Mignot, J.-P. (1987) *Precis de Protistologie*. Boubée et Fondation Singer Polignac, Paris.

Raikov, I.B. (1958) Der Formwechsel des Kernapparates einiger niederer Ciliaten. I. Die Gattung *Trachelocerca*. *Archiv für Protistenkunde*, **103**, 129–92.

Raikov, I.B. (1971) Bactéries épizoiques et mode de nutrition du cilié psammophile *Kentrophoros fistulosum* Fauré-Fremiet (étude au microscope électronique). *Protistologica*, **7**, 365–78.

Raikov, I.B. (1978) Ultrastructure du cytoplasme et des nématocystes du cilié *Remanella multinucleata* Kahl (Gymnostomata, Loxodidae). Existence de nématocystes chez les ciliés. *Protistologica*, **14**, 413–32.

Raikov, I.B. (1982) The protozoan nucleus. Morphology and evolution. *Cell Biology Monographs*, **9**, XV + 1–474.

Raikov, I.B. (1994) The nuclear apparatus of some primitive ciliates, the karyorelictids: structure and divisional reorganization. *Bollettino di Zoologia*, **61**, 19–28.

Raikov, I.B. Gerassimova-Matvejeva, Z.P. and Puytorac, P. de (1975) Cytoplasmic fine structure of the marine psammobiotic ciliate *Tracheloraphis dogieli* Raikov. I. Somatic infraciliature and cortical organelles. *Acta Protozoologica*, **14**, 17–42.

Remane, A. (1933) Verteilung und Organisation der benthonischen Mikrofauna der Kieler Bucht. *Wissenschaftliche Meeresuntersuchungen Kiel (Neue Folge)* **21**, 162–221.

Small, E.B. (1984) An essay on the evolution of ciliophoran oral cytoarchitecture based on descent from within a karyorelictean ancestry. *Origins of Life*, **13**, 217–28.

Small, E.B. and Lynn, D.H. (1985) Phylum Ciliophora, in *Illustrated Guide to the Protozoa* (eds J.J. Lee, S.J. Hutner and E.C. Bovee), Lawrence, Kansas, pp. 393–575.

Wichtermann, R. (1986) *The Biology of Paramecium*, 2nd edn, Plenum Press, New York.

Wilbert, N. (1986) Die orale Infraciliatur von *Tracheloraphis dogieli* Raikov, 1957 (Ciliophora, Gymnostomata, Karyorelictida). *Archiv für Protistenkunde*, **132**, 191–5.

18

Molecular and cellular evolution of ciliates: a phylogenetic perspective

Robert P. Hirt, Mark Wilkinson and T. Martin Embley*

*Department of Zoology, The Natural History Museum, Cromwell Road, London SW7 5BD. E-mail: rch@nhm.ac.uk

ABSTRACT

The ciliated protozoa are characterized by an enormous diversity expressed at every level of their biology and from this perspective they represent a paradigm for the study of eukaryotic cellular–molecular evolution. The rationalization of this diversity within a systematic framework has a long history and the evolution of ciliate systematics has paralleled advances in microscopic methodologies. More recently molecular sequence data have been used to complement the large body of morphological data. Ribosomal RNA genes have been the central molecules used to investigate phylogenetic relationships between ciliates. In this paper we discuss two questions of ciliate evolution using the tools of molecular systematics: (1) the origin and evolution of the features of the karyorelictid macronucleus, namely its non-capacity to divide and its nearly diploid status, which has been hypothesized to be a "primitive" trait for ciliates and (2) the origin and evolution of a hydrogen-producing organelle called the hydrogenosome found in both free-living and rumen ciliates.

18.1 INTRODUCTION

Ciliates have attracted the interest of biologists since the first microscopes were developed and the intense search for such "animalcules" in the environment for nearly three centuries has clearly established that

Evolutionary Relationships Among Protozoa. Edited by G.H. Coombs, K. Vickerman, M.A. Sleigh and A. Warren. Published in 1998 by Chapman & Hall, London. ISBN 0 412 79800 X

the ciliated protozoa are one of the most successful groups of protists (Corliss, 1979; Hausmann and Hülsmann, 1996; Lynn and Corliss, 1991). They have colonized virtually every aquatic and humid environment, including anaerobic niches, where they represent a substantial fraction of the protist biomass (Fenchel, 1987; Fenchel and Finlay, 1991). In addition to free-living species, many ciliates are symbionts or parasites and they are abundant in the digestive tracts of ruminants (Williams and Coleman, 1992). Ciliates are involved in the dynamics of many ecosystems and are of direct practical importance for human activities such as waste water treatments (Curds, 1992), soil fertility (Darbyshire, 1994) and the fisheries industry (Ragan *et al.*, 1996; Wright and Lynn, 1995). In addition to their ecological importance, ciliates have been used as model systems to address molecular, cell biology, genetic and developmental questions (Frankel, 1989; Nanney, 1980, 1986; see Madireddi, Smothers and Allis, 1995 and Wheatley, Rasmussen and Tiedtke, 1994 for recent reviews).

In relation to their ecological diversity, ciliates are characterized by an enormous biological diversity manifested at every cellular level, with external morphology being only the most conspicuous. The rationalization of this morphological diversity into different phylogenetic hypotheses has paralleled advances in microscopic methodologies, including electron microscopy and new staining-fixation protocols (Corliss, 1979; Lynn and Corliss, 1991). Some ciliate classifications have made extensive use of kinetid ultrastructure revealed by electronic microscopy, and are thus based on epigenetic characters (Small and Lynn, 1985). The cortical ciliature of ciliates involves the phenomenon of cytotaxy (Sonneborn, 1970), in which supramolecular cell structures provide templates for new similar structures, and the ability of environmentally-induced changes to these structures to be inherited without any change in the cell's DNA, are all well characterized (Frankel, 1989; Nanney, 1982). Epigenetic inheritance systems are apparently important in the development and evolution of multicellular organisms (Edelman, 1988; Jablonka and Lamb, 1995) and this underlines the potential for ciliates as model systems for studying these phenomena (Frankel, 1989). The use of such epigenetic characters for phylogenetics and classification has, however, been questioned or prohibited by some authors: "The stable inheritance of characteristics is mediated by the genome. Differences due to epigenetic or environmental factors do not provide useful phylogenetic information and must be specifically avoided; all characters of interest are genetically mediated." (Swofford *et al.*, 1996, pp. 409). In contrast, we see no reason why the use of epigenetically-inherited characters, rather than ecophenotypic variation, should be prohibited in phylogenetic studies. What is important is stable heritability, whether it be genetic, epigenetic or a combination of both.

More recently, molecular sequences have been used to complement the large body of morphological data for ciliate systematics, with the small subunit ribosomal RNA gene (SSUrDNA) the main source of such sequence data (Patterson and Sogin, 1993). There is also a substantial dataset of partial large subunit (LSU) rDNA sequences which have been used to investigate ciliate phylogeny (Baroin-Tourancheau *et al.*, 1992; 1995).

It is the purpose of this chapter to discuss some of these morphological and molecular data by focusing on phylogenetic hypotheses concerning two of the important features of ciliates, namely their peculiar nuclear organization and their capacity to adopt an anaerobic lifestyle based upon hydrogenosomes.

18.2 CILIATE MOLECULAR PHYLOGENIES

18.2.1 The relationship of ciliates to other protists

Ultrastructural features have been used to propose a relationship between ciliates and dinoflagellates (Corliss, 1988; Lee and Kugrens, 1992) and molecular sequences (Gajadhar *et al.*, 1991) have identified a monophyletic group comprising ciliates, dinoflagellates and apicomplexans. The clade was subsequently named the alveolates to reflect the presence of a putative synapomorphy, the cortical alveoli (Cavalier-Smith, 1993). This result has been supported by sampling additional taxa (reviewed in Sogin *et al.*, 1996).

The position of the alveolar clade in the SSUrDNA tree suggests a close relationship with the stramenopiles, metazoans, fungi and plants which constitute the so-called crown taxa (Sogin *et al.*, 1996). In contrast, protein sequences have not confirmed this position for ciliates and they raise the commonly discussed issue of the level of confidence one can place in a single gene tree (Philippe and Adoutte, 1995) (chapter 2). For instance ciliates are recovered below, rather than above, the kinetoplastic-euglenid lineage in actin (Drouin, Moniz de Sa and Zucker, 1995) and EF-1alpha trees (Hashimoto *et al.*, 1995). Moreover, the ciliates (as well as the alveolates) are not monophyletic in the actin trees, which suggests caution in the interpretation of these particular data. Further sampling of genes and taxa may be needed before (if ever) any consensus can be reached regarding these issues.

18.2.2 Ciliate morphological and molecular phylogenies

There have been a number of different morphological-based taxonomies proposed for ciliates (Corliss, 1979; de Puytorac *et al.*, 1993; Small and Lynn, 1985) and the incongruence between them can be explained by emphasis being placed upon different morphological or ultrastructural characters. A major problem for estimating phylogenetic relationships between organisms such as ciliates, which exhibit highly divergent morphologies, are the difficulties of identifying homologous characters (see Patterson and Sogin, 1993). More recently, developmental data have also been used to infer evolutionary relationships (Bardele, 1989; Eisler and Fleury, 1995). The ability to obtain ultrastructural and ontogenetic data is often dependent on sophisticated microscopical techniques which makes their general use relatively difficult (see chapter 17). In contrast, the development of molecular phylogenetics has allowed the systematic comparison of the same set of characters (SSU or LSUrDNA) between all sampled taxa. Molecular sequence data also have the convenient feature of making transparent any hypothesis of homology based upon columns of aligned nucleotides. Sequences generated by different laboratories can easily be retrieved from databases, facilitating new analyses as further data or methods become available. The sampling effort from different laboratories for ciliate SSUrDNA sequences has covered a broad taxonomic range with nearly 70 sequences described so far (Bernhard *et al.*, 1995; Embley *et al.*, 1995; Hammerschmidt *et al.*, 1996; Hirt *et al.*, 1995) and there are 43 sequences for the LSUrDNA dataset (Eisler and Fleury, 1995). Although none of the previously published taxonomies based on morphology, ultrastructure and developmental data are in complete accord with the molecular phylogenies, there is generally good agreement at lower taxonomic levels (Eisler and Fleury, 1995). It is worth noting that the apparent disagreements between phylogenies has greatly stimulated the accumulation of additional molecular data, and the re-examination of morphological features, promoting an ongoing process of reciprocal illumination (c.f. Hennig, 1966) which greatly benefits the study of ciliate biology.

18.2.3 Phylogeny of karyorelictids and evolution of nuclear dualism

Ciliates are characterized by nuclear dualism, that is their cytoplasm contains simultaneously two types of nuclei, a generative micronucleus and a somatic macronucleus (MAC) (Raikov, 1982). One point of general agreement in pre-molecular ciliate systematics, was that the ancestor of all ciliates was similar to present-day karyorelictids, a relatively small group with about 135 named species (Corliss, 1979; Corliss, 1988; Raikov, 1985) (chapter 17). The key feature found in karyorelictids (as their name implies – nuclear relics), was their supposed ancestral (or plesiomorphic) type of nuclear dualism. This is characterized by a paradiploid and non-dividing MAC in contrast to the polyploid and dividing MAC found in the relatively few other ciliates that have been well studied (Raikov, 1985). In order to test the hypothesis of whether the karyorelictid type is plesiomorphic, we have sequenced the SSUrDNA from *Loxodes magnus* and a *Tracheloraphis* sp. and analysed their phylogenetic relationships (Hirt *et al.*, 1995). In addition, we have accumulated additional heterotrich sequences to investigate a possible relationship with karyorelictids, as suggested by some ultrastructural data (Small and Lynn, 1985; de Puytorac, Grain and Legendre, 1994), and by analysis of a LSUrDNA partial sequence from *Loxodes striatus* (Baroin-Tourancheau *et al.*, 1992). Recently published karyorelictid and heterotrich SSUrDNA sequences (Hammerschmidt *et al.*, 1996) were also integrated into the analysis.

The major features of the ciliate SSUrDNA tree are shown in Figure 18.1 and can be summarized as follows: the ciliates sampled are split into two major monophyletic groups: (1) clade **a**, composed of karyorelictids and aerobic heterotrichs, (2) clade **b** containing all of the other ciliates. This basic split was observed with all phylogenetic inference methods investigated (including least-squares, neighbour-joining, maximum parsimony and maximum likelihood (see Embley *et al.*, 1995; Hammerschmidt *et al.*, 1995; Hirt *et al.*, 1995 for more details). Clade **c** which lies within clade **b**, was also consistently recovered and supported by high bootstrap proportions.

Thus phylogenetic analyses of the karyorelictid sequences unambiguously support the hypothesis that they are closely related to the aerobic heterotrichs (Hirt *et al.*, 1995) (Figure 18.1). The distribution of ciliates with a dividing or non-dividing MAC mapped onto this tree, also reveals that it is most parsimonious to infer a dividing MAC, rather than a non-dividing MAC, as the ancestral state. This hypothesis requires only one evolutionary step, the loss of the capacity to divide the MAC on the karyorelictid lineage. In contrast, two independent acquisitions of the capacity to divide the MAC are required if, as Raikov (1985, 1994) has proposed, the non-dividing MAC is the ancestral state. The hypothesis of an ancestral dividing MAC is compatible with Fauré-Fremiet's proposition that the nuclear dualism present in karyorelictids is a derived trait (Fauré-Fremiet, 1961, 1970). However, Fauré-Fremiet (1961) also suggested that the non-dividing MAC was an adaptation to an interstitial life style rather than a synapomorphy for karyorelictids whose monophyly he questioned. Sampling of all named groups is required to test this hypothesis further, but at least for *Loxodes* and *Tracheloraphis* the non-dividing MAC can be regarded as homologous (Figure 18.1). See also chapter 17 for a morphological perspective on these issues.

Hammerschmidt *et al.* (1996) also characterized the ciliate *Protocruzia* SSUrDNA but, like our analysis here, they could not unambiguously resolve its position (Figure 18.1). However, the molecular data decisi-

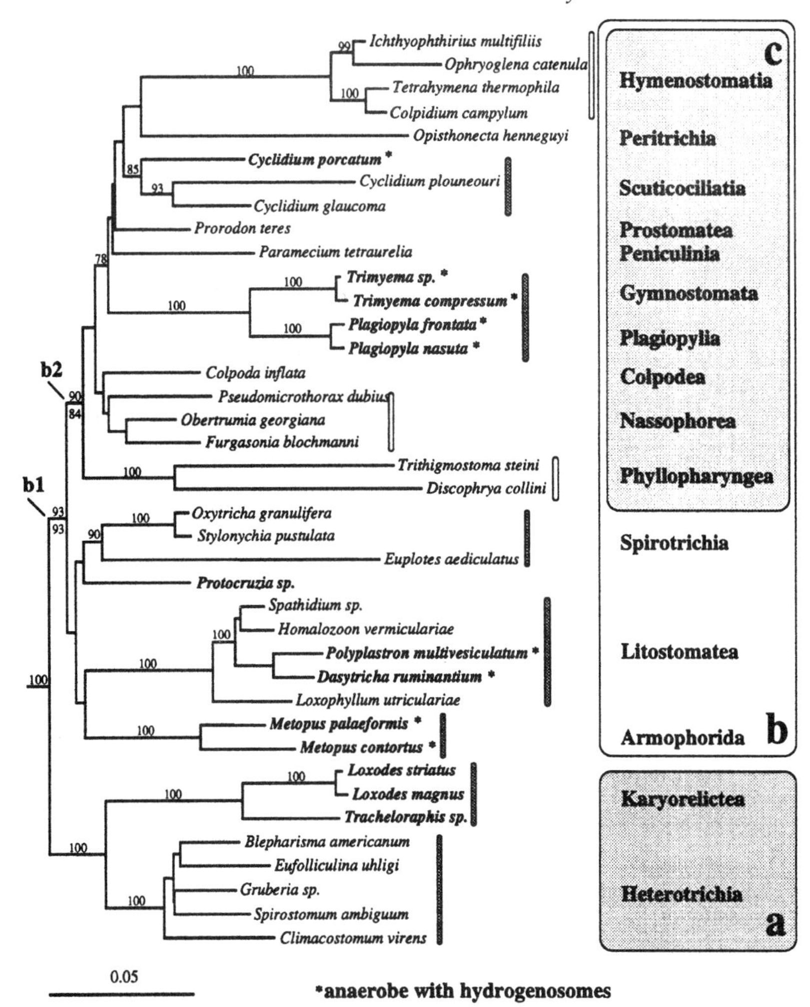

Figure 18.1 Ciliate molecular phylogeny based on SSUrDNA sequences. A representative selection of ciliate and outgroup taxa (outgroup taxa not shown, see Embley *et al.* 1995) SSUrDNA sequences were aligned and a phylogenetic tree estimated using a least squares distance method (Fitch and Margoliash, 1967) with pairwise distances estimated by the Jukes and Cantor model, bootstrap proportions (BP) were also estimated (all calculations used programs implemented in PHYLIP v3.5: DNADIST, FITCH, SEQBOOT, CONSENSE). Similar topologies are recovered using maximum parsimony and maximum likelihood analysis, see for example Embley *et al.* (1995) and Hirt *et al.* (1995) for additional details. For the sake of simplicity we have named the three major ciliate clades recovered most consistently by single letters: a, b and c. Branches leading to clade b and c are indicated by b1 and b2, see text. BP $\geqslant$70% are

vely reject the hypothesis that *Protocruzia,* which has a dividing and paradiploid MAC, is related to the karyorelictids which have non-dividing MAC (Figure 18.1) with which it was previously classified (Small and Lynn, 1985).

Interestingly, Raikov (1994) has recently pointed out that the mode of MAC division in the aerobic heterotrichs involves external microtubules, instead of internal microtubules as in other ciliates. This prompted Hammerschmidt *et al.* (1996) to propose an alternative interpretation of the evolution of nuclear dualism over the SSUrDNA tree. They analysed MAC division as an unordered three character state (i.e. 1: non-dividing, 2: dividing MAC with internal microtubules and 3: dividing MAC with external microtubules). This complicates the interpretation of MAC evolution relative to the SSUrDNA tree. Five different hypothesis, including four which would accommodate an ancestral dividing MAC, imply two evolutionary changes and additional criteria are needed to decide between them. Two possible interpretations are illustrated in Figure 18.2A. Outgroup comparison does not provide any resolution of ancestral state because MACs do not exist in dinoflagellates and apicomplexans, and these taxa display a collection of different nuclear division mechanisms (Raikov, 1982). Clearly additional comparative molecular and mechanistic data on the two MAC division processes are needed to support any of these hypotheses.

The SSUrDNA tree (Figure 18.1) can perhaps provide some useful guidance. For instance there is strong support that heterotrichs are not monophyletic (Figure 18.1). The anaerobic armophorid heterotrichs (Small and Lynn, 1985) of the genus *Metopus* which group in clade **b**, are clearly unrelated to the aerobic heterotrichs. This tree contrasts with morphological taxonomies where the complexity of heterotrich oral ciliature has always been seen as an advanced and likely monophyletic feature (Corliss, 1979). Non-monophyly of heterotrichs has recently been confirmed by additional armophorid SSUrDNA sequence data from

---

shown on the branches. BP from a neighbour-joining analysis are also shown for b1 and b2 below the branches for comparison with previous analyses, see text. The previously reported *Entodinium simplex* SSUrDNA sequence (Genbank accession number: U27815) was subsequently shown to originate from *Polyplastron multivesiculatum* (Andre-Denis Wright, Department of Zoology, University of Guelph, personal communication and Robert Hirt, unpublished data). All taxa discussed in the text are highlighted by bold type or dark vertical bars. Names for ciliate higher taxa are from Corliss (1994) except for Gymnostomata from Serrano *et al.* (1988) and Armophorida and Plagiopylia from Small and Lynn (1985).

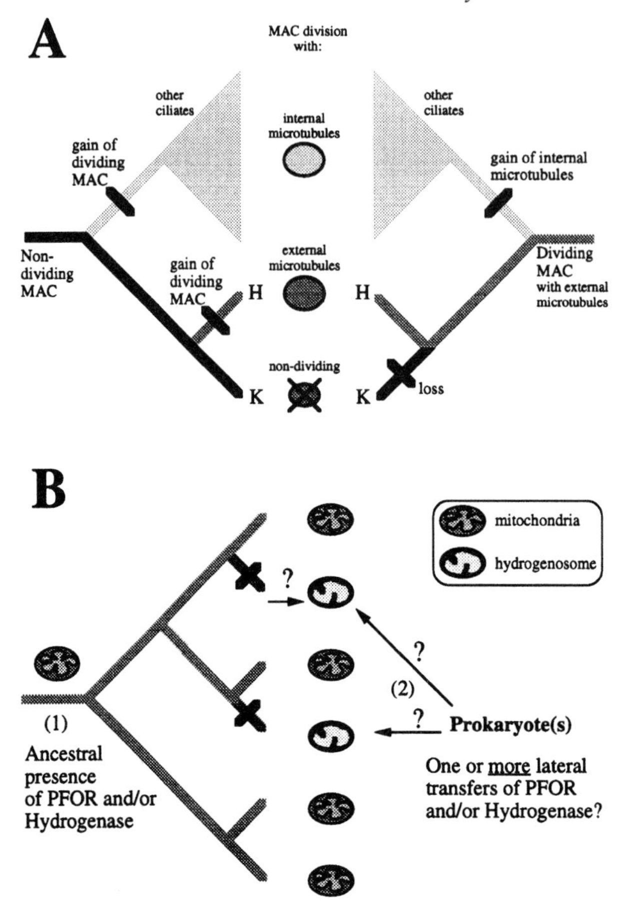

Figure 18.2 Mapping of cellular features on the ciliate SSUrDNA phylogeny. (A) Nuclear dualism diversity found in ciliates, with two possible evolutionary hypotheses for the ancestral state, left: non-dividing macronucleus (MAC); right: dividing MAC with external microtubules (selected from five possible scenarios with two steps). Character changes are shown by thick lines. H, aerobic heterotrichs; K, karyorelictids. (B) Multiple origins of hydrogenosomes in ciliates (only two shown for simplicity) with some alternative hypotheses for the origin(s) of enzymes (not necessarily related to the origin of the organelle) specific for hydrogenosomes, namely hydrogenase and pyruvate:ferredoxin oxidoreductase (PFOR). These enzymes could have been gained ancestrally (1) or transferred from an unknown donor organism – probably a prokaryote (2). The phylogeny permits the identification of the close relatives of hydrogenosome containing ciliates, which should be included for comparative sampling of genes to test these hypotheses.

Caenomorpha uniserialis (R.P. Hirt, B.J. Finlay, T.M. Embley unpublished) and LSUrDNA from a *Brachonella* sp. (Eisler and Fleury, 1995). This topology highlights the need to determine how anaerobic armophorids divide their MAC, i.e. whether they use internal or external microtubules. It also suggests that the proposed morphological homologies which previously united the two heterotrich groups need to be reexamined.

18.2.4 Origin and evolution of ciliate hydrogenosomes

It is commonly thought that all eukaryotes are aerobes with mitochondria. However, secondary loss of functional mitochondria has occurred in a number of protist lineages (see chapter 6) and is associated with the colonization of anaerobic niches where food bacteria are common and competition is limited (Fenchel and Finlay, 1995). Ciliates are probably the best group for studying this phenomenon because they seem to have lost functional mitochondria on a number of occasions, and some have gained a redox organelle which produces hydrogen and is thus called a hydrogenosome (Müller, 1993; Fenchel and Finlay, 1995). Members of at least seven orders (according to the taxonomy of Corliss, 1979) have been described as possessing hydrogenosomes (Fenchel and Finlay, 1995). However, given the difficulties in constructing hypotheses of higher relationships for the ciliates based on morphology, there have previously been no robust hypotheses of relationships among hydrogenosome-containing ciliates. Molecular data can provide this phylogeny and have the additional advantage of being entirely independent of the structure whose evolutionary distribution is being investigated.

Nine hydrogenosome-containing ciliates representing at least five higher groups (Corliss, 1979) have so far been investigated for their nuclear SSUrDNA (Embley *et al.*, 1995) (Figure 18.1). All unconstrained analyses of the taxa sampled support four independent origins of hydrogenosomes during ciliate evolution. Detailed investigation of hypotheses of fewer origins using constrained trees and the Kishino–Hasegawa maximum likelihood difference test (Kishino and Hasekawa, 1989) reveal that one hypothesis of three origins falls within the 95% confidence set (Embley *et al.*, 1995). This places the scuticociliate *Cyclidium porcatum* as sister group to *Plagiopyla* and *Trimyema* and renders *Cyclidium* paraphyletic. In contrast, a monophyletic *Cyclidium* was recovered in all unconstrained analyses and there are also good morphological reasons for believing this to be an accurate reflection of the evolution of the genus (Esteban *et al.*, 1993) (Figure 18.1). *Cyclidium porcatum* is also of interest because its hydrogenosome looks very much like mitochondria in some aerobic *Cyclidium* species (Esteban *et al.*, 1993;

Fenchel and Finlay, 1995), lending some support to the hypothesis that ciliate hydrogenosomes may represent modified mitochondria (Finlay and Fenchel, 1989).

Morphological investigations have continued to detect new anaerobic ciliates (Esteban, Finlay and Embley, 1993) reinforcing the impression that among ciliates the transformation from aerobe with mitochondria, to anaerobe with hydrogenosomes, is relatively easy. The big questions that remain are 'From whence do the genes encoding anaerobic biochemistry originate? and 'How many times has this occurred?' Once again the molecular tree can provide useful guidance on taxon sampling to most efficiently address the question of the origin and evolution of the hydrogenosomal enzymes (Embley *et al.*, 1995) (Figure 18.2B). Of particular interest are the hydrogenosomal enzymes pyruvate:ferredoxin oxidoreductase and hydrogenase which are typical of anaerobic prokaryotes (Müller, 1993). Ciliates thus represent a paradigm for the study of hydrogenosome evolution and detailed comparative studies between ciliates and other anaerobic protists are likely to contribute to the understanding of a major puzzle of eukaryote evolution, the origin and evolution of anaerobic metabolism (Müller, 1996).

18.3 PERSPECTIVES

18.3.1 Molecules and morphology

There is fairly good concordance between the taxonomy proposed by Small and Lynn (1985) which uses epigenetically-based ciliate characters and the SSUrDNA gene tree. For example, the monophyly of the litostomes, defined by their shared similarity in kinetid ultrastructure (Small and Lynn, 1985), is confirmed by molecular studies (Figure 18.1 and see LSUrDNA tree in Eisler and Fleury, 1995). However, there are also conflicts: the stichotrichs form a clade with the hypotrichs in the molecular tree (Figure 18.1) whereas the Small and Lynn (1985) system splits them into two distinct classes (implying non-monophyly). In the case of ciliates, epigenetic characters would seem to be neither better nor worse than other characters in that they may be homologous or homoplastic, and may provide reliable as well as misleading evidence of phylogenetic relationships. Frankel (1989: 75–76) gives an example of transfer of ciliary rows between two conjugating ciliates demonstrating lateral "plasma gene" transfer for ultrastructural units, a problem well known for DNA encoded genes (Sprague, 1991), which further illustrates the similarity between the two types of heritable information. We believe that studies of molecules and morphology are highly complementary. As we have shown, molecular data provide independent estimates of phylogeny that can be put to work in interpreting the evolution of ciliate morphology. Conflicts between molecular and morphological-based hypothesis of relationships, such as the relationships of aerobic and anaerobic heterotrichs, identify areas in need of further study and thus help direct research efforts. Ultimately, the dialogue promoted by these two sources of information will contribute to a better understanding of ciliate diversification, phylogeny, and general biology.

18.3.2 The future of ciliate molecular phylogenetics

Both the SSU and LSUrDNA phylogenies are characterized by a lack of resolution between major taxa in clade **b** (Figure 18.1). Experience shows that the inclusion of additional taxa is one important means of improving resolution and support for some groups. For example, bootstrap proportions for nodes **b1** and **b2** in Figure 18.1 are much more impressive than in previous analyses (Embley *et al.*, 1995; Hirt *et al.*, 1995; Leipe *et al.*, 1994). Sampling of additional taxa may contribute some further resolution within clade **b**, as might combined analysis of the complementary SSU and LSUrDNA dataset. New sequence data from alternative genes will also be important in providing both additional resolution and independent estimates of ciliate phylogeny.

Acknowledgements

This work was supported by a NERC grant GR3/8146 to TME and RPH. Collaboration with B.J. Finlay, G. Esteban and P.L. Dyal on many aspects of this project is gratefully acknowledged.

18.4 REFERENCES

Bardele, C.F. (1989) From ciliate ontogeny to ciliate phylogeny: a program. *Bollettino di Zoologia*, **56**, 235–43.

Baroin-Tourancheau, A., Delgado, P., Perasso, R. and Adoutte, A. (1992) A broad molecular phylogeny of ciliates: identification of major evolutionary trends and radiations within the phylum. *Proceedings of the National Academy of Sciences USA*, **89**, 9764–8.

Baroin-Tourancheau, A., Tsao, N., Klobutcher, L.A. *et al.* (1995) Genetic code deviation in the ciliates, evidence for multiple and independent events. *EMBO Journal*, **14**, 3262–7.

Bernhard, D., Leipe, D.D., Sogin, M.L. and Schlegel, M. (1995) Phylogenetic relationships of Nassulida within the phylum Ciliophora inferred from the complete small subunit rRNA gene sequences of *Furgasonia blochmanni*, *Obertrumia georgiana*, and *Pseudomicrothorax dubius*. *Journal of Eukaryotic Microbiology*, **42**, 126–31.

Cavalier-Smith, T. (1993) Kingdom Protozoa and its 18 phyla. *Microbiological Reviews*, **57** 953–94.

Corliss, J.O. (1979) *The Ciliated Protozoa*, 2nd edn, Pergamon Press, Oxford.

Corliss, J.O. (1988) The quest for the ancestor of the Ciliophora: a brief review of the continuing problem. *BioSystems*, **21**, 323–31.

Corliss, J.O. (1994) An interim utilitarian ("user-friendly") hierarchical classification and characterization of the protists. *Acta Protozoologica*, **33**, 1–51.

Curds, C.R. (1992) *Protozoa and the Water Industry*, Cambridge University Press, Cambridge.

Darbyshire, J.F. (ed) (1994) *Soil Protozoa*, CAB International, Oxon.

Drouin, G., Moniz de Sá, M. and Zucker, M. (1995) The *Giardia lamblia* actin gene and the phylogeny of eukaryotes. *Journal of Molecular Evolution*, **41**, 841–9.

Edelman, G.M. (1988) *Topobiology* Basic Books, Harper-Collins, New York.

Eisler, K. and Fleury, A. (1995) Morphogenesis and evolution in ciliates, in *Protistological Actualities, Proceedings of the Second European Congress of Protistology* (eds G. Brugerolle and J.-P. Mignot), Clermont-Ferrand, pp. 102–27.

Embley, T.M., Finlay, B.J., Dyal, P.L. *et al.* (1995). Multiple origins of anaerobic ciliates with hydrogenosomes within the radiation of aerobic ciliates. *Proceedings of the Royal Society London, Series B*, **262** 87–93.

Esteban, G., Finlay, B.J. and Embley, T.M. (1993) New species double the diversity of anaerobic ciliates in a Spanish lake. *FEMS Microbiology Letters*, **109**, 93–9.

Esteban, G., Guhl, B.E., Clarke, K. *et al.* (1993) *Cyclidium porcatum* n.sp.: a free-living anaerobic scuticociliate containing a stable complex of hydrogenosomes, eubacteria and archaeobacteria. *European Journal of Protistology*, **29**, 262–70.

Fauré-Fremiet, E. (1961) Quelques considérations sur les ciliés mésopsammiques a propos d'un récent travail de J. Dragesco. *Cahiers de Biologie Marine*, **2**, 177–86.

Fauré-Fremiet, E. (1970) A propos de la note de M. Thomas Njiné sur le cilié *Loxodes magnus. Comptes Rendus de L'Academie des Sciences, Paris*, **270**, 523–4.

Fenchel, T. (1987) *Ecology of Protozoa: The Biology of Free-living Phagotrophic Protists*, Science Tech Publisher/Springer-Verlag, Madison.

Fenchel, T. and Finlay, B.J. (1991) The biology of free-living anaerobic ciliates. *European Journal of Protistology*, **26**, 201–15.

Fenchel, T. and Finlay, B.J. (1995) *Ecology and Evolution in Anoxic Worlds*, Oxford University Press, New York.

Finlay, B.J. and Fenchel, T. (1989) Hydrogenosomes in some anaerobic protozoa resemble mitochondria. *FEMS Microbiology Letters*, **65**, 311–14.

Fitch, W.M. and Margoliash, E. (1967) Construction of phylogenetic trees. *Science*, **155**, 279–84.

Frankel, J. (1989) *Pattern Formation*, Oxford University Press, New York.

Gajadhar, A.A., Marquardt, W.C., Hall, R. *et al.* (1991) Ribosomal RNA sequences of *Sarcocystis muris, Theileria annulata* and *Crypthecodinium cohnii* reveal evolutionary relationships among apicomplexans, dinoflagellates, and ciliates. *Molecular and Biochemical Parasitology*, **45**, 147–54.

Hammerschmidt, B., Schlegel, M., Lynn, D.H. *et al.* (1996) Insights into the evolution of nuclear dualism in the ciliates revealed by phylogenetic analysis of rRNA sequences. *Journal of Eukaryotic Microbiology*, **43**, 225–30.

Hashimoto, T., Makamura, Y., Kamaishi, T. *et al.* (1995) Phylogenetic place of kinetoplastid protozoa inferred from a protein phylogeny of elongation factor 1 α. *Molecular and Biochemical Parasitology*, **70**, 181–5.

Hausmann, K. and Hülsmann, N. (1996) *Protozoology*, 2nd edn. Thieme, Stuttgart.

Hennig, W. (1966) *Phylogenetic Systematics*, University of Illinois Press, Urbana.

Hirt, R.P., Dyal, P.L., Wilkinson, M. *et al.* (1995) Phylogenetic relationships among karyorelictids and heterotrichs inferred from small subunit rRNA sequences: resolution at the base of the ciliate tree. *Molecular Phylogenetics and Evolution*, **4**, 77–87.

Jablonka, E. and Lamb, M.J. (1995) *Epigenetic Inheritance and Evolution*, Oxford University Press, New York.

Kishino, H. and Hasegawa, M. (1989) Evaluation of the maximum likelihood estimate of the evolutionary tree topologies from DNA sequence data, and the branching order in Hominoidea. *Journal of Molecular Evolution*, **29**, 170–9.

Lee, R.E. and Kugrens, P. (1992) Relationship between the flagellates and ciliates. *Microbiological Reviews*, **56**, 529–42.

Leipe, D.D., Bernhard, D., Schlegel, M. and Sogin, M.L. (1994) Evolution of 16S-like ribosomal RNA genes in the ciliophoran taxa Litostomatea and Phyllopharyngea. *European Journal of Protistology*, **30**, 354–61.

Lynn, D.H. and Corliss, J.O. (1991) Ciliophora, in *Microscopic Anatomy of Invertebrates, Protozoa*, Vol. 1 (eds F.W. Harrison and J.O. Corliss), Wiley-Liss, New York, pp. 333–67.

Madireddi, M.T. Smothers, J.F. and Allis, C.D. (1995) Waste not, want not. Does DNA elimination fuel gene amplification during development in ciliates? *Seminars in Developmental Biology*, **6**, 305–15.

Müller, M. (1993) The hydrogenosome. *Journal of General Microbiology*, **139**, 2879–89.

Müller, M. (1996) Energy metabolism of amitochondriate protists, an evolutionary puzzle, in *Christian Gottfried Ehrenberg-Festschrift* (eds M. Schlegel and K. Hausmann), Leibzig Universitätsverlag, Leibzig, pp. 63–76.

Nanney, D.L. (1980) *Experimental Ciliatology*, Wiley, New York.

Nanney, D.L. (1982) Genes and phenes in *Tetrahymena. BioScience*, **32**, 783–8.

Nanney, D.L. (1986) Introduction, in *The Molecular Biology of Ciliated Protozoa* (ed. J.G. Gall), Academic Press, London, pp. 1–26.

Patterson, D.J. and Sogin, M.L. (1993) Eukaryotic origins and protistan diversity, in *The Origin and Evolution of the Cell* (eds H. Hartman and K. Matsuno), World Scientific, Singapore, pp. 13–46.

Philippe, H. and Adoutte, A. (1995) How reliable is our current view of eukaryotic phylogeny? in *Protistological Actualities, Proceedings of the Second European Congress of Protistology* (eds G. Brugerolle and J.-P. Mignot), Clermont-Ferrand, pp. 17–32.

Puytorac, P. de Batisse, A., Deroux, G. *et al.* (1993) Proposition d'une nouvelle classification du phylum des protozoaires Ciliophora Doflein 1901. *Comptes Rendus de L'Academie des Sciences, Paris*, **316**, 716–29.

Puytorac, P. de, Grain, J and Legendre, P. (1994) An attempt at reconstructing a phylogenetic tree of the ciliophora using parsimony methods. *European Journal of Protistology*, **30**, 1–17.

Ragan, M.A., Cawthorn, R.J., Despres, B. *et al.* (1996) The lobster parasite *Ano-*

phryoides haemophila (Scuticociliatida, Orchitophrydae), nuclear 18S rDNA sequence, phylogeny and detection using oligonucleotide primers. *Journal of Eukaryotic Microbiology*, **43**, 341–6.

Raikov, I.B. (1982) *The Protozoan Nucleus*, Springer-Verlag, Wien and New York.

Raikov, I.B. (1985) Primitive never-dividing macronucleus of some lower ciliates. *International Review of Cytology*, **95**, 267–325.

Raikov, I.B. (1994) The nuclear apparatus of some primitive ciliates, the karyorelictids: structure and divisional reorganization. *Bollettino di Zoologia*, **61**, 19–28.

Serrano, S., Martin-Gonzales, A., and Fernandez-Galiano, D. (1988) *Trimyema compressum* Lackey 1925: morphology, morphogenesis and systematic implications. *Journal of Protozoology*, **35**, 315–20.

Small, E.B. and Lynn, D.H. (1985) Phylum Ciliophora, in *Illustrated Guide to the Protozoa* (Eds S.H. Hunter, J.J. Lee and E.C. Bovee), Allen Press, Lawrence, pp. 393–575.

Sogin, M.L., Morrison, H.G., Hinkle, G. and Silberman, J.D. (1996) Ancestral relationships of the major eukaryotic lineages. *Microbiologià SEM*, **12**, 17–28.

Sonneborn, T.M. (1970) Gene action in development. *Proceedings of the Royal Society London, Series B*, **176**, 347–66.

Sprague, G.F. (1991) Genetic exchange between kingdoms. *Current Opinion in Genetics and Development*, **1**, 530–3.

Swofford, D.L., Olsen, G.J., Waddel, P.J. and Hillis, D.M. (1996) Phylogeny inference, in *Molecular Systematics* (eds D.M. Hillis, C. Moritz and B.K. Mable), Sinauer Associates, Massachusetts, pp. 407–514.

Wheatley, D.N., Rasmussen, L. and Tiedtke, A. (1994) *Tetrahymena*, a model for growth, cell cycle, with biotechnological potential. *BioEssays*, **16**, 367–72.

Williams, A.G. and Coleman, G.S. (1992) *The Rumen Protozoa*, Springer-Verlag, New York.

Wrights, A.-D. G. and Lynn, D.H. (1995) Phylogeny of the fish parasite *Ichthyophthirius* and its relatives *Ophryoglena* and *Tetrahymena* (Ciliophora, Hymenostomatia) inferred from 18S ribosomal RNA sequences. *Molecular Biology and Evolution*, **12**, 285–90.

19

Phylogenetic relationships of the Myxozoa

Cort L. Anderson

Department of Biology, Imperial College at Silwood Park, Ascot, Berkshire, SL5 7PY.
E-mail: c.l.anderson@ic.ac.uk

ABSTRACT

The phylum Myxozoa is a group of organisms whose biology and taxonomic placement are not well established. This phylum contains some 1200 described species, all obligate parasites, but the evolutionary relationship of the phylum to other major groups is obscure, as are the systematic relationships within the phylum. However, molecular data on the Myxozoa are beginning to accumulate, and these data are beginning to resolve the phylogenetic position of this phylum. Three molecular studies have been published, in which whole or partial 18S rRNA sequences have been used to construct phylogenetic hypotheses (Schlegel *et al.*, 1996; Siddall *et al.*, 1995; Smothers *et al.*, 1994). All three agree in identifying the Myxozoa as metazoan. However, these same studies reach different conclusions regarding the group's evolutionary relationships within the Metazoa. Using a combination of DNA sequence data and some morphological and developmental characters, Siddall *et al.* (1995) conclude that myxozoans are extremely reduced cnidarians. In contrast, Schlegel *et al.* (1996) and Smothers *et al.* (1994) would group the Myxozoa with the bilateral animals. Further DNA sequence data – from genes other than the 18S rRNA gene – would be useful to resolve differences between these competing hypotheses. Moreover, molecular data could also have application to taxonomic questions within the phylum, helping to sort out the confused systematics of this group and to demonstrate genetic identity between life cycle stages from different hosts.

Evolutionary Relationships Among Protozoa. Edited by G.H. Coombs, K. Vickerman, M.A. Sleigh and A. Warren. Published in 1998 by Chapman & Hall, London. ISBN 0 412 79800 X

19.1 INTRODUCTION

The Myxozoa (phylum Myxozoa Grassé, 1970) (formal creation of the phylum Myxozoa was by Grassé and Lavette (1978)) are an intriguing group of parasitic organisms whose biology, taxonomy, and evolutionary relationships are not well established. The phylum contains well over 1100 described species, most of which infect teleost fishes, with new species continually being added. A number of these infect commercially important fishes (reviewed in El-Matbouli, Fischer-Scherl and Hoffmann, 1992b), and indeed, much of the information available on these organisms is to be found in the parasitology literature. Myxozoa have also been found in amphibians, reptiles, and a variety of invertebrates, most commonly in oligochaetes (Lom, 1990). Because of the logistical difficulties entailed in studying these minuscule parasites however, very little is known of their life histories, distribution, or host range.

19.2 TAXONOMIC POSITION

First described in the 19th century (see Lom, 1990 for general account), the Myxozoa were long grouped with the Microsporidia, in a class Cnidosporidia Doflein 1901, in large part because both possess extrusible polar filaments. Subsequently, as the differences in structure and organization between the two groups became apparent, the Microsporidia were placed in a separate protistan phylum Microspora Sprague 1977 (see also chapter 4). The phylum Myxozoa, while still frequently described as protist, should more appropriately be considered a metazoan phylum.

Traditionally, the phylum Myxozoa was thought to contain two classes, Myxosporea Bütschli, 1881, found mainly in poikilothermic vertebrates, and the Actinosporea Levine, 1980, found in marine and freshwater aquatic invertebrates (see Levine *et al.*, 1980). However, Markiw and Wolf (1983) and Wolf and Markiw (1984) demonstrated that the life cycle of *Myxobolus cerebralis* Hofer, 1903, includes an actinosporean stage in a tubificid host. Myxosporean spores were infective for tubificid worms, but not able to propagate infection in the fish host, while actinosporean spores could infect fish, but were not able to infect tubificid worms. Initially received with scepticism, this obligate alternation of hosts has been independently confirmed for *M. cerebralis*, the organism responsible for whirling disease in salmonids (El-Matbouli and Hoffmann, 1989). Alternation of hosts has since been shown to occur in multiple genera (Kent, Margolis and Corliss, 1994, and references therein)[1]. This has given rise to new guidelines for the taxonomy of this

[1] Andree *et al.* (1997) have recently demonstrated identity of SSUrRNA gene sequences from the myxosporean and actinosporian stages of *Myxosoma cerebralis* (Eds).

group, rules which do away with the class Actinosporea altogether (Kent, Margolis and Corliss, 1994).

The discovery that at least some, if not all, of the Actinosporea are disguised myxosporeans leaves the taxonomy of this phylum in an awkward condition. It is clear that traditional, morphology-based taxonomic methods are inadequate. It is not possible to use morphology alone to determine which actinosporean stage(s) correspond to which myxosporean stage. Transmission studies, in which potential fish hosts are exposed to myxozoan-infected oligochaetes, can be used to establish links between myxosporean and actinosporean life stages, but these are time and labour intensive undertakings (El-Matbouli and Hoffman, 1989; El-Matbouli, Fischer-Scherl and Hoffmann, 1992a), and do not preclude the possibility that a given parasite may utilize multiple hosts. Finally, the sheer number of possible host–parasite combinations renders this approach impractical. Moreover, many of the myxozoan genera have numerous species, e.g. *Myxobolus* has more than 450 named species, based on small differences in spore structure and on occurrence in new hosts. These small differences in spore structure between parasites in different hosts may, however, be host-related rather than indicative of different species, making absolute species identification in this phylum extremely difficult.

19.3 ASPECTS OF MYXOZOAN BIOLOGY

While almost any host tissue may be a site of infection, most myxosporeans exhibit a degree of tissue specificity. Some myxosporean species are coelozoic, producing plasmodia or pseudoplasmodia which attach to the lining of hollow organs (urinary bladder, gall bladder), or are free floating within the lumen. Histozoic species are found in solid tissues, typically in the form of multinucleate plasmodia embedded in host connective tissue. These histozoic infections can take the form of cysts, or more rarely, there may be a diffuse distribution of small pseudoplasmodia (Lom, 1990). While the majority of myxosporean infections occur in vertebrates, there are a very few infections reported in invertebrates which exhibit characteristic myxosporean-type development. Among these are *Myxobolus* species found in the oligochaete *Nais* (Gurley, 1894), *Chlormyxum diploxys* infections in the lepidopteran *Tortrix viridana* (Gurley, 1893), and *Myxosporidium bryozoides*, in the aquatic bryozoan *Plumatella* (*Alcyonella*) *fungosa* (Korotneff, 1892). Actinosporean infections typically occur in the coelomic cavity and intestinal lining of the host organism.

Spore formation in Myxozoa involves the enclosure of cells or nuclei and surrounding cytoplasm by other cells of parasite origin culminating in the diagnostic valved spore with polar capsules. Sporoplasms are enveloped by valvogenic cells and capsulogenic cells, which subsequently differentiate into spore valves and polar capsules. These structures vary in number and arrangement in the mature spore, and provide the basis for traditional taxonomy of this group. Infection is thought to occur after extrusion of the polar filaments which anchor the spore to the prospective host tissue, whereupon the valves open, releasing the infective sporoplasm.

For most species of Myxozoa, the details of transmission of infection are completely unknown. *M. cerebralis* provides one of the few instances where a complete life cycle has been established (Markiw, 1992; El-Matbouli and Hoffmann, 1991; El-Matbouli, Hoffmann and Mandok, 1995). Myxosporean-type spores are released upon the death and decomposition of the fish host. These spores are ingested by tubificid worms, and initiate infection in the gut of the worms. Mature actinosporean spores appear after ca. 3.5 months, and are continually released for several weeks. Host fish can acquire infection by ingestion of the tubificid host, or by encountering free floating mature actinosporean spores. The *M. cerebralis* will then begin to produce mature myxosporean spores after two to three months maturation in the fish host.

19.4 PHYLOGENETIC POSITION OF THE MYXOZOA

Because of their apparently simple organization, the Myxozoa were initially considered to be protists. However, as more of their biology has become known, the situation has inevitably become more complicated. In particular, the similarity of polar capsules to cnidarian nematocysts, both during ontogeny and as mature structures, and the resemblance between sporoblasts and parasitic cnidarian Narcomedusae larvae, have been frequently remarked upon (Weill, 1938; Lom and de Puytorac, 1965; Mitchell, 1977). The myxozoan polar capsule primordium is a small vesicle which extends as a long tube into the cytoplasm of the capsulogenic cell, vesicle and tube being wrapped in microtubules. The polar filament is assembled in the external tube, and as the tube shortens, the filament is withdrawn into the bulbous vesicle in a coiled arrangement. The capsule wall is composed of an inner, alkali-resistant, chitinous layer, covered with a proteinaceous outer layer (Lukes, Volf and Lom, 1993). This process is in all essentials the same as the development of cnidarian nematocysts (Lom and de Puytorac, 1965; Siddall *et al.*, 1995). A hypothesis which does not accommodate an evolutionary link between Cnidaria and Myxozoa would require that the developmental pathway and structures common to polar capsules and nematocysts have been independently acquired by convergent evolution, or else lost by other metazoan groups. This has naturally led to speculation that the myxozoans are grossly simplified Cnidaria. While most authors have conceded the striking similarities, some have been less willing to conclude that this qualifies Myxozoa for inclusion in the Cnidaria, or that the two derive from a common ancestor. Arguing for their retention in the kingdom Protozoa, Cavalier-Smith (1993) maintained that a hypothesis of myxozoans derived from an ancestral cnidarian 'involves too great a degree of parasitic reduction to be contemplated seriously'.

However, more recent data from studies of several myxozoan species indicates that the Myxozoa are indeed metazoans, and might in fact be extremely derived cnidarians. Siddall *et al.* (1995) describe a detailed morphological study of the developmental stages of a representative myxozoan, *Thelohanellus nikolskii* Akhmerov, 1955, in which they were able to demonstrate the presence of several metazoan-defining properties. These included multicellularity, terminal cell differentiation and specialization, collagen production, and the presence of cellular junctions. In addition, molecular data in the form of novel full length 18S rDNA sequences from a myxozoan, *Henneguya doori* Guilford, 1963, and a narcomedusan cnidarian, *Polypodium hydriforme* Ussov, 1885, were analysed, together with two other myxozoan 18S sequences from the database, three other cnidarian sequences, 14 metazoan sequences, and two protistan sequences, likewise from the database. A second analysis used ca. 800bp 18S partial sequences of an additional 12 cnidarian genera, and also incorporated the morphological data, scored as the presence or absence of multicellularity, cellular junctions, terminal differentiation, nematocysts, and radial symmetry. In both these analyses, sequence alignment was accomplished using Malign (Wheeler and Gladstein, 1993), and the resultant data matrix was subjected to parsimony analysis. A third set of analyses was carried out employing OPTALIGN (Wheeler, 1996). All three analyses unambiguously placed the myxozoan taxa among the Metazoa, and furthermore, as a sister group of the parasitic narcomedusan cnidaria (Figure 19.1).

A somewhat different phylogenetic hypothesis had been published by Smothers *et al.* (1994). They amplified and sequenced five 18S sequences, two of *Myxobolus*, two of *Henneguya*, and one of *Myxidium*. Also included in their analyses were 24 18S sequences from a variety of taxa, including: Chordata, Arthropoda, Mollusca, Platyhelminthes, Nematoda, Cnidaria (2 spp. represented), Ctenophora, Porifera, and a variety of more distant taxa. Sequence alignment was accomplished with the GCG Pileup alignment program (GCG Sequence Analysis Software 7.0). In this analysis, those sequences whose alignments were ambiguous were excluded, resulting in a somewhat truncated data set. Parsimony analysis and neighbour-joining (Saito and Nei, 1987) methods were used to infer phylogeny. Both methods of analysis con-

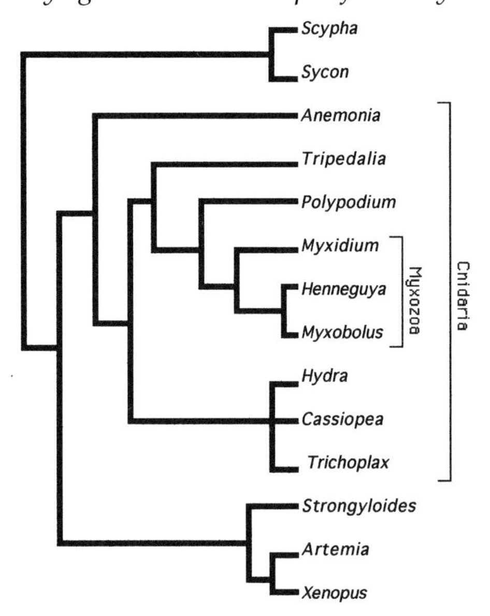

Figure 19.1 Dendrogram based upon parsimony analysis of 18S ribosomal sequences and morphological characters, grouping Myxozoa with cnidarian taxa. Branch lengths are not to scale. Adapted from Siddall *et al.* (1995).

verged upon the same topology. Again, the myxozoan taxa are clustered with the Metazoa. However, rather than forming a monophyletic group with the Cnidaria, this analysis groups the Myxozoa with the Nematoda, and as a sister taxon to the other bilateral animals included in the analysis (Figure 19.2). This clustering is supported by the recent work of Schlegel *et al.* (1996), whose phylogeny likewise shows the Myxozoa as more closely related to the Bilateria.

The differences between these two phylogenies could arise from a number of causes. In their analysis, Smothers *et al.* (1994) opted to leave out stretches of sequence whose proper alignment was ambiguous. By doing so they may have discarded phylogenetically informative data which, if retained in the analysis, might have resulted in a different tree topology. And while the node connecting bilateral animals with the Myxozoa is strongly supported, the affinity with the nematode lineage is not. A further interesting feature is provided by the relationships within the Myxozoa: the data suggest that the two genera, *Henneguya* and *Myxobolus* are paraphyletic.

While both phylogenies are based upon ribosomal sequences, they were not the same sequences; Siddall *et al.* (1995) incorporated sequen-

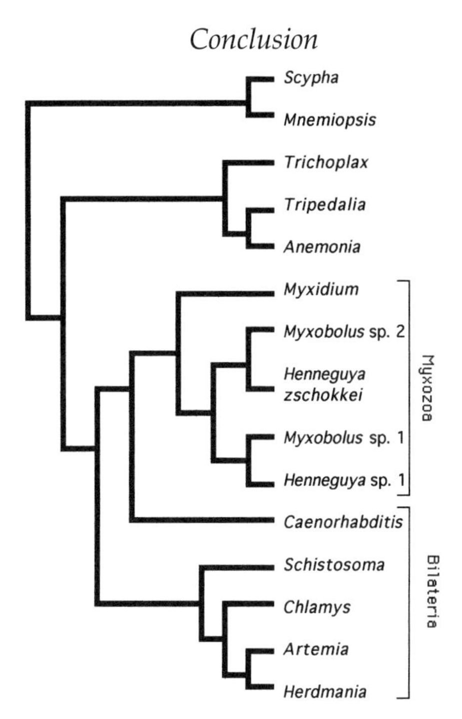

Figure 19.2 Dendrogram based upon parsimony and neighbour-joining analysis of 18S ribosomal sequences, grouping myxozoan taxa with *Caenorhabditis* (Nematoda). Branch lengths are not to scale. Adapted from Smothers *et al.* (1994).

ces from more, and different cnidarian taxa in their analyses, including the narcomedusan species. In a perfect world this would not be a problem; all sets of data should converge upon a single topology and provide a measure of confidence in the resultant phylogeny. Instead, these competing phylogenies demonstrate the limits of phylogenetic resolution of ribosomal sequence data, and indicate the need for further sequence data, in order to establish the phylogenetic position of this phylum.

19.5 CONCLUSION

Notwithstanding the difficulties involved in obtaining material, no mean feat when the source organism's cells are usually embedded in host tissue, reliable sequence data on the Myxozoa are beginning to accumulate. Already the handful of available 18S sequences have unambiguously placed phylum Myxozoa within the Metazoa. Whether the

Myxozoa are highly derived parasitic cnidarians or are relatives of the Bilateria, the lineage has nevertheless undergone an extreme reduction, losing those features which are common to bilateral animals and cnidarians. However, further sequence data are needed to augment that which already exists. In particular, comparison of sequences **other** than 18S rRNA could help resolve the still problematic position of the Myxozoa vis-a-vis the Cnidaria. Equally important, sequence data and molecular markers could prove extremely useful in sorting out the systematics of these organisms.

Addendum: Recent investigation of *Hox* genes in the Myxozoa has revealed *Hox* sequences in myxozoan species which resemble classes of *Hox* genes found in triploblasts, but not in diploblasts (C.L. Anderson, B. Okamura, and E.U. Canning, 1997, *Nature,* in press.). *Hox* genes confirm that Myxozoa are extremely reduced triploblasts (Anderson *et al.*, in press). If borne out by further investigation, this would lend support to a phylogenetic hypothesis which places the Myxozoa with the Bilateria.

REFERENCES

Anderson, C.L., Okamuta, B. and Canning, E.U. (1998) *Hox* genes confirm that Myxozoa are extremely reduced triploblasts. *Nature* (in press).

Andree, K.B., Gresoviac, S.J. and Hedrick, R.P. (1997) Small subunit ribosomal RNA sequences unite alternate actinosporean and myxosporean stages of *Myxobolus cerebralis*, the causative agent of whirling disease in salmonid fish, *Journal of Eukaryotic Microbiology*, **44**, 208–15.

Cavalier-Smith, T. (1993) Kingdom Protozoa and its 18 phyla. *Microbiological Reviews*, **57**, 953–94.

El Matbouli, M. and Hoffmann, R.W. (1989) Experimental transmission of two *Myxobolus* spp. developing bisporogeny via tubificid worms. *Parasitology Research*, **75**, 461–4.

El Matbouli, M. and Hoffmann, R.W. (1991) Effects of freezing, aging, and passage through the alimentary canal of predatory animals on the viability of *Myxobolus cerebralis* spores. *Journal of Aquatic Animal Health*, **3**, 260–2.

El Matbouli, M., Hoffmann, R.W. and Mandok, C. (1995) Light and electron microscopic observations on the route of the triactinomyxon-sporoplasm of *Myxobolus cerebralis* from epidermis into rainbow trout cartilage. *Journal of Fish Biology*, **49**, 919–35.

El Matbouli, M., Fischer-Scherl, T. and Hoffmann, R.W. (1992a) Transmission of *Hoferellus carassii* Achmerov 1960 to goldfish *Carassius auratus* via an aquatic oligochaete. *Bulletin of the European Association of Fish Pathologists*, **12**, 54–6.

El Matbouli, M., Fischer-Scherl, T. and Hoffmann, R.W. (1992b) Present knowledge of the life cycle, taxonomy pathology and therapy of some Myxosporea spp. important for freshwater fish. *Annual Review of Fish Diseases*, 367–402.

Grassé, P.-P. and Lavette, A. (1978) La myxosporidie *Sphaeromyxa sabrazesi* et le nouvel embranchement des Myxozoaires (Myxozoa). Recherches sur l'etat pluricellulaire primitif et considerations phylogenetiques *Annales des Sciences Naturelles*, Zoologie (Paris), **20**, 193–285.

Gurley, R.R. (1893) On the classification of Myxosporidia, a group of protozoan parasites infecting fishes. *Bulletin of the United States Fish Commission*, **XI**, 407–20.

Gurley, R.R. (1894) The Myxosporidia or psorosperms of fishes and the epidemic produced by them. *Report of the United States Commissioner of Fish and Fisheries*, **26**, 65–304.

Kent, M.L., Margolis, L. and Corliss, J.O. (1994) The demise of a class of protists: taxonomic and nomenclatural revisions proposed for the protist phylum Myxozoa Grassé 1970. *Canadian Journal of Zoology*, **72**, 932–7.

Korotneff, A. (1892) *Myxosporidium bryozoides*. *Zeitschrift für Wissenschaftliche Zoologie*, **53**, 591–6.

Levine, N.D., Corliss, J.O., Cox, F.E.G. *et al.* (1980) A newly revised classification of the Protozoa. *Journal of Protozoology*, **27**, 37–58.

Lom, J. (1990) Phylum Myxozoa, in *Handbook of Protoctista* (eds L. Margulis, J.O. Corliss, M. Melkonian and D.J. Chapman), Jones and Bartlett, Boston, pp. 36–52.

Lom, J. and Puytorac, P. de (1965) Studies on the myxosporidian ultrastructure and polar capsule development. *Protistologica*, **1**, 53–65.

Lukes, J., Volf, P. and Lom, J. (1993) Detection of chitin in spores of *Myxobolus muelleri* and *M. subepithelialis* (Myxosporea, Myxozoa). *Parasitology Research*, **79**, 439–40.

Markiw, K. (1992) *Salmonid Whirling Disease*. U.S. Fish and Wildlife Service, Fish and Wildlife Leaflet 17.

Markiw, M.E. and Wolf, K. (1983) *Myxosoma cerebralis* (Myxozoa, Myxosporea) etiologic agent of salmonid whirling disease requires tubificid worms (Annelida: Oligochaeta) in its life cycle. *Journal of Protozoology*, **30**, 561–4.

Mitchell, L.G. (1977) Myxosporida, in *Parasitic Protozoa*, vol. 4, (ed. J.P. Kreier), Academic Press, London, pp. 115–54.

Mitchell, L.G. (1989) Myxobolid parasites (Myxozoa: Myxobolidae) infecting fishes of western Montana, with notes on histopathology, seasonality and intraspecific variation. *Canadian Journal of Zoology*, **67**, 1915–22.

Saito, N. and Nei, M. (1987) The neighbor-joining method: a new method for reconstructing trees. *Molecular Biology and Evolution*, **4**, 406–25.

Schlegel, M., Lom, J., Stechmann, A. *et al.* (1996): Phylogenetic analysis of complete small subunit ribosomal RNA coding region of *Myxidium lieberkuehni*: Evidence that Myxozoa are Metazoa and related to Bilateria. *Archiv für Prosistenkunde*, **147**, 1–9.

Siddall, M.E., Martin, D.S., Bridge, D. *et al.* (1995) The demise of a phylum of protists: phylogeny of Myxozoa and other parasitic Cnidaria. *Journal of Parasitology*, **81**, 961–7.

Smothers, J.F., von Dohlen, C.D., Smith, L.H. and Spall, R.D. (1994) Molecular evidence that myxozoan protists are metazoans. *Science*, **265**, 1719–21.

Weill, R. (1938) 'L'interprétation des Cnidosporidies et le valeur taxonomique de leur cnidôme. Leur cycle comparé à phase larvaire des Narcomeduses Cunindes. *Travaux de la Station Zoologique de Wimereaux*, **13**, 727–44.

Wheeler, W. (1995) Optimization alignment: The end of multiple sequence alignment in phylogenetics? *Cladistics*, **12**, 1–9.

Wheeler, W., and Gladstein, D. (1993) *Malign. Version 1.85*. Department of Invertebrates, American Museum of Natural History, New York.

Wolf, K. and Markiw, M. (1984) Biology contravenes taxonomy in the Myxozoa: new discoveries show alternation of invertebrate and vertebrate hosts. *Science*, **225**, 1449–52.

20

Relationships between lower fungi and protozoa

Gordon W. Beakes

Department of Biological and Nutritional Sciences, The Agriculture Building, The University, Newcastle upon Tyne, NE1 7RU. E-mail: G.W.Beakes@ncl.ac.uk

ABSTRACT

This review discusses the relationships between lower fungi and protozoa. One complication of interpreting the relationships between fungi and protists has been the use of different, sometimes conflicting, classification schemes by protozoologists and mycologists which are here summarized for comparison. The organisms traditionally studied by mycologists, although mostly falling within the kingdom Fungi, are now known to include representatives which are phylogenetically within the kingdoms Protozoa and Chromista. Although the amoeboid 'slime moulds' have been studied by mycologists they are clearly protozoan. The biflagellate plasmodial plasmodiophoromycete plant pathogens have now been placed in a phylum with a number of amoebo-flagellate protozoans, euglyphid testate amoebae and the photosynthetic amoeboid chlorarachniophytes. It also seems likely that a number of rotifer parasites in the genus *Haptoglossa,* which are currently classified as oomycetes, are also plasmodiophoromycetes. Molecular sequencing has confirmed that the biflagellate oomycetes, labyrinthulids (slime net fungi and thraustochytrids) and uniflagellate hyphochytrids are all members of the heterokont (infrakingdom) clade of the kingdom Chromista. The oomycetes and hyphochytrids cluster as sister groups, whereas the labyrinthulids branch more deeply and appear to be more closely related to the protist bicoecids. The uniflagellate Chytridiomycota and non motile Zygomycota are the remaining 'lower fungal' representatives of the true kingdom Fungi. The Fungi form part of a monophyletic clade within the crown eukaryotes which also include the protozoan collar flagellates and their allies

Evolutionary Relationships Among Protozoa. Edited by G.H. Coombs, K. Vickerman, M.A. Sleigh and A. Warren. Published in 1998 by Chapman & Hall, London. ISBN 0 412 79800 X

(in the phylum Choanozoa) as well as the metazoan animals. Recent molecular sequencing has shown, unexpectedly, that the phylum Choanozoa also encompasses a group of fish pathogens which had been variously classified as both fungi and protists and the semi-amoeboid marine organism, *Corallochytrium*, which had been formerly assigned to the thraustochytrids. Other problematic protists which might be related to this clade are also briefly described together with a speculative discussion as to the likely flagellate ancestors of these organisms.

20.1 INTRODUCTION

The likely phylogenetic relationships between primitive single-celled organisms falling within the protozoa, fungi and algae have been the subject of much speculation. Although ultrastructural and biochemical data have extended the available phylogenetic characters (Cavalier-Smith 1995) it has undoubtedly been the systematic comparison of DNA base sequences for genes encoding conserved cell molecules (e.g. ribosomal RNAs, tubulin, cytochromes) that has provided an objective phylogenetic framework for these simple organisms (Hasegawa, Hashimoto and Adachi, 1992; Matheson 1992). Whilst it will be apparent from other chapters in this volume that trees generated from such information are also subject to artifacts and misinterpretation (see chapter 2) they do provide testable phylogenetic hypotheses against which structural, developmental and biochemical information can be evaluated. Once an organism finds its place in a DNA-based phylogenetic tree, many of its apparently anomalous or inconsistent structural features can be reinterpreted in the context of those found in likely sister or progenitor groups.

20.1.1 What is a fungus?

Fungi have been considered to constitute a eukaryote kingdom in their own right, based initially upon their distinctive mode of nutrition (Whittaker, 1969). However, even amongst mycologists there has been debate as to exactly which organisms should be considered as 'fungi'. It had long been recognized that some of the major groups of organisms traditionally studied by mycologists were not closely related to others. In particular the biflagellate oomycetes were recognized as having ultrastructural and biochemical similarities to the chromophyte algae (Beakes, 1987, 1989).

So what are the main defining characteristics of fungi? The biological feature which characterizes fungi above all else is their osmotrophic (absorptive) mode of nutrition. In addition, the majority of fungi have a walled thallus composed of characteristic branching filaments (hyphae –

Figure 20.1) which grow by apical (tip) extension (Bartnicki-Garcia, 1996). Furthermore, the phyla which have been shown to share a common evolutionary lineage (and as such represent the true 'Fungi', see Section 20.4) all have chitin (a polymer of N-acetyl glucosamine) in their cell walls and mitochondria with flat plate-like cristae (Figures 20.6, 20.9). With the exception of chytrids, which have a single smooth posterior flagellum, the 'true Fungi' lack flagellate stages. However, in one of the most lucid recent discussions of fungal phylogeny, Barr (1992) suggests that fungi could be considered as a phylogenetically diverse but biologically coherent 'Union' of organisms. Thus, the oomycetes would be biological fungi within this union but are not part of the kingdom Fungi.

The 'lower fungi' have been characterized by having undivided (aseptate or coenocytic) hyphae (Figure 20.1) or ovoid or spherical thalli (Figures 20.10, 20.11). Many are aquatic and produce either uni- (Figure 20.7) or bi- (Figure 20.2) or polyflagellate zoospores (Figure 20.8). Traditionally the term 'lower fungi' also encompasses the largely terrestrial coenocytic Zygomycota, which includes the familiar mucoraceous 'pin-moulds' as well as many entomopathogenic fungi. Ribosomal gene DNA sequences have revealed clear evolutionary links between the non-motile terrestrial zygomycetes and *Allomyces* within the Chytridiomycota (Nagahama *et al.*, 1995; Paquin *et al.*, 1995).

The organisms covered in this review, together with their classification as presented in recent protozoological (Corliss, 1994) and mycological (Alexopoulos, Mims and Blackwell, 1996) literature are given in Table 20.1. One of the difficulties and inconsistencies in the classification of lower organisms has been the use of different nomenclature schemes by mycologists, phycologists and protozoologists, which is complicated by the different conventions underpinning the zoological and botanical codes (Patterson and Larsen, 1992). The classification of these primitive organisms is currently undergoing constant re-evaluation as new molecular sequencing data become available. Readers are also directed towards chapter 21 for the very latest revisions of the schemes summarized in Table 20.1. As a mycologist I have elected to adopt the accepted mycological terminology in this account and will describe the phylogenetic relationships under the Kingdom headings outlined in Table 20.1.

20.2 KINGDOM PROTOZOA

20.2.1 Acrasiomycota, Dictyosteliomycota and Myxomycota (Slime moulds)

The amoeboid slime moulds are a complex group which because of their amoeboid vegetative stages and phagotrophic mode of nutrition

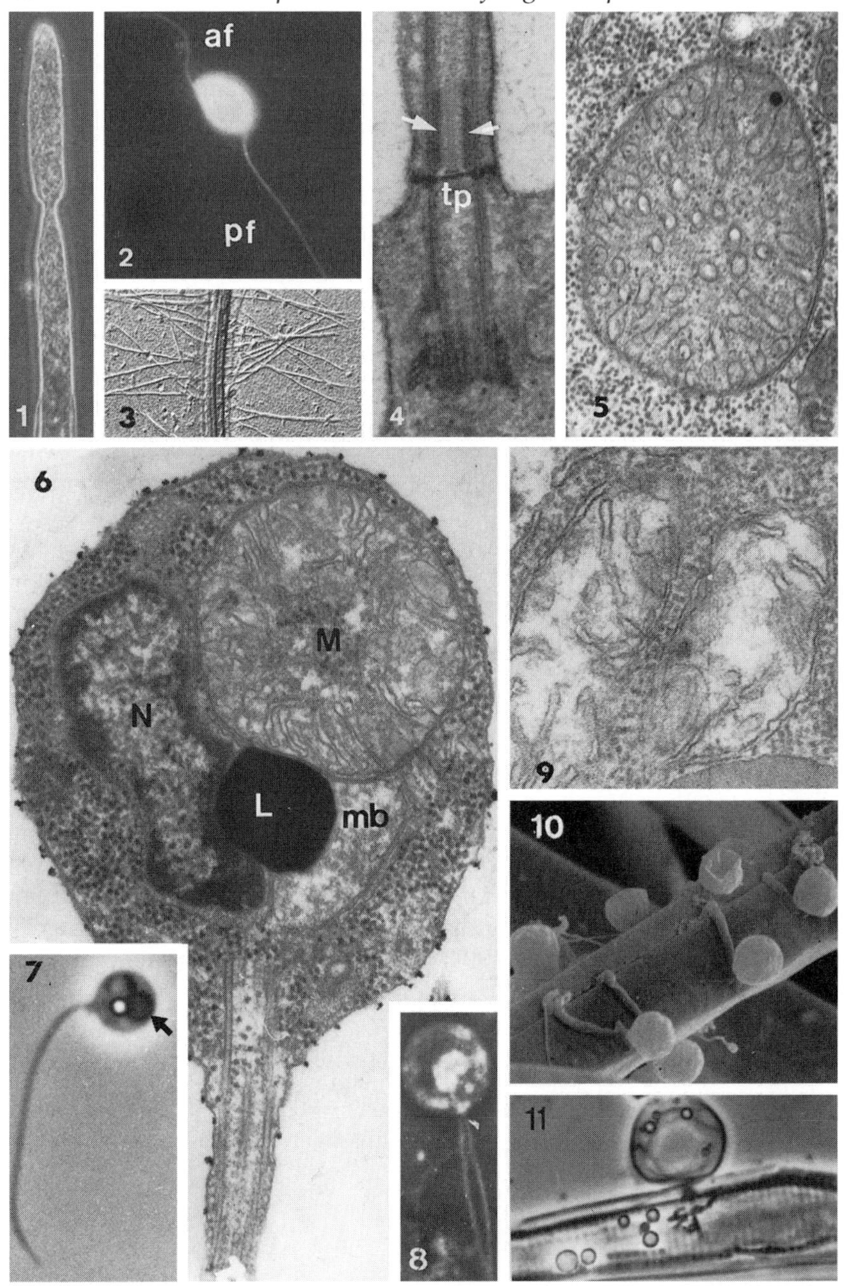

Plate 20.1 Photographs illustrating some key morphological and cytological features of lower fungi within the Kingdoms Chromista and Fungi.

do not conform to the generally accepted biological concept of 'fungi' defined above. A review of their general biology is given in Alexopoulos, Mims and Blackwell (1996). The plasmodial slime moulds belonging to the phylum Mycetozoa/Myxomycota (Table 20.1) have

---

Figure 20.1 Apical region of a sporulating mycelium of the oomycete water mould *Saprolegnia*, showing hyphoid structures which characterize the thallus of most fungi. Phase contrast light micrograph (LM) × 450.

Figure 20.2 Epi-fluorescence LM of a secondary type zoospore of the oomycete *Aphanomyces* showing characteristic bean shape and shorter anterior flagellum (af) and trailing posterior flagellum (pf) typical of these heterokont organisms. Stained with a FITC-labelled polyclonal antibody. × 2500.

Figure 20.3 Whole mount transmission electron micrograph (TEM) showing detail of the anterior flagellum of a secondary type zoospore of the oomycete *Plasmopara* showing double row of tripartite mastigoneme/stramenopile hairs which characterize the Chromistan Kingdom. × 45 000.

Figure 20.4 Higher power TEM profile through the base of an anterior flagellum of a secondary zoospore of the oomycete fungus, *Achlya,* showing prominent terminal plate (tp) and double transitional helix structure above it (arrowed). This structure is also phylogenetically significant in the chromistan lineage shared by a number of slopalinid protists. × 50 000.

Figure 20.5 TEM of profile of a mitochondrion from an encysted zoospore of *Saprolegnia* showing the rounded tubular cristae which characterize chromistan 'fungi'. × 70 000.

Figure 20.6 TEM of a zoospore of the chytrid *Zygorhizidium planktonicum* showing single posterior flagellum, and cluster of organelles (nucleus N, mitochondrion M and microbody mb) around the central electron dense lipid globule. Note the simple structure of the flagellar basal region in this species. × 60 000.

Figure 20.7 Phase contrast LM of corresponding zoospore of *Z. planktonicum,* showing long tapering posterior flagellum which is one of the characteristic features of the superkingdom Opisthokonta. The single optically dense mitochondrion is arrowed. × 2730.

Figure 20.8 Phase contrast LM of polyflagellate zoospore of the rumen chytrid, *Neocallimastix.* × 1650.

Figure 20.9 TEM showing part of a single reticulate mitochondrion in the chytrid *Zygorhizidum affluens.* This shows clearly the flattened plate-like cristae which characterize the true Fungi. × 60 000.

Figure 20.10 Scanning electron micrograph (SEM) of infected cells of the diatom *Synedra* showing settled spores of *Z. planktonicum* with their single infection rhizoids penetrating the cell between the upper and lower valves. × 3800.

Figure 20.11 Phase contrast LM of a discharged sporangium of *Z. planktonicum* on *Synedra* showing branched coralloid absorptive rhizoid system within the host cell. × 1600.

Table 20.1 A comparison of recent protistological (Corliss, 1994) and mycological (Alexopoulos, Mims and Blackwell, 1996) classification schemes of organisms which have been categorized as 'lower fungi' or related to them

Protozoological perspective of classification Corliss 1994	*Mycological perspective of classification Alexopoulos et al. 1996*
KINGDOM PROTOZOA	KINGDOM PROTOZOA
Phylum: Percolozoa	
Class: Heterolobosea	
Order: Acrasida	Phylum: Acrasiomycota[1,2]
Phylum: Mycetozoa	
Class : Dictyostelea	Phylum: Dictyosteliomycota[1,2]
	Phylum: Myxomycota[1,2]
Class: Myxogastrea	Order: Physarales etc.
Class: Protostelea	Order: Ceratiomyxales[1]
Phylum: Opalozoa[4]	
Order: Plasmodiophorida	Phylum: Plasmodiophoromycota[1]
Phylum: Choanozoa	
Order: Choanoflagellida	
KINGDOM CHROMISTA	KINGDOM STRAMENOPILA[3]
Subkingdom: HETEROKONTA	
Phylum: Labyrinthulomorpha	Phylum: Labyrinthulomycota[1]
Class: Labyrinthulea	Order: Labyrinthulales
Class: Thraustochytriacea	Order: Thraustochytridiales
Phylum: Pseudofungi	
Class: Hyphochytriomycetes	Phylum: Hyphochytriomycota[1]
Class: Oomycetes	Phylum: Oomycota[1]
KINGDOM FUNGI	KINGDOM FUNGI
Phylum: Chytridiomycota	Phylum: Chytridiomycota[1]
Class: Chytridiomycetes	Order: Neocallimasticales
	Order: Blastocladiales
	Order: Spizellomycetales
	Order: Chytridiales
	Order: Monoblepharidales
Phylum: Zygomycota	Phylum: Zygomycota[1]
Phylum: Ascomycota	Phylum: Ascomycota
Phylum: Basidiomycota	Phylum: Basidiomycota

NOTE
[1] Organisms which have been classified as 'lower fungi' and will be discussed in this review.
[2] These are the myxomycete Classes defined in the Dictionary of Fungi (Hawksworth *et al.* 1995).
[3] Dick (pers. comm.) states that this group should be correctly spelt Straminipila; Cavalier-Smith (1989) uses the name Chromista to encompass this group.
[4] Cavalier-Smith (this volume) has replaced Opalazoa with Neomonada.

extraordinarily complex life histories. They produce three types of uninucleate amoeboid cells (one of which is biflagellate), a multinucleate somatic 'plasmodial phase' and a fungus-like sporophore containing walled spores which are dispersed by wind, water or arthropods. These moulds exhibit two quite different patterns of nuclear division in which the nuclear envelope either breaks down or remains intact until late anaphase/early telophase.

Two other groups of amoeboid phagotrophic organisms have also been traditionally placed in the 'slime moulds'. These are the dictyostelid cellular slime moulds (Dictyostela/Dictyosteliomycota, Table 20.1) and the acrasid cellular slime moulds (Acrasida/Acrasiomycota, Table 20.1). The dicytostelids have amoebae with filose pseudopodia and spindle pole bodies associated with their nuclei but lack a flagellate stage and appear to be closely related to the true slime moulds. The acrasids have lobose pseudopodia and lack spindle pole bodies but produce biflagellate swarmers. The mitochondria of acrasid slime moulds have flattened plate-like cristae (Alexopoulos, Mims and Blackwell, 1996) and are related to the amoeboflagellates within the Heterolobosea in the phylum Percolozoa (Corliss, 1994). Both of these groups have stages in which their amoebae aggregate into coherent 'multicellular' phases, exemplified by the slug like pseudoplasmodia of *Dictyostelium* and multicellular sorocarps of *Acrasis*.

In phylogenetic trees based upon 18S ribosomal DNA sequences the three 'mycological phyla' of slime moulds often appear widely separated. However, the precise position of these phyla is not particularly well supported and their relative positioning often shows considerable variation from one tree to another. A recent study using the EF-1 alpha protein synthesis elongation factor gene sequences has shown that representatives of the Myxomycota and Dictyosteliomycota cluster closely together within the crown eukaryotes (S. Baldauf personal communication) and supports their inclusion by protozoologists within the single protozoan phylum Mycetozoa. Although not yet sequenced, it seems likely that the acrasids will cluster with the Heterolobosea outside of the crown eukaryotes. Since it is clear from both phylogenetic evidence and their general biology that these organisms are unequivocally in the kingdom Protozoa it might be sensible for mycologists to adopt current protistan terminology in the future (Table 20.1).

20.2.2 Plasmodiophoromycota (endoparasitic slime moulds)

The plasmodiophoromycetes are biflagellate zoosporic organisms which have intracellular plasmodial thalli which can exhibit amoeboid movement (Mithen and Magrath, 1992) and are generally placed by mycologists with the slime moulds (Table 20.1). They include a number of plant pathogens, the most economically important of which is probably *Plasmodiophora brassicae* which causes club root of cabbages. However, the classification of these organisms has been problematic and they have even been placed within the biflagellate oomycetes. Their taxonomic position has been critically discussed in a number of recent reviews (Barr, 1983, 1992; Buczaki, 1983). Their zoospores have a short anterior flagellum and a long posterior flagellum both of which are smooth. They have a distinctive flagellar base with long kinetosomes extending deep into the cytoplasm and a basal structure similar to that of protists such as *Naegleria* (Barr and Allan, 1982; Barr, 1992). This character clearly separates them from the chromistan lineage (see Section 20.3). *Plasmodiophora* produces long-lived resting cysts which have chitin in their walls which Buczaki (1983) suggests points to their fungal ancestry. However, their characteristic cruciate pattern of mitosis, where the metaphase chromosomes straddle a prominent axial nucleolus (Dylewski and Miller, 1983), and the elaborate penetration apparatus produced by the encysted spores (Aist and Williams, 1971) are both features strongly indicative of a protozoan ancestry (Barr, 1983, 1992).

One of the problematic characters which have made this group so difficult to place has been the precise configuration of their mitochondrial cristae. They generally have been placed within groups containing flat cristae (see mitochondrial profiles in *Plasmodiophora*, Aist and Williams, 1971). However in other genera the mitochondrial cristae could just as easily be interpreted as being tubular-vesiculate (e.g. *Polymyxa* Barr and Allan 1982 and *Woronia* Dylewski and Miller 1983). Preliminary analysis of their 18S rRNA gene sequence apparently indicated that they clustered with the alveolate ciliates (Castlebury and Domier, 1994). However, subsequent studies apparently show them to cluster within a clade together with the cercomonadid and thaumatomonad flagellates and euglyphid testate amoebae, as well as the photosynthetic chlorarachniophytes, to which Cavalier-Smith (chapter 21) has given the phylum designation Cercozoa.

Finally, there are a number of biflagellate parasites of Ascheleminths (e.g. *Haptoglossa*), which have been traditionally classified as lagenidialian oomycetes, whose cysts contain a complex penetration apparatus (Robb and Banen, 1982). Their so called 'gun cells' have an almost identical fine-structure and mode of action to the elaborate infection structures described in *Plasmodiophora* (Aist and Williams, 1971). It will be necessary to confirm that the zoospores of these organisms have smooth flagella, although Dick (pers. comm.) has pointed out that the unusual swimming behaviour of the zoospores of *Haptoglossa* is consistent with their lacking mastigonemes.

20.2.3 Relationships with other Protista

Finally, before leaving 'fungi' which have been grouped within the protozoa, some brief mention must be made to *Perkinsus marinus*. This species is a devastating pathogen of oysters and other shellfish and was originally classified as the fungus *Dermatocystidium marinum* before being briefly transferred to *Labyrinthulomyxa* within the chromistan Labyrinthulales (see Dick 1996). However the distinctive structure of its biflagellate zoospores clearly showed that it had apicomplexan affiliations, a conclusion that has been subsequently supported by molecular sequencing (Goggin and Barker, 1993). A detailed account of this organism has recently been given by Perkins (1996).

20.3 KINGDOM CHROMISTA (STRAMENOPILES)

20.3.1 Oomycota (water moulds)

Molecular sequencing (Gunderson *et al.*, 1987; Förster *et al.*, 1990; Van der Auwera, Chapelle and De Wachter, 1994) has confirmed that the biflagellate (Figure 20.2) oomycete fungi are members of the infrakingdom Heterokonta within the kingdom Chromista (Cavalier-Smith, 1989). They consequently show evolutionary affiliations with photosynthetic chlorophyll c-containing algae such as the diatoms and xanthophytes. All of these groups have flagellate zoosporic stages in which the anterior flagellum is decorated with distinctive tubular hairs (Figure 20.3). Indeed, Patterson (1989) used this character to group these organisms under the 'stramenopile' banner (meaning straw hair) to allow for the inclusion of protist groups without ornamented cilia but with similar hairs decorating their body surface. However, recently Dick (1996) has pointed out that this is an incorrect Latin derivation of this name and the correct spelling should be straminipila. Although 'Stramenopila' has never been formally published as a valid name it has been widely used as a synonym for Haptophyta/Chromista in much contemporary literature (Table 20.1).

The oomycetes are typically organisms of aquatic ecosystems. They include a number of economically important pathogens of marine and freshwater animals within the orders Saprolegniales and Lagenidiales as well as significant plant pathogens in the order Peronosporales (which includes the genera *Pythium* and *Phytophthora*). The biochemical and fine-structural similarities between oomycetes, hyphochytridiomycetes and chromophyte algae have already been comprehensively reviewed (Beakes 1987, 1989). Key features include the presence of cellulose (α1-4 glucan) microfibrils in their walls, soluble β1-3 linked glucan storage reserves (laminarins), a coiled transitional helix (TH) in their flagellar base (Figure 20.4; see Patterson, 1989; Barr and Désaulniers, 1989a) and mitochondria with tubular cristae (Figure 20.5). However it is worth noting that only a relatively small number of genera of these biflagellate fungi have been critically examined ultrastructurally (e.g. *Saprolegnia* and *Phytophthora*; Beakes, 1989). The small number of studies which have been carried out on members of the Lagenidiales reveal a much more varied flagellar base structure. *Lagena radicola* (Barr and Désaulniers, 1986, 1989b) for instance completely lacks a TH, whilst in *Olpidiopsis saprolegniae* (Bortnick, Powell and Barstow, 1985; Barr, 1992) there is only a single rather than the double TH coil typical of most oomycetes. The latter genus is also of note in that it is the only oomycete so far described to have a semi-open pattern of mitosis (Martin and Miller, 1986). Many lagenidialian genera, including the two mentioned above, would be well worth including in future phylogenetic gene sequencing studies.

20.3.2 Hyphochytridiomycota (hyphochytrids)

The holocarpic (non mycelial) Hyphochytridiomycota, although once classified with the Chytridiomycota, are now known to possess a single anterior flagellum ornamented with tripartite hairs and the double TH characteristic of oomycetes (Cooney, Barr and Barstow, 1985). These fungi have both chitin and cellulose in their cell walls (Fuller and Clay, 1993) and undergo semi-open mitosis with polar fenestrae (Barstow, Freshour and Fuller, 1989), which again sets them apart from most other oomycetes (Beakes, 1981). Although frequently considered as having close taxonomic affiliations with the order Saprolegniales in the oomycetes (Beakes, 1987, 1989; Dick, personal communication), recent molecular studies seem to indicate that the hyphochytrids are a sister group to the oomycetes (Van der Auwera *et al.*, 1995).

20.3.3 Labyrinthulomycota (slime net fungi and thraustochytrids)

The final group of heterokont aquatic fungi encompass two small marine orders constituting the phylum Labyrinthulomycota (Table 20.1) which have been variously classified with both the oomycetes and slime moulds (Moss 1991). Most are saprophytes of marine plants but some (e.g. *Labyrinthula*) can cause diseases of sea grasses. Morphologically the Labyrinthulids are unique in forming a curious netlike thallus through which their non-flagellate spindle-like cells migrate. The Thraustochytrids have typically spherical or ovoid thalli with a rhizoid-like basal net (Moss, 1991). In both of these groups the absorptive net is isolated from the body cells by a complex, endoplasmic reticulum-derived, plugging organelle – the sagenogenetosome or bothrosome. Their thallus is coated in adpressed scales which are rich in fucose (Darley, Porter and Fuller, 1973). The non motile spherical marine protist *Diplophrys marina* is coated with similar scales and also produces rhizoid-like net structures, but apparently lacks the specialized plugging bothrosomes (Dykstra and Porter, 1984).

Molecular sequencing of their SSUrRNA genes clearly shows that members of the Thraustochytridiales (*Thaustochytrium kinnei* and *Ulkenia profunda*: Cavalier-Smith *et al.* 1994 and *Labyrinthuloides minuta* Leipe *et al.* 1994) cluster within the Heterokonta infrakingdom of the Chromista. In some trees they appear to be amongst the first diverging groups in this clade (Leipe *et al.*, 1994) and it has been suggested that they split off, together with the biflagellate bisoecids, before the evolution of the TH (which they lack).

20.3.4 Other chromistan flagellates

Possible relationships between chromistan organisms and other flagellate protozoa have been discussed by Patterson (1989). It is clear that the colourless biflagellate bisoecids (Bisoecales), such as the marine *Cafeteria,* share many structural features (e.g. their flagella rootlet structure, peripheral extrusive organelles) with the Oomycota and other chromists (Fenchel and Patterson, 1988). Large subunit ribosomal DNA analysis reveals that *Cafeteria* often appears to branch either as a sister clade to the thraustochytrids or immediately below them (Leipe *et al.*, 1994; Cavalier-Smith, Allsopp and Chao, 1994).

Other protists associated with this clade are to be found within the multiflagellate Slopalinida (syn. Opalinata) order of Patterson (1989). These include the cigar-shaped flattened flagellates such as *Opalina* which inhabit the intestinal tracts of cold-blooded vertebrates and the ridged proteromonad flagellates such as *Karatomorpha* and *Proteromonas*. Although all these flagellates lack the characteristic tripartite hairs on their flagella, they both have a double TH in their flagellar bases (Patterson, 1989). The positioning of the proteromonads within the heterokont chromistans has recently been confirmed by rRNA phylogeny (Silberman *et al.* 1996). Remarkably the body surface of proteromonads is decorated by bipartite hairs (incorrectly referred to as tripartite) and it has been suggested that this character may have preceded flagellar ornamentation (Patterson, 1989), although the reverse has also been claimed. Perhaps the most unexpected protist to be included with these heterokont chromistans has been the non-motile human parasite *Blastocystis hominis*, whose recently published 18S ribosomal DNA sequence reveals that it clusters closest to *Proteromonas* (Silberman *et al.*, 1996). At present molecular phylogeny has not been able to resolve the precise branching order within the heterokont chromistan clade. If the early branching of the opalinid protists proves to be well supported it could imply that groups such as the labyrinthulids, bicoecids and lagenidialian fungi such as *Lagena* all have lost their TH during evolutionary history.

20.4 KINGDOM FUNGI

20.4.1 Chytridiomycota, Zygomycota (true fungi)

Although the grouping of osmotrophic hyphal organisms into their own separate Kingdom has been suggested for some time, the rationale for this has been given additional support by recent molecular sequencing studies. The Chytridiomycota, Zygomycota, Ascomycota and Basidiomycota form a coherent evolutionary lineage (Bowman *et al.*, 1992) evolving as a sister clade to the metazoan animals plus choanoflagellates (Baldauf and Palmer, 1993; Wainwright *et al.*, 1993). From these studies it seems hardly justifiable for Margulis (1996) to continue to include uniflagellate chytrids within the Protoctista. In addition to the characters already outlined in Section 20.1.1, all of these true fungi synthesize lysine via the alpha amino adipic acid (AAA) pathway (see Cavalier-Smith, 1987) and have closed mitosis with non-centriolar spindle pole bodies (Heath, 1986). Ribosomal RNA gene sequences indicate that the orders within the Chytridiomycota (Table 20.1) show significant phylogenetic divergence (Nagahama *et al.*, 1995). The Blastocladiales appear to be the evolutionary ancestors of most of the terrestrial fungi in the Zygomycota including the mycorrhizal Glomaceae. The Chytridiales, in contrast, are apparently only ancestral to the anomalous pathogenic zygomycete, *Basidiobolus*. Berbee and Taylor (1992) have used nucleotide substitutions in the 18S ribosomal RNA sequence to calibrate the likely evolutionary rate in fungi. Based on an estimation of a substitution rate of 1% per lineage per 100 million years, they have suggested that the terrestrial fungi diverged from the chytrids approximately 550 million years ago.

20.4.2 Neocallimasticales (rumen chytrids)

No account of fungal–protist relationships would be complete without some mention of the anaerobic rumen fungi. These free swimming poly- and biflagellate inhabitants of the rumen of herbivorous artiodactyls (Figure 20.8) were originally assigned to the protozoan genus *Callimastix* (see Heath, Bauchop and Skipp, 1983). The polyflagellate species were later transferred to their own genus *Neocallimastix*. However, it was not until the mid 1970s that Colin Orpin observed that these flagellate organisms were in fact liberated from rhizoidal sporangia, usually attached to plant particles in the rumen, and suggested that they were chytridiomycete fungi (see Heath, Bauchop and Skipp, 1983). Although originally assigned to the order Spizellomycetales (Heath, Bauchop and Skipp, 1983) they have recently been reassigned to their own order the Neocallimasticales (Li and Heath, 1992). These were the first obligately anaerobic fungi to be described and are genuinely polyflagellate with a characteristic kinetosome structure (Heath, Bauchop and Skipp, 1983). They also have reductive hydrogenosomes in place of mitochondria. 18S ribosomal DNA analysis confirms that these rumen fungi cluster with other chytrids and possibly represent a basal branch within the phylum (Li and Heath, 1992). Significantly they have a closed pattern of mitosis (Heath and Bauchop, 1985) which is a feature shared with the Blastocladiales and the majority of terrestrial fungi (Heath 1986). It would be of interest to discover if there are either any free-living members of this order in other anaerobic environments or any aerobic representatives of this order.

20.4.3 Protozoan 'cousins' of fungi

Cavalier-Smith (1987) speculated that because of their shared flattened mitochondrial cristae and uniflagellate cells, the collar flagellates (Choanoflagellatea, Table 20.1) were the most likely extant protist group to be ancestral to fungi and animals. He grouped these uniflagellate organisms into the 'super kingdom' Opisthokonta. Although this proposed evolutionary relatedness of the choanoflagellates and the chytrids was greeted with a certain amount of incredulity by mycologists at the time, the grouping of fungi, choanozoa and animals into a single clade is now well supported by molecular studies (see Cavalier-Smith and Chao, 1995). Although choanoflagellates do not share many obvious structural features with fungi, the complex striate rootlet structure in the freshwater choanoflagellate *Codosiga* (Hibberd, 1975) is remarkably similar to the striate fan-like basal bodies connecting the flagellar base to their nucleus in spizellomycete chytrids such as *Karlingia* (Barr, 1980). Furthermore the double plate which occurs in the flagellar base of many choanoflagellates is similar to that seen in some members of the Monoblepharidales (Barr, 1992). Both members of the Spizellomycetales and choanoflagellates have a semi-open pattern of mitosis with the development of an invasive spindle (Beakes, 1981; Heath, 1986; Leadbeater, 1994). Certainly, there is a superficial morphological similarity between epiphytic freshwater choanoflagellates such as *Salpingoeca* (Figure 20.12) which are associated with diatoms and many chytrid parasites of these algae (Figures 20.10, 20.11). Again the transformation of the basal pseudopodia which attach these flagellate protists (Figure 20.12) to their substrates (Hibberd, 1975) into infective

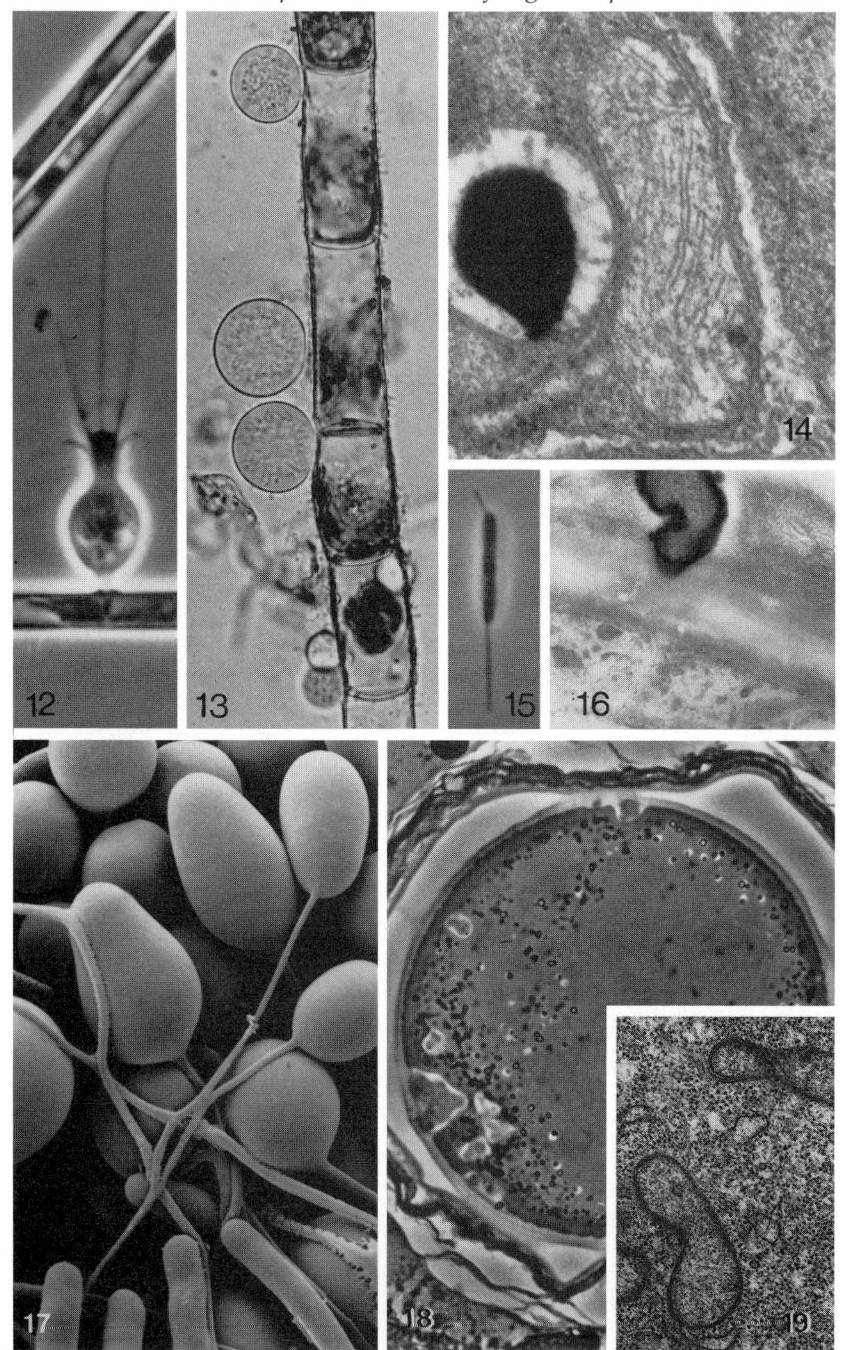

Plate 20.2 Illustrations of protist organisms which may be ancestral to the Fungi.

rhizoids (Figure 20.11) appears conceptually to be a fairly straightforward evolutionary step (Cavalier-Smith, 1987).

Recently, Cavalier-Smith and Allsopp (1996) have also shown that the free-living non-motile organism, *Corallochytrium limacisporum* clusters as a sister group to the choanoflagellates in the phylum Choanozoa. This organism was discovered in tropical lagoons and was placed in the thraustochytrids because of its apparently scaly wall (Raghu-Kumar, 1987). However, *Corallochytrium* lacks a flagellate stage but does produce a slow moving amoeboid phase and apparently has flattened mitochondrial cristae (Cavalier-Smith and Allsopp, 1996) although these have not been clearly illustrated.

Another surprise has been the recent discovery that a number of morphologically unexciting, walled non-flagellate parasites of marine fish (Figure 20.18) and crustacea also fit in this basal Choanozoan clade on the basis of their 18S rRNA gene sequences (Kerk *et al.*, 1995; Ragan

---

Figure 20.12 Phase contrast LM of the freshwater choanoflagellate *Salpingoeca* attached to the surface of the freshwater diatom *Asterionella*. Note the single flagellum, and tentacle like projections around the neck of this flask shaped cell. Note that these organisms are remarkably similar in size to the sporangia of their diatom-associated chytrid counterparts (Figure 20.11). × 1600.

Figure 20.13 LM of a filament of the green alga *Oedogonium* with its associated undescribed flagellate protist. Note the spherical cells which appear superficially very chytrid like. × 640.

Figure 20.14 TEM of mitochondria of the unidentified *Oedogonium* protist showing its 'Choanozoan-like' flattened cristae. × 50 000.

Figure 20.15 Phase contrast LM of a biflagellate zoospore of the *Oedogonium* protist. Note short anterior flagellum and stiff rod like posterior flagellum of uniform diameter (compare with Figures 20.2 and 20.7). × 1600

Figure 20.16 TEM showing detail of the electron dense attachment thread which anchors *Oedogonium* protist to its host cells. Note the host cell wall has thickened in the region adjacent to this attachment structure. × 32 000.

Figure 20.17 SEM of cultured cells of the fish pathogenic protist *Ichthophonus hoferi*. Note swollen terminal cells on the tips of tubular thread like structures. × 350.

Figure 20.18 LM of sectioned spherical infective thallus of *Ichthyophonus* in the muscle tissue of fish. Such rounded trophocysts are typical of many of the pathogenic organisms in the DRIP clade. × 900.

Figure 20.19 TEM of peripheral cytoplasm of cultured cell of *Ichthyophonus* showing mitochondria with sparse tubular-vesiculate cristae. × 7500.

Figures 20.7, 20.11, 20.12, 20.13 and 20.15 supplied by Dr H M Canter and Figures 20.17–20.19 by Dr Bettina Spanggaard.

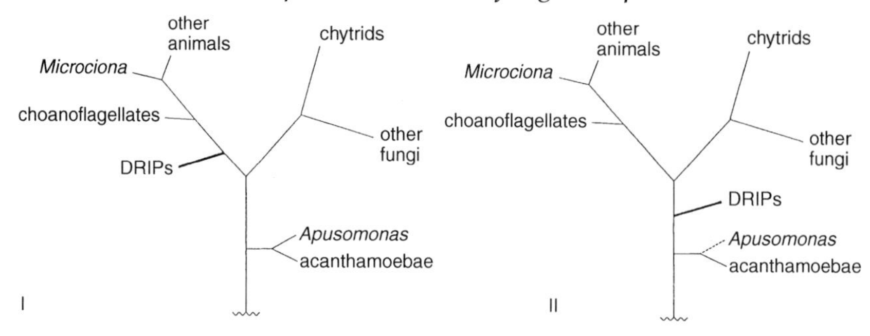

Figure 20.20 Alternative possible tree relationships between the DRIPs clade (Ichthyosporea) and other eukaryote lineages based upon comparison of their small subunit ribosomal RNA gene sequences. From Ragan *et al.* 1996 with permission.

et al., 1996; Spanggaard *et al.*, 1996). Many of these organisms have long puzzled both mycologists and protozoologists as to their exact taxonomic position. This confusion is perhaps best exemplified by the important parasite of marine fish, *Ichthyophonus hoferi* (Figures 20.17, 20.18). This organism has been variously assigned to the Entomophthorales (related to *Basidiobolus*), Chytridiales and Saprolegniales within the fungi and under the guise of *Ichthyosporidium* to microsporidian protists (see Spanggaard *et al.*, 1996). However, most recent reviewers of marine parasites of fish have preferred to assign this problematic organism to those of *incertae sedis* pending more evidence (Alderman, 1982). In liquid culture at low pH (3.5) *Ichthyophonus* adopts a filamentous growth form (Spanggaard *et al.* 1994; Spanggaard, Huss and Bresciani, 1995) in which empty filaments are terminated by the swollen cytoplasm-filled, often bifurcated, structures (Figure 20.17). At higher pH (7.0) and *in situ* the organism forms mainly spherical bodies (Figure 20.18) which generate small amoeboid cells, which are thought to disperse the organism in the host blood vessels (Spanggaard, Huss and Bresciani, 1995). The cell walls of the Canadian strain of *Ichthyophonus* did not appear to contain cellulose or chitin but are apparently rich in sulphated and acid mucopolysaccharides (Rand, 1994). In contrast, using biochemical rather than histochemical protocols, Spanggaard *et al.* (1996) report that their strain of *Ichthyophonus* does have a chitin component to its walls. What is unexpected about an organism within this clade is that it appears to have mitochondria (Figure 20.19) of a tubular-vesicular type (Ragan *et al.*, 1996; Spanggaard *et al.*, 1996). Recent analyses of the 18S rRNA gene sequences have shown that *Ichthyophonus* clusters as a sister group to a number of other taxonomically problematic pathogens of aquatic organisms, namely *Dermocystidium* spp. *Psorospermium* and the enigmatic 'rosette agent' pathogen of Chinook salmon (Ragan *et al.*, 1996; Spanggaard *et al.*, 1996). Interestingly the mitochondria of *Dermocystidium* sp. from brook trout have flattened cristae (Ragan *et al.*, 1996). The salmonid 'rosette agent' clusters as a sister group to the choanoflagellates in 18S ribosomal DNA trees (Kerk *et al.*, 1995). Although not specifically referred to by the authors their micrographs of the vegetative bodies of the rosette agent also seem to show mitochondria with flattened cristae (Harrell *et al.*, 1986: see their Figure 9). Thus *Dermocystidium* and the 'rosette agent' appear to be linked by having flattened mitochondrial cristae.

The crustacean parasite *Psorospermium* clusters within this clade on the same branch as *Ichthyophonus* (Ragan *et al.*, 1996) and thus might be predicted to have tubular-vesiculate cristae, although it has not been possible to confirm this. The close relationship between these two organisms is supported by the similar morphology of their encysted resting stage and their amoeboid dispersive stages (see Ragan *et al.*, 1996). Ragan *et al.* (1996) have used the initials of these assorted organisms (DRIPs) to define this clade although Cavalier-Smith (chapter 21) has proposed the class name Ichthyosporea for them.

The precise branching position of these DRIPs in relation to the choanoflagellate, fungal and animal branches has not been completely resolved (Figure 20.20). The best supported tree suggests that this clade diverges subsequent to the animal–fungus dichotomy, as the most basal branch within the animal lineage and is a sister group to the animals and choanoflagellates (Ragan *et al.*, 1996). However, these authors also acknowledge the possibility that the DRIPs clade may have split immediately before the animal–fungal dichotomy (see Figure 20.20). However, it should be noted that Spanggaard *et al.* (1996) suggest that their data indicate that *Ichthyophonus* does not appear to be older than the earliest chytrid fungi or even the zygomycete family Glomaceae. They suggest that this organism evolved at about the same time as its fish hosts, around 390 million years ago, which again suggests the Ichthyosporea could represent a related (parallel) rather than an ancestral group to the fungi. This topic is also discussed in some detail by Cavalier-Smith in chapter 21.

The apparent discrepancy between organisms with tubular and flattened cristae appearing within the DRIPs/Ichthyosporea clade has been explained as an apparent reversion of flattened cristae to tubular form. However, this may genuinely reflect a pivotal basal position of this group within the animal/choanozoa branch since it seems likely that these organisms have evolved from biflagellate ancestors with tubular cristae such as *Apusomonas* (Cavalier-Smith and Chao, 1995). Note that Cavalier-Smith has now placed this apparently basal biflagellate group in a newly created phylum Neomonada since its former phylum name of Opalozoa is no longer appropriate (see chapter 21). The variation in mitochondrial morphology may reflect a variable acquisition of flattened cristae from tubular ancestral forms rather than the reversion of morphology previously suggested (Ragan *et al.*, 1996).

20.4.4 Other possible protist relatives of fungi

To conclude, it is worth remembering that there may be many missing link organisms which may help clarify the link between fungi and protists and which have not been studied in detail or even discovered. One such organism may be the enigmatic protist *Nephromyces*, which lives in the renal sac of molgulid tunicates and was orginally thought to be an anomalous chytrid (Saffo, 1981) but was placed by Corliss (1994) in the Opalozoa. However, it does seem to possess many of the morphological and life cycle characteristics exhibited by members of the newly created Ichthyosporea, including an amoeboid infective stage (Saffo and Nelson, 1983). However, this organism also produces a hyphal-like filamentous phase which apparently has chitin in its cell wall (Saffo and Fultz, 1986) and has a biflagellate swarmer stage. Therefore, this organism does appear to show characters which could place it at the base of the fungal/choanozoan/animal clade and it certainly deserves attention from the molecular phylogenists. It will also be interesting to see where molecular phylogeny places biflagellate colonial flagellates such as *Rhipidodendron* and *Spongomonas* since these have a striated basal root structure to their flagella (Hibberd, 1976) similar to those described in many choanoflagellates and chytrids.

Finally, Dr H. Canter (personal communication) has recently come across an unusual protist which superficially resembled a chytrid and is associated with filaments of the green alga *Oedogonium* (Figure 20.13). This protist produces spherical bodies which are physically attached to the algal wall by an electron dense rootlike structure (Figure 20.16). The mitochondria have flattened plate-like cristae (Figure 20.14) which may associate it with the animal/choanozoan/fungal clade. It produces biflagellate rod-like zoospores with a short anterior flagellum and stiff rod-like trailing flagellum (Figure 20.15). These are somewhat reminiscent of flagellates of the genus *Cercomonas* which is currently within the Neomonada. Again it would be of interest to determine where this organism ends up on the 18S rDNA tree.

20.5 CONCLUSIONS

In this review I have discussed protist organisms which have commonly been studied by mycologists. It is clear that 'fungi' are a polyphyletic assemblage, principally falling into three kingdoms: Protozoa, Chromista and Fungi. Molecular phylogeny clearly supports the concept of a separate kingdom Fungi, which as well as the terrestrial fungi should also include its basal uniflagellate phylum, the Chytridiomycota. Molecular phylogeny has shown that the Fungi appear to share a monophyletic ancestor with the Choanozoa and the animals within the crown eukaryotes. It seems likely that the fungi and animals have phylogenetic affiliations with the protist choanoflagellates and a number of previously problematic organisms now grouped in the Ichthyosporea. Furthermore it seems likely that all of these groups evolved from biflagellate protistan ancestors with tubular mitochondrial cristae now placed in the Neomonada. However the most likely flagellate protist ancestor and precise order of branching at the base of this well defined clade both remain unresolved. There are clearly still plenty of exciting challenges ahead for the evolutionary phylogenist.

Acknowledgements

I would like to thank Dr Alan Warren for inviting me to write this review and acknowledge the help given by Dr Hilda Canter, Dr Thomas Rand and Dr Bettina Spanggaard, all of whom have allowed me to make use of their published or unpublished material as illustrations.

20.6 REFERENCES

Aist, J.R. and Williams, P.H. (1971) The cytology and kinetics of cabbage root hair penetration by *Plasmodiophora brassicae*. *Canadian Journal of Botany*, **49**, 2023–34.

Alderman, D.J. (1982) Fungal diseases of aquatic animals, in *Microbial Diseases of Fish* (ed R.J. Roberts), Academic Press, London, pp. 189–242.

Alexopoulos, C.J., Mims, C.W. and Blackwell, M. (1996) *Introductory Mycology*, 4th edn. Wiley, New York.

Baldauf, S. and Palmer, J.D. (1993) Animals and fungi are each other's closest relatives: congruent evidence from multiple proteins. *Proceedings of the National Academy of Sciences USA*, **90**, 1158–62.

Barr, D.J.S. (1980) An outline for the reclassification of the Chytridiales, and for a new order, the Spizellomycetales. *Canadian Journal of Botany*, **58**, 2380–94.

Barr, D.J.S. (1983) The zoosporic grouping of plant pathogens. Entity or non-entity, in *Zoosporic Plant Pathogens: A Modern Perspective* (ed. S.T. Buczacki), Academic Press, London, pp. 43–83.

Barr, D.J.S. (1992) Evolution and kingdoms of organisms from the perspective of a mycologist. *Mycologia*, **84**, 1–11.

Barr, D.J.S. and Allan, P.M.E. (1982) Zoospore ultrastructure of *Polymyxa graminis* (Plasmodiophoromycetes). *Canadian Journal of Botany*, **60**, 2496–504.

Barr, D.J.S. and Désaulniers, N.L. (1986) Ultrastructure of the *Lagena radicola* zoospore, including a comparison of primary and secondary *Saprolegnia zoospores. Canadian Journal of Botany*, **65**, 2161–76.

Barr, D.J.S and Désaulniers, N.L. (1989a) The flagellar apparatus in zoospores of *Phytophthora, Pythium* and *Halophytophthora. Canadian Journal of Botany*, **70**, 2163–9.

Barr, D.J.S and Désaulniers, N.L. (1989b) The flagellar apparatus of the Oomycetes and Hyphochytridiomycetes, in *The Chromophyte Algae: Problems and Perspectives* (eds J.C. Green, B.S.C. Leadbeater and W.I Diver), Clarendon Press, Oxford, pp. 342–55.

Barstow, W.E., Freshour, G.D, and Fuller, M.S. (1989) The ultrastructure of mitosis during zoosporogenesis in *Rhizidiomyces apophysatus. Canadian Journal of Botany*, **67**, 3401–9.

Bartnicki-Garcia, S. (1996) The hypha: unifying thread of the fungal kingdom, in *A Century of Mycology* (ed. B.C. Sutton), Cambridge University Press, Cambridge, pp. 105–33.

Beakes, G.W. (1981) Ultrastructure of the phycomycete nucleus, in *The Fungal Nucleus* (eds K. Gull and S.G. Oliver), Cambridge University Press, Cambridge, pp. 1–35.

Beakes, G.W. (1987) Oomycete phylogeny: ultrastructural perspectives, in *Evolutionary Biology of Fungi* (eds A.D.M. Rayner, C.M. Brasier and D. Moore), Cambridge University Press, Cambridge, pp. 405–21.

Beakes, G.W. (1989) Oomycete fungi: their phylogeny and relationship to chromophyte algae, in *The Chromophyte Algae: Problems and Perspectives* (eds J.C. Green, B.S.C. Leadbeater and W.I Diver), Clarendon Press, Oxford, pp. 325–42.

Berbee, M.L. and Taylor, J.W. (1992) Dating the evolutionary radiations of the true fungi. *Canadian Journal of Botany*, **71**, 1114–27.

Bortnick, R.N., Powell, M.J. and Barstow, W.E. (1985) Zoospore fine structure of the parasite *Olpidiopsis saprolegniae* variety *saprolegniae* (Oomycetes, Lagenidiales). *Mycologia*, **77**, 861–79.

Bowman, B.H., Taylor, J.W., Brownlee, A.G. et al.(1992) Molecular evolution of the fungi: relationships of the Basidiomycetes, Ascomycetes, and Chytridiomycetes. *Molecular Biology and Evolution*, **9**, 285–96.

Buczaki, S.T. (1983) *Plasmodiophora*: An inter-relationship between biological and practical problems, in *Zoosporic Plant Pathogens: A Modern Perspective* (ed. S.T. Buczacki), Academic Press, London, pp. 162–95.

Castlebury, L.A. and Domier, L.L. (1994) Small-subunit ribosomal RNA gene phylogeny of *Plasmodiophora brassicae. Abstracts of 5th International Mycological Congress*, Vancouver, p. 32.

Cavalier-Smith, T. (1987) The origin of Fungi and pseudofungi, in *Evolutionary Biology of the Fungi* (eds A.D.M. Rayner, C.M. Brasier and D. Moore), Cambridge University Press, Cambridge, pp. 339–53.

Cavalier-Smith, T. (1989). The Kingdom Chromista, in *The Chromophyte Algae: Problems and Perspectives* (eds J.C. Green, B.S.C. Leadbeater and W.I Diver), Clarendon Press, Oxford, pp. 381–407.

Cavalier-Smith, T. (1995) Evolutionary protistology comes of age: biodiversity and molecular cell biology. *Archiv für Protistenkunde*, **145**, 145–54.

Cavalier-Smith, T. and Chao, E.E. (1995) The opalozoan *Apusomonas* is related to the common ancestor of animals, fungi and choanoflagellates. *Proceedings of the Royal Society of London Series B*, **261**, 1–6.

Cavalier-Smith, T. and Allsopp, M.T.E.P. (1996) *Corallochytrium*, an enigmatic non-flagellate protozoan related to Choanoflagellates. *European Journal of Protistology*, **32**, 1–9.

Cavalier-Smith, T. Allsopp, M.T.E.P. and Chao, E.E. (1994) Thraustochytrids are chromists, not Fungi.: 18s rRNA signatures of Heterokonta. *Philosophical Transactions of the Royal Society, Series B*, **346**, 387–97.

Cooney, E.W., Barr, D.J.S. and Barstow, W.E. (1985) The ultrastructure of the zoospores of *Hyphochytrium catenoides*. *Canadian Journal of Botany*, **63**, 497–505.

Corliss, J.O. (1994) An interim utilitarian ('user-friendly') hierarchical classification and characterization of the protists. *Acta Protozoologica*, **33**, 1–51.

Darley, W., Porter, D. and Fuller, M.S. (1973) Cell wall composition and synthesis via Golgi-directed scale formation in the marine eukaryote, *Schizochytrium aggregatum* with a note on *Thraustochytrium* sp. *Archives für Mikrobiologie*, **90**, 89–106.

Dykstra, M.J. and Porter, D. (1984) *Diplophrys marina*, a new scale-forming marine protist with labyrinthulid affinities. *Mycologia*, **76**, 626–32.

Dylewski. D.P. and Miller, C.E. (1983) Cruciform nuclear division in *Woronia pythii* (Plasmodiophoromycetes). *American Journal of Botany*, **70**, 1325–39.

Fenchel, T. and Patterson, D.J. (1988) *Cafeteria roenbergensis* nov. gen., nov. sp. a heterotrophic microflagellate from marine plankton. *Marine Microbial Food Webs*, **3**, 9–19.

Förster, H., Coffey, M.D., Ellwood H., and Sogin, M.L. (1990) Sequence analysis of the small subunit ribosomal RNAs of three zoosporic fungi and implications for fungal evolution. *Mycologia*, **82**, 306–12.

Fuller, M.S. and Clay, R.P. (1993) Observations of *Gonyapoda* in pure culture: growth, development and cell wall characterization. *Mycologia*, **85**, 38–45.

Goggin, C.L. and Barker, S.C. (1993) Phylogenetic position of the genus *Perkinsus* (Protista, Apicocomplexa) based on small subunit ribosomal RNA. *Molecular and Biochemical Parasitology*, **60**, 65–70.

Gunderson, J.H., Elwood, H., Ingold, A. *et al.* (1987) Phylogenetic relationships between chlorophytes, chrysophytes and oomycetes. *Proceedings of the National Academy of Sciences USA*, **84**, 5823–7.

Hasegawa, M., Hashimoto, T. and Adachi, J. (1992) Origin and evolution of eukaryotes as inferred from protein sequence data, in *The Origin and Evolution of the Cell* (eds H. Hartman and K. Matsuno), World Scientific, Singapore, pp. 107–30.

Harrell, L.W., Elston, R.A., Scott, T.M. and Wilkinson, M.T. (1986). A significant new systematic disease of net-pen reared Chinook salmon (*Oncorhynchus tshawytscha*) brood stock. *Aquaculture*, **55**, 249–62.

Hawksworth, D.L., Kirk, P.M., Sutton, B.C. and Pegler D.N. (1995) *Ainsworth and Bisby's Dictionary of Fungi*, 8th Edn, CAB International, Wallingford.

Heath, I.B. (1986) Nuclear division: a marker for protist phylogeny? *Progress in Protistology*, 1, 115–62.

Heath, I.B. and Bauchop, T. (1985) Mitosis and phylogeny in the genus *Neocallimastix*. *Canadian Journal of Botany*, **63**, 1595–604.

Heath, I.B., Bauchop, T. and Skipp, R.A. (1983) Assignment of the rumen anaerobe *Neocallimastix frontalis* to the Spizellomycetales (Chytridiomycetes) on the basis of its polyflagellate zoospore ultrastructure. *Canadian Journal of Botany*, **61**, 295–307.

Hibberd, D.J. (1975) Observations on the ultrastructure of the choanoflagellate *Codosiga botrytis* (Ehr.) Saville-Kent with special reference to the flagellar apparatus. *Journal of Cell Science*, **17**, 191–219.

Hibberd, D.J. (1976) The fine-structure of the colonial colourless flagellates *Rhipidodendron splendidum* Stein and *Spongomonas uvella* Stein with special reference to the flagellar apparatus. *Journal of Protozoology*, **23**, 374–85.

Kerk, D., Gee, A., Standish, M. *et al.* (1995) The rosette agent of Chinook salmon (*Oncorhynchus tshawytscha*) is closely related to choanoflagellates, as determined by the phylogenetic analyses of its small ribosomal subunit RNA. *Marine Biology*, **122**, 187–92.

Leadbeater, B.S.C. (1994) Developmental studies on the loricate choanoflagellate *Stephanoeca diplocostata* Ellis. *European Journal of Protistology*, **30**, 171–83.

Leipe, D.D., Wainwright, P.O., Gunderson, J.H. *et al.* (1994) The stramenopiles from a molecular perspective: 16S-like rRNA sequences from *Labyrinthuloides minuta* and *Cafeteria roenbergensis*. *Phycologia*, **33**, 369–77.

Li, J. and Heath, I.B. (1992) The phylogenetic relationships of the anaerobic chytridiomycetous gut fungi (Neocallimasticaceae) and the Chytridiomycota. I. Cladistic analysis of rRNA sequences. *Canadian Journal of Botany*, **70**, 1738–46.

Margulis, L. (1996) Archaeal-eubacterial mergers in the origin of eukarya: phylogenetic classification of life. *Proceedings of the National Academy of Sciences USA*, **93**, 1071–6.

Martin, R.W. and Miller, C.E. (1986) Ultrastructure of mitosis in the endoparasite *Olpidiopsis varians*. *Mycologia*, **78**, 11–21.

Matheson, A.T. (1992) Ribosomal protein structure: A probe into the evolution of the cell, in *The Origin and Evolution of the Cell* (eds H. Hartman and K. Matsuno), World Scientific, Singapore, pp 311–31.

Mithen, R. and Magrath, R. (1992) A contribution to the life history of *Plasmodiophora brassicae*: secondary plasmodia development in root galls of *Arabidopsis thaliana*. *Mycological Research*, **96**, 877–85.

Moss, S.T. (1991). Thraustochytrids and other zoosporic marine fungi, in *The Biology of Free-living Heterotrophic Flagellates* (eds D.J. Patterson and J. Larsen), Clarendon Press, Oxford, pp. 415–25.

Nagahama, T., Sato, H., Shimazu, M. and Sugiyama, J. (1995) Phylogenetic divergence of the entomophthoralean fungi: evidence from nuclear 18S ribosomal RNA gene sequences. *Mycologia*, **87**, 203–9.

Paquin, B., Roewer, I., Wang, Z. and Lang, B.F. (1995) A robust fungal phylogeny using the mitochondrially encoded NAD5 protein sequence. *Canadian Journal of Botany*, **73** (Suppl. 1), S180–5.

Patterson, D.J. (1989) Stramenopiles: chromophytes from a protistan perspective, in *The Chromophyte Algae: Problems and Perspectives* (eds J.C. Green, B.S.C. Leadbeater and W.I Diver), Clarendon Press, Oxford, pp. 357–79.

Patterson, D.J. and Larsen, J. (1992) A perspective on protistan nomenclature. *Journal of Protozoology*, **39**, 125–31.

Perkins, F.O. (1996) The structure of *Perkinsus marinus* (Mackin, Owen and

Collier, 1950) Levine, 1978 with comments on taxonomy and phylogeny of *Perkinsus* spp. *Journal of Shellfish Research*, **15**, 67–87.

Ragan, M.A., Goggin, C.L., Cawthorn, R.J. *et al.* (1996) A novel clade of fungus-like parasites near the animal–fungal divergence. *Proceedings of the National Academy of Sciences USA*, **93**, 11907–12.

Rand, T.G. (1994) An unusual form of *Ichthyophonus hoferi* (Icthyophonales: Ichthyophonaceae) from yellowtail flounder *Limanda ferruginea* from the Nova Scotia shelf. *Diseases of Aquatic Organisms*, **18**, 21–8.

Raghu-Kumar, S. (1987) Occurrence of the Thraustochytrid, *Corallochytrium limacisporum* gen. et sp. nov. in the Coral Reef Lagoons of the Lakshadweep Islands in the Arabian Sea. *Botanica Marina*, **30**, 83–9.

Robb, J. and Barron, G.L. (1982) Nature's ballistic missile. *Science*, **218**, 1221–2.

Saffo, M.B. (1981) The enigmatic protist *Nephromyces*. *BioSystems*, **14**, 487–90.

Saffo, M.B. and Nelson, R. (1983) The cells of *Nephromyces*: developmental stages of a single life cycle. *Canadian Journal of Botany*, **61**, 3230–9.

Saffo, M.B. and Fultz, S. (1986) Chitin and the symbiotic protist *Nephromyces*. *Canadian Journal of Botany*, **64**, 1306–10.

Silberman, J.D., Sogin, M.L., Leipe, D.D. and Clark, C.G. (1996) Human parasite finds taxonomic home. *Nature*, **380**, 398.

Spanggaard, B., Gram, N., Okamoto, N. and Huss, H.H. (1994) Growth of the fish-pathogenic fungus, *Ichthyophonus hoferi*, measured by conductimetry and microscopy. *Journal of Fish Diseases*, **17**, 145–53.

Spanggaard, B., Huss, H.H. and Bresciani, J. (1995) Morphology of *Ichthyophonus hoferi* assessed by light and scanning electron microscopy. *Journal of Fish Diseases*, **18**, 567–77.

Spanggaard, B., Shouboe, P., Rossen, L. and Taylor, J.W. (1996) Phylogenetic relationships of the intercellular fish pathogen *Ichthyophonus hoferi* and fungi, choanoflagellates and the rosette agent. *Marine Biology*, **126**, 109–15.

Van der Auwera, G., Chapelle, S. and De Wachter, R. (1994) Structure of the large ribosomal subunit RNA of *Phytophthora megasperma*, and phylogeny of the oomycetes. *FEBS Letters*, **338**, 133–6.

Van der Auwera, G., De Baere, R. de Peer, Y. *et al.* (1995) The phylogeny of the Hyphochytridiomycota as deduced from ribosomal RNA sequences of *Hyphochytrium catenoides*. *Molecular Biology and Evolution*, **12**, 671–8.

Wainwright, P.O., Hinkle, G., Sogin, M.L. and Stickel, S.K. (1993) Monophyletic origins of the metazoa: an evolutionary link with fungi. *Science*, **260**, 340–1.

Whittaker, R.H. (1969) New concepts of kingdoms of organism. *Science*, **163**, 150–60.

21

Neomonada and the origin of animals and fungi

T. Cavalier-Smith

Department of Botany, University of British Columbia, Vancouver, B.C. Canada, V6T 1Z4.
E-mail: tom@tcs.botany.ubc.ca

ABSTRACT

Zooflagellate protists are very diverse and are spread across 12 phyla in two kingdoms. Three protozoan phyla (Metamonada, Trichozoa, and Neomonada) consist entirely or almost entirely of zooflagellates, while two others (Euglenozoa, Dinozoa) are a mixture of zooflagellates and phytoflagellates. A few zooflagellates are also found in three predominantly amoeboid protozoan phyla (the largely amoeboflagellate Percolozoa, the Cercozoa and the Amoebozoa) and in four phyla of the primarily photosynthetic kingdom Chromista. Neomonada, my most recently proposed major zooflagellate assemblage, are pivotally important for understanding the origin of the kingdoms Animalia and Fungi and the diversification of advanced protozoa of the subkingdom Neozoa.

I recently established the phylum Neomonada by grouping the once separate phylum Choanozoa, now a subphylum, with what remained of the former phylum Opalozoa after I transferred Opalinata from the kingdom Protozoa into the kingdom Chromista. Other formerly opalozoan flagellates, the phytomyxans, amoeboflagellate sarcomonads, and proteomyxids are now not grouped with the neomonads but with certain filose and reticulofilose sarcodines in the phylum Cercozoa as a result of recent molecular sequence evidence for this novel amoeboflagellate phylum.

Except for a few secondarily non-flagellate walled species, neomonads are naked phagotrophic and aerobic zooflagellates. They have mitochondria with well-developed cristae, which are seldom discoid (unlike those of Percolozoa and Euglenozoa), and Golgi dictyosomes. Neomonads typically have one or two flagella, rarely four or none.

Evolutionary Relationships Among Protozoa. Edited by G.H. Coombs, K. Vickerman, M.A. Sleigh and A. Warren. Published in 1998 by Chapman & Hall, London. ISBN 0 412 79800 X

They are distinguished from Dinozoa by the absence of cortical alveoli and from most chromistan zooflagellates by lacking tubular ciliary hairs. They have no chloroplasts and are typically non-amoeboid. Neomonada are classified here in 25 orders (one, Dermocystida, new), 10 classes (one, the mainly non-flagellate Ichthyosporea, new) and three subphyla: Choanozoa, Apusozoa, and Isomita.

Subphylum Choanozoa is predominantly uniflagellate with usually flat mitochondrial cristae; in addition to the familiar choanoflagellates it includes two groups of walled, usually non-flagellate protists formerly confused with fungi: the free-living marine *Corallochytrium*, and the parasitic ichthyosporeans (*Dermocystidium*, rosette agent, *Ichthyophonus, Psorospermium*) here placed in two orders within a new class Ichthyosporea. Ribosomal RNA sequences show that the non-flagellate *Corallochytrium* is more closely related to choanoflagellates than are ichthyosporeans and that all three choanozoan classes are specifically related to each other and to the kingdoms Animalia and Fungi. This finding supports the thesis that animals and fungi each arose by radical transformations of choanozoan-like uniflagellate ancestors.

The formerly opalozoan neomonads, usually with tubular cristae, are divided into 17 orders in seven classes and two subphyla: Apusozoa and Isomita, which differ in extrusome type and flagellar arrangement. Apusozoa, named after the thecomonad *Apusomonas*, are anisokont biflagellates with strongly diverging centrioles; they are rather diverse, their 12 orders being grouped into five classes and two infraphyla; one infraphylum has extrusomes of three differing non-ejectisome types that are used to delineate classes, while the other lacks extrusomes. Isomita (classes Cyathobodonea and Nephromycea) are isokont biflagellates (rarely uniflagellate), with parallel centrioles, and ejectisome-like extrusomes or none. A new family, Adriamonadidae, is created here for the recently described genus *Adriamonas*, placed within the order Pseudodendromonadida of the Cyathobodonea. Ribosomal RNA sequencing shows that *Apusomonas* is specifically related to the clade Opisthokonta comprising Choanozoa, Animalia, and Fungi. Sequence information is needed also for the other neomonad groups to test the validity of grouping them all in a single phylum.

Many, possibly even most, of the roughly 100 zooflagellate genera of uncertain taxonomic position may prove to belong to the Neomonada or Cercozoa when they are characterized by electron microscopy and/or rRNA sequencing.

21.1 INTRODUCTION

During the past three decades our perception of the kingdoms of life and their interrelationships has changed dramatically. Thirty years ago most biologists thought simply of two kingdoms: animals and plants.

During the 1970s variants of the five kingdom system of Whittaker (1969) became increasingly widely adopted by elementary textbooks. However research biologists never wholeheartedly accepted any of them, partly because some of Whittaker's eukaryotic kingdoms seemed unduly heterogeneous and/or polyphyletic and partly because his demarcation between the eukaryotic kingdoms entirely ignored the burgeoning new ultrastructural data. Therefore several authors in the 1970s proposed a larger number of kingdoms (Leedale, 1974; Edwards, 1976; Cavalier-Smith, 1978). When first presenting a multikingdom system that comprehensively took account of ultrastructural diversity, I stated: 'I do not see how biological diversity can be taught in a clear way even to schoolchildren with fewer than the six kingdoms'. Those six kingdoms included a new kingdom Chromista, comprising not only all the chlorophyll c-containing algae except the dinoflagellates, but also the various heterotrophic heterokonts such as the oomycetes. The major differences between the six kingdoms are shown in Figure 21.1.

The higher level classification of Protozoa, the most basal eukaryotic kingdom, has undergone numerous changes since the widely used system of Levine *et al.* (1980) was published. These include ranking the group as a kingdom, as first suggested by Owen (1858), not as a subkingdom, as had prevailed for more than a century, and an increase in the number of phyla from seven to 13. The grouping of these 13 phyla into three subkingdoms, and their subdivision into 26 subphyla, are summarized in the appendix. All four of the higher eukaryote kingdoms evolved independently from the most advanced protozoan subkingdom, Neozoa. The precise protozoan ancestry of the predominantly photosynthetic Plantae and Chromista is not yet known. Both the derived heterotrophic kingdoms, Animalia and Fungi, however, probably evolved from the same neozoan group, Choanozoa, which I have recently included as a subphylum of a broader zooflagellate phylum Neomonada (Cavalier-Smith, 1997a). This chapter summarizes the recent evidence for the choanozoan origins of animals and fungi and discusses the classification of Neomonada.

I shall present evidence that Choanozoa should include not only the familiar choanoflagellates, which are structurally similar to the trophic cells of the first animals (namely sponges), but also two phyletically distinct classes of protist that are walled in the trophic state and therefore were previously confused with Fungi, i.e. Corallochytrea Cavalier-Smith, 1995 and Ichthyosporea cl. nov. I shall also explain why I recently created the new zooflagellate phylum Neomonada by grouping Choanozoa with the major part of the former protozoan phylum Opalozoa (Cavalier-Smith, 1997a), and discuss the major lines of evolution within the new phylum. The classification of the Neomonada, as slightly revised here, is summarized in Table 21.1.

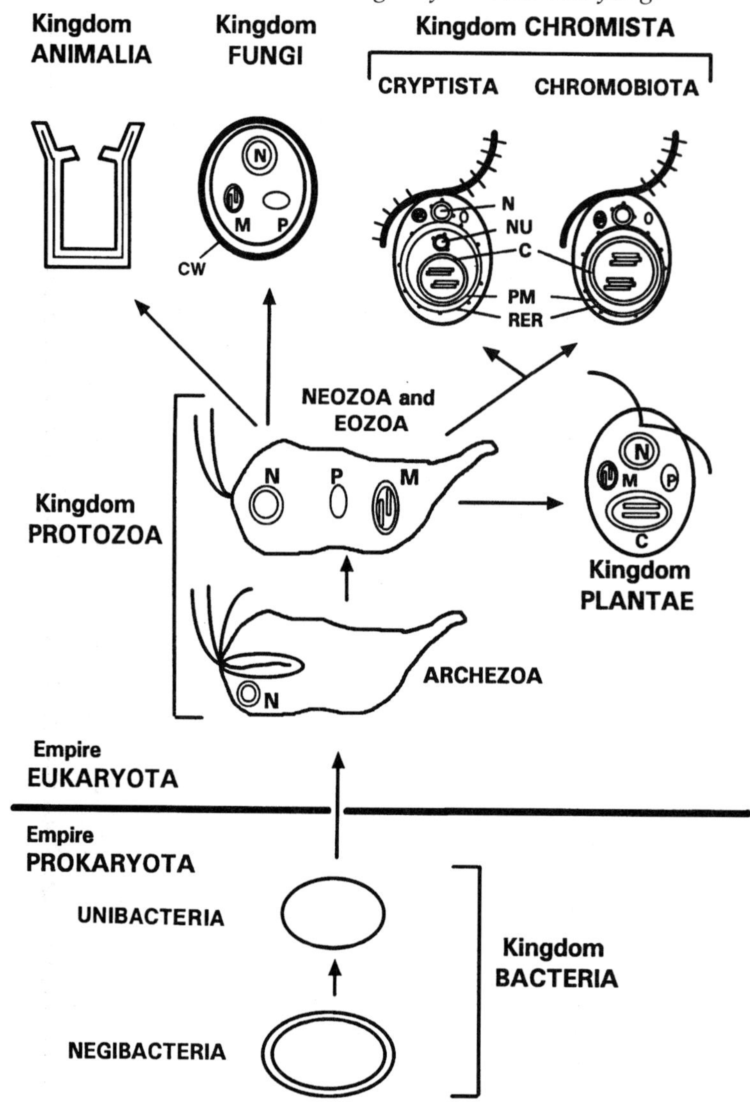

Figure 21.1 Diagrammatic outline of the six kingdoms of life. The animal kingdom is represented by a cnidarian polyp and the other five kingdoms by one or two generalized unicells, even though some organisms in each of them (many in Plantae, Fungi and Chromista) independently became multicellular. The ancestral kingdom Bacteria is divided into two subkingdoms by the number of envelope membranes. Negibacteria (also known as Gracilicutes) are the majority, and have two bounding membranes, which were both inherited by mitochondria and chloroplasts when they evolved independently from different endosymbiotic negibacteria. Unibacteria, comprising the infrakingdoms Archaebacteria and Posibacteria (Cavalier-Smith 1998), have only one bounding

Because some readers may be a little overwhelmed by the dizzying speed with which major new protozoan taxa have been created during the past decade, or may not be familiar with the idea that there are several radically different zooflagellate phyla, I shall preface this discussion by some historical remarks. The present revolution in protozoan systematics stems primarily from two major technical innovations: high quality electron microscopy, which gave the initial impetus; and rapid techniques for DNA sequencing coupled with powerful computers and refined algorithms for calculating molecular phylogenetic trees. Also important have been a growing rigour in cladistic analysis, and major advances in molecular cell biology that have given us a better biological understanding of cell organelles, the structure and arrangement of which form the various body plans of Protozoa, and therefore the fundamental basis for a sound protozoan taxonomy.

---

membrane, and gave rise to eukaryotes by the autogenous evolution of phagotrophy, the endomembrane system, cytoskeleton, nucleus and 9+2 flagella. The earliest eukaryotes, of the protozoan subkingdom Archezoa, lack mitochondria (M), peroxisomes (P), and Golgi stacks in trophic cells, and were probably ancestrally quadriflagellate. Eozoa also were probably ancestrally quadriflagellate, but the ancestors of Euglenozoa and Neozoa became biflagellate. Biflagellate Neozoa gave rise to Plantae (i.e. green plants, red algae, and glaucophytes) by transforming an endosymbiotic cyanobacterium into obligate chloroplasts (C), and to Chromista by the independent symbiogenetic incorporation of a red alga. Chromista differ from Plantae in the location of their chloroplasts within the rough endoplasmic reticulum (RER), where they are surrounded by a periplastid membrane (PM) that is the relic of the plasma membrane of the symbiotic red alga; the red algal nucleus was lost by the subkingdom Chromobiota (e.g. brown algae, diatoms, haptophytes, oomycetes, Opalinata) but remains as a tiny nucleomorph (NU) with three minute chromosomes in cryptomonads (subkingdom Cryptista, which also includes the zooflagellate *Goniomonas*). Uniflagellate Neozoa gave rise to animals by the evolution of multicellular epithelia and collagenous connective tissue and to fungi by the origin of chitinous walls (CW) and the loss of phagotrophy. Chloroplasts have been lost several times within the Chromista and Dinoflagellata, mitochondria have been converted to hydrogenosomes or totally lost in several protozoan phyla and in rumen fungi, while peroxisomes have been lost in the same organisms and also in the Opalinata; flagella have been lost even more frequently, in all kingdoms. These and other organellar losses greatly confused earlier classifications; the very simple cell structure of Microsporidia (now in the Kingdom Fungi) probably results from similar losses but was once thought to be a primitive trait justifying their current position within the Archezoa (see appendix 21A) is very unclear.

Table 21.1 Classification of the infrakingdom Neomonada Cavalier-Smith 1997

Phylum Neomonada Cavalier-Smith 1997
- Subphylum 1. Apusozoa Cavalier-Smith 1997
  - Infraphylum 1. Extrumonada Cavalier-Smith 1997
    - Class 1. Thecomonadea Cavalier-Smith 1993 (as subclass) stat. nov. 1995
      - Order 1. Apusomonadida Karpov and Mylnikov 1989 em. Cavalier-Smith (*Apusomonas*)
      - Order 2. Cryomonadida Cavalier-Smith 1993 (*Cryothecomonas*)
    - Class 2. Jakobea Cavalier-Smith 1997
      - Order 1. Jakobida Cavalier-Smith 1993 (*Jakoba*)
      - Order 2. Discocelida Cavalier-Smith 1997 (*Discocelis*)
    - Class 3. Kinetomonadea Cavalier-Smith 1993
      - Subclass 1. Varicristia Cavalier-Smith 1997
        - Order 1. Histionida Cavalier-Smith 1993 (e.g. *Histiona, Reclinomonas, Ancyromonas*)
        - Order 2. Ancyromonadida ord. nov. (Diagnosis: flatcristae; dense cortical layer beneath plasma membrane)
      - Subclass 2. Ramicristia Cavalier-Smith 1997
        - Order 1. Heliomonadida Cavalier-Smith 1993 (*Dimorpha, Tetradimorpha*)
        - Order 2. Commatiida Cavalier-Smith 1997 (*Commation*)
  - Infraphylum 2. Eurymonada Cavalier-Smith 1997
    - Class 1. Anisomonadea Cavalier-Smith 1993 emend.
      - Order 1. Diphylleida Cavalier-Smith 1993 (*Diphyllea*)
      - Order 2. Phagodinida Cavalier-Smith 1993 (*Phagodinium*)
      - Order 3. Caecitellida Cavalier-Smith 1997 (*Caecitellus* Patterson *et al.* 1993)
    - Class 2. Ebridea Lemmerman 1901 emend. Deflandre 1936 stat. nov. Loeblich III 1970
      - Order Ebriida Deflandre 1936 (e.g. *Ebria, Hermesinum*)
- Subphylum 2. Isomita Cavalier-Smith 1997
  - Class 1. Cyathobodonea Cavalier-Smith 1993
    - Order 1. Pseudodendromonadida Hibberd 1985
      - Family 1. Pseudodendromonadidae Cavalier-Smith 1993 (*Cyathobodo, Pseudodendromonas*)
      - Family 2. Adriamonadidae fam. nov. (diagnosis: like Pseudodendromonadidae have a gullet and two anterior flagella with parallel centrioles, but unlike them have no stalk or scales: type genus *Adriamonas* Verhagen *et al.* 1994)
    - Order 2. Phalansteriida Hibberd 1983
      - Family Phalansteriidae Kent 1880 (*Phalansterium*)
    - Order 3. Kathablepharida Cavalier-Smith 1993 (*Kathablepharis*)
  - Class 2. Telonemea Cavalier-Smith 1993
    - Order 1. Telonemida Cavalier-Smith 1993 (*Telonema*)
    - Order 2. Nephromycida Cavalier-Smith 1993 (*Nephromyces*)

Table 21.1 Continued

Subphylum 3. Choanozoa Cavalier-Smith 1981 emend. 1983 stat. nov. 1997
Class 1. Choanoflagellatea Kent 1880 (as order Choanoflagellida) stat. nov. (syn. Choanomonadea Krylov *et al.* 1980)
Order 1. Craspedida Cavalier-Smith 1997 (e.g. *Monosiga, Sphaeroeca*)
Families Codonosigidae Kent 1880 and Salpingoecidae Kent 1880
Order 2. Acanthoecida Cavalier-Smith 1997 (e.g. *Diaphanoeca*)
Family Acanthoecidae Norris 1965
Class 2. Corallochytrea Cavalier-Smith 1995
Order Corallochytrida Cavalier-Smith 1995 (*Corallochytrium*)
Class 3. Ichthyosporea cl. nov. (diagnosis: unicellular parasites of animals; trophic cells with walls that lack chitin; zoospores, if present, uniflagellate, without a collar or cortical alveoli)
Order 1. Dermocystida ord. nov. (diagnosis: with flat mitochondrial cristae; parasites of vertebrates: *Dermocystidium,* rosette agent)
Order 2. Ichthyophonida Rand 1994 orth. mut. (*Ichthyophonus* with vesicular cristae, *Psorospermium*)
Class 4. Cristidiscoidea, Page 1987 stat. nov.
Order 1. Nucleariida Cavalier-Smith 1993 (*Nuclearia*)
Order 2. Fonticulida Cavalier-Smith 1983 (*Fonticula*)
Order 3. Ministeriida Cavalier-Smith 1997 (*Ministeria*)

21.2 CHANGING VIEWS OF ZOOFLAGELLATE CLASSIFICATION

Improved light microscopes led to the discovery of the vast diversity of protozoa during the 19th century. During the same period they revealed the fundamental differences in microanatomy and embryology by which the body plans of the various multicellular animal phyla differ, and therefore enabled zoologists to divide animals into well defined phyla, most of which are still valid today. But the resolution of the light microscope was insufficient to define the body plans of the unicellular protozoa with equal precision. The simple late-19th century classification of protozoa into sporozoa, amoebae, flagellates and ciliates, with which many of us grew up and which still dominates many elementary biological textbooks, relied on superficial characters and has been recognized as profoundly unsatisfactory for many decades. Only when electron microscopy, with its 1000-fold greater resolution, was applied to protozoa in the 1950s and 1960s was it possible to discern the fundamental body plans of the different types of protozoa.

Protozoologists were however rather slow in applying this important new information to reforming protozoan macrosystematics. By the

1970s there were, however, several explicit discussions of the phylogenetic significance of these new ultrastructural characters (Manton, 1965; Taylor, 1976, 1978; Hibberd, 1979; Cavalier-Smith, 1978). The 1980 classification of the Society of Protozoologists (Levine *et al.*, 1980) made an important step forward in splitting the former 'sporozoans' into several, ultrastructurally radically different phyla. But it was archaic in retaining zooflagellates and phytoflagellates as formal taxa and in treating zooflagellates as a single class. Moreover, flagellates and Sarcodina, which even the non-specialist Singer (1959) had ranked (like ciliates) as separate phyla within the subkingdom Protozoa were grouped together as the single phylum Sarcomastigophora, not for positive reasons but simply because of the difficulty in separating them and the likelihood that sarcodines had arisen many times from flagellates by losing flagella.

It was well established long before 1980 that ultrastructural diversity within zooflagellates was much greater than that which separated multicellular animals and fungi (Taylor, 1976). Phylogeneticists had by then recognized that the body plan of major flagellate groups could be defined in terms of the presence or absence of major cell organelles (notably mitochondria, chloroplasts, flagella, Golgi dictyosomes) and their differences in ultrastructure, e.g. in the topology of chloroplast membranes, the structure of the flagellar transition region and roots, and the shape of mitochondrial cristae (Taylor, 1976, 1978; Hibberd, 1979; Cavalier-Smith, 1975, 1978). Even in comparison with the classification of Grassé (1952) that was too early to incorporate any insights from electron microscopy, the 1980 classification seemed retrogressive in omitting altogether some key groups (e.g. bisoecids and proteromonads) and in reducing the rank of many others. In consequence both the general protozoologist and the general biologist have remained largely unaware of the tremendous ultrastructural, evolutionary and ecological diversity of the zooflagellates, and their study has been impeded considerably in comparison with better known groups.

Zooflagellates were first divided into several phyla according to their ultrastructural body plans by Cavalier-Smith (1981a) who proposed seven new phyla to include zooflagellates: Euglenozoa, Dinozoa, Parabasalia, Metamonadina (now Metamonada), Choanozoa, Proterozoa, and Heterokonta, the last of which was placed outside the Protozoa in the new botanical kingdom Chromista. All except Proterozoa, which was later (Cavalier-Smith, 1991, 1993a) subsumed in the broader phylum Opalozoa, persisted in the recent major re-evaluation of the kingdom Protozoa and of protist classification (Cavalier-Smith, 1993b). Since then Heterokonta have been increased in rank to infrakingdom and subdivided into three separate phyla (Ochrophyta,

Sagenista and Bigyra), each of which contains some zooflagellates (Cavalier-Smith, 1995a,b, 1997b). Corliss (1994) advocated seven heterokont phyla, but most of these have not been validly published under the International Code of Botanical Nomenclature. Heterokonta are grouped with the algal phylum Haptophyta as the chromist subkingdom Chromobiota.

Later I proposed that the amitochondrial Metamonada and Parabasalia diverged substantially before the other mitochondria-containing zooflagellate taxa (Cavalier-Smith, 1983), a view amply confirmed by rRNA phylogeny. Metamonada and Microsporidia are grouped as the subkingdom Archezoa. Molecular sequence trees have also confirmed not only the phyletic unity and distinctiveness of the Euglenozoa, Dinozoa, Choanozoa, and Heterokonta (Cavalier-Smith, 1993b), but also the distinctness from all of them of the phylum Percolozoa that was proposed more recently to encompass zooflagellates, amoeboflagellates, and amoebae that have discoid mitochondrial cristae but which lack Golgi dictyosomes (Cavalier-Smith, 1993c). Percolozoa may be the earliest diverging organisms with mitochondria (Cavalier-Smith, 1993b,d; Cavalier-Smith and Chao, 1996; Sogin *et al.*, 1996). The recent ultrastructural characterization of the amitochondrial flagellate *Trimastix*, which despite its name has four not three flagella, shows that it has hydrogenosomes and Golgi dictyosomes like Parabasala but unlike them lacks a striated parabasal filament (Brugerolle, 1995). I therefore recently reduced Parabasala in rank to subphylum in order to group them with *Trimastix* (in its own subphylum Anaeromonada) as a new protozoan phylum Trichozoa (Cavalier-Smith, 1997a). Percolozoa and Euglenozoa, both of which have distinctive discoid mitochondrial cristae, constitute the Infrakingdom Disaicristata. They have been grouped with Trichozoa as the subkingdom Eozoa. All Protozoa more advanced than Archezoa and Eozoa are grouped together in the subkingdom Neozoa. The subdivision of Neozoa into four infrakingdoms and their pivotal position in eukaryotic phylogeny are shown in Figure 21.2 (see Note 1, p. 407).

The phylum Opalozoa was accepted by Corliss (1994), but the zooflagellates included in it are so varied (despite almost all having tubular mitochondrial cristae) that their fundamental unity was always more doubtful than for other major zooflagellate groups. Because of this we have started to sequence ribosomal RNA genes from as many such flagellates as we can in order to understand their relationships better. Our results so far show a fundamental difference between sarcomonad flagellates (*Cercomonas, Heteromita, Thaumatomonas*), which prove to be closely related to euglyphid filose testate amoebae (Cavalier-Smith and Chao, 1997), and the thecomonad *Apusomonas*, which instead is related, though rather distantly, to Choanozoa, animals and fungi

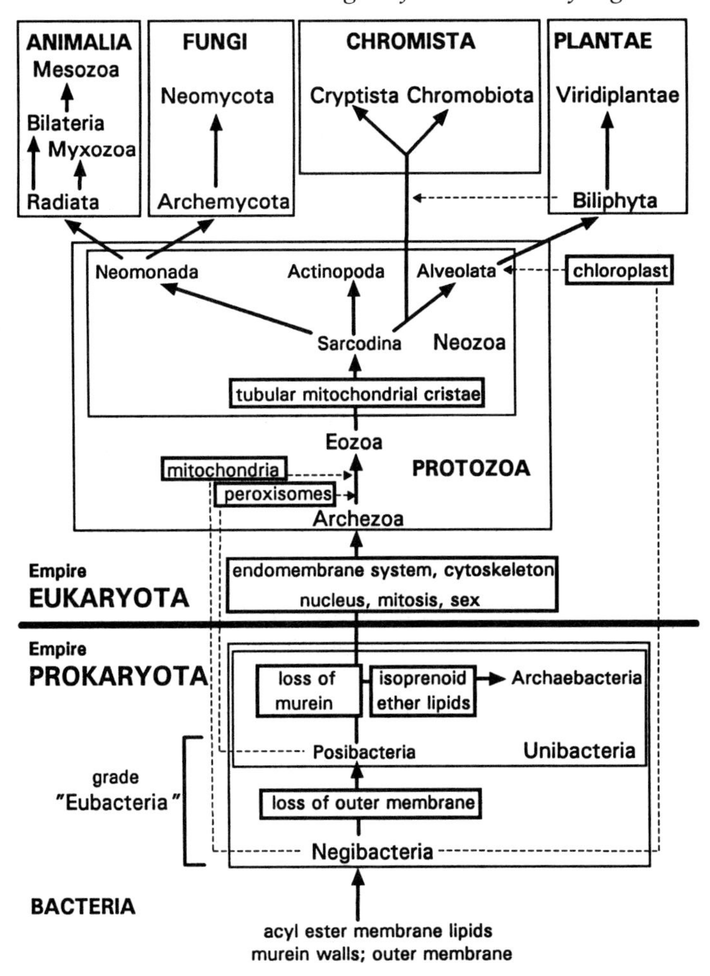

Figure 21.2 Postulated phylogenetic relationships between the six kingdoms (upper case) and their subkingdoms. Infrakingdoms are also shown (in smaller type) for the two paraphyletic subkingdoms (Neozoa and Unibacteria) that contain the ancestors of higher kingdoms, as are some key innovations (within heavy boxes). The four major symbiogenetic events in the history of life are shown with dashed arrows: (1) the symbiogenetic origin of mitochondria from an α-proteobacterium, (2) the probable origin of peroxisomes from a posibacterium (Cavalier-Smith, 1990), (3) the monophyletic origin of chloroplasts from a cyanobacterium, and (4) the origin of the ancestral chromist as a eukaryote–eukaryote chimaera between a red algal endosymbiont and a biflagellated protist host; whether the host was a protozoan, as assumed here, or a very early plant is unclear. For clarity infrakingdoms are not shown for the four higher holophyletic kingdoms (Animalia, Fungi, Plantae, and Chromista) or for

(Cavalier-Smith and Chao, 1995). Ultrastructural evidence has long suggested that proteromonads are more closely related to the chromistan heterokonts than to any of the preceding groups, a conclusion recently confirmed by rRNA phylogeny (Silberman *et al.*, 1996). Silberman *et al.* (1996) have also shown that the enigmatic non-flagellate parasitic protist *Blastocystis* is more closely related to *Proteromonas* than to any other organisms for which rRNA sequences are available, which was entirely unexpected. I have therefore placed *Blastocystis* in a new order, Blastocystales, within the class Proteromonadea (Cavalier-Smith, 1998). This combination of molecular and ultrastructural evidence makes it likely that the former Opalozoa included at least three phyletically disparate groups.

The first of these is the Proteromonadea plus the ultrastructurally related Opalinea, which have both been transferred into the Heterokonta as the subphylum Opalinata of a new phylum Bigyra characterized by a double flagellar transition helix (Cavalier-Smith, 1997b, 1998). Bigyra also include the Pseudofungi (Oomycetes and Hyphochytrea) and the zooflagellate *Developayella elegans* (see Cavalier-Smith, 1997b for a more detailed treatment of this phylum and of the Sagenista, which include the zooflagellate bicoecids as well as the Labyrinthulea).

The second group comprises the amoeboid sarcomonad flagellates, proteomyxids and phytomyxans, which were recently transferred to a revised sarcodine phylum Rhizopoda (Cavalier-Smith, 1997a), on account of the molecular evidence for a specific relationship between sarcomonads and euglyphid amoebae and also, but more distantly, with amoeboid chlorarachnean algae (Cavalier-Smith and Chao, 1997). This modified use of the term Rhizopoda (Cavalier-Smith 1995a; 1997a) is closer to that of Dujardin's (1841) 'rhizopodes', which was restricted to testate amoebae and foraminifera, than it is to von Siebold's (1845) later formalization Rhizopoda, which soon (von Siebold, 1848) lumped

---

Negibacteria. The diagnoses of the taxa and a complete listing of their phyla and subphyla are given by Cavalier-Smith (1998). Chloroplasts are assumed to have originated in the latest common ancestor of Plantae and the alveolate dinoflagellate protozoa (Cavalier-Smith, 1982), but the alternative possibility that dinoflagellates obtained their chloroplasts secondarily by lateral transfer from a chromobiote, though considered distinctly less likely, cannot currently be excluded (Cavalier-Smith, 1995b). For simplicity two secondary symbiogenetic lateral transfers of green algal chloroplasts into Protozoa are not shown: the certain transfer to a cercozoan (see p. 386) host to create the class Chlorarachnea (e.g. *Chlorarachnion*) and the possible transfer to a euglenoid host to create the euglenoid subclass Euglenia.

Amoebaea (naked lobose amoebae) together with Dujardin's rhizopodes. The term 'rhizopodes' ('root feet') was initially devised for sarcodines with filiform root-like pseudopods (Dujardin, 1841); it did not then include, and has always been descriptively inappropriate for, the Amoebaea or Heliozoa. However, as we have now found by rRNA sequencing (Cavalier-Smith and Chao, in preparation) that several entirely non-amoeboid flagellates also belong in Rhizopoda, *sensu* Cavalier-Smith, 1995, I have now created the new name Cercozoa for this important phylum and discontinued the name Rhizopoda for any taxon (Cavalier-Smith, 1998).

The third and most diverse group of former Opalozoa comprises *Apusomonas* and a large number of non-amoeboid phagotrophic zooflagellates that have now been grouped with Choanozoa to form the new phylum Neomonada (Cavalier-Smith, 1997a). The composition of the Choanozoa has also been broadened recently by the inclusion of the walled non-flagellate protist *Corallochytrium* in a new class Corallochytrea alongside the choanoflagellates, which was necessitated by the close relationship shown by their ribosomal RNA sequences (Cavalier-Smith and Allsopp, 1996).

Thus, the very recent changes necessitated by the new molecular trees have required the dismemberment not only of the former Opalozoa, but also of the Rhizopoda: first the lobose amoebae were removed from the Rhizopoda as the phylum Amoebozoa (Cavalier-Smith, 1995a); later the secondarily amitochondriate Archamoebae and the Mycetozoa (Cavalier-Smith, 1997a) were placed within the Amoebozoa as distinct subphyla. Protozoologists have long accepted the exclusion of Foraminifera from Rhizopoda (e.g. Hartog, 1906; Margulis *et al.*, 1990), even though they were central to and constituted the majority of species in both Dujardin's and von Siebold's rhizopods. The recent placement of Athalamea in the subphylum Reticulofilosa of the Cercozoa (Cavalier-Smith, 1998) instead of grouping them with the Foraminifera (Cavalier-Smith, 1993b), must be thoroughly tested by molecular phylogeny. My infrakingdom Sarcodina includes most organisms other than Heliozoa that were included in von Siebold's Rhizopoda, but also includes some flagellates in the Cercozoa as well as *Multicilia* in the Amoebozoa. My present classification of the kingdom Protozoa into 13 phyla and 26 subphyla is summarized in the appendix.

21.3 SUBPHYLUM CHOANOZOA: RELATIONSHIP OF *CORALLOCHYTRIUM* AND THE ICHTHYOSPOREANS TO CHOANOFLAGELLATES

Molecular phylogenetic analysis has unexpectedly identified two groups of largely non-flagellate protozoa that are more closely related to choanoflagellates than are any other known groups (Kerk *et al.*, 1995; Cavalier-Smith and Allsopp, 1996; Ragan *et al.*, 1996). Both groups, unlike almost all other protozoa, have walled trophic stages, which previously led to the mistaken classification of many of these protists as fungi rather than protozoa. Kerk *et al.* (1995) showed that the enigmatic rosette agent, an intracellular parasite of salmonid fish, is specifically related to choanoflagellates. We have confirmed this, and have shown that *Corallochytrium*, a free-living walled saprotrophic non-flagellate protist previously treated as a thraustochytrid (Raghu-Kumar, 1987), is even more closely related to choanoflagellates than is the rosette agent (Cavalier-Smith and Allsopp, 1996). As *Corallochytrium* apparently has flat mitochondrial cristae, like choanoflagellates, I now classify it in the Choanozoa, but in a separate class (Corallochytrea: Cavalier-Smith, 1995a) from the choanoflagellates (class Choanoflagellata). Ragan *et al.* (1996) have shown that the rosette agent is part of a much broader clade of parasitic protozoa that includes also at least two species of *Dermocystidium, Ichthyophonus hoferi,* and *Psorospermium haeckeli.* All these are parasites of fish, amphibians or crustaceans, and their life cycles are complex but not fully worked out. Because their interrelationship was not previously recognized no formal name has been given to this clade, which Ragan *et al.* (1996) refer to simply as the DRIP clade. As the group is sufficiently phylogenetically coherent, and also morphologically different from all other protozoa, to be treated as a distinct class, I here create a new class, Ichthyosporea, for the DRIP clade: its diagnosis is given in Table 21.1. I chose the name Ichthyosporea because these organisms resemble unicellular walled spores, and mostly infect fish (*Ichthys* in Greek).

The demonstration by Ragan *et al.* (1996) that the ichthyosporeans form a single clade is very robust and undoubtedly correct. However, the exact position shown for the group is open to question since it disagrees with the conclusion of Kerk *et al.* (1995) that the rosette agent is the sister of choanoflagellates and the conclusion of Cavalier-Smith and Allsopp (1996) that the rosette agent is a sister to choanoflagellates plus *Corallochytrium*. Instead Ragan *et al.* (1996) found ichthyosporeans to be a sister to choanoflagellates plus animals (or more rarely to choanoflagellates plus animals and fungi. Ragan *et al.* (1996) attributed this slight contradiction to the results of Kerk *et al.* (1995) to the fact that their trees included not only rosette but also four other ichthyosporeans. To test this interpretation I have carried out a new maximum likelihood analysis that includes not only all the ichthyosporean sequences studied by Ragan *et al.* (1996) but also our *Corallochytrium* sequence and a third choanoflagellate sequence not available to them. As Figure 21.3 shows, this analysis shows ichthyosporeans as a sister group to choanoflagellates plus *Corallochytrium*, not to choanoflagellates plus both *Corallochytrium* and animals as would be expected from Ragan *et al.*'s interpretation. A separate parsimony analysis of the same data set as in Figure 21.3 yielded a single most parsimonious tree with ichthyosporeans as the sister group to choanoflagellates plus *Corallochytrium*. A neighbour-joining distance analysis also showed the Choanozoa as holophyletic (i.e. as a single clade) as in the parsimony and distance analysis, but bootstrap support for this was not high.

The albeit poorly-known morphological characters of ichthyosporeans are also consistent with the position shown in Figure 21.3. A flagellate stage is known only for some species of *Dermocystidium*, which have uniflagellate zoospores (Olson, Dungan and Holt, 1991), which is consistent with their being related to the uniflagellate choanoflagellates. At least one species of *Dermocystidium* has flat cristae (Ragan *et al.*, 1996). The statement that *Dermocystidium* has tubular cristae (Cavalier-Smith and Allsopp, 1996) was an error based on confusion with the tubulicristate apicomplexan flagellate *Perkinsus*, which was once classified as a *Dermocystidium*, but is morphologically and phylogenetically distinct from any of the approximately 20 species of genuine *Dermocystidium*. *Ichthyophonus* has vesicular not flat cristae, but the shape of those of the rosette agent and *Psorospermium* is unknown. Ichthyosporea can reasonably be placed as a third class within the Choanozoa, if one accepts the reasonable suggestion of Ragan *et al.* (1996) that the vesicular cristae of *Ichthyophonus* are a unique derived character and not an indication of a close relationship with the very few other protists (certain filose amoebae) with vesicular cristae. On this interpretation the ancestral choanozoan had flat cristae.

A fungal order, Ichthyophonales, was previously established (Rand, 1994) for *Ichthyophonus*. As the molecular results clearly show that it is not specifically related to fungi, but is closer to choanoflagellate protozoa, I here change the ending of the order to Ichthyophonida to accord with zoological practice. I also place *Psorospermium* in the Ichthyophonida since the two genera are very close together on Figure 21.3 and in the tree of Ragan *et al.* (1996) and there is no apparent morphological evidence that would argue against their placement in the same order. The molecular trees also show that *Dermocystidium* spp. are so closely related to the rosette agent that the two should be placed in the same order as each other. In view of the fact that *Dermocystidium* has flat cristae and *Ichthyophonus* has vesicular ones, I create a new order Dermocystida for *Dermocystidium* plus the rosette agent. If it turns out that rosette has flat cristae and *Psorospermium* vesicular ones this would corroborate such a subdivision of the class Ichthyosporea.

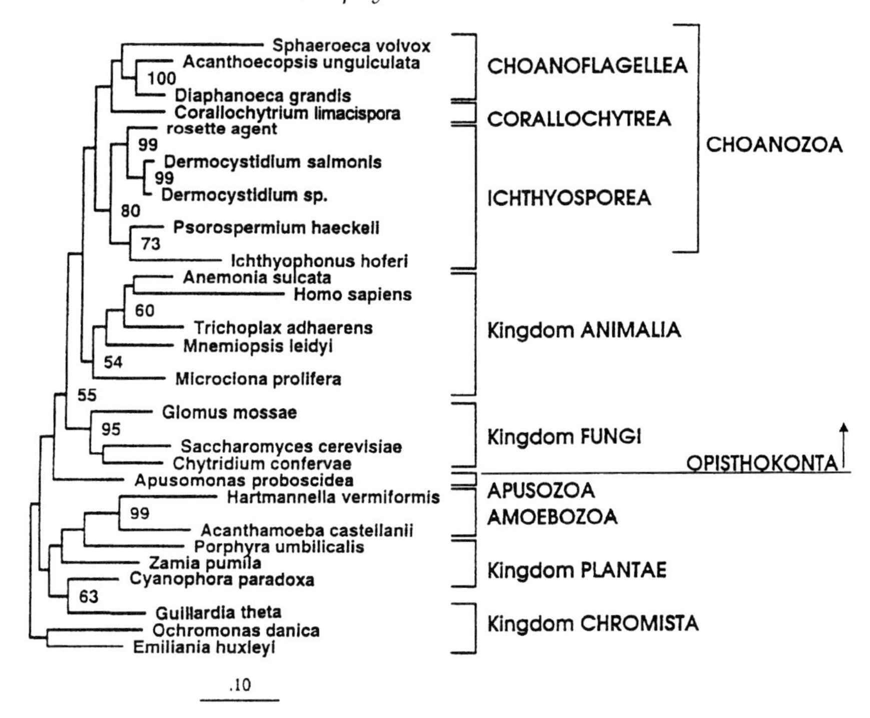

Figure 21.3 Maximum likelihood and parsimony analysis of all available neomonad 18S ribosomal RNA sequences together with representatives of the four higher kingdoms (Animalia, Fungi, Plantae and Chromista) and of the Amoebozoa, the only other protozoan phylum that tends to branch near the Neomonada. The tree shown is the maximum likelihood tree, calculated by the fastDNAml v.1.0.6 (Olsen *et al.*, 1994), with quick add option, empirical base frequencies, and global rearrangements (transition/transversion parameter 1.514676). Three independent fastDNAml calculations using different species input orders (randomly chosen with the jumble option) gave the same tree (with a log likelihood of -20828.69024). Figures at the nodes are all the bootstrap percentages for a separate maximum parsimony analysis using the PHYLIP (Version 3.5) program DNAPARS (Felsenstein, 1992) (100 pseudoreplicates) that were 50% or greater. Both analyses used equal weighting for all the nucleotide positions; only the most conservative and best aligned 1758 alignment positions were included.

21.4 OPISTHOKONTA AND THE CHOANOZOAN ORIGINS OF ANIMALS AND FUNGI

From Figure 21.3, and from Figure 21.4 which includes a much greater variety of outgroup taxa, it can be seen that Choanozoa, animals, and fungi form a clade, as was first shown by Wainright *et al.* (1993). This clade was predicted by me (Cavalier-Smith, 1987a) and named opisthokonta (Gk. opistho = posterior, kont = cilium or flagellum) because I considered that a posteriorly uniflagellate motile state was a shared derived character for the group (posteriorly uniflagellate zoospores in most chytridiomycete fungi, posteriorly uniflagellate sperm in the most primitive animal phyla, the swimming of most – not all – choanoflagellates with the flagellum at the rear when they are not attached to a substratum). I also postulated that flat non-discoid mitochondrial cristae were a shared derived character for the opisthokont clade, since most protozoa other that choanoflagellates have either tubular or discoid cristae. The validity of the opisthokont clade and the two shared derived characters postulated for it is strongly supported by the recent demonstration that the tubulicristate anisokont biflagellated zooflagellate *Apusomonas* is the sister to the opisthokonts. This means that posterior uniflagellation and flat cristae are not only shared by but are also a derived character for the opisthokonts (Cavalier-Smith and Chao, 1995).

A choanoflagellate ancestry for sponges has long been accepted because of the similarity in structure and mode of feeding of sponge choanocytes and choanoflagellate protozoa. Because the triploblastic tissue organization of sponges is so fundamentally similar to that of Cnidaria and other higher animals (as shown for example by De Cecatty, 1986), I contended in my unpublished contribution to the 1984 meeting of the Systematics Association (see preface of Conway Morris *et al.*, 1985) that the two could not possibly have arisen independently and therefore that all animals had a choanoflagellate ancestry. This then controversial position, which was contrary to the dogmas of generations of textbook writers who had grossly exaggerated the differences between sponges and other animals, has now been vindicated by molecular sequence evidence for the monophyly of the animal kingdom including sponges (Wainright *et al.*, 1993; Cavalier-Smith *et al.*, 1996).

In my 1984 Systematics Association contribution I argued that sponges arose from a member of the choanoflagellate family Codonosigidae similar to *Proterospongia* (Leadbeater, 1983), that Cnidaria evolved from a calcareous sponge, that Ctenophora and Bilateria evolved independently from different Cnidaria; and that the first bilateral animals were bryozoan-like coelomates that evolved by the radical transformation of an anthozoan polyp and all acoelomate bilateria are secondarily so. Recent molecular trees neither refute nor directly support these postulates, partly because many relevant branches are so close together that their actual order is uncertain, and partly because molecular trees cannot reveal the phenotypes at ancestral nodes. It therefore remains necessary to infer these indirectly.

This uncertainty about the phenotype of ancestral nodes, and a neglect of an explicit consideration of where on molecular trees the boundaries between the phenotypes used to define taxa really lie, often leads molecular phylogeneticists into a rather loose and potentially confusing terminology. For example Ragan *et al.* (1996) refer to choanoflagellates and ichthyosporeans as 'within the animal lineage', or even 'within the animal kingdom' and to ichthyosporeans as being 'the most basal branch of the metazoa'. But ichthyosporeans are not metazoans and do not become so merely because on molecular trees they are a sister group either to choanoflagellates or to choanoflagellates plus metazoans. Admittedly I myself once did classify choanoflagellates within the kingdom Animalia and outside the kingdom Protozoa (Cavalier-Smith, 1981a). But I soon realized that this, though cladistically correct, is taxonomically undesirable since it gives too much weight to the cladistic relationship and too little to the considerable phenetic gulf between Choanozoa and Animalia.

Since 1983 (Cavalier-Smith, 1983) therefore, I have consistently used the term animal as a synonym for metazoan and to refer only to members of the kingdom Animalia; this usage excluded from the term animal all those protists classified in the kingdom Protozoa, and therefore excludes Choanozoa. If one wishes to refer to an animal/choanozoan clade one should use some such compound phrase or, if the situation merits it, invent a specific new single term, such as opisthokont for the animal/choanozoan/fungal clade, and not use a term like 'animal lineage' for a clade that includes non-animals. The title of Wainright *et al.*'s (1993) paper (monophyletic origins of the Metazoa: an evolutionary link with Fungi) was similarly misleading: their tree showed Choanozoa, not fungi, as a sister group to metazoa, and also showed opisthokonta as a clade. Their statement that I suggested 'a specific evolutionary link between fungi and animals' is equally misleading. What I actually suggested was that animals and fungi evolved entirely independently from different choanoflagellate ancestors (Cavalier-Smith, 1987a). The only published ribosomal RNA trees including choanoflagellate sequences that show animals and fungi as sister groups are those of Philippe and Adoutte (1996) and Cavalier-Smith *et al.*, (1996b); virtually all the others (e.g. Wainright *et al.*, 1993; Cavalier-Smith and Allsopp, 1996; Cavalier-Smith and Chao, 1995, 1996; Cavalier-Smith *et al.*, 1996a,c; Ragan *et al.*, 1996) show Choanozoa as

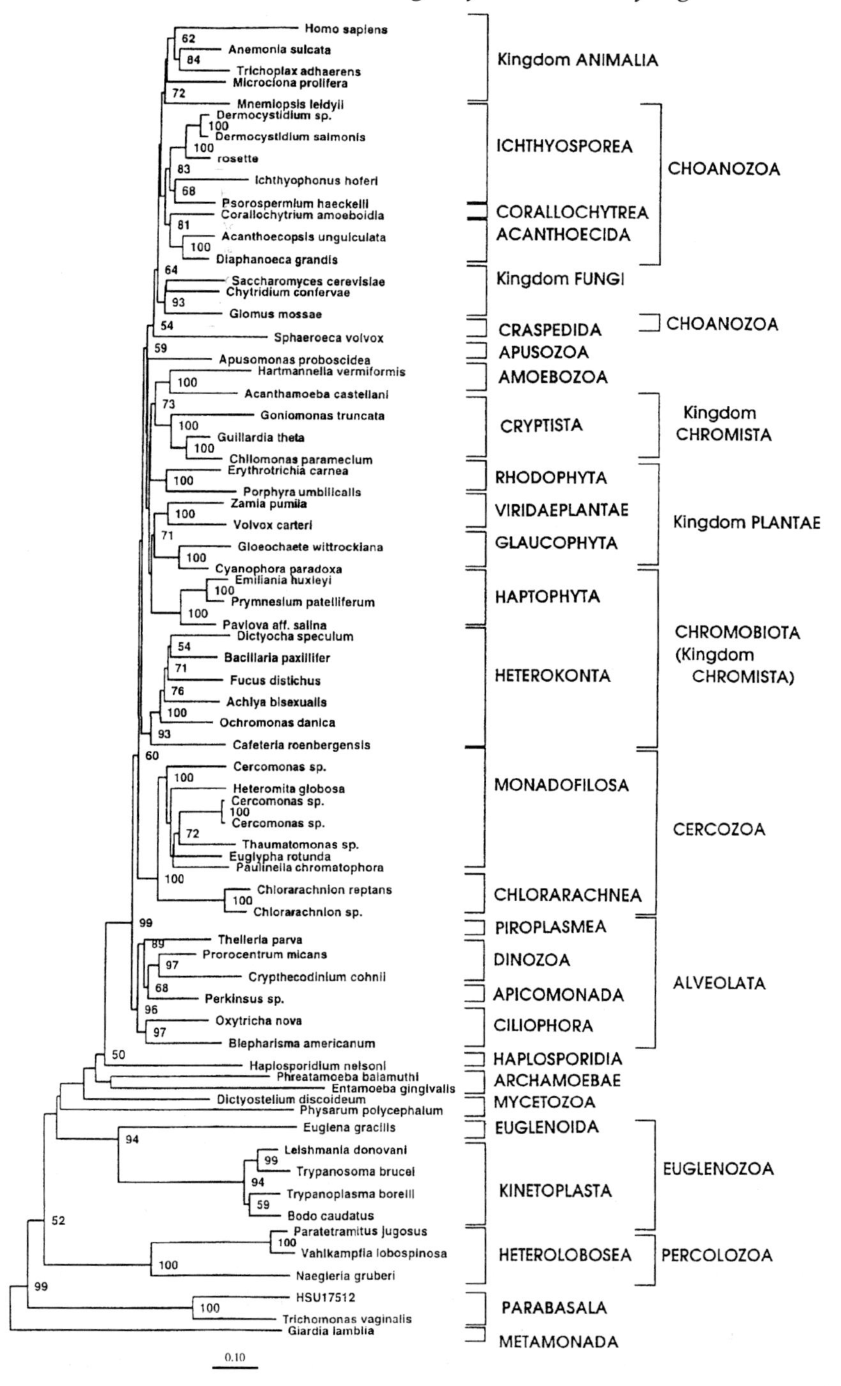
Homo sapiens
Anemonia sulcata
Trichoplax adhaerens
Microciona prolifera
Mnemiopsis leidyii
Dermocystidium sp.
Dermocystidium salmonis
rosette
Ichthyophonus hoferi
Psorospermium haeckelii
Corallochytrium amoeboidia
Acanthoecopsis unguiculata
Diaphanoeca grandis
Saccharomyces cerevisiae
Chytridium confervae
Glomus mossae
Sphaeroeca volvox
Apusomonas proboscidea
Hartmannella vermiformis
Acanthamoeba castellani
Goniomonas truncata
Guillardia theta
Chilomonas paramecium
Erythrotrichia carnea
Porphyra umbilicalis
Zamia pumila
Volvox carteri
Gloeochaete wittrockiana
Cyanophora paradoxa
Emiliania huxleyi
Prymnesium patelliferum
Pavlova aff. salina
Dictyocha speculum
Bacillaria paxillifer
Fucus distichus
Achlya bisexualis
Ochromonas danica
Cafeteria roenbergensis
Cercomonas sp.
Heteromita globosa
Cercomonas sp.
Cercomonas sp.
Thaumatomonas sp.
Euglypha rotunda
Paulinella chromatophora
Chlorarachnion reptans
Chlorarachnion sp.
Theileria parva
Prorocentrum micans
Crypthecodinium cohnii
Perkinsus sp.
Oxytricha nova
Blepharisma americanum
Haplosporidium nelsoni
Phreatamoeba balamuthi
Entamoeba gingivalis
Dictyostelium discoideum
Physarum polycephalum
Euglena gracilis
Leishmania donovani
Trypanosoma brucei
Trypanoplasma borelli
Bodo caudatus
Paratetramitus jugosus
Vahlkampfia lobospinosa
Naegleria gruberi
HSU17512
Trichomonas vaginalis
Giardia lamblia
0.10
Kingdom ANIMALIA
ICHTHYOSPOREA
CORALLOCHYTREA
ACANTHOECIDA
CHOANOZOA
Kingdom FUNGI
CRASPEDIDA
CHOANOZOA
APUSOZOA
AMOEBOZOA
CRYPTISTA
Kingdom CHROMISTA
RHODOPHYTA
VIRIDAEPLANTAE
Kingdom PLANTAE
GLAUCOPHYTA
HAPTOPHYTA
HETEROKONTA
CHROMOBIOTA (Kingdom CHROMISTA)
MONADOFILOSA
CERCOZOA
CHLORARACHNEA
PIROPLASMEA
DINOZOA
ALVEOLATA
APICOMONADA
CILIOPHORA
HAPLOSPORIDIA
ARCHAMOEBAE
MYCETOZOA
EUGLENOIDA
EUGLENOZOA
KINETOPLASTA
HETEROLOBOSEA
PERCOLOZOA
PARABASALA
METAMONADA

sisters to Animalia as in the present analysis, though in one analysis Choanozoa appeared as sisters of Fungi (Cavalier-Smith, Allsopp and Chao, 1994). In Cavalier-Smith (1993b) Choanozoa were sisters to the radiate animals (not within them as stated by Ragan *et al.*, 1996), and bilateral animals were sisters to the Fungi. Since the bootstrap support for any of these three positions is never high, the fact that not all trees agree is not surprising. Although the ribosomal RNA trees show that Choanozoa are the most closely related protozoa (of the groups so far sequenced) to both fungi and animals, the branching order of these three taxa clearly cannot be determined unambiguously from rRNA trees.

We still know too little about choanoflagellate molecular diversity to test my specific proposals that animals evolved from a non-loricate codonosigid choanoflagellate, and that fungi evolved from a possibly chitinous loricate salpingoecid choanoflagellate (Cavalier-Smith, 1987a). The only choanoflagellates on most published trees are from the family Acanthoecidae, which having siliceous loricas seemed to me improbable ancestors for either kingdom. A sequence now available for the codonosigid *Sphaeroeca* is very highly divergent from the two acanthoecid sequences, as can be seen on Figure 21.3. Though this divergence may be a little exaggerated by some obvious sequencing errors in this particular sequence, it is significant for two reasons. First, it provides reasonable evidence, given that there is no fossil record for the class, that the choanoflagellate phenotype is an ancient one, an important point if one is arguing that choanoflagellates were ancestral to both the animal and the fungal kingdoms. Second it shows that the deep divergence between the siliceous loricate choanoflagellates in body structure and in

Figure 21.4 Maximum likelihood phylogenetic tree for all available neomonad 18S ribosomal RNA sequences and representatives of all major groups of metakaryotes. Metakaryotes are defined as all eukaryotes that have mitochondria or which are known to have secondarily lost them (Cavalier-Smith, 1987b), i.e. all eukaryotes other than the possibly primitively amitochondrial subkingdom Archezoa – Metamonada. The metamonad archezoan *Giardia lamblia* was also included in order to root the tree. Methods were as for Figure 21.3 except that only one maximum likelihood calculation was done. Figures at the nodes are bootstrap percentages for the separate parsimony analysis. HSU 17512 is an unidentified gut symbiont of termites. A similar tree in Sogin *et al.* (1996 Figure 1b) includes also the unpublished sequence of the radiolarian *Thalassicola nucleata*, showing it within a very broad array of deeply branching sarcodine taxa, consistent with the assumption of Figure 21.2 that Actinopoda, Neomonada and Alveolata all evolved independently from a paraphyletic Sarcodina.

the dynamics of the feeding structure is mirrored at the genetic level. I have therefore established a new order Acanthoecida to separate the siliceous loricate Acanthoecidae from the non-siliceous choanoflagellates (i.e. families Codonosigidae and Salpingoecidae), and have grouped the two latter families in a new order, for which I adopted a variant (Craspedida) of the old name Craspedomonadida (Cavalier-Smith, 1997a).

Sequences are needed from each of the choanoflagellate genera in order to test this dichotomy further and to identify the most deeply branching lineages. Only when this is done will we be able to say whether animals and fungi evolved from within the crown choanoflagellates, as I suggested (Cavalier-Smith, 1987a), or whether instead they arose from a stem choanoflagellate (here crown and stem are used in the original sense of Jefferies, 1979). One argument favouring an origin from a stem choanoflagellate is that sex is unknown in choanoflagellates; if this absence is not just a result of our ignorance of their habits, it might mean that sex was lost in their common ancestor, in which case only an earlier stem choanoflagellate can have been ancestral to animals or fungi. In fact Neomonada is the only one of the 50 currently recognized eukaryotic phyla (Cavalier-Smith, 1998) in which sex has never been observed. I conjecture that this is only because they are mostly very small flagellates that have never been thoroughly studied in nature, and that it would be impossible to distinguish between a mating pair and a dividing cell in a fixed sample or by casual observation of a live one.

If, as I have argued, both Fungi and Animalia evolved from choanoflagellates, then Choanozoa must be paraphyletic, though crown choanoflagellates need not be if both Fungi and Animalia evolved from very early stem choanoflagellates. However, the appearance of choanozoan paraphyly on Figure 21.3 is probably an artefact caused by a too low position on that tree for *Sphaeroeca*, which tends to be unstable in its position on molecular trees. For various technical reasons its position on the smaller tree of Figure 21.3 (and in Philippe and Adoutte, 1996) in which crown Choanozoa appear to be holophyletic is probably more accurate.

The discovery that *Corallochytrium* and ichthyosporeans also are choanozoans in no way alters the choanoflagellate theory of animal origins, since neither is structurally or functionally suitable to have served as an animal ancestor. Indeed ichthyosporeans, being exclusively parasites of animals, probably did not evolve their special phenotypes until after their common ancestor became a parasite in some aquatic animal. I suggest that their common ancestor was most likely a stem choanoflagellate that lost its collar and evolved wall stages as adaptations to parasitism. Because of their wide separation from each other on the molecular trees I argue that these two groups (Cavalier-Smith and Allsopp, 1996) evolved cell walls and gave up the phagotrophic choanoflagellate mode of feeding independently of each other and independently of the ancestor of fungi. Since at least some *Dermocystidium* species have flat cristae, and some are posteriorly uniflagellate, one must regard these features as general properties of Choanozoa rather than of choanoflagellates alone. This means that in principle fungi could have evolved from a uniflagellate choanozoan other than a choanoflagellate. Unlike sponges, which almost certainly evolved from a collared monad, there is no necessary reason why the immediate flagellate ancestor of fungi need have had a collar. Given these uncertainties it would be safer to say that fungi probably evolved from a uniflagellate choanozoan rather than specifically from a choanoflagellate. One structural feature other than flat cristae and posterior uniflagellation also pointed to a choanoflagellate ancestry for fungi (Cavalier-Smith, 1987a): this was the resemblance in flagellar root structure between certain chytridiomycete fungi and choanoflagellates, first noted by Barr (1981, 1992). However, until all the various uniflagellates of uncertain taxonomic position (Patterson and Zölfell, 1991) have been studied by electron microscopy, we cannot rule out the possibility that choanozoan zooflagellates other than choanoflagellates may actually exist and have similar flagellar roots. They would be equally good candidates for fungal ancestors.

21.5 NON-CHOANOZOAN NEOMONADA

The fact that *Apusomonas* appears as the sister group of the opisthokonts on molecular trees, or sometimes even tends to intrude slightly within the Choanozoa (Cavalier-Smith and Allsopp, 1996; Ragan *et al.*, 1996), means that *Apusomonas* is the closest known protozoan relative of Choanozoa. The question immediately arises whether these two taxa are similar enough in morphology to be included in the same phylum or not. I have come to the view that the differences between the tubular cristae and biflagellate state of *Apusomonas* and the flat or vesicular cristae and uniflagellate state of Choanozoa are insufficient to justify their continued separation in separate phyla. I have therefore reduced Choanozoa in rank to a subphylum and placed the two taxa together in the new phylum Neomonada (Cavalier-Smith, 1997a). The prefix Neo- was chosen to emphasize that these taxa branch much higher up the phylogenetic tree than the majority of zooflagellate phyla (i.e. than Metamonada, Trichozoa, Percolozoa, Euglenozoa) and also higher up than the various sarcodine phyla; the latter suggests that their non-amoeboid nature is derived compared with the amoeboflagellate character of many sarcodines. Although neomonads do not undergo amoeboid movement, one group of them, the thecomonads, does have pseudopods projecting from a ventral groove (*Apusomonas*) or ventrolateral grooves (*Cryothecomonas*) for feeding. Pseudopodia, in contrast to amoeboid locomotion, are therefore not entirely absent from Neomonada as they are from Euglenozoa – it seems inappropriate to call the unique slow rocking motion of *Corallochytrium* (Raghu-Kumar, 1987) amoeboid.

The fact that a protozoan phylum containing both Choanozoa and Apusozoa is obviously paraphyletic is no impediment to its adoption; any protozoan phylum containing the ancestors of animals and fungi must be paraphyletic, and it is impossible to create a comprehensive phylogenetic classification of organisms without having some paraphyletic taxa (Mayr and Ashlock, 1991; Cronquist, 1987; Cavalier-Smith, 1993b, 1998) despite the dogmatic wishes of Hennig (1966).

A new subphylum Apusozoa (Cavalier-Smith, 1997a) was created for tubulicristate and anisokont flagellate (formerly opalozoan) monads which cannot be included in Dinozoa (because they lack cortical alveoli) or in Heterokonta (because they lack tubular ciliary hairs or any morphological evidence that they were derived from ancestors that had them), and which cannot be included in the Sarcodina because they are not obviously amoeboid. Apusozoa comprise five distinct classes: Thecommadea, which include *Apusomonas* and also *Cryothecomonas*, which has long extrusomes and a theca which is probably not homologous to that of *Apusomonas*; Jakobea, which includes *Jakoba* and *Discocelis* with isodiammetric extrusomes; Kinetomonadea (histionids, heliomonads and *Commation*) that have kinetocyst extrusomes; Ebridea with an internal siliceous skeleton; and Anisomonadia with neither siliceous skeleton nor extrusomes (i.e. *Phagodinium, Diphyllea* and *Caecitellus*). The validity of these five classes and their placement in the Apusozoa need testing by sequencing, but there is no compelling morphological reason to think that any of the taxa included are phyletically closer to any other group so far placed on the ribosomal RNA tree than to Apusomonadida. However, some of these taxa might eventually turn out to be evolutionarily closer to Cercozoa or Heliozoa (some of which have kinetocysts similar to those of Kinetomonadea) than to *Apusomonas* and Choanozoa.

A second new subphylum Isomita has been created to group the two isokont former opalozoan classes Telonemea and Cyathobodonea together. Despite being isokont, these groups resemble Apusozoa more closely than they do any other group. However, as most of these resemblances (tubular cristae, biflagellate condition, absence of cortical alveoli and of tubular hairs) are characters that would have been present in the ancestors of neomonads they cannot be used to support their monophyly (here I use monophyly in the classical sense to include both paraphyly and holophyly); since however the immediate ancestors of neomonads were probably amoeboflagellates (Cavalier-Smith, 1997a), the loss of amoeboid movement by a stabilization of the cell cortex by a mechanism not involving cortical alveoli can be argued to be a shared derived character uniting Choanozoa, Apusozoa and Isomita. But as the molecular basis for this cortical stabilization is unknown, and as the assumption that Isomita will prove to branch above the sarcodine phyla (Cavalier-Smith, 1997a) remains to be tested, the grouping of Isomita, Choanozoa and Apusozoa together in a single phylum needs rigorous testing by molecular phylogeny. Though at present the evidence for the inclusion of all the apusozoan and isomitan orders is relatively weak, there would be even less justification for placing any of them in other established phyla or as separate phyla in their own right. It remains unclear whether the superficial similarities of the isomitan *Phalansterium* to choanoflagellates, that led to their former classification in the same order, are convergent or not.

The Spongomonadidae (*Spongomonas*, *Rhipidodendron*), which I formerly classified in the Isomita (Cavalier-Smith, 1997a) have now been transferred to the Cercozoa in the Subphylum Monadofilosa. I recently placed the order Hemimastigida within the Protalveolata (Cavalier-Smith, 1998), rather than as a fourth neomonad subphylum Hemimastigophora (Cavalier-Smith, 1997a).

The tubulicristate neomonads are sufficiently diverse ultrastructurally that they may eventually need to be placed in more than one phylum. The boundary that I have drawn between the Neomonada and Cercozoa (formerly Rhizopoda, Cavalier-Smith 1997a) is not very sharp; the neomonad *Apusomonas*, which forms pseudopods for feeding, and the sarcomonads, currrently placed in the Cercozoa because of molecular sequence evidence, are sufficiently similar morphologically that their placement in the same phylum (Cavalier-Smith, 1993a) was justifiable prior to the molecular sequence evidence that they are more closely related to other groups than to each other. It is therefore possible that some, or even many, taxa now included in the Neomonada may really belong with the sarcomonads in the Cercozoa or elsewhere. The classificatory problem will be particularly acute if, as preliminary evidence suggests (Cavalier-Smith and Chao, unpublished) some tubulicristate neomonads turn out to branch more deeply in the tree than either Apusozoa or Cercozoa.

21.6 ZOOFLAGELLATES NOT YET ASSIGNED TO PHYLA

There are several scores of zooflagellate genera not yet assigned to phyla because they have not been studied in the electron microscope, and lack characters visible in the light microscope that would enable them to be identified as euglenoids, dinoflagellates, diplozoans, parabasalans, heterokonts or choanoflagellates, i.e. the zooflagellate groups that are most readily recognizable in the light microscope. I suggest that, when these are studied by electron microscopy and/or molecular sequencing, most of them will be assignable to the Neomonada or Cercozoa. Unfortunately, with the exception of choanoflagellates, neomonads lack shared characters visible in the light microscope that would enable them to be assigned to the phylum. At present over half the 25 recognized neomonad orders are monogeneric; but a thorough study of these unassigned flagellates may greatly change this state of affairs by allowing many of them to be assigned to these orders, which are probably mostly richer in genera than they now appear. Of course some of these homeless flagellates may need to be placed in new orders or even classes, since there is no reason to think that electron microscopists are ceasing to discover very distinctive new body plans among the zooflagellates.

21.7 DID NEOMONAD ZOOFLAGELLATES HAVE A PHOTOSYNTHETIC ANCESTRY?

Ever since the acceptance of the symbiotic origin of chloroplasts (Cavalier-Smith, 1980, 1981a, b; Doolittle, 1980; Gray and Doolittle, 1982) it has been widely assumed that all zooflagellates are primitively non-photosynthetic. But this assumption has now been disproved for the non-photosynthetic pedinellids (Cavalier-Smith, Chao and Allsopp, 1995) and *Oikomonas* (Cavalier-Smith *et al.*, 1996a), which evolved by three separate losses of chloroplasts from heterokont algal ancestors. The recent demonstration of a relict plastid in Sporozoa (syn. Apicomplexa) (McFadden *et al.*, 1996), not only shows that more protozoa than is commonly thought may have had a photosynthetic ancestry, but also makes it likely that the common ancestor of Dinoflagellata and Sporozoa (grouped together with Protalveolata in the superphylum Miozoa: Cavalier-Smith, 1987c, 1993b) was photosynthetic. Because glaucophytes have cortical alveoli, I proposed long ago that chloroplasts first evolved in a common ancestor of dinoflagellates and the kingdom Plantae furnished with cortical alveoli (Cavalier-Smith, 1982; see Cavalier-Smith (1995b) for a more recent discussion of this). If this is correct, which is made increasingly likely by the discovery of the sporozoan plastid, then it would follow from the rRNA trees that the common ancestor of dinoflagellates, sporozoans and Ciliophora (i.e. the ancestral alveolate) was also photosynthetic, and that a common ancestor of Ciliophora had also lost plastids. Such loss was postulated by both Taylor (1978) and Cavalier-Smith (1978) before they wholeheartedly accepted the symbiogenetic origin of chloroplasts.

Although, unlike the case for heterokonts and alveolates, there is no specific reason to postulate a photosynthetic ancestry for neomonads, there is also no strong argument against such an origin. The basic problem, as discussed in detail by Philippe and Adoutte (1996), Cavalier-Smith *et al.* (1996c, d) and Cavalier-Smith and Chao (1997), is that the relative branching order of Plantae, Alveolata and the opisthokont/apusozoan clade is uncertain. These three groups form part of a major radiation, the megakaryote radiation (Cavalier-Smith, 1995b), which also included the origins of the Chromista, Amoebozoa and Cercozoa, and which was so sudden that the relative branching order of most of the major megakaryote taxa cannot be reliably resolved by ribosomal RNA phylogeny. Some ribosomal RNA trees group together the Plantae, Chromista and Alveolata as a sister group to the purely non-photosynthetic opisthokont/apusozoan clade, whereas others, as in Figure 21.4 place the alveolates below the plant/animal divergence. (With both topologies the exact position of the two neosarcodine phyla, Cercozoa and Amoebozoa is variable). If the former topology is correct then it would be reasonable to regard Neomonada as primitively non-photosynthetic. If the latter topology is correct, and if also the dinoflagellate/plant common ancestor was indeed photosynthetic, then it would follow that the Neomonada arose ultimately by plastid loss from a very early alga. Probably only the serendipitous discovery of a number of separate cladistically unambiguous molecular markers will enable one to determine the branching order of the megakaryote taxa and thus dispel this uncertainty.

21.8 ENVOI

At present so little is known of the phylogeny of the tubulicristate neomonads that we cannot predict whether they will turn out to be, like *Apusomonas,* simply a sister group to the opisthokonta, or a much broader group with affinities also with the Chromista and/or Plantae. If the latter is the case, then a better understanding of neomonads may have as much to contribute to an understanding of the origins of these kingdoms, as to those of fungi and animals; a better understanding of neomonad phylogeny and systematics will also be important for clarifying their relationship with the neosarcodine phyla Cercozoa and Amoebozoa. I hope that the establishment of the phylum Neomonada will also stimulate the study of these fascinating organisms for their own intrinsic interest. Neomonads are relatively small but remarkably diverse flagellates, which occur abundantly in soil, freshwater, and the oceans, and must be of great ecological importance as primary consumers, mainly of bacteria and small protozoa.

Acknowledgements

I thank NSERC for a research grant; the Canadian Institute for Advanced Research Evolutionary Biology Program for Fellowship support; M.A. Ragan, C. L. Goggin and their co-authors for providing the ichthyosporean sequences and a pre-print of their paper prior to publication; and G. Beakes and K. Vickerman for helpful advice on the manuscript.

21.9 APPENDIX. CLASSIFICATION OF THE KINGDOM PROTOZOA

The arrangement of the 13 phyla and 26 subphyla follows that of my recent revisions (Cavalier-Smith, 1995a, 1997a, 1998) of the system of Cavalier-Smith, 1993a. These publications (and Corliss, 1994) give a more detailed classification of Protozoa to the level of subclass and discuss its phylogenetic basis. Major changes in the circumscription of the Protozoa are the transfer of Microsporidia to the kingdom Fungi (Cavalier-Smith, 1998, and see Edlin's chapter in this volume) and of myxozoa to the kingdom Animalia (Cavalier-Smith, 1998 and see Anderson, this volume). The only classificatory innovations in the present paper are the creation of the neomonad class Ichthyosporea, the orders Dermocystida, and Ancyromonadida, and the family Adriamonadidae for the reasons given in the text; in addition, the Cristidiscoidea (see p. 381) have been transferred from Cercozoa to Neomonada (based on one rRNA sequence evidence for *Ministeria* (Cavalier-Smith and Chao, in prep.) The taxa marked with an asterisk are probably (certainly in the case of the kingdom Protozoa) paraphyletic. Some others must be also, but in many cases we do not know which are paraphyletic and which are holophyletic. (see note on p. 407)

Kingdom Protozoa* Goldfuss 1818 stat. nov. Owen 1858/9 em. Cavalier-Smith 1983, 1997

Subkingdom 1. Archezoa* Cavalier-Smith 1983 em. 1997

Phylum. Metamonada Grassé 1952 stat. nov. et em. Cavalier-Smith 1981

Subphylum 1. Eopharyngia Cavalier-Smith 1993 (e.g. *Giardia, Trepomonas, Chilomastix, Retortamonas*)

Subphylum 2. Axostylaria Grassé 1952 stat. nov. em. Cavalier-Smith 1993 (e.g. *Oxymonas, Pyrsonympha*)

[Before it was transferred to the kingdom Fungi, the phylum Microsporidia was included here]

Subkingdom 2. Eozoa* Cavalier-Smith 1997

Phylum. Trichozoa Cavalier-Smith 1997

Subphylum 1. Anaeromonada Cavalier-Smith 1997 (*Trimastix*)
Subphylum 2. Parabasala Honigberg 1973 stat. nov. Cavalier-Smith 1997 (e.g. *Trichomonas, Trichonympha*)

Infrakingdom. Discicnstata Cavalier-Smith 1998
Phylum 1. Percolozoa Cavalier-Smith 1991
Subphylum 1. Tetramitia Cavalier-Smith 1993 (e.g. *Percolomonas, Lyromonas;* heteroloboseans e.g., *Naegleria*)
Subphylum 2. Pseudociliata Cavalier-Smith 1993 (*Stephanopogon*)
Phylum 2. Euglenozoa Cavalier-Smith 1981
Subphylum 1. Plicostoma Cavalier-Smith 1998 [superclasses Diplonemia Cavalier-Smith 1993 (e.g. *Diplonema*) and Euglenoida Bütschli 1884 stat. nov. Cavalier-Smith 1998 (e.g. *Euglena, Petalomonas, Peranema*)]
Subphylum 2. Kinetoplasta Honigberg 1963 stat. nov. Cavalier-Smith 1993 (e.g. *Bodo, Trypanosoma, Leishmania*)

Subkingdom 3. Neozoa* Cavalier-Smith 1993 stat. nov. 1997
Infrakingdom 1. Sarcodina* Schmarda 1874 orthog. em. Lütken 1876 stat. nov. Cavalier-Smith 1997
Superphylum 1. Eosarcodina Cavalier-Smith 1997 emend. 1998
Phylum. Foraminifera (D'Orbigny 1926) Eichwald 1830 stat. nov. Margulis 1974 (e.g. *Allogromia, Ammonia, Globigerina*)

Superphylum 2. Neosarcodina* Cavalier-Smith 1993 em. 1997
Phylum 1. Cercozoa Cavalier-Smith 1998
Subphylum 1. Phytomyxa Cavalier-Smith 1997 (e.g. *Plasmodiophora*)
Subphylum 2. Reticulofilosa Cavalier-Smith 1997 (e.g. *Chlorarachnion, Cryptochlora, Leucodiotyon*
Subphylum 3. Monadofilosa Cavalier-Smith 1997 (e.g. *Cercomonas, Heteromita, Thaumatomonas, Euglypha, Paulinella, Spongomonas, Reticulomyxa, Biomyxa, Gymnophrys*)
Phylum 2. Amoebozoa Lühe 1913 stat. nov. Corliss 1984 em. Cavalier-Smith 1998
Subphylum 1. Lobosa Carpenter 1861 stat. nov. Cavalier-Smith 1997 (e.g. *Amoeba, Acanthamoeba, Arcella, Difflugia*)
Subphylum 2. Conosa Cavalier-Smith 1998 [infraphyla Archamoebae Cavalier-Smith 1983 stat. nov. 1998 (e.g. *Pelomyxa, Mastigamoeba, Phreatamoeba, Entamoeba*) and Mycetozoa De Bary 1859 stat.nov. (e.g. *Protostelium, Dictyostelium, Physarum*)]

Infrakingdom 2. Alveolata Cavalier-Smith 1991

Superphylum 1. Miozoa Cavalier-Smith 1987

Phylum 1. Dinozoa* Cavalier-Smith 1981 em. 1987, 1997

Subphylum 1. Protalveolata* Cavalier-Smith 1991 (e.g. *Colponema, Oxyrrhis, Hemimastix, Ellobiopsis, Colpodella, Perkinsus*)

Subphylum 2. Dinoflagellata Bütschli 1885 stat. nov. Cavalier-Smith 1991 (e.g. *Noctiluca, Crypthecodinium, Amphidinium*)

Phylum 2. Sporozoa Leuckart 1879 (syn. Apicomplexa Levine 1970) stat. nov. Cavalier-Smith 1981

Subphylum 1. Gregarinia Dufour 1828 stat. nov. Cavalier-Smith 1998 (e.g. *Monocystis*)

Subphylum 2. Coccidiomorpha Doflein 1901 stat. nov. Cavalier-Smith 1998 (e.g. *Eimeria, Sarcocystis, Paramyxa, Haplosporidium, Plasmodium, Babesia*)

Subphylum 3. Manubrispora Cavalier-Smith 1998 (e.g. *Metchnikovella*)

Superphylum 2. Heterokaryota Hickson 1903 stat. nov. Cavalier-Smith 1993

Phylum Ciliophora Doflein 1901 stat. nov. Copeland 1956 em. auct.

Subphylum 1. Tubulicorticata de Puytorac *et al.* 1992 (e.g. *Loxodes, Stylonychia, Colpoda*)

Subphylum 2. Epiplasmata de Puytorac *et al.* 1992 (e.g. *Tetrahymena, Paramecium, Vorticella*)

Subphylum 3. Filocorticata de Puytorac *et al.* 1992 (e.g. *Spathidium*)

Infrakingdom 3. Actinopoda Calkins 1902 stat. nov. Cavalier-Smith 1997

Phylum 1. Heliozoa Haeckel 1886 stat. nov. Margulis 1974 (e.g. *Actinophrys, Acanthocystis*)

Phylum 2. Radiozoa Cavalier-Smith 1987

Subphylum 1. Spasmaria Cavalier-Smith 1993 (e.g. *Acanthometra, Sticholonche*)

Subphylum 2. Radiolaria Müller 1858 emend. stat. nov. Cavalier-Smith 1993 (e.g. *Thalassicolla, Aulacantha*)

Infrakingdom 4. Neomonada* Cavalier-Smith 1997

Phylum Neomonada* Cavalier-Smith 1997

Subphylum 1. Apusozoa* Cavalier-Smith 1997 (e.g. *Apusomonas, Jakoba, Histiona, Ebria, Tetradimorpha, Caecitellus*)

Subphylum 2. Isomita Cavalier-Smith 1997 (e.g.

Cyathobodo, Phalansterium, Kathablepharis, Nephromyces)
Subphylum 3. Choanozoa* Cavalier-Smith 1981 em. 1983 stat. nov.
(e.g. *Monosiga, Diaphanoeca, Corallochytrium, Dermocystidium, Ichthyophonus, Ministeria*)

21.10 REFERENCES

Barr, D.J. (1981) The phylogenetic and taxonomic implications of the flagellar rootlet morphology among zoosporic fungi. *BioSystems*, **14**, 359–70.

Barr, D.J. (1992) Evolution and kingdoms of organisms from the perspective of a mycologist. *Mycologia*, **84**, 1–11.

Brugerolle, G. (1995) *Trimastix convexa* a free-living amitochondriate flagellate without close relationships with Percolozoa, retortamonad or trichomonad flagellates. *European Journal of Protistology*, **31**, 410.

Cavalier-Smith, T. (1975) The origin of nuclei and of eukaryote cells. *Nature, London*, **256**, 463-8.

Cavalier-Smith, T. (1978) The evolutionary origin and phylogeny of microtubules, mitotic spindles and eukaryote flagella. *BioSystems*, **10**, 93–113.

Cavalier-Smith, T. (1980) Cell compartmentation and the origin of eukaryote membranous organelles, in *Endocytobiology: Endosymbiosis and Cell Biology, a Synthesis of Recent Research* (eds W. Schwemmler and H.E.A. Schenk), de Gruyter, Berlin, pp. 893–916.

Cavalier-Smith, T. (1981a) Eukaryote kingdoms, seven or nine? *BioSystems*, **14**, 461–81.

Cavalier-Smith, T. (1981b) The origin and early evolution of the eukaryotic cell, in *Molecular and Cellular Aspects of Microbial Evolution*, Society for General Microbiology Symposium **32** (eds M.J. Carlile, J.F. Collins and B.E.B. Moseley), Cambridge University Press, Cambridge, pp. 33–84

Cavalier-Smith, T. (1982) The origins of plastids. *Biological Journal of the Linnean Society*, **17**, 289–306.

Cavalier-Smith, T. (1983) A 6-kingdom classification and a unified phylogeny. In *Endocytobiology II* (eds H.E.A. Schenk and W. Schwemmler), De Gruyter, Berlin, pp. 1027–34.

Cavalier-Smith, T. (1986) The kingdom Chromista, origin and systematics, in *Progress in Phycological Research*, Vol. 4 (eds F. E. Round and D. J. Chapman), Biopress, Bristol, pp. 309–47.

Cavalier-Smith, T. (1987a) The origin of Fungi and pseudofungi, in *Evolutionary Biology of the Fungi*. Symposium of the British Mycological Society 13 (eds A.D.M. Rayner, C. M. Brasier and D. M. Moore), Cambridge University Press, Cambridge, pp. 339–53.

Cavalier-Smith, T. (1987b) The origin of cells, a symbiosis between genes, catalysts and membranes. *Cold Spring Harbor Symposia on Quantitative Biology*, **52**, 805–24.

Cavalier-Smith, T. (1987c) The origin of eukaryote and archaebacterial cells. *Annals of the New York Academy of Sciences*, **503**, 17–54.

Cavalier-Smith, T. (1990) Symbiotic origin of peroxisomes. In *Endocytobiology* IV.

(eds P. Nardon, V. Gianinazzi-Pearson, A.M. Grenier, L. Margulis and D.C. Smith). Institut National de la Recherche Agronomique, Paris, pp. 515–521.

Cavalier-Smith, T. (1991) Cell diversification in heterotrophic flagellates, in *The Biology of Free-living Heterotrophic Flagellates* (eds D.J. Patterson and J. Larsen), Clarendon Press, Oxford, pp. 113–31.

Cavalier-Smith, T. (1993a) The protozoan phylum Opalozoa. *Journal of Eukaryotic Microbiology*, **40**, 609–15.

Cavalier-Smith, T. (1993b) Kingdom Protozoa and its 18 phyla. *Microbiological Reviews*, **57**, 953–94.

Cavalier-Smith, T. (1993c) Percolozoa and the symbiotic origin of the metakaryote cell, in *Endocytobiology V* (eds H. Ishikawa, M. Ishida, and S. Sato), Tübingen University Press, Tübingen, pp. 399–406.

Cavalier-Smith, T. (1993d) Evolution of the eukaryotic genome, in *The Eukaryotic Genome, Organization and Regulation* (eds P.M.A. Broda, S.G. Oliver and P.F.G. Sims), Cambridge University Press, Cambridge, pp. 333–85.

Cavalier-Smith, T. (1995a) Zooflagellate phylogeny and classification. *Cytology* (St. Petersburg), **37**, 1010–29.

Cavalier-Smith, T. (1995b) Membrane heredity, symbiogenesis, and the multiple origins of algae, in *Biodiversity and Evolution* (eds R. Arai, M. Kato and Y. Doi), The National Science Museum Foundation, Tokyo, pp. 75–114.

Cavalier-Smith, T. (1998) A revised six-kingdom system of life. *Biological Reviews* **73**, (in press).

Cavalier-Smith, T. (1997a) Amoeboflagellates and mitochondrial cristae in eukaryote evolution: megasystematics of the new protozoan subkingdoms Eozoa and Neozoa. *Archiv für Protistenkunde*, **147**, 237–58.

Cavalier-Smith, T. (1997b) Sagenista and Bigyra, two phyla of heterotrophic heterokont chromists. *Archiv für Protistenkünde* **148**, 253–267.

Cavalier-Smith, T. and Allsopp, M.T.E.P. (1996) *Corallochytrium*, an enigmatic non-flagellate protozoan related to choanoflagellates. *European Journal of Protistology*, **32**, 306–10.

Cavalier-Smith, T. and Chao, E.E. (1995) The opalozoan *Apusomonas* is related to the common ancestor of animals, fungi, and choanoflagellates. *Proceedings of the Royal Society of London Series B*, **261**, 1–6.

Cavalier-Smith, T. and Chao, E.E. (1996) Molecular phylogeny of the free-living archezoan *Trepomonas agilis* and the nature of the first eukaryote. *Journal of Molecular Evolution*, **43**, 551–62.

Cavalier-Smith, T. and Chao, E.E. (1997) Sarcomonad ribosomal RNA sequences, rhizopod phylogeny, and the origin of euglyphid amoebae. *Archiv für Protistenkunde*, **147**, 227–36.

Cavalier-Smith, T., Allsopp, M.T.E.P. and Chao, E.E. (1994) Chimeric conundra: are nucleomorphs and chromists monophyletic or polyphyletic? *Proceedings of the National Academy of Sciences USA*, **91**, 11368–72.

Cavalier-Smith, T., Chao, E.E. and Allsopp, M.T.E.P. (1995) Ribosomal RNA evidence for chloroplast loss within Heterokonta, pedinellid relationships and a revised classification of ochristan algae. *Archiv für Protistenkunde*, **145**, 209–20.

Cavalier-Smith, T., Allsopp, M.T.E.P., Chao, E.E. *et al.* (1996a) Sponge phylogeny, animal monophyly and the origin of the nervous system: 18S rRNA evidence. *Canadian Journal of Zoology*, **74**, 2031–2045.

Cavalier-Smith, T., Chao, E. E., Thompson, C. and Hourihane, S. (1996b) *Oikomonas*, a distinctive zooflagellate related to chrysomonads. *Archiv für Protistenkunde*, **146**, 273–9.

Cavalier-Smith, T., Couch, J.A., Thorsteinsen, K.E. *et al.* (1996c) Cryptomonad nuclear and nucleomorph 18S rRNA phylogeny. *European Journal of Phycology*, **31**, 315–28.

Cavalier-Smith, T., Allsopp, M.T.E.P., Häuber, M.M. *et al.* (1996d) Chromobiote phylogeny: the enigmatic alga *Reticulosphaera japonensis* Grell is an aberrant haptophyte, not a heterokont. *European Journal of Phycology*, **31**, 255–63.

Conway Morris, S., George, J.D., Gibson, R. and Platt, H.M. (eds) (1985) *The Origins and Relationships of Lower Invertebrates*, Clarendon Press, Oxford.

Corliss, J.O. (1994) An interim utilitarian ('user friendly') hierarchical classification and characterization of the protists. *Acta Protozoologica*, **33**, 1–51.

Cronquist, A. (1987) A botanical critique of cladism. *The Botanical Review*, **53**, 1–52.

De Ceccatty, M.P. (1986) Cytoskeletal organization and tissue patterns of epithelia in the sponge *Ephydatia mulleri*. *Journal of Morphology*, **189**, 45–65.

Dujardin, F. (1841) *Histoire Naturelle des Zoophytes Infusoires*, Roret, Paris.

Edwards, P. (1976) A classification of plants into higher taxa based on cytological and biochemical criteria. *Taxon*, **25**, 529–42.

Felsenstein, J. (1992) *Phylip Manual (Version 3.5)* University of Washington, Seattle.

Grassé, P-P. (ed.) (1952) *Traité de Zoologie*, Vol. 1. Fasc. 1, Masson, Paris.

Gray, M.W. and Doolittle, W.F. (1982) Has the endosymbiont hypothesis been proven? *Microbiological Reviews*, **46**, 1–42.

Hartog, M. (1906) Protozoa, in *The Cambridge Natural History* (eds S.F. Harmer and A.E. Shipley), Macmillan, London, pp.1–162.

Hennig, W. (1966) *Phylogenetic Systematics*, University of Illinois Press, Urbana.

Hibberd, D.J. (1979) The structure and phylogenetic significance of the flagellar transition region in the chlorophyll *c*-containing algae. *BioSystems*, **11**, 243–61.

Jefferies, R.S. (1979) The origin of chordates: a methodological essay, in *The Origin of Major Invertebrate Groups* (ed. M.R. House), Academic Press, London, pp. 443–77.

Karpov, S.A. and Mylnikov, A.P. (1989) Biology and ultrastructure of colourless flagellates Apusomonadida Ord. N. *Zoologicheskii Zhurnal*, **68**, 5–17 (in Russian).

Kerk, D., Gee, A., Standish, M. *et al.* (1995) The rosette agent of chinook salmon. (*Oncorhynchus tshawytscha*) is closely related to choanoflagellates, as determined by the phylogenetic analysis of its small ribosomal subunit RNA. *Marine Biology*, **122**, 187–92.

Leadbeater, B.S.C. (1983) Life-history and ultrastructure of a new marine species of *Proterospongia* (Choanoflagellida). *Journal of the Marine Biological Association*, **63,** 135–60.

Leedale, G.F. (1974) How many are the kingdoms of organisms? *Taxon*, **32**, 261–70.

Levine, N.D., Corliss, J.O., Cox, F.E.G. *et al.* (1980) A newly revised classification of the Protozoa. *Journal of Protozoology*, **27,** 37–58.

Manton, I. (1965) Some phyletic implications of flagellar structure in plants. *Advances in Botanical Research*, **2**, 1–34.

Margulis, L., Corliss, J.O., Melkonian, M. and Chapman, D.J. (eds.) (1990) *Handbook of Protoctista*. Jones and Bartlett, Boston.

Mayr, E. and Ashlock, P.D. (1991) *Principles of Systematic Zoology*, 2nd Edn, McGraw Hill, New York.

McFadden, G.I., Reith, M., Munholland, J. and Lang-Umasch, N. (1996) Plastid in human parasites. *Nature*, **381**, 482.

Olsen, G.L., Matsuda, H., Hagstrom, R. and Overbeek, R. (1994) FastDNAml: a tool for construction of phylogenetic trees of DNA sequences using maximum likelihood. *Computer Applications in the Biosciences*, **10**, 41–8.

Olson, R.E., Dungan, C.F. and Holt, R.A. (1991) Water-borne transmission of *Dermocystidium salmonis* in the laboratory. *Diseases of Aquatic Organisms*, **12**, 41–8.

Owen, R. (1858) Palaeontology, in *Encyclopedia Britannica*, 8th edn (ed. T.S. Traill), Black, Edinburgh, Vol. **17**, 91-176.

Patterson, D.J. and Zölffel, M. (1991) Heterotrophic flagellates of uncertain taxonomic position, in *The Biology of Free-living Heterotrophic Flagellates* (eds D. J. Patterson and J. Larsen), Clarendon Press, Oxford, pp. 427–75.

Philippe, H. and Adoutte, A. (1996) How reliable is our current view of eukaryotic phylogeny? in *Protistological Actualities* (eds G. Brugerolle and J-P. Mignot), G.P.L.F, Clermont-Ferrand, pp. 17–33.

Ragan, M.A., Goggin, C.L., Cawthorn, R.J. *et al.* (1996) A novel clade of protistan parasites near the animal fungal divergence. *Proceedings of the National Academy of Sciences USA*, **93**, 11907–12.

Raghu-Kumar, S. (1987) Occurrence of the thraustochytrid, *Corallochytrium limacisporum* gen. et sp. nov. in the coral reef lagoons of the Lakshadweep Islands in the Arabian Sea. *Botanica Marina*, **30**, 83–9.

Rand, T.G. (1994). An unusual form of *Ichthyophonus hoferi* (Ichthyophonales, Ichthyophonaceae) from yellowtail flounder *Limanda ferruginea* from the Nova Scotia shelf. *Diseases of Aquatic Organisms*, **18**, 21–8.

Silberman, J.D., Sogin, M.L., Leipe, D.D. and Clark, C.G. (1996) Human parasite finds taxonomic home. *Nature*, **380**, 398.

Singer, C. (1959) *A History of Biology*, 3rd and revised edn, Abelard-Schuman, London and New York, pp. 202–3.

Sogin, M.L., Silberman, J.D., Hinkle, G. and Morrison, H.G. (1996) Problems with molecular diversity in the Eukarya, in *Evolution of Microbial Life* (eds D.M Roberts, P. Sharp, G. Alderson and M.A. Collins), Cambridge University Press, Cambridge, pp. 167–84.

Taylor, F.J.R. (1976) Flagellate phylogeny: a study in conflicts. *Journal of Protozoology*, **23**, 28–40.

Taylor, F.J.R. (1978) Problems in the development of an explicit hypothetical phylogeny of the lower eukaryotes. *BioSystems*, **10**, 67–89.

Verhagen, F.J.M., Zölffel, M., Brugerolle, G. and Patterson, D.J. (1994) *Adriamonas peritocrescens* gen. nov., sp. nov., a new free-living soil flagellate (Protista, Pseudodendromonadidae incertae sedis). *European Journal of Protistology*, **30**, 295–308.

Von Siebold, C.T.E. (1845) Bericht über die leistungen in der Naturgeschichte der Würmer, Zoophyten und Protozoen während des Jahres 1843 und 1844. *Archio für Naturgeschicte* **11**, 256–296.

Von Siebold, C.T. (1848) Lehrbuch der Vergleichenden Anatomie der Wirbellosen Thiere. *In* Von Siebold, C.T. and Stannius, H. Lehrbuch der Vergleichenden Anatomie, Vol. 1, Berlin, pp. 1–679.

Wainright, P.O., Hinkle, G., Sogin, M.L. and Stickel, S.K. (1993) Monophyletic origins of the Metazoa, an evolutionary link with Fungi. *Science*, **260**, 340–2.

Whittaker, R.H. (1969) New concepts of kingdoms of organisms. *Science*, **163**, 150–60.

NOTES ADDED IN PROOF:

1. [p. 383] The phylogenetic position of Discicrisitata is very uncertain, because protein trees place them higher up than do rRNA trees. Their classification in Neozoa (Cavalier-Smith, 1998) and placement of Trichozoa in Archezoa may be preferable to their inclusion in a possibly polyphyletic Eozoa.
2. [p. 400] A somewhat simpler system, which I now prefer to that adopted when the present chapter was written, merges infrakingdoms Sarcodina and Neomonada into a single infrakingdom Sarcomastigota, includes Discicristata in the Neozoa and Trichozoa in the Archezoa, the taxa Eozoa, Eosarcodina and Neosarcodina are abandoned (Cavalier-Smith, 1988).

22

Classification of protozoa and protists: the current status

John O. Corliss

P.O. Box 2729, Bala Cynwyd, Pennsylvania 19004, USA. E-mail: Jochezmoi@aol.com

ABSTRACT

In tracing the history of classifications of protozoa/protists, one is struck by two outstanding facts. First, as is true for past systems proposed for all living things, humans seem to possess an innate, compelling desire to recognize a natural orderliness among the objects of their study. The inevitable result of this is that organisms typically end up in a hierarchical taxonomic scheme, one thought to reflect their evolutionary/phylogenetic relationships as well as establishing the systematic juxtaposition of groups one to another. Comparative morphology has faithfully served as the principal source of characters used to justify workers' conclusions. The second fact is that, with continuous advances and improvements in techniques, which allow protozoologists to amass more and more data relevant to their comparative taxonomic studies, resulting classifications over the years have shown a tremendous inflation in the numbers of ranks/categories deemed necessary to contain all the new (and old) species involved. Increased appreciation of the undeniably great diversity among protists has supported this awesome expansion and accompanying rearrangements of groups recognized as unique among these widely distributed forms. Against this background, the present paper concludes that the current status of the megasystematics of the protozoa and their nearby relatives is unsettled, in a state of flux. But it is considered worthwhile to discuss in some detail the strong points and the drawbacks of major schemes of classification proposed in the literature of the past few years. Singled out for such treatment are the innovative approaches being suggested by a trio of workers in modern protozoan phylogenetics and systematics, viz Lynn Margulis, Tom Cavalier-Smith and David

Evolutionary Relationships Among Protozoa. Edited by G.H. Coombs, K. Vickerman, M.A. Sleigh and A. Warren. Published in 1998 by Chapman & Hall, London. ISBN 0 412 79800 X

J. Patterson. Finally, a number of challenging points to be kept in mind are listed for consideration by any aspirant hoping to produce the ideal systematic arrangement of protozoa *sensu lato* by the turn of the 21st century or shortly thereafter.

22.1 INTRODUCTION

Whether the arranging of groups of organisms into some kind of classificatory system is considered an art or a science, the process has been carried out by humans for many centuries. Even primitive peoples often had a taxonomy of sorts, recognizing and giving vernacular names, for example, to different species of birds and mammals common to their own living ranges. All major taxonomic schemes in wide usage today, for both contemporary and extinct organisms, stem, of course, from the binomial system proposed by Carolus Linnaeus in his celebrated tomes of the mid-18th century. With subsequent advances in our detailed knowledge of the world's plants and animals, large and small, expansion and refinement have naturally resulted. With comparative anatomy serving as the principal backbone of taxonomy, biologists have long maintained that their schemes of classification are not arbitrary but, rather, reflect the natural or 'true' evolutionary relationships that exist among the included groups (Magr, 1982; Ragan, 1997).

Early workers with protists – essentially the algae and protozoa of the classical literature – felt obliged to give taxonomic weight also to physiological characteristics and to features of the life cycle, because of the dearth of characters of a strictly morphological nature. Nonetheless, for the conventional protozoan groups (taxa considered to be composed of primitive or 'first animals': proto-zoa), the traditional procedures in taxonomy long ago established by vertebrate zoologists have been used automatically until quite recent times. That is, protozoologists have primarily utilized the comparative morphological approach in determining a presumably appropriate (i.e. as natural as seemed possible) hierarchical classification of their 'lower' eukaryotic organisms.

Today, protistologists (students, from various fields, of the mostly unicellular, microscopic protozoa, algae, and 'lower' fungi) can – and do – take advantage of a vast array of characters, following especially the advent of more sophisticated microscopical and cytological techniques and of greatly refined physiological/biochemical/molecular approaches. In the present Symposium, we have heard ample indication of such methods of investigation and of the sometimes spectacular data resulting from their application. Electron microscopy has probably provided the greatest single boon in all these advances, since we must have robust identifiable characters in hand both to construct phylogenetic trees and to erect our schemes of classification. Hopefully, we have not forgotten the continued importance of whole organism biology in constructing the 'new systematics' of the highly diverse groups of 'lower' eukaryotes, including the geological perspective (Knoll, 1992).[1]

Before considering the tremendous expansion of – and multiple changes in – the classifications of protozoa, algae, and protists, as found in recently proposed overall taxonomic schemes, let us recall very briefly some of the now classical arrangements of such organisms in the published literature of the past 120 years. We may well wistfully note the simplicity of some of these much earlier taxonomic schemes while admiring – in view of the woeful lack of relevant data available in those times – the often insightful ideas concerning possible phylogenetic interrelationships of major groups.

22.2 CLASSICAL ARRANGEMENTS OF THE PAST 120 YEARS

Kent (1880) considered amoeboid forms as the progenitors of all other groups of protozoa (Figure 22.1). This view has not ever been a popular one, yet there have never been grounds for total rejection of it. He was the first to postulate a close evolutionary relationship between choanoflagellates and sponges. He understandably (for the times) separated the suctorians widely from (other) ciliates, misleading subsequent authors to do the same for the next 75 years. But his wise separation of opalinids from ciliates was ignored over the same length of time. Kent (1880–1882) omitted the very recently erected Sporozoa (see Leuckart, 1879) from his taxonomic classification, although he included some gregarines in his overall scheme.

In the very same year, independently, Otto Bütschli (1880) – that great architect of systematic protozoology (Corliss, 1978; Dobell, 1951) – published his equally unique evolutionary arrangement of the Protista known at that time (Figure 22.2). He gave more attention to (some of the) algal lines and considered primitive flagellates to be at the base of the eukaryotic groups of his tree. Bacteria branched off from the same stem of early life forms that led to his flagellates. In Bütschli's (1880–1889) classification scheme (Table 22.1), which endured for many decades with only slight expansion and refinement, the microsporidians were appended to his Sporozoa and the trichonymphid flagellates to his Ciliata. Note that four classes and nine orders were deemed sufficient to represent the protozoan diversity recognized in the latter part of the 19th century.

Not until 80 years later (Corliss, 1960) was another 'phylogenetic tree' proposed for the major higher taxa of the protozoa *sensu lato*, a drawing

[1] For a concise history of the emergence of the field of modern protistology see Corliss (1986) and the recent 'up-date' treatment by Cavalier-Smith (1995a).

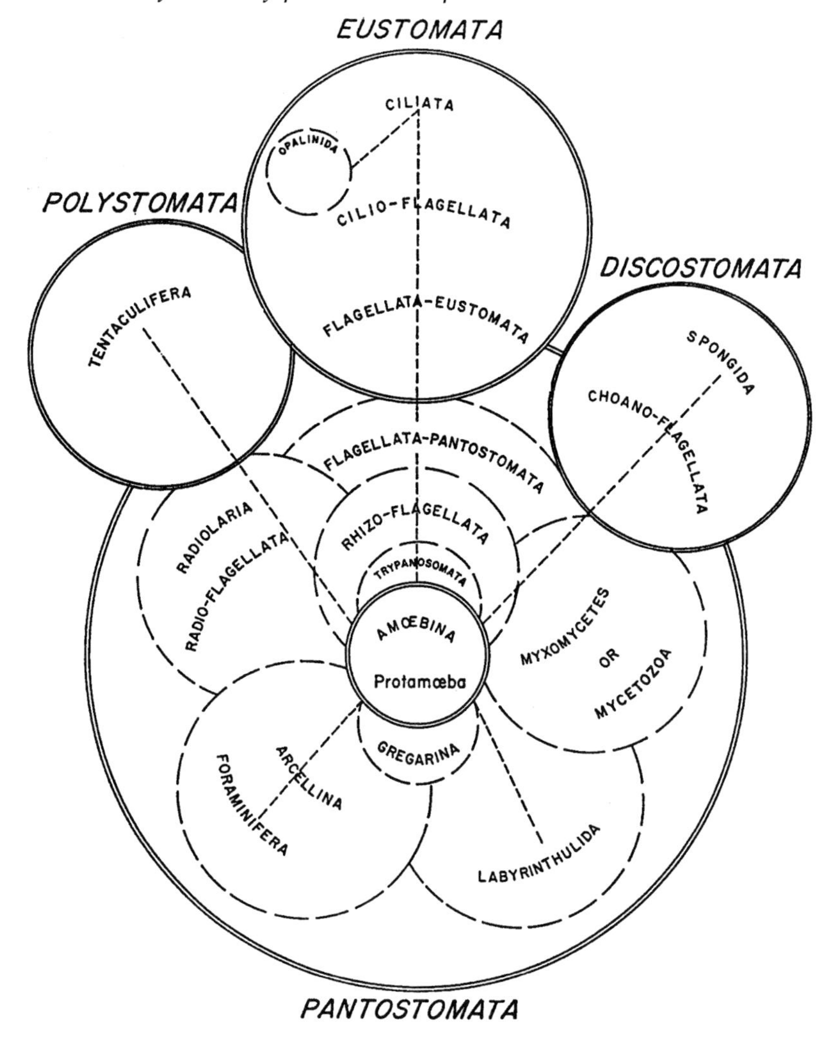

Figure 22.1 A redrawing of Kent's (1880) phylogenetic tree for the Protozoa (from Corliss, 1960).

of mine based largely on the prevailing conventional wisdom of that time (Figure 22.3). The taxonomic scheme endorsed was essentially that of the Society of Protozoologists' sponsored Honigberg Report (Honigberg *et al.*, 1964), to which I was an active contributor, which appeared in print five years later (Table 22.2). As in the cases of the much earlier trees and classifications, both the tree and the hierarchical scheme may be considered classical in the sense that they have now become com-

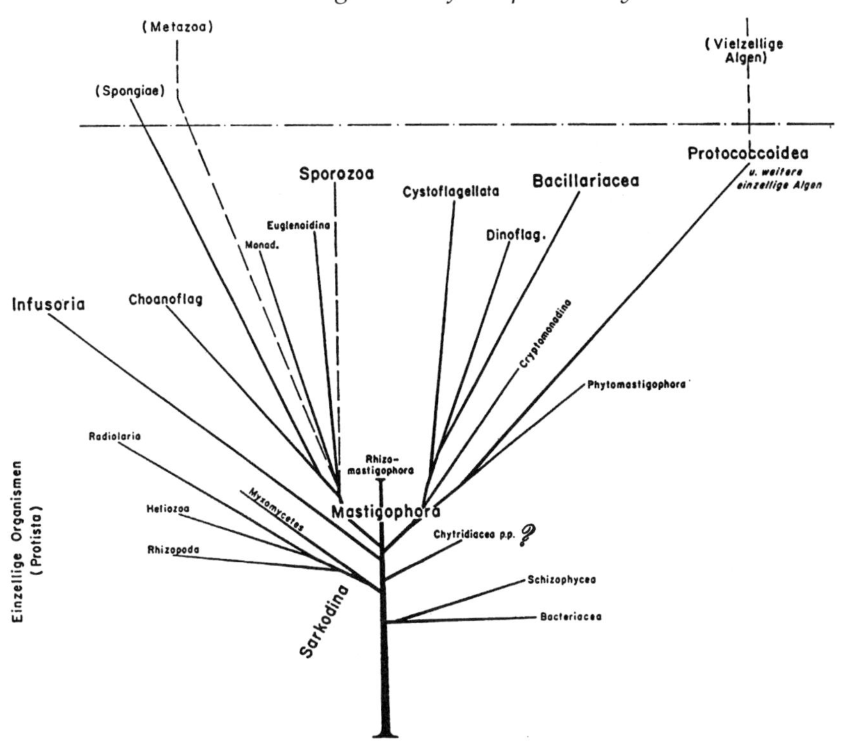

Figure 22.2 A redrawing of Bütschli's (1880) phylogenetic tree for the Protozoa (plus bacteria) and their relationship to other eukaryotes (from Corliss, 1960).

pletely outmoded, while nevertheless reflecting the state (including the limitations) of our knowledge well into the second half of the 20th century. But some 11 classes and 66 orders appear to have been needed by that time, a significant increase over the numbers of the 1880's (cf. Tables 22.1 and 22.2).

During the past 25 years, the burgeoning literature on the protists has included thousands of papers containing contributions to the systematics and/or the evolution of the implicated groups, be they strains and species, genera and families, or orders, classes, and phyla. All relevant works can no longer be cited in single, even circumscribed, reviews for lack of space. New overall classifications have ranged from those suggested by individual protistologists and by textbook writers/ editors (de Puytorac *et al.*, 1987; Sleigh, 1989; Hausmann and Hülsmann, 1996) to schemes representing the combined efforts of a special group or committee, often international in composition, of

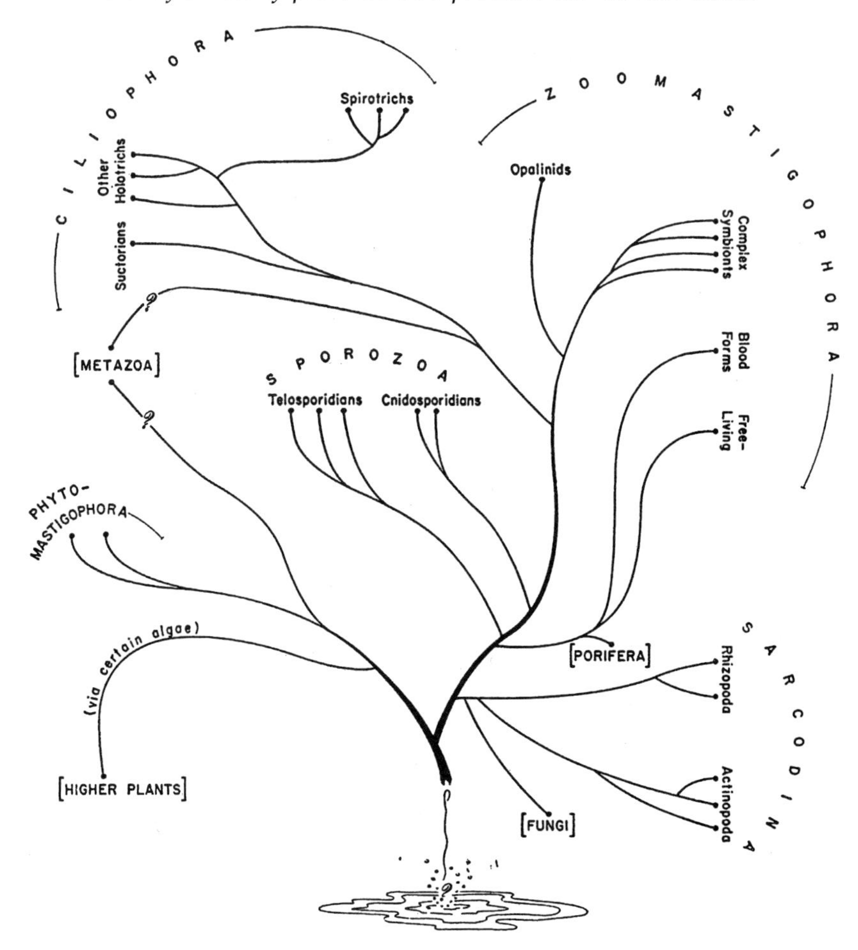

Figure 22.3 The phylogenetic tree of Corliss (1960) for the Protozoa, with an indication of other major eukaryotic lineages (from Corliss, 1960), based on the limited knowledge of those times (effectively prior to the Age of Ultrastructure and certainly to the advent of rRNA gene sequencing).

several protozoologists and/or phycologists and/or mycologists. The kinds of characters used in proposing these new taxonomies have become much more exacting, moving from gross morphological or generalized physiological/biochemical and ecological attributes to precise configurations of cytoarchitectural details (now possible to obtain by application of more sophisticated electron microscopical techniques) and to significant comparisons of refined biochemical or molecular biological properties. Numerous major examples of such

Table 22.1 Suprafamilial taxonomic classification of the Protozoa according to Bütschli (from Bütschli, 1880–1889)

Class I. **SARKODINA**
- Subclass 1. RHIZOPODA
  - Order (1) **Rhizopoda**
    - Suborder (i) Amoebaea
    - (ii) Testacea
    - (iii) Perforata
- Subclass 2. HELIOZOA
- 3. RADIOLARIA

Class II. **SPOROZOA**
- Subclass 1. GREGARINIDA
  - Order (1) **Monocystidea**
  - (2) **Polycystidea**
- Subclass 2. COCCIDIA
- 3. MYXOSPORIDIA
- 4. SARCOSPORIDIA

Class III. **MASTIGOPHORA**
- Order (1) **Flagellata**
  - Suborder (i) Monadina
  - (ii) Euglenoidina
  - (iii) Heteromastigoda
  - (iv) Isomastigoda
- Order (2) **Choanoflagellata**
- (3) **Dinoflagellata**
  - Suborder (i) Adinida
  - (ii) Dinifera
- Order (4) **Cystoflagellata**

Class IV. **INFUSORIA**
- Subclass 1. CILIATA
  - Order (1) **Gymnostomata**
  - (2) **Trichostomata**
    - Suborder (i) Aspirotricha
    - (ii) Spirotricha
      - Section (a) Heterotricha
      - (b) Oligotricha
      - (c) Hypotricha
      - (d) Peritricha
- Subclass 2. SUCTORIA

advanced approaches have been exposed by other speakers participating in the present Symposium.

In 1980, the Levine Report (Table 22.3) appeared (Levine *et al.*, 1980), another Society of Protozoologists' special committee effort whose membership included three of the same protozoologists (Corliss, Honigberg, Levine) who had been most involved in preparing the Honigberg Report of 1964 (see above). The popular *Illustrated Guide to the Protozoa* (Lee, Hutner and Bovee, 1985) in large measure endorsed the 'by-consensus' taxonomic proposals of that second Society-sponsored committee. Although hailed as a radical revision and embraced by numerous workers, the Levine Report system was, in fact, also immediately criticized for not going far enough in its changes over preceding conventional classifications, although seven separate phyla were recognized and the numbers of classes and orders were expanded to 27 and 98 (117 in Lee, Hutner and Bovee), respectively. But botanically long-accepted divisions or classes of algae were still reduced to ordinal level (all jammed into the identical 10 orders of the same single class Phytomastigophorea of the earlier Honigberg Report) or neglected altogether. The 'animal' nature of protozoa was still tacitly followed. In short, the

Table 22.2 Taxonomic classification of the Protozoa in the Honigberg Report down through subclasses, with indication of numbers of included orders (taken from Honigberg *et al.*, 1964). The total number of suprafamilial taxa would be 140, counting suborders as well; contrast this with Bütschli's total of 37 (see Table 22.1)

- Subphylum I. **SARCOMASTIGOPHORA**
  - Superclass *I.* Mastigophora
    - Class 1. PHYTOMASTIGOPHOREA [with 10 orders]
    - Class 2. ZOOMASTIGOPHOREA [9 orders]
  - Superclass *II.* Opalinata [1 order]
  - Superclass *III.* Sarcodina
    - Class 1. RHIZOPODEA
      - Subclass *1.* Lobosia [2 orders]
      - Subclass *2.* Filosia [2 orders]
      - Subclass *3.* Granuloreticulosia [3 orders]
      - Subclass *4.* Mycetozoia [3 orders]
      - Subclass *5.* Labyrinthulia [1 order]
    - Class 2. PIROPLASMEA [1 order]
    - Class 3. ACTINOPODEA
      - Subclass *1.* Radiolaria [2 orders]
      - Subclass *2.* Acantharia [2 orders]
      - Subclass *3.* Heliozoia [2 orders]
      - Subclass *4.* Proteomyxidia [1 order]
- Subphylum II. **SPOROZOA**
  - Class 1. TELEOSPOREA
    - Subclass *1.* Gregarinia [3 orders]
    - Subclass *2.* Coccidia [2 orders]
  - Class 2. TOXOPLASMEA [1 order]
  - Class 3. HAPLOSPOREA [1 order]
- Subphylum III. **CNIDOSPORA**
  - Class 1. MYXOSPORIDEA [3 orders]
  - Class 2. MICROSPORIDEA [1 order]
- Subphylum IV. **CILIOPHORA**
  - Class 1. CILIATEA
    - Subclass *1.* Holotrichia [7 orders]
    - Subclass *2.* Peritrichia [1 order]
    - Subclass *3.* Suctoria [1 order]
    - Subclass *4.* Spirotrichia [6 orders]

exciting protist perspective Corliss, 1986 that was enthusiastically being embraced and pursued in an increasing number of major laboratories around the world was conspicuous by its absence. Considering assemblages of conventional protozoa, algae, and 'lower' fungi as being inextricably commingled taxonomically and phylogenetically was already becoming the wave of the future as early as the late 1970s (see reviews in Corliss, 1986, 1987), a fact reflected in neither the Levine Report nor the 1985 edition of the *Illustrated Guide*.

Table 22.3 Taxonomic classification of the Protozoa in the Levine Report down only through the class rank, thus not including 16 subclasses, 5 superorders, 98 orders, and 71 suborders (from Levine *et al.*, 1980). The total number of suprafamilial taxa, 229, may be contrasted with the number of 140 found in the Honigberg Report of 16 years earlier (see Table 22.2)

Phylum I. **SARCOMASTIGOPHORA**
Subphylum *I.* **Mastigophora**
Class 1. PHYTOMASTIGOPHOREA
2. ZOOMASTIGOPHOREA
Subphylum *II. Opalinata*
Class 1. OPALINATEA
Subphylum *III.* **Sarcodina**
Superclass (1) Rhizopoda
Class 1. LOBOSEA
2. ACARPOMYXEA
3. ACRASEA
4. EUMYCETOZOEA
5. PLASMODIOPHOREA
6. FILOSEA
7. GRANULORETICULOSEA
8. XENOPHYOPHOREA
Superclass (2) Actinopoda
Class 1. ACANTHAREA
2. POLYCYSTINEA
3. PHAEODAREA
4. HELIOZOEA

Phylum II. **LABYRINTHOMORPHA**
Class 1. LABYRINTHULEA

Phylum III. **APICOMPLEXA**
Class 1. PERKINSEA
2. SPOROZOEA

Phylum IV. **MICROSPORA**
Class. 1. RUDIMICROSPOREA
2. MICROSPOREA

Phylum V. **ASCETOSPORA**
Class 1. STELLATOSPOREA
2. PARAMYXEA

Phylum VI. **MYXOZOA**
Class 1. MYXOSPOREA
2. ACTINOSPOREA

Phylum VII. **CILIOPHORA**
Classes:
1. KINETOFRAGMINOPHOREA
2. OLIGOHYMENOPHOREA
3. POLYHYMENOPHOREA

Although treating overlapping taxa of protozoa and algae as components of a protistan system tended to bring together workers from formerly different areas and fields of research, the old barriers of tradition, territoriality, and authoritarianism delayed progress – true, still, even today – in appreciating the importance of and the differences in taxonomic schemes being proposed separately by phycologists, protozoologists, or mycologists. Even in such areas as cell and evolutionary biology – quite distinct from taxonomy – there was/is not complete elimination of a considerable degree of bias, preconceived notions, weighted interpretations of hard data, preferences in selection of the characters studied, influence of specific leaders of diverse schools of thought, and the like (Corliss, 1986, 1994; Ragan, 1997, 1998; Silva, 1984, 1993). Subjectivity cannot be completely eliminated!

While many indispensable studies (too numerous to cite here individually) have been carried out in recent years the results of which impinge significantly on revisions in our concepts of both protistan phylogeny and protistan megasystematics, three major highly influential overall works (better, series of works) may serve as leading examples of current detailed taxonomic treatments involving all groups of the highly diversified protists. Following my critical comments on this trio of unique and quite dissimilar schemes, I shall conclude that the present status of the high-level systematics of the protists *sensu lato* is still in a state of flux. But I shall then suggest what it is, in my opinion, that we ought to keep in mind as bases for ultimate production of a truly robust, evolutionarily compatible, single classification embracing all groups of these fascinating 'lower' eukaryotes.

A couple of words of explanation may be in order at this point, if I may digress for a moment. Since the goal of this particular paper is to elucidate the present status of protozoan/protistan classification (most other speakers in the Symposium are specifically covering evolutionary relationships; and see the excellent very recently published review by Adoutte *et al.* (1996) and the important notes of caution concerning the pitfalls of phylogenetic reconstruction by Philippe and Adoutte (1995)), our attention will not be focused here on the advances in our understanding of protistan phylogenetics, linked though the two approaches obviously are. Protistan evolutionary trees and cladograms are of great importance, but I am considering them below only when their creators have used them directly in production of a revised taxonomy of the multiple high-level groups comprising the huge assemblage of 'lower' eukaryotes now known to populate the earth. Incidentally, many studies by phycologists may appear to have been slighted in this review; but, interestingly enough, to date seldom have students of the algae proposed overall schemes embracing conventional protozoan groups as well as their own organisms of research investigation, whereas protozoological protistologists have been bolder in freely incorporating algal (and fungal) taxa in their overviews.

Finally, in this digression, a brief but necessary comment must be made concerning the taxonomic relationship of protozoa and protists, or even of Protozoa with a capital 'P' (both in the conventional older sense and in the sense of Cavalier-Smith (1993b) and Corliss (1994)) and Protoctista or Protista (in the sense of Margulis (1996; Margulis *et al.*, 1990; Margulis and Schwartz, 1982, 1988) and of Corliss (1984) and numerous others). Regarding this potentially confusing topic, I am, as a generalist, in sympathy with the need for biological terms and names that can be widely applied without undue stress in comprehending them. The old phylum Protozoa was deficient mostly in **including** and downgrading those groups of **algal** protists many of which are now known to be quite unrelated to major **protozoan** taxa and, indeed, often to one another. The new kingdom Protozoa (its upgrading in rank necessitated by the great evolutionary diversity of its many included phyletic groups) has been purged of various major algal lines and of primitive forms, the former now assigned more appropriately to other kingdoms and the latter to a separate kingdom Archezoa. And now we must transfer the phylum Myxozoa to the animal kingdom! Thus the present (kingdom) Protozoa is both more refined and more precisely definable, while still containing a large number of high-level protistan taxa. The (former) single kingdom for **all** protists, whether it be called the Protoctista (Proto-ctista) or, more simply, the Protista, is now considered by many workers – on the basis of our current knowledge – to be unwieldy, undefinable, and too all-embracing of unique groups widely separable (both taxonomically and evolutionarily) at very high (including kingdom) levels. Over the years, I have changed my own mind on this controversial subject, turning from strong support of a kingdom Protista with elimination/dropping of the phylum Protozoa (Corliss, 1981, 1984, 1991) to an equally strong endorsement of a refined kingdom Protozoa as one of several kingdoms needed taxonomically to contain all of the protists (Corliss, 1994, 1995b). (End of digression.)

22.3 INNOVATIVE MAJOR SCHEMES OF CLASSIFICATION

Let us first consider the high-level taxonomic contributions of Lynn Margulis of Boston, now Amherst, a prodigious worker interested in microbial genetics, ecology, and evolution and a founding parent of the field of modern protistology (Corliss, 1986) and specifically of the International Society for Evolutionary Protistology. She has long been actively promoting the idea of a five-kingdom system to represent all life on Earth (Margulis, 1974a,b, 1996; Margulis *et al.*, 1990; Margulis and Schwartz, 1982, 1988, 1998). Margulis's taxonomic inspirations were probably mostly derived originally from her reading of Copeland's (1956) novel book and her association (see Whittaker and Margulis, 1978) with the late Robert H. Whittaker, ecologist from Cornell University, who stressed the seeming importance of modes of nutrition (photosynthetic, absorptive, ingestive) in recognizing multiple kingdoms of organisms (Whittaker, 1959, 1969 (and see Figure 22.4, reproduced from this second provocative essay of his)). Margulis's own tree of life on Earth and her high-level taxonomic scheme have basically varied little over the period from 1974 to the present time, the main – but highly significant – addition to Whittaker's contributions being her incorporation of postulated symbiotic events (Figure 22.5).

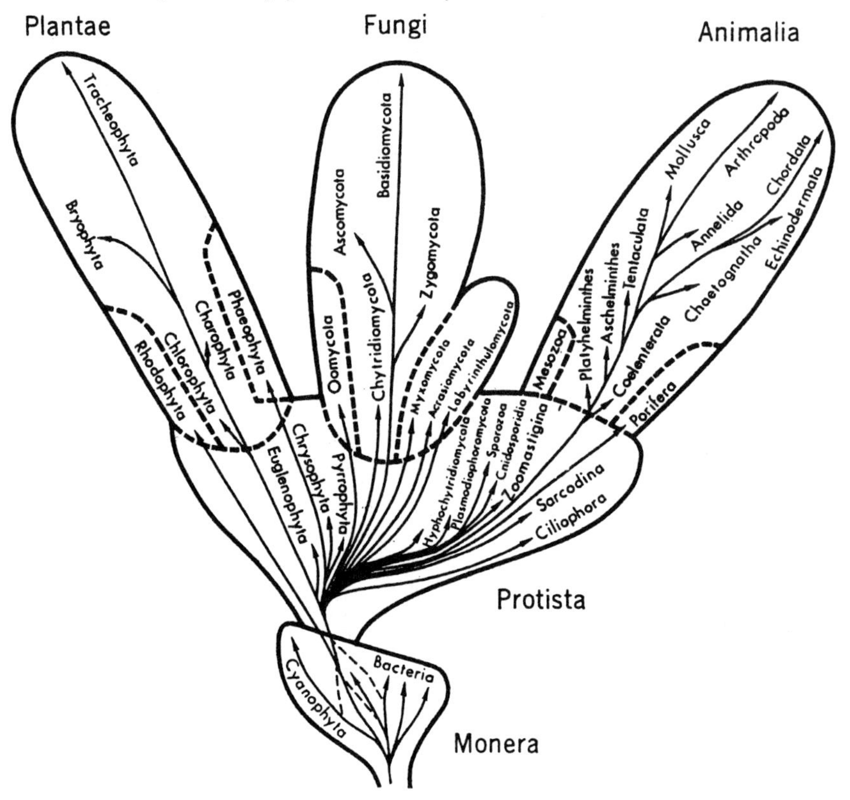

Figure 22.4 Tree of life according to Whittaker (1969), showing interrelationships of his five kingdoms based in large measure on differing modes of nutrition, with his Protista, however, showing all three kinds: photosynthesis, absorption, ingestion (redrawn from Whittaker, 1969).

The latest Margulis classification of her kingdom (now called by her merely a 'taxon,' to remove any socio-political implications) Protoctista embraces some 29 phyla of protists, as seen in Table 22.4 (Margulis, 1996; and see Margulis and Schwartz, 1998), representing some consolidation over her former number of 35 (Margulis *et al.*, 1990; Margulis, McKhann and Olendzenski, 1993). Five of the 29 are designated by vernacular names only; names of the others are mostly conventional latinized labels taken from the zoological or botanical literature; while three are original with Margulis: Archaeprotista, Dinomastigota, and Discomitochondriates. Her phyla contain the more or less classically established classes, orders, and lower taxa, as revised/updated by

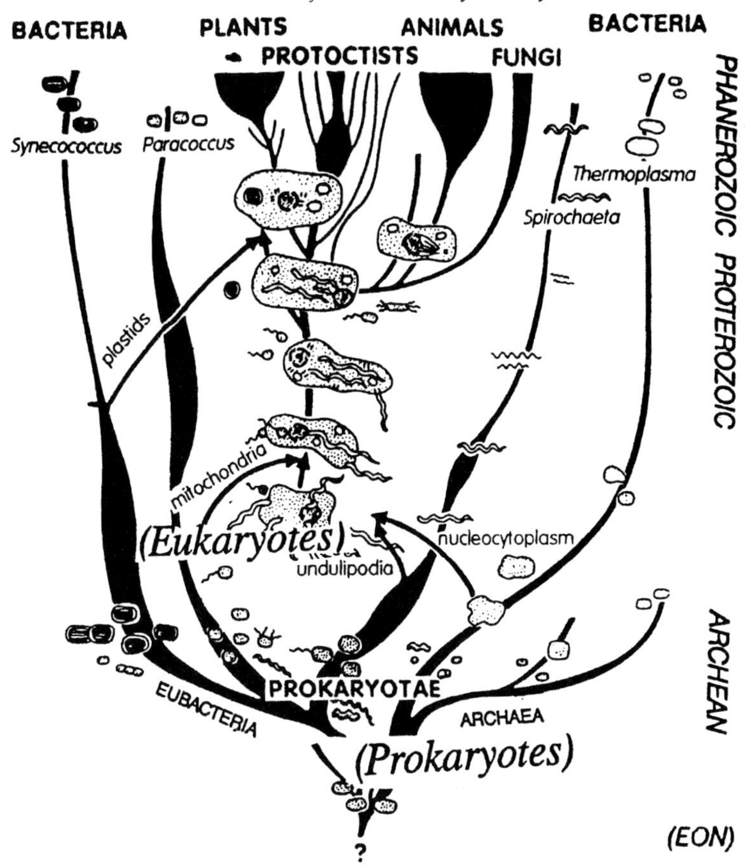

Figure 22.5 Tree of life according to Margulis (1996), showing interrelationships and juxtapositions of the five kingdoms and also times and places of major symbiotic events: acquisition of mitochondria, plastids, and undulipodia (eukaryotic flagella) (redrawn from Figure 2 in Margulis, 1996).

current specialists on the groups involved (see the individually authored chapters by phycologists, protozoologists, and mycologists in Margulis *et al.*, 1990). Classes are shown in Table 22.5, the data taken mostly from Margulis, McKhann and Olendzenski (1993), a composite work in which the number of phyla was 35 (some given only colloquial names), exclusive of two uncertain groups. The total number of classes reaches 75, and the number of orders 285, nearly tripling the figures of the preceding Levine Report (Levine *et al.*, 1980) and the Lee, Hutner and Bovee (1985) tome for those particular taxonomic categories.

Table 22.4 List of ungrouped protistan phyla (after Margulis, 1996), arranged as in Table 3 of her paper. Note that there are only 29 phyla in this (also see the forthcoming book by Margulis and Schwartz, 1998) most recent consideration by Margulis of the phyletic composition of her kingdom Protoctista. The number is larger (35) in a slightly earlier treatment (see Margulis, McKhann and Olendzenski 1993), and Table 22.5)

Archaeprotista	**Discomitochondriates**	**Oomycota**
Rhizopoda	**Zoomastigina**	**Hyphochytriomycota**
Granuloreticulosa	**Chrysomonads**	**Haplosporidia**
Xenophyophora	**Xanthophyta**	**Paramyxea**
Myxomycota	**Eustigmatophyta**	**Myxozoa**
Dinomastigota	**Bacillariophyta**	**Rhodophyta**
Ciliophora	**Phaeophyta**	**Gamophyta**
Apicomplexa	**Labyrinthulids/ Thraustochytrids**	**Chlorophyta**
Haptomonads	**Plasmodiophorids**	**Actinopoda**
Cryptomonads		**Chytridiomycota**

Criticisms of the Margulis scheme – aside from the minor running battles over terminology such as 'Protoctista' and 'undulipodium' (see remarks in papers by Cavalier-Smith, 1993b; Corliss, 1984, 1994; Hülsmann, 1992; and references within those reviews) – have been mostly concerned with her maintenance of a single kingdom for all protists; the bases used for her groupings of included phyla (presence or absence of cilia/flagella and/or complex sexual cycles); and her distribution of subtaxa within her phyla. Phylogeneticists/cladists decry her sometime seeming neglect of molecular (e.g. rRNA sequencing) data – which may identify monophyletic assemblages – in postulating high-level interrelationships. In past years, several workers have followed her general outline, including wholehearted support of the five-kingdom concept. For an example, in my own treatment of the protists a dozen years ago (Corliss, 1984), I presented a classification differing principally from that of Margulis's in its expansion in the number of phyla (to 45), in assignment of these groups to 18 vernacularly named supraphyletic assemblages (Table 22.6), in the (evolutionary-based) juxtaposition of some taxa, and in matters of terminology and nomenclature. Although a number of specialists have not followed Margulis's outline, it nevertheless, quite understandably, became, and still is, very popular with textbook writers, educators, and students around the whole world. (Incidentally, her long-held, imaginative insights into the significance of symbiosis in the evolution of all present-day life forms (Margulis, 1970, 1976, 1981, 1993, 1996;

Margulis and Kester, 1991) remain most heuristic, quite apart from her taxonomic contributions *per se*.)

Let us turn next to the contributions of Tom Cavalier-Smith, of London and now of Vancouver, who became interested in the protists through his early researches in cell biology, recognizing that many of these minute eukaryotes could serve as ideal material for investigations into membrane systems and the like (Cavalier-Smith, 1975, 1978, 1981a; and see 1987a, 1991a,b, 1993c, 1995a-c). His keen interest in the origin and evolution of such cellular inclusions led him to recognize/appreciate the need for revision of the current taxonomies of many of the algae, protozoa, and fungi under intensive study in numerous research centres around the world. His resulting bold attempts to straighten out the misconceptions/misunderstandings of the past led to his creation of many innovative high-level taxa (Cavalier-Smith, 1981b, 1983, 1986, 1989), causing considerable consternation among practicing taxonomists many of whom seemed reasonably content with more cautious, more or less conventional – and ever so much simpler! – schemes of classification (as embodied in the Levine Report of 1980 and the *Illustrated Guide* of 1985). Cavalier-Smith continues to think along his original, if now refined, lines (see Cavalier-Smith, 1997a,b, and chapter 21 in this volume), with the perhaps inevitable result of generally a seemingly greater multiplicity of high-level names in his most recent (and future!) presentations of multi-kingdom systems for all organisms (Cavalier-Smith, 1993a,b, 1995b,c; Cavalier-Smith, Allsopp and Chao, 1994; Cavalier-Smith and Chao, 1995; Cavalier-Smith, Chao and Allsopp, 1995).

Considering the many publications of Cavalier–Smith through the year 1995, his megasystematics of organisms has embraced some 8-9 kingdoms (two for the bacteria *sensu lato*). The protists are distributed among five (six, if one includes the new kingdom Cryptista erected in Cavalier-Smith, 1995c) eukaryotic kingdoms, but are found alone in great abundance in three of these: the Archezoa, the Chromista, and the huge Protozoa. These three contain some 3, 4, and 18 phyla, respectively, based principally on data in Cavalier-Smith (1993b). Looking a bit ahead, we find that, in Cavalier-Smith (1997a), the kingdom Protozoa is compacted and its contained taxa rearranged: for example, many subunits of the phylum Opalozoa are transferred to other phyla. Then, in Cavalier-Smith (1997b), the kingdom Archezoa is reduced to a subkingdom of Protozoa (the rank it held originally in Cavalier-Smith, 1983), and revised numbers of included phyla read 5 for the Chromista and 16 for the Protozoa. Most significantly, as part of the reduction of the latter's overall size, the ill-fated Opalozoa disappears completely, with its last vestige, the Opalinata, being transferred to a new position among the heterokonts of the neighboring kingdom Chromista.

Table 22.5 List of 35 *protoctistan* phyla and their included classes (sometimes named only in the vernacular) according to information in the composite work of Margulis, McKhann and Olendzenski (1993). Some 76 classes are given here; contrast this number with the 27 of the Levine Report (Levine *et al.*, 1980) and the 12 of the still earlier Honigberg Report (Honigberg *et al.*, 1964), although the latter two works were limited primarily to consideration of just so-called *protozoan* taxa

Names of Phyla	*Names of Included Classes*
Rhizopoda	Lobosea, Filosea
Haplosporidia	Haplosporea
Paramyxea	Parayxidea, Marteiliidea
Myxozoa	Myxosporea, Actinosporea
Microspora	Rudimicrosporea, Microsporea
Acrasea	Acrasids
Dictyostelida	Dictyostelids
Rhodophyta	Rhodophyceae
Conjugaphyta	Conjugatophyceae
Xenophyophora	Psamminida, Stannominida
Cryptophyta	Cryptophyceae
Glaucocystophyta	Glaucocystophyceae
Karyoblastea	Karyoblastea
Zoomastigina	Amebomastigota, Bicosoecids, Choanomastigota, Diplomonadida, Pseudociliata, Kinetoplastida, Opalinata, Proteromonadida, Parabasalia, Retortamonadida, Pyrsonymphida
Euglenida	Euglenophyceae
Chlorarachnida	Chlorarachniophyceae
Prymnesiophyta	Prymnesiophyceae
Raphidophyta	Raphidophyceae
Eustigmatophyta	Eustigmatophyceae
Actinopoda	Polycystina, Phaeodaria, Heliozoa, Acantharia
Hyphochytriomycota	Hyphochytrids
Labyrinthulomycota	Labyrinthulids, Thraustochytrids
Plasmodiophoromycota	Plasmodiophorids
Dinomastigota	Dinophyceae
Chrysophyta	Chrysophyceae, Synurophyceae, Pedinellophyceae, Dictyochophyceae, Silicomastigota
Chytridiomycota	Chytridiomycetes
Plasmodial slime molds	Protostelids, Myxomycetes
Ciliophora	Karyorelictea, Spirotrichea, Prostomatea, Litostomatea, Phyllopharyngea, Nassophorea, Oligohymenophorea, Colpodea
Granuloreticulosa	Athalamea, Foraminiferea
Apicomplexa	Gregarinia, Coccidia, Hematozoa
Bacillariophyta	Coscinodiscophyceae, Fragilariophyceae, Bacillariophyceae
Chlorophyta	Prasinophyceae, Chlorophyceae, Ulvophyceae, Charophyceae
Oomycota	Saprolegniomycetidae, Peronosporomycetidae
Xanthophyta	Xanthophyceae
Phaeophyta	Phaeophyceae

Table 22.6 The 45 phyla of the kingdom Protista arranged within 18 vernacularly named supraphyletic assemblages (after Corliss, 1984)

RHIZOPODS
- **Karyoblastea**
- **Amoebozoa**
- **Acrasia**
- **Eumycetozoa**
- **Plasmodiophorea**
- **Granuloreticulosa**
- **Xenophyophora**

MASTIGOMYCETES
- **Hyphochytridiomycota**
- **Oomycota**
- **Chytridiomycota**

CHLOROBIONTS
- **Chlorophyta**
- **Prasinophyta**
- **Conjugatophyta**
- **Charophyta**
- **Glaucophyta**

EUGLENOZOA
- **Euglenophyta**
- **Kinetoplastidea**
- **Pseudociliata**

RHODOPHYTES
- **Rhodophyta**

CRYPTOMONADS
- **Cryptophyta**

CHOANOFLAGELLATES
- **Choanoflagellata**

CHROMOBIONTS
- **Chrysophyta**
- **Haptophyta**
- **Bacillariophyta**
- **Xanthophyta**
- **Eustigmatophyta**
- **Phaeophyta**
- **Proteromonadea**

LABYRINTHOMORPHS
- **Labyrinthulea**
- **Thraustochytriacea**

POLYMASTIGOTES
- **Metamonadea**
- **Parabasalia**

PARAFLAGELLATES
- **Opalinata**

ACTINOPODS
- **Heliozoa**
- **Taxopoda**
- **Acantharia**
- **Polycystina**
- **Phaeodaria**

DINOFLAGELLATES
- **Peridinea**
- **Syndinea**

CILIATES
- **Ciliophora**

SPOROZOA
- **Sporozoa**

MICROSPORIDIA
- **Microsporidia**

HAPLOSPORIDIA
- **Haplosporidia**

MYXOSPORIDIA
- **Myxosporidia**

The lowness of figures for phyla is somewhat misleading because of Cavalier-Smith's common insertion of intermediate ranks even at the highest taxonomic levels of his classifications. For example, he has used – often created as new – the ranks of subkingdoms, branches, infrakingdoms, and parvkingdoms as well as superphyla, subphyla, and infraphyla, categories the employment of most of which is typically eschewed by taxonomic protozoologists. So, he has a lot more ranks, and thus his systems become lengthy and unwieldy (top-heavy!) at supraordinal levels. For his kingdom Protozoa, Cavalier-Smith (1993b) employed 12 distinct ranks above his ordinal and familial taxa, involving a total of 175 often unfamiliar names within this assemblage alone. He stoutly defends such actions, however, as quite necessary to show proper evolutionary/phylogenetic separateness of the top-level groups involved, and, at the same time, to keep down (from excessive) the numbers of kingdoms and phyla that otherwise might be required. He also claims that he is thus enabled to employ a goodly number of conventional/classical names (for some of these intermediate ranks) that will aid the reader/user of his schemes in tying past classifications to his modern ones.

While basing his taxonomic arrangements on molecular phylogenetic data, along with evidence from ultrastructural and biochemical investigations of others and of his own, Cavalier-Smith nonetheless does endorse a traditional hierarchical arrangement of his taxa at all (their many) levels; and he carefully publishes new names in accordance with provisions of both the zoological and the botanical codes of nomenclature. Figure 22.6, reproduced from Cavalier-Smith (1995b), portrays a kind of evolutionary chart or diagram of phylogeny among all the eukaryotes, as the scenario was viewed at that time by this prolific worker. All groups of protists not included in the five labelled kingdoms belong to the sprawling (unidentified on the chart) kingdom Protozoa. In Table 22.7 are the names of the 18 phyla (and their 63 classes) assigned to the single kingdom Protozoa by Cavalier-Smith (1993b) in his most comprehensive overall paper on this large group of protists, which he first identified as a separate major eukaryotic assemblage some 15 years ago (Cavalier-Smith, 1981b). Notice the general similarity of many of the names in Table 22.7 with a number of those appearing in publications of other workers (e.g., in Corliss (1984), Lee, Hutner and Bovee (1985), Margulis *et al.* (1990)) and, of course, in earlier papers by Cavalier-Smith himself.

Critics bemoan the proliferative/inflationary aspects of his classifications, as I have pointed out above, and perhaps find even more distasteful Cavalier-Smith's annual or even semi-annual revisions of his 'latest' schemes, which suggests that some of his taxonomic decisions have perhaps been made prematurely. Various specialists also quarrel

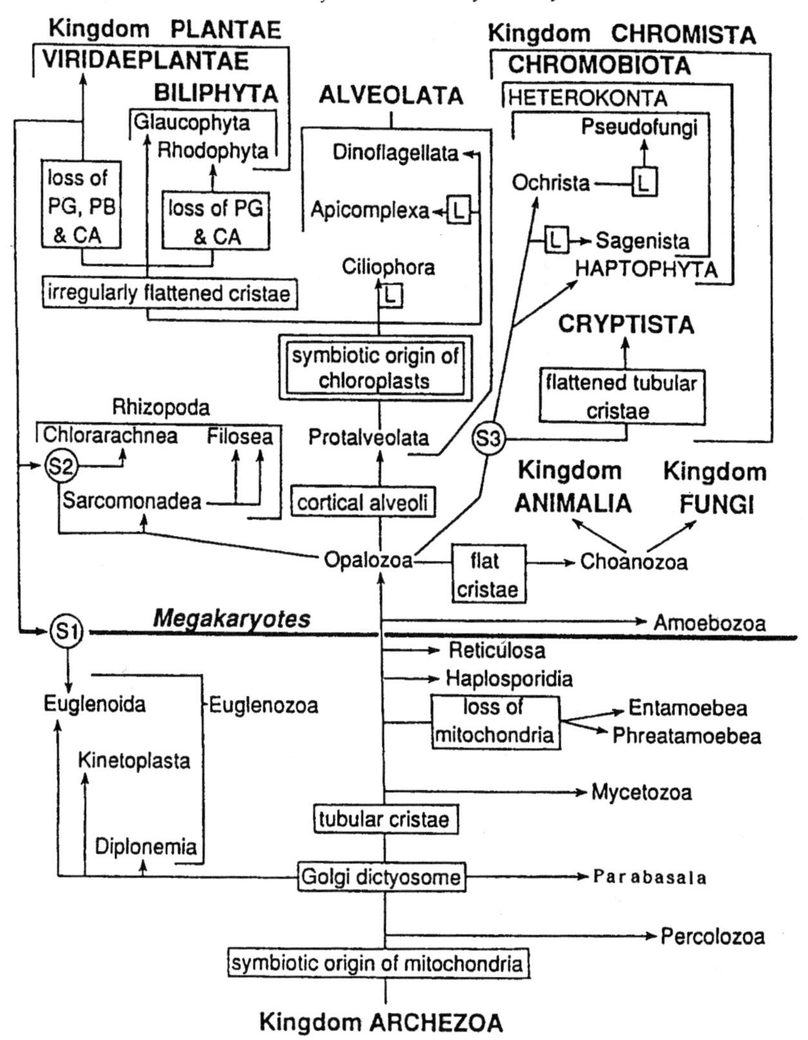

Figure 22.6 Diagram of evolutionary transitions during phylogeny of eukaryotes, according to Cavalier-Smith, with emphasis on protistan groups (after Cavalier-Smith, 1995b). The only kingdom not labelled as such is the Protozoa, which embraces the majority of taxa shown in the chart. For further explanation and a key to abbreviations used, the reader is referred to the original paper.

with several of his conclusions concerning interrelationships within specific higher taxa, claiming that some of his proposals are too speculative in nature, considering the dearth of relevant hard data available at the time of their promulgation. It is difficult for others of us to accept

Table 22.7 List of the 18 phyla assigned to the kingdom Protozoa by Cavalier-Smith (1993b), including their associated classes (supra- and infra-taxa not shown at any of the ranks). Some changes have been introduced in papers by Cavalier-Smith since the year 1993 (see text). The order (arrangement) of the names here follows that of their appearance in Table 4 of Cavalier-Smith (1993b)

Names of Phyla	*Names of Included Classes*
Percolozoa	Percolomonadea, Heterolobosea, Lyromonadea, Pseudociliatea
Parabasalia	Trichomonadea, Hypermastigea
Euglenozoa	Diplonemea, Petalomonadea, Peranemea, Aphagea, Kinetoplastea
Opalozoa	Heteromitea, Telonemea, Cyathobodonea, Ebridea, Phytomyxea, Opalinea, Kinetomonadea, Hemimastigea
Mycetozoa	Protostelea, Myxogastrea, Dictyostelea
Choanozoa	Choanomonadea
Dinozoa	Colponemea, Oxyrrhea, Ellobiopsea, Syndinea, Noctilucea, Haplozooidea, Peridinea, Bilidinea
Apicomplexa	Apicomonadea, Eogregarinea, Neogregarinea, Coelotrophea, Eucoccidea, Haemosporea, Piroplasmea
Ciliophora	Spirotrichea, Prostomatea, Litostomatea, Phyllopharyngea, Nassophorea, Oligohymenophorea, Colpodea, Karyorelictea
Heliozoa	Nucleohelea, Centrohelea
Radiozoa	Acantharea, Sticholonchea, Polycystinea, Phaeodarea
Rhizopoda	Lobosea, Filosea
Reticulosa	Athalamea, Monothalamea, Polythalamea
Entamoebia	Entamoebea
Myxosporidia	Myxosporea/Actinomyxea
Haplospiridia	Haplosporea
Paramyxia	Paramyxea
Mesozoa	Rhombozoa, Orthonectea

'cilia' for cilia + flagella (although to describe dinoflagellates as 'biciliated flagellates' is not quite as awkward as to call them 'biundulipodiated mastigotes'!). His determined resurrection of the name 'Protozoa' has caused some concern among modern protistologists who are, understandably, trying to eliminate terms that harken back to divisive systems (that, for example, seem to imply – although his Protozoa does not do this – maintenance, taxonomically and evolutionarily, of false separations of certain algal and protozoan groups). Still, a number of

Table 22.8 Assignment of 34 phyla of **protists** to six eukaryotic kingdoms (after Corliss, 1994, 1995b). Phyletic names are arranged alphabetically

Kingdoms	*Included Protistan Phyla*
ARCHEZOA	**Archamoebae, Metamonada, Microspora**
PROTOZOA	**Apicomplexa, Ascetospora, Choanozoa, Ciliophora, Dinozoa, Euglenozoa, Heliozoa, Mycetozoa, Opalozoa, Parabasala, Percolozoa, Radiozoa, Rhizopoda**
CHROMISTA	**Bicosoecae, Diatomae, Dictyochae, Chlorarachniophyta, Cryptomonada, Haptomonada, Labyrinthomorpha, Phaeophyta, Pseudofungi, Raphidophyta**
PLANTAE	**Charophyta, Chlorophyta, Glaucophyta, Prasinophyta, Rhodophyta, Ulvophyta**
FUNGI	**Chytridiomycota**
ANIMALIA	**Myxozoa**

workers have accepted the main themes of his schemes, notably Corliss (1994, 1995b). In my own 'interim user-friendly' hierarchical classification of the protists (Corliss, 1994; and see Table 22.8), I have attempted to justify the major conclusions drawn by Cavalier-Smith (1993b) and to show how the resulting positions of most given groups in the overall scheme may be constructively linked nomenclaturally as well as taxonomically to more or less corresponding parts of time-honoured, more conventional, classifications of the older literature (points which cannot be detailed in this brief overview).

Finally, let us review the relevant contributions of David J. ('Paddy') Patterson (of Bristol, now Sydney), who was trained as a zoologist and cytologist (Patterson, 1979, 1980; Patterson and Fenchel, 1985), and/but was among the first young biologists to participate wholeheartedly in the exciting 'protist revolution.' He was a co-founder of the *Progress in Protistology* series (see Corliss and Patterson, 1986, 1987; Patterson and Corliss, 1989), co-editor of (and an author in) the thorough and most useful book on free-living heterotrophic flagellates (Patterson and Larsen, 1991), and proposer of the now very popular group which he named the stramenopiles (better, straminopiles), essentially a refined/redefined assemblage of heterokont protists the composition of which is largely identical to the Heterokonta moiety of Cavalier-Smith's kingdom

Chromista (Patterson, 1989a). An ecologically oriented taxonomist (see Fenchel and Patterson, 1986, 1988; Larsen and Patterson, 1990; Patterson, Larsen and Corliss, 1989) as well as an accomplished electron-microscopist, Patterson soon became absorbed in the evolutionary problems that were presented by the protists overall, quickly perceiving the potential conflict between the implications of cladistically derived 'phylogenetic trees,' on the one hand, and standard hierarchical arrangements of ranked and named (high-level) taxa, on the other (see Brugerolle and Patterson, 1990; Patterson, 1986a; Patterson and Brugerolle, 1988; Delvinquier and Patterson, 1992; Leipe *et al.*, 1994; Smith and Patterson, 1986).

Patterson's ideas are clearly the most revolutionary – as they affect top-level protozoan/protistan taxonomic systems – yet to appear on the scene (Patterson, 1989a,b, 1994; Patterson and Sogin, 1993; and see Figure 22.7). His direct translation of robust eukaryotic evolutionary trees into phylogenetic classifications leaves many suprafamilial groups of protists unranked and without formal names. Some taxa at all levels are considered to be of uncertain status, and quite a number of (less well known) genera are left entirely to one side, as Patterson equates speculation with pure ignorance, neither allowed in his schemes. In the long awaited forthcoming second edition of the well known *Illustrated Guide to the Protozoa* (edited by Lee *et al.*, 1998), the superstructure of the classificatory system is being prepared by a committee of the Society of Protozoologists chaired by D. J. Patterson. Predictably, some five dozen monophyletic groups of protists will be recognized there as assignable to the protozoa *sensu lato*. They may correspond roughly to the list published in Patterson (1994); the groups seem to range from families to orders or higher, with ranks not always indicated and with some names apparently to be given in the vernacular only.

Patterson's insistence on rigour is laudable, but critics of his taxonomic approaches regret the rather complete departure from standard procedures of recognizing named hierarchical levels in a system that can be related, to some degree, with published classification schemes of the past. However, it is admittedly true that the usual traditional taxonomies of yesteryear were rather authoritarian in nature, arbitrarily constructed for convenience, and probably followed often because of resistance to change *per se*. Yet lack of data practically obliged earlier synthesizers to engage in speculation concerning group relationships.

Has the time perhaps indeed arrived for a major, iconoclastic turnaround, in which the looser, relatively undisciplined taxonomic practices of the past are to be discarded wholesale and replaced with a classificatory system based on phylogenetic trees rigorously produced,

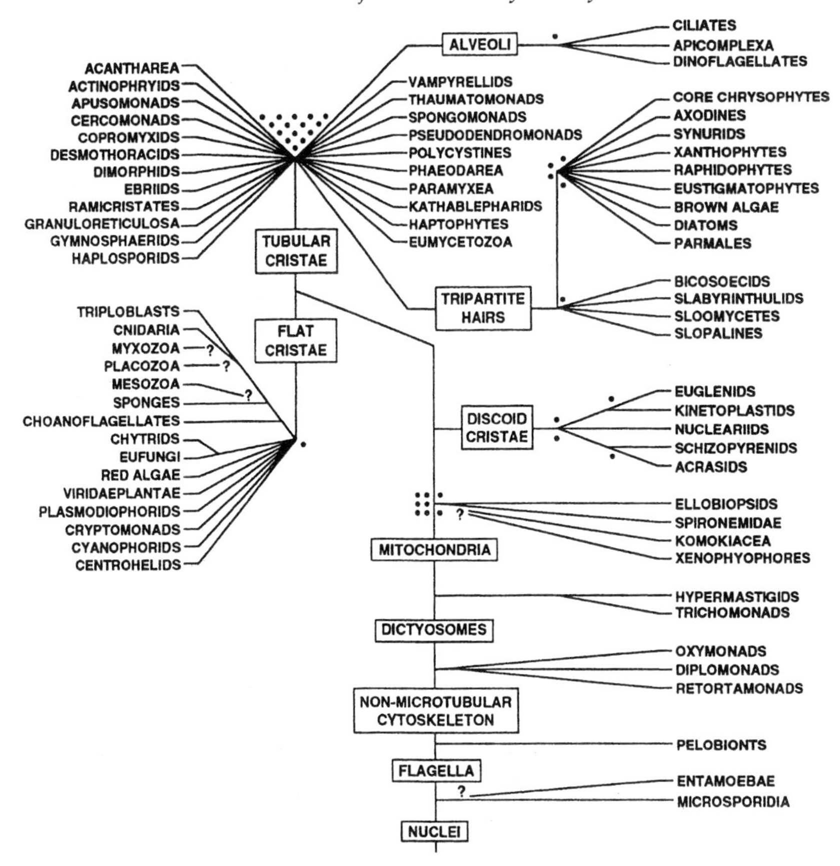

Figure 22.7 A skeletal tree of the eukaryotes, according to Patterson, directly usable also, by appropriate transformation, to depict the superstructure of a phylogenetic classification for these organisms, with emphasis on protistan groups (after Patterson, 1994). (See original paper for further explanation.)

with focused attention on monophyly and the points of emergence of carefully selected innovative (synapomorphic, often ultrastructural) characters? This essentially seems to be Patterson's goal, as I understand it. Is the objective too idealistic to be realizable at this time? Figure 22.7, based on data in Patterson (1994), shows his skeletal phylogenetic tree which reveals the times and places of emergence of key ultrastructurally detectable synapomorphic characters and the branching patterns of major monophyletic groups of protists known to Patterson at that time.

Unlike many other taxonomic protistologists, Patterson also has a keen interest in nomenclatural problems, especially as they affect the 'ambiregnal' species of protists. His pleas for appropriate revision of existing Codes of Nomenclature (Patterson, 1986b; Patterson and Larsen, 1992), like those proposed by Corliss (1984, 1986), have been echoed and extended by others (see reviews in Corliss, 1990, 1993, 1995a).

22.4 AIMING FOR THE IDEAL

What, then, is the current status of the classification of the protozoa (or, if you will, of the protists *sensu lato*)? I believe that the expression 'in a state of flux' is probably more accurate than to consider it discouragingly to be 'in a state of chaos'! More importantly, what can be done, in any case, to bring about a significant degree of welcome stability?

Several years ago (Corliss, 1994), I attempted to produce an overall classification of the protists based on information of all kinds then available. Since that time, I have made some improvements in this 'interim utilitarian user-friendly' scheme, but my new macrosystem is hardly ready for unveiling at this time. Admittedly, it suffers from acceptance of some paraphyletic (even polyphyletic) groups, from incorporation of some 'educated guesses' about group interrelationships (speculation masking ignorance?), and from our incomplete understanding, to date, of the impact of such very recent findings as, for example, those concerning possible mitochondrial/hydrogenosomal homologies plus the basic question of using the shape/type of mitochondrial cristae as a robust differential character at high levels, matters of great potential significance (Adoutte *et al.*, 1996; Biazini, Finlay and Lloyd, 1997; Cavalier-Smith, 1997a; Cavalier-Smith and Chao, 1995; Seravin, 1993).

In conclusion – and I am thinking of the future – let me mention the major matters which ought to be kept clearly in mind by all of us who may be addicted to or committed to production of still more acceptable high-level schemes of classification of protozoa and, indeed, of all protists or even of all eukaryotes. We should continue to carry out the indispensable research projects that result in accumulation of much needed new hard data of relevance, but the following points might serve as guidelines, if you will, to be taken into account when thinking about overall protistan megasystematics:

1. Because of the widespread natural occurrence of symbioses in past aeons, the chimaeric evolutionary origin of a number of protistan lineages must be accepted and dealt with. Multiple invasions/engulfments and/or reinfections further complicate the sorting out process. Workers ask, what is taxonomically the more important, the 'host' cytoplasm, or the (persisting remnant of the) symbiont, or the entire newly constituted organism? Surely the last; but all such

factors must be taken into consideration. The general notions of Lynn Margulis (see citations on p. 422) and others (e.g. see Bandele (1997), Sapp (1994, Sogin *et al.* (1996), Taylor (1987) on the subject are sound and, today, some of them are indisputable – above all, those concerned with the origins of plastids and mitochondria.

2. Species or groups of protists ideally should be studied not only ultrastructurally and molecularly but also ecologically and behaviorally: whole organism biology must not be neglected. And findings should be thoroughly compared with results obtained on other species in the same or different taxa. Taxonomists are sometimes negligent with regard to carrying out comprehensive *comparative* studies in their cytological work; and molecular biologists should not be satisfied with considering rRNA sequencing data from only a few species as unconditionally representative of the position of a large and perhaps internally diverse group/line/clade of protists.
3. Although once highly stimulating, the concept of a single neoHaeckelian 'kingdom Protista/Protoctista' is no longer tenable.[1] The protists *sensu lato* need to be distributed among multiple kingdoms or some other kind of very high-level assemblages of the eukaryotic organisms overall. Because of their truly incredible diversity – morphological, ecological, biochemical, molecular, genetic, phylogenetic – the protists require such assignment to *multiple* high-level taxa in any scheme of classification. Proliferation and taxonomic inflation of groups and names are, to a degree, the *inevitable* consequences of rigorous comparative studies. Some 20 new classes (a high rank for botanists) of algae have been erected and accepted in the past 30 years (Norton, Melkonian and Anderson, 1996); and even more high taxa of protozoa have emerged during that same period (and not solely from the pen of Cavalier-Smith). In Table 22.9, notice the great jumps in numbers of suprafamilial protozoan taxa, correlatable with increases in our knowledge, even in decades before molecular data were available. The solution is surely not to ignore such new groups but to find the most acceptable (i.e. supportable) way of absorbing/incorporating them satisfactorily in a single overall taxonomic system.
4. Megasystematics of the protists **should** reflect their evolutionary relationships, but there still needs to be a degree of internal consistency in resulting taxonomic juxtapositions; hierarchies, ranks, and names are important and need be neither misleading, irrelevant, nor overly speculative. Assigning/equating some monophyletic clusterings to appropriate previously or newly named higher protistan

[1] Incidentally, the idea of a third kingdom between plants and animals far predates the time of Haeckel, Hogg, and other 19th-century natural philosophers: see the scholarly essay by Ragan, 1997.

Table 22.9 Numbers of higher taxa (at selected ranks) of **protozoa** or **protists** according to various authors through the year 1996[a]

Sources of data	*Kingdoms*	*Phyla*	*Classes*	*Orders*	*Suborders*
Bütschli (1880–1889)			4	9	11
Honigberg Report (1964)		1	12	66	41
Levine Report (1980)		7	27	98	71
Margulis/Schwartz (1982)	1	27			
Corliss (1984)	1	45			
Lee *et al.* (1985)		6	27	117	72
de Puytorac *et al.* (1987)	1	13	53	135	79
Margulis *et al.* (1990)	1	35	76	285	54
Cavalier-Smith (1993b, 1995b)	5	31	101		
Corliss (1994, 1995b)	6	34	83		
Hausmann/Hülsmann (1996)	2	18	45		
Margulis (1996)	1	29			

[a]The first three sources and the sixth work have treated only (their concepts of) the PROTOZOA; the 4th, 5th, 7th, 8th, and last entry, a single but all-embracing kingdom PROTISTA/PROTOCTISTA; the remaining three (9th–11th), essentially the **protists** as distributed throughout multiple kingdoms – two, five, or six – of 'lower' or all eukaryotes. See original works for details.

taxa, however, remains a vexatious challenge often unresolved satisfactorily to date.

5. Nomenclature (names for the accepted higher taxa) must eventually, somehow, be stabilized. The problem is quite complex because of the constant shifting in the concept, taxonomic boundaries, and rank of many groups; the changes in spellings of names used in the literature over the years (and to what extent should strict priority be recognized?); and, last but not least, the inescapable influence of (past and present!) authoritarianism and of the background training (e.g. botanical or zoological, taxonomic or nontaxonomic) of the users of familiar names as well as of the proposers of new ones. Nevertheless, one may take hope from the fact – as an examination of the tables and figures of this paper will reveal – that there is already considerable overlapping in names favoured by different taxonomists for essentially the same groups of protists: often agreement needs to be found only on the relatively minor matter of endings/suffixes.
6. Finally, let me mention very briefly a dozen selected examples of proposed taxonomic changes within the past few years (some as recently as six months ago at the time of writing) that ought to be readily accepted, for the most part, now:

(i) Various green algal groups (e.g. prasinophytes, chlorophytes, ulvophytes, charophytes, and conjugatophytes – by whatever names and ranks given to these taxa in the future) must be assigned/returned to the plant kingdom (or at least classified *with* plants). There is increasing agreement among phycologists and protozoologists that the 'greens' are, in effect, 'lower' plants – or perhaps, better, that the 'higher' plants are a minor recent evolutionary offshoot from the green algae! (Plants have had their turn in the sun?)

(ii) Numerous algal groups (at ordinal ranks and below) must be shifted out of their conventional classes and into other ones (or serve themselves as nuclei for the formation of new classes), primarily on the basis of rRNA sequencing patterns (see the series of papers emanating from R. A. Andersen's laboratory, with more in press and preparation: Andersen, 1987, 1989, 1991, 1992; Ariztia, Andersen and Sogin, 1991; Bhattacharya *et al.*, 1992; Andersen *et al.*, 1993; Saunders *et al.*, 1995; Karlson *et al.*, 1996; see also the very recent review by Ragan, 1998).

(iii) The chytridiomycotes, phylogenetically far separated from the oomycetes and hyphochytriomycetes, must be considered as 'true' fungi: mycologists widely support this view (Barr, 1992; Bruns, 1991; Powers, 1993).

(iv) The parasitic myxozoa/myxosporidians of the literature, always attached to the protozoan protists (although as a highly unique group), must now be considered as a distinct group of **animals** (Smothers *et al.*, 1994; Corliss, 1995b; Schlegel *et al.*, 1996), with cnidarians – but **not** bilaterians – likely the most closely related taxon of contemporary metazoan groups (Siddall *et al.*, 1995).

(v) The celebrated ciliate evolutionary tree of yore (last seen in Corliss, 1979) has been turned at least partially upsidedown by modern ultrastructural and molecular studies (Bardele, 1987; Lynn and Small, 1988; Lynn and Sogin, 1988; Lynn and Corliss, 1991; Baroin-Tourancheau *et al.*, 1992; Hammerschmidt *et al.*, 1996).

(vi) Heterotrophic free-living flagellates – diverse, numerous, and understudied – provide the ultimate ancestry for all eukaryotic taxa (Patterson and Larsen, 1991; Vickerman, 1992; Patterson, 1993; Cavalier-Smith, 1995b,c; Sleigh, 1995); thus, they are fully deserving of more attention from both taxonomists and phylogeneticists in the future.

(vii) Ciliates, dinoflagellates, and sporozoa (apicomplexans) – superficially quite strange bedfellows – turn out to be unitable under the caption of 'alveolates' (Gajadhar *et al.*, 1991; Wolters,

1991; Cavalier-Smith, 1991a, b, 1993b; Patterson and Sogin, 1993; Patterson, 1994); Several studies in progress indicate that some other protists may be eligible for addition to this now surprisingly clearcut assemblage, all members of which possess or have possessed cortical alveoli in their structural make-up (perhaps the very curious haplosporidians (Siddall, Stokes, and Burreson, 1995), or even the foraminifers (see Wray *et al.*, 1995; but also Merle *et al.*, 1994; and Pawlowski *et al.*, 1994), or maybe the glaucophytes (as mentioned in Corliss, 1994)?).

(viii) Fungi are phylogenetically much more closely related to animals than to plants (Baldauf and Palmer, 1993; Cavalier-Smith, 1987b; Nikoh *et al.*, 1994; Wainright *et al.*, 1993). (Vegetarians might keep this in mind.)

(ix) Red algae, once considered primitive or totally aberrant forms, may be distantly related to plants, as well as the green algae mentioned above (Cavalier-Smith, 1987c; Ragan, 1998; Ragan and Gutell, 1995).

(x) Euglenids, lumped with the 'phytoflagellates' in older classifications, are closely related to the trypanosomatid group of (former) 'lower zooflagellates,' and this must be reflected in modern overall protistan classificatory schemes (Kivic and Walne, 1984; Triemer and Farmer, 1991). The marine ciliate-turned-flagellate, *Stephanopogon* (Lipscomb and Corliss, 1982), has sometimes been assigned to this group as well. But whether the Euglenozoa are properly placed in a 'middle zone' of the phylogenetic tree of the eukaryotes has been questioned (Philippe and Adoutte, 1995).

(xi) Candidates for the most primitive of protists (and thus of eukaryotes in general: Schlegel, 1991; Sogin, 1991, 1994) have varied during the past 10 years. All of them have been members of groups formerly assigned to very widely separate/disparate protistan taxa: red algae, fungi, rhizopod amoebae, zooflagellates, dinoflagellates, and cnidosporidians. There is still uncertainty concerning this matter; and the parallel problem of the true phylogenetic relationship of contemporary species sometimes placed together in the kingdom Archezoa (mostly on the basis of their apparent current common lack of mitochondria) remains to be resolved satisfactorily (see the helpful critical discussion in Philippe and Adoutte, 1995). The unique microsporidians of the protozoological literature (shown by rRNA and other data to branch very early on the eukaryotic evolutionary tree: Vossbrinck *et al.*, 1987; Sogin *et al.*, 1996), appear to be the most likely truly primitive forms, although their long existence as minute and obligate intracellular symbionts (mainly in insects and fishes) confounds the issue to some extent. Worthy of note: Hausmann and Hülsmann (1996) have assigned the protists to two kingdoms (within the empire Eukaryota), the Microspora and the Mastigota.

(xii) *Blastocystis hominis*, considered both a fungus and a (most often rhizopod) protozoon, turns out to be evolutionarily related to the heterokont algae (Patterson's stramenopiles, part of Cavalier-Smith's kingdom Chromista): see Silberman *et al.* (1996). *Pneumocystis carinii*, however, does appear to remain best classified as a fungus (Edman and Sogin, 1994), although more work on its taxonomic position still could be carried out.

From the point of view of users – be they students, teachers, information retrieval systems, medical doctors, naturalists, historians of science, general biologists/scientists or professional phycologists, mycologists, protozoologists, parasitologists, taxonomists, paleontologists, ecologists, physiologists, biochemists, or cell, molecular, or evolutionary biologists – the ideal classification system for protists (and for all eukaryotes) will be one that, while accurately reflecting known phylogenetic interrelationships, also is reasonably compact, clear, uncomplicated, and understandable to readers/users without numerous explanatory footnotes.

Can this goal be achieved by the beginning of the 21st century?

NOTE ADDED IN PROOF

I am grateful to the editors and publisher for giving me an opportunity, at proofreading time, to update this chapter a bit by insertion of several more citations in the text and references section and by addition of a few new summarizing comments here.

Although there has been a continuing avalanche of papers in the literature on topics involving protists as ideal research material (Corliss, 1998), relatively few works that have come to my attention during the past two years have treated the subject of the overall classification or phylogeny of these 'lower' eukaryotes in any depth. These are, as James Thurber once whimsically put it, practically countable on the fingers of one thumb: Bardele (1997), Cavalier-Smith (1997a, b), Margulis and Schwartz (1998), Ragan (1997, 1998), Sogin *et al.* (1996), and the several chapters in the present volume (see especially those by Beakes, Cavalier-Smith, Corliss, Philippe and Adoutte, and Vickerman).

Whether it is directly so stated or not, it is satisfying to note that a 'protist perspective' (Corliss, 1986) has pervaded the above-cited contributions (and many others in recent years), helping to break the century-old hold of botany and zoology on the taxonomic placement of major groups of the eukaryotic microorganisms and their close relatives. A second point on which there is widespread agreement is recognition of the value of studying protists in attacking a great diversity of significant biological and biomedical problems today. Finally, there seems also to be a growing tacit understanding that sometimes it may be sensible to consider acceptance of different ways of viewing protist relationships overall, depending on the needs or purposes of the different clientele being served at a given time. With respect to the last-made point, let me briefly remind the reader of the five ways in which protists *sensu lato* have been or may be conveniently treated with differing objectives in mind.

1. They may be classified from a largely ecological (including nutritional) point of view. This is yet to be developed on a large scale, but perhaps it has been hinted at by some marine protistologists (e.g. Sieburth, 1979). Also, might such an approach possibly become helpful to workers on biodiversity (such as Finlay, Maberly and Cooper, 1997) in the future? And recall that it was partially used in the original systematic schemes on kingdoms (including one for the protists) that were published by Whittaker (1959, 1969).
2. Protistan assemblages may be treated solely/merely as evolutionary grades, with no formally recognized top-level taxonomic status. While there is general agreement that certain protists have, in effect, served as a step in the evolutionary transition from prokaryotes to multicellular eukaryotes (Corliss, 1989), the hypothesis does not lend itself well to taxonomic or phylogenetic treatment of all protistan groups in existence today and thus would appear to be of limited usefulness, especially from pragmatic or didactic point of view. But cytologically, protists – eukaryotic organisms the majority of which display a unicellular nature – may be thought of as commanding our attention in terms of representing a specific and abiding type or level of biological organization (Bardele, 1997).
3. Protists may be recognized as constituting clades or diverging evolutionary lines or lineages, many of which may have no easily discernible direct taxonomic relationship to each other. If strict monophyly of a line can be unequivocably established, we are then provided with a unique phylogenetically circumscribed evolutionary group, a fact of considerable importance. But it seems increasingly clear that such independent clades are often difficult to rank taxonomically (at least, above familial or ordinal levels) and to fit into hierarchically arranged overall classificatory systems, as the cladists themselves readily concede. Beautiful molecular phylogenetic trees can be generated (see examples in such works already cited elsewhere in this paper as those by Adoutte, Andersen, Bardele, Cavalier-Smith, Leipe, Lynn, Patterson, Schlegel, Siddall, Sogin, and many coworkers), and these, undeniably, are of great evolutionary – if not directly high-level systematic – value.
4. All protists (however exactly defined!) may be conveniently lumped together into a single kingdom (be it named Protista or Protoctista), as has persistently been done by Margulis and associates (see references cited in preceding sections plus, now, Margulis and Schwartz (1998)). This option has obvious appeal, and it certainly gives the protists a conspicuous position on the tree of life. It has received enthusiastic support from teachers of biology, from laypersons generally, and from textbook writers all of whom can easily understand the (theoretical) concept of a simple five-kingdom biotic world.

However, unfortunately, as research-oriented protistologists are increasingly pointing out with clarity (add Bardele (1997) and Ragan (1998) to the list of workers already including Andersen, Cavalier-Smith, Corliss, Patterson, Sogin, and numerous other colleagues), the notion is perhaps too simplistic today, failing as it does to meet a number of both phylogenetic (cladistic) and evolutionary precepts of established significance and wide acceptance.

5. Finally, let us consider very briefly what I believe is the distinctly supportable concept that the protists (spelled with a small 'p') may be distributed taxonomically among multiple kingdoms (a major one of which may be known formally as the 'Protozoa'). To adopt this viewpoint, one needs only (1) to acknowledge the fact (supplied by the phylogeneticists, mostly with the aid of rRNA sequencing data) that some major groups of protists appear to have closer evolutionary affinities with members of other kingdoms (outstanding example: green algae with conventional plants) than with each other; and (2) to accept the assumption that certain other high-ranking assemblages, by virtue of their uniquenesses, warrant taxonomic separation from one another at the kingdom level. It is my personal hope, as is evident from this chapter, that such a megasystematic scheme of classification for the protists (which can also nomenclaturally bridge well the gap separating modern from long-established conventional systems in algal and protozoan taxonomy: see Corliss (1994)) can be refined in due time to the point of wide general acceptance and preferred usage in teaching, research, and information retrieval, as well as in meeting still further demands of a diverse clientele.

22.5 REFERENCES

Adoutte, A., Germot, A., Le Guyader, H. and Philippe, H. (1996) Que savons-nous de l'histoire évolutive des Eucaryotes? 2. De la diversification des protistes à la radiation des multicellulaires. *Medecine/Sciences*, **12**, I–XVII.

Andersen, R.A. (1987) Synurophyceae classis nov., a new class of algae. *American Journal of Botany*, **74**, 337–53.

Andersen, R.A. (1989) The Synurophyceae and their relationship to other golden algae. *Beihefte zur Nova Hedwigia*, **95**, 1–26.

Andersen, R.A. (1991) The cytoskeleton of chromophyte algae. *Protoplasma*, **164**, 143–59.

Andersen, R.A. (1992) Diversity of eukaryotic algae. *Biodiversity and Conservation*, **1**, 267–92.

Andersen, R.A., Saunders, G.W., Paskind, M.P. and Sexton, J.P. (1993) Ultrastructure and 18S rRNA gene sequence for *Pelagomonas calceolata* gen. et sp. nov. and the description of a new algal class, the Pelagophyceae classis nov. *Journal of Phycology*, **29**, 701–16.

Ariztia, E.V., Andersen, R.A. and Sogin, M.L. (1991) A new phylogeny for chromophyte algae using 16S-like rRNA sequences from *Mallomonas papillosa* (Synurophyceae) and *Tribonema aequale* (Xanthophyceae). *Journal of Phycology*, **27**, 428–36.

Baldauf, S.L. and Palmer, J.D. (1993) Animals and fungi are each other's closest relatives: congruent evidence from multiple proteins. *Proceedings of the National Academy of Sciences USA*, **90**, 11558–62.

Bardele, C.F. (1987) Auf neuen Wegen zu alten Zielen-moderne Ansätze zur Verwandtschaftsanalyse der Ciliaten. *Verh. Dtsch. Zool. Ges.*, **80**, 59–75.

Bardele, C.F. (1997) On the symbiotic origin of protists, their diversity, and their pivotal role in teaching systematic biology. *Italian Journal of Zoology*, **64**, 107–13.

Baroin-Tourancheau, A., Delgado, P., Perasso, R. and Adoutte, A. (1992) A broad molecular phylogeny of ciliates: identification of major evolutionary trends and radiations within the phylum. *Proceedings of the National Academy of Sciences USA*, **89**, 9764–8.

Barr, J.D.S. (1992) Evolution and kingdoms of organisms from the perspective of a mycologist. *Mycologia*, **84**, 1–11.

Bhattacharya, D., Medlin, L., Wainright, P.O. *et al.* (1992) Algae containing chlorophylls *a* + *c* are paraphyletic: molecular evolutionary analysis of the Chromophyta. *Evolution*, **46**, 1801–17.

Biagini, G.A., Finlay, B.J. and Lloyd, D. (1997) Evolution of the hydrogenosome. *FEMS Microbiology Letters*, **155**, 133–40.

Brugerolle, G. and Patterson, D.J. (1990) A cytological study of *Aulacomonas submarina* Skuja 1939, a heterotrophic flagellate with a novel ultrastructural identity. *European Journal of Protistology*, **25**, 191–9.

Bruns, T.D. (1991) Fungal molecular systematics. *Annual Review of Ecology and Systematics*, **22**, 525–64.

Bütschli, O. (1880) (Page xii of Abt. I of Bütschli, 1880–89: see following reference.)

Bütschli, O. (1880–89) Protozoa. Abt. I. Sarcodina und Sporozoa, Abt. II. Mastigophora, Abt. III. Infusoria und System der Radiolaria, in *Klassen und Ordnung des Thier-Reichs* (ed. H.G. Bronn), C.F. Winter, Leipzig, **1**, 1–2035.

Cavalier-Smith, T. (1975) The origin of nuclei and eukaryotic cells. *Nature*, **256**, 463–8.

Cavalier-Smith, T. (1978) The evolutionary origin and phylogeny of microtubules, mitotic spindles and eukaryote flagella. *BioSystems*, **10**, 93–114.

Cavalier-Smith, T. (1981a) The origin and early evolution of the eukaryotic cell. *Society of General Microbiology, Symposium*, **32**, 33–84.

Cavalier-Smith, T. (1981b) Eukaryote kingdoms: seven or nine? *BioSystems*, **14**, 461–81.

Cavalier-Smith, T. (1983) A 6-kingdom classification and a unified phylogeny, in *Endocytobiology II* (eds H.E.A. Schenk and W. Schwemmler), Walter de Gruyter, Berlin, pp. 1027–34.

Cavalier-Smith, T. (1986) The kingdom Chromista: origin and systematics. *Progress in Phycological Research*, **4**, 309–47.

Cavalier-Smith, T. (1987a) The origin of cells: a symbiosis between genes, catalysts, and membranes. *Cold Spring Harbor Symposia on Quantitative Biology*, **52**, 805–24.

Cavalier-Smith, T. (1987b) The origin of Fungi and pseudofungi. *Symposium of the British Mycological Society*, **13**, 339–53.

Cavalier-Smith, T. (1987c) Glaucophyceae and the origin of plants. *Evolutionary Trends in Plants*, **2**, 75–8.

Cavalier-Smith, T. (1989) The kingdom Chromista, in *The Chromophyte Algae: Problems and Perspectives* (eds J.C. Green, B.S.C. Leadbeater and W.L. Diver), Clarendon Press, Oxford, pp. 381–407.

Cavalier-Smith, T. (1991a) Evolution of eukaryotic and prokaryotic cells, in *Foundations of Medical Cell Biology* (ed. G.E. Bittar), J.A.I. Press, Greenwich, Conn., **1**, 217–72.

Cavalier-Smith, T. (1991b) Cell evolution, in *Evolution of Life* (eds S. Osawa and T. Honjo), Springer-Verlag, Tokyo, pp. 271–304.

Cavalier-Smith, T. (1993a) The protozoan phylum Opalozoa. *Journal of Eukaryotic Microbiology*, **40**, 609–15.

Cavalier-Smith, T. (1993b) Kingdom Protozoa and its 18 phyla. *Microbiological Reviews*, **57**, 953–94.

Cavalier-Smith, T. (1993c) The origins, losses and gains of chloroplasts, in *Origins of Plastids: Symbiogenesis, Prochlorophytes, and the Origins of Chloroplasts* (ed. R.A. Lewin), Chapman and Hall, New York, pp. 291–349.

Cavalier-Smith, T. (1995a) Evolutionary protistology comes of age: biodiversity and molecular cell biology. *Archiv für Protistenkunde*, **145**, 145–54.

Cavalier-Smith, T. (1995b) Membrane heredity, symbiogenesis, and the multiple origins of algae, in *Biodiversity and Evolution* (eds R. Arai, M. Kato and Y. Doi), Tokyo, pp. 75–114.

Cavalier-Smith, T. (1995c) Zooflagellate phylogeny and classification. *Cytology*, **37**, 1010–29.

Cavalier-Smith, T. (1997a) Amoeboflagellates and mitochondrial cristae in eukaryote evolution: megasystematics of the new protozoan subkingdoms Eozoa and Neozoa. *Archiv für Protistenkunde* **147**, 237–58.

Cavalier-Smith, T. (1997b) A revised six-kingdom system of life. *Biological Reviews* (in press).

Cavalier-Smith, T., Allsopp, M.T.E.P. and Chao, E.E. (1994) Chimeric conundra: are nucleomorphs and chromists monophyletic or polyphyletic? *Proceedings of the National Academy of Sciences USA*, **91**, 11368–72.

Cavalier-Smith, T. and Chao, E.E. (1995) The opalozoan *Apusomonas* is related to the common ancestor of animals, fungi, and choanoflagellates. *Proceedings of the Royal Society of London, Series B*, **261**, 1–6.

Cavalier-Smith, T., Chao, E.E. and Allsopp, M.T.E.P. (1995) Ribosomal RNA evidence for chloroplast loss within Heterokonta: pedinellid relationships and a revised classification of ochristan algae. *Archiv für Protistenkunde*, **145**, 209–20.

Copeland, H.F. (1956) *The Classification of Lower Organisms*, Pacific Books, Palo Alto, California.

Corliss, J.O. (1960) Comments on the systematics and phylogeny of the Protozoa. *Syst. Zool.*, 8(4) (year 1959), 169–90.

Corliss, J.O. (1978) A salute to fifty-four great microscopists of the past: a pictorial footnote to the history of protozoology. Part I. *Transactions of the American Microscopical Society*, **97**, 419–58.

Corliss, J.O. (1979) *The Ciliated Protozoa: Characterization, Classification, and Guide to the Literature*, 2nd edn, Pergamon Press, Oxford and New York.

Corliss, J.O. (1981) What are the taxonomic and evolutionary relationships of the protozoa to the Protista? *BioSystems*, **14**, 445–59.

Corliss, J.O. (1984) The kingdom Protista and its 45 phyla. *BioSystems*, **17**, 87–126.

Corliss, J.O. (1986) Progress in protistology during the first decade following reemergence of the field as a respectable interdisciplinary area in modern biological research. *Progress in Protistology*, **1**, 11–63.

Corliss, J.O. (1987) Protistan phylogeny and eukaryogenesis. *International Review of Cytology*, **100**, 319–70.

Corliss, J.O. (1989) Protistan diversity and origins of multicellular/multitissued organisms. *Bollettino di Zoologia*, **56**, 227–34.

Corliss, J.O. (1990) Toward a nomenclatural protist perspective, in *Handbook of Protoctista* (eds L. Margulis, J.O. Corliss, M. Melkonian and D.J. Chapman), Jones and Bartlett, Boston, pp. xxv–xxx. (reprinted, essentially without change, in Margulis, Mckhann and Olendzenski, 1993, pp. xxvii–xxxii)

Corliss, J.O. (1991) Introduction to the Protozoa, in *Microscopic Anatomy of Invertebrates* (eds F.W. Harrison and J.O. Corliss), Wiley-Liss, New York, **1**, 1–12.

Corliss, J.O. (1993) Should there be a separate code of nomenclature for the protists? *BioSystems*, **28**, 1–14.

Corliss, J.O. (1994) An interim utilitarian ('user-friendly') hierarchical classification and characterization of the protists. *Acta Protozoologica*, **33**, 1–51.

Corliss, J.O. (1995a) The ambiregnal protists and the codes of nomenclature: a brief review of the problem and of proposed solutions. *Bulletin of Zoological Nomenclature*, **52**, 11–17.

Corliss, J.O. (1995b) The need for a new look at the taxonomy of the protists. *Revista de la Sociedad Mexicana de Historia Natural*, 45 (year 1994), 27–35.

Corliss, J.O. (1998) The protists deserve attention: what are the outlets providing it? *Protist*, **149**, 3–6.

Corliss, J.O. and Patterson, D.J. (eds) (1986) *Progress in Protistology*, Vol. 1, Biopress, Bristol.

Corliss, J.O. and Patterson, D.J. (eds) (1987) *Progress in Protistology*, Vol. 2, Biopress, Bristol.

Delvinquier, B.L.J. and Patterson, D.J. (1992) The opalines, in *Parasitic Protozoa*, 2nd edn (eds J.P. Kreier and J.R. Baker), Academic Press, New York and London, **3**, 247–325.

Dobell, C. (1951) In memoriam. Otto Bütschli (1848–1920), 'Architect of Protozoology.' *Isis*, **42**, 20–2.

Edman, J.C. and Sogin, M.L. (1994) Molecular phylogeny of *Pneumocystis carinii*, in Pneumocystis carinii *Pneumonia*, 2nd edn (ed. P.D. Walzer), Marcel Dekker, New York, pp. 91–105.

Fenchel, T. and Patterson, D.J. (1986) *Percolomonas cosmopolitus* (Ruinen) n. gen., a new type of filter feeding flagellate from marine plankton. *Journal of the Marine Biological Association of the United Kingdom*, **66**, 465–82.

Fenchel, T. and Patterson, D.J. (1988) *Cafeteria roenbergensis* nov. gen., nov. sp., a heterotrophic microflagellate from marine plankton. *Marine Microbial Food Webs*, **3**, 9–19.

Finlay, B.J., Maberly, S.C. and Cooper, J.I. (1997) Microbial diversity and ecosystem function. *Oikos*, **80**, 209–13.

Gajadhar, A.A., Marquardt, W.C., Hall, R. *et al.* (1991) Ribosomal RNA sequences of *Sarcocystis muris, Theileria annulata* and *Crypthecodinium cohnii* reveal evolutionary relationships among apicomplexans, dinoflagellates, and ciliates. *Molecular and Biochemical Parasitology*, **45**, 147–54.

Hammerschmitt, B., Schlegel, M., Lynn, D.H. *et al.* (1996) Insights into the evolution of nuclear dualism in the ciliates revealed by phylogenetic analysis of rRNA sequences. *Journal of Eukaryotic Microbiology*, **43**, 225–30.

Hausmann, K. and Hülsmann, N. (1996) *Protozoology*, 2nd edn, Georg Thieme Verlag, Stuttgart and New York.

Honigberg, B.M., Balamuth, W., Bovee, E.C. *et al.* (1964) A revised classification of the phylum Protozoa. *Journal of Protozoology*, **11**, 7–20.

Hülsmann, N. (1992) Undulipodium: end of a useless discussion. *European Journal of Protistology*, **28**, 253–7.

Karlson, B., Potter, D., Kuylenstierna, M. and Andersen, R.A. (1996) Ultrastructure, pigment composition, and 18S rRNA gene sequence for *Nannochloropsis granulata* sp. nov. (Monodopsidaceae, Eustigmatophyceae), a marine ultraplankter isolated from the Skagerrak, northeast Atlantic Ocean. *Phycologia*, **35**, 253–60.

Kent, W. S. (1880) (Page 37 of Kent, 1880–82: see following reference.)

Kent, W.S. (1880–82) *A Manual of the Infusoria*, Vols. I–III, David Bogue, London.

Kivic, P.A. and Walne, P.L. (1984) An evaluation of a possible phylogenetic relationship between the Euglenophyta and Kinetoplastida. *Origins of Life*, **13**, 269–88.

Knoll, A.H. (1992) The early evolution of eukaryotes: a geological perspective. *Science*, **256**, 622–7.

Larsen, J. and Patterson, D.J. (1990) Some flagellates (Protista) from tropical marine sediments. *Journal of Natural History*, **24**, 801–937.

Lee, J.J., Hutner, S.H. and Bovee, E.C. (eds) (1985) *An Illustrated Guide to the Protozoa*, Society of Protozoologists, Lawrence, Kansas.

Lee, J.J., Leedale, G.F., Patterson, D.J. and Bradbury, P.C. (eds) (1998) *An Illustrated Guide to the Protozoa*, 2nd edn, Society of Protozoologists, Lawrence, Kansas (in press).

Leipe, D.L., Wainright, P.O., Gunderson, J.H. *et al.* (1994) The stramenopiles from a molecular perspective: 16S-like rRNA sequences from *Labyrinthuloides minuta* and *Cafeteria roenbergensis*. *Phycologia*, **33**, 369–77.

Leuckart, K. (1879) *Allgemeine Naturgeschichte der Parasiten*, G.F. Winter, Leipzig.

Levine, N.D., Corliss, J.O., Cox, F.E.G. *et al.* (1980) A newly revised classification of the Protozoa. *Journal of Protozoology*, **27**, 37–58.

Lipscomb, D.L. and Corliss, J.O. (1982) *Stephanopogon*, a phylogenetically important 'ciliate,' shown by ultrastructural studies to be a flagellate. *Science*, **215**, 303–4.

Lynn, D.H. and Corliss, J.O. (1991) Ciliophora, in *Microscopic Anatomy of Invertebrates* (eds F.W. Harrison and J.O. Corliss), Wiley-Liss, New York, **1**, 333–467.

Lynn, D.H. and Small, E.B. (1988) An update on the systematics of the phylum Ciliophora Doflein, 1901: the implications of kinetid diversity. *BioSystems*, **21**, 317–22.

Lynn, D.H. and Sogin, M.L. (1988) Assessment of phylogenetic relationships among ciliated protists using partial ribosomal RNA sequences derived from reverse transcripts. *BioSystems*, **21**, 249–54.

Margulis, L. (1970) *Origin of Eukaryotic Cells*, Yale University Press, New Haven.
Margulis, L. (1974a) The classification and evolution of prokaryotes and eukaryotes, in *Handbook of Genetics* (ed. R.C. King), Plenum Press, New York, **1**, 1–41.
Margulis, L. (1974b) Five-kingdom classification and the origin and evolution of cells. *Evolutionary Biology*, **7**, 45–78.
Margulis, L. (1976) Genetic and evolutionary consequences of symbiosis. *Experimental Parasitology*, **39**, 277–349.
Margulis, L. (1981) *Symbiosis in Cell Evolution: Life and its Environment on the Early Earth*, W. H. Freeman, San Francisco.
Margulis, L. (1993) *Symbiosis in Cell Evolution*, 2nd edn, W.H. Freeman, San Francisco.
Margulis, L. (1996) Archaeal-eubacterial mergers in the origin of Eukarya: phylogenetic classification of life. *Proceedings of the National Academy of Sciences USA*, **93**, 1071–6.
Margulis, L., Corliss, J.O., Melkonian, M. and Chapman, D.J. (eds) (1990) *Handbook of Protoctista*, Jones and Bartlett, Boston.
Margulis, L. and Fester, R. (eds) (1991) *Symbiosis as a source of Evolutionary Innovation*, MIT Press, Cambridge, Massachussetts.
Margulis, L., McKhann, H.I. and Olendzenski, L. (eds) (1993) *Illustrated Glossary of Protoctista*, Jones and Bartlett, Boston.
Margulis, L. and Schwartz, K.V. (1982) *Five Kingdoms: an Illustrated Guide to the Phyla of Life on Earth*, 1st edn, W. H. Freeman, San Francisco and New York.
Margulis, L. and Schwartz, K.V. (1988) *Five Kingdoms: an Illustrated Guide to the Phyla of Life on Earth*, 2nd edn, W. H. Freeman, San Francisco and New York.
Margulis, L. and Schwartz, K.V. (1998) *Five Kingdoms: an Illustrated Guide to the Phyla of Life on Earth*, 3rd edn, W. H. Freeman and Company, New York.
Mayr, E. (1982) *The Growth of Biological Thought: Diversity, Evolution, and Inheritance*. Belknap Press, Cambridge, Massachusetts.
Merle, C., Moullade, M., Lima, O. and Perasso, R. (1994) Essai de caractérisation phylogénétique de Foraminifères planctoniques à partir de séquences partielles d'ARNr 28S. *Compt. rend. Acad. Sci., Paris*, **319**, 149–53.
Nikoh, N., Hayase, N., Iwabe, N. *et al.* (1994) Phylogenetic relationship of the kingdoms Animalia, Plantae, and Fungi, inferred from 23 different protein species. *Molecular and Biochemical Evolution*, **11**, 762–8.
Norton, T.A., Melkonian, M. and Andersen, R.A. (1996) Algal biodiversity. *Phycologia*, **35**, 308–26.
Patterson, D.J. (1979) On the organization and classification of the protozoon *Actinophrys sol* Ehrenberg, 1830. *Microbios*, **26**, 165–208.
Patterson, D.J. (1980) Contractile vacuoles and associated structures: their organization and function. *Biological Reviews*, **55**, 1–46.
Patterson, D.J. (1986a) The fine structure of *Opalina ranarum* (family Opalinidae): opalinid phylogeny and classification. *Protistologica*, 21(year 1985), 413–28.
Patterson, D.J. (1986b) Some problems of ambiregnal taxonomy and a possible solution. *Symposia Biologica Hungarica*, **33**, 87–93.
Patterson, D.J. (1989a) Stramenopiles: chromophytes from a protistan perspective, in *The Chromophyte Algae: Problems and Perspectives* (eds J.C. Green, B.S.C. Leadbeater and W.I. Diver), Clarendon Press, Oxford, pp. 357–79.

Patterson, D.J. (1989b) The evolution of protozoa. *Mems. Inst. Oswaldo Cruz,* 83 (Suppl. 1, year 1988), 580–600.

Patterson, D.J. (1993) The current status of the free-living heterotrophic flagellates. *Journal of Eukaryotic Microbiology,* **40**, 606–9.

Patterson, D.J. (1994) Protozoa: evolution and systematics, in *Progress in Protozoology, Proceedings of the IX International Congress of Protozoology,* Berlin, 1993 (eds K. Hausmann and N. Hülsmann), Gustav Fischer Verlag, Stuttgart, pp. 1–14.

Patterson, D.J. and Brugerolle, G. (1988) The ultrastructural identity of *Stephanopogon apogon* and the relatedness of the genus to other kinds of protists. *European Journal of Protistology,* **23**, 279–90.

Patterson, D.J. and Corliss, J.O. (eds) (1989) *Progress in Protistology,* Vol. 3, Biopress, Bristol.

Patterson, D.J. and Fenchel, T. (1985) Insights into the evolution of heliozoa (Protozoa, Sarcodina) as provided by ultrastructural studies on a new species of flagellate from the genus *Pteridomonas. Biological Journal of the Linnean Society,* **34**, 381–403.

Patterson, D.J. and Larsen, J. (eds) (1991) *The Biology of Free-living Heterotrophic Flagellates,* Clarendon Press, Oxford.

Patterson, D.J. and Larsen, J. (1992) A perspective on protistan nomenclature. *Journal of Protozoology,* **39**, 125–31.

Patterson, D.J., Larsen, J. and Corliss, J.O. (1989) The ecology of heterotrophic flagellates and ciliates living in marine sediments. *Progress in Protistology,* **3**, 185–277.

Patterson, D.J. and Sogin, M.L. (1993) Eukaryote origins and protistan diversity, in *The Origin and Evolution of the Cell* (eds H. Hartman and K. Matsuno), World Scientific Publishing, Singapore, pp. 13–46.

Pawlowski, J., Bolivar, I., Guiard-Maffia, J. and Gouy, M. (1994) Phylogenetic position of foraminifera inferred from LSU rRNA gene sequences. *Molecular Biology and Evolution,* **11**, 929–38.

Philippe, H. and Adoutte, A. (1995) How reliable is our current view of eukaryotic phylogeny?, in *Protistological Actualities* (Proceedings of the Second European Congress of Protistology, Clermont-Ferrand, France) (eds G. Brugerolle and J.-P. Mignot), pp. 17–33.

Powers, M.J. (1993) Looking at mycology with a Janus face: a glimpse at Chytridiomycetes active in the environment. *Mycologia,* **85**, 1–20.

Puytorac, P. de, Grain, J. and Mignot, J.-P. (1987) *Précis de Protistologie,* Société Nouvelle des Editions Boubée, Paris.

Ragan, M.A. (1997) A third kingdom of eukaryotic life: history of an idea. *Archiv für Protistenkunde* **148**, 225–43.

Ragan, M.A. (1998) On the delineation and higher-level classification of algae. *European Journal of Phycology* (in press).

Ragan, M.A. and Gutell, R.R. (1995) Are red algae plants? *Botanical Journal of the Linnean Society,* **118**, 81–105.

Sapp, J. (1994) *Evolution by Association: A History of Symbiosis.* OUP Inc., New York.

Saunders, G.W., Potter, D., Paskind, M.P. and Andersen, R.A. (1995) Cladistic analyses of combined traditional and molecular data sets reveal an algal lineage. *Proceedings of the National Academy of Sciences USA,* **92**, 244–8.

Schlegel, M. (1991) Protist evolution and phylogeny as discerned from small subunit ribosomal RNA sequence comparisons. *European Journal of Protistology,* **27**, 207–19.

Schlegel, M., Lom, J., Stechmann, A. *et al.* (1996) Phylogenetic analysis of complete small subunit ribosomal RNA coding region of *Myxidium lieberkuehni*: evidence that Myxozoa are Metazoa and related to the Bilateria. *Archiv für Protistenkunde,* **147**, 1–9.

Seravin, L.N. (1993) The main fine structural types and forms of mitochondrial cristae; the extent of their evolutionary conservation (ability to undergo morphological transformation). *Tsitologia,* **35**, 3–34.

Siddall, M.E., Martin, D.S., Bridge, D. *et al.* (1995) The demise of a phylum of protists: phylogeny of Myxozoa and other parasitic Cnidaria. *Journal of Parasitology*, **81**, 961–7.

Siddall, M.E., Stokes, N.A. and Burreson, E.M. (1995) Molecular phylogenetic evidence that the phylum Haplosporidia has an alveolate ancestry. *Molecular Biology and Evolution*, **12**, 573–81.

Sieburth, J.McN. (1979) *Sea Microbes*. Oxford University Press, New York.

Silberman, J.D., Sogin, M.L., Leipe, D.D. and Clark, C.G. (1996) Human parasite finds taxonomic home. *Nature*, **380**, 398.

Silva, P.C. (1984) The role of extrinsic factors in the past and future of green algal systematics, in *Systematics of the Green Algae* (eds D.E.G. Irvine and D.M. John), Academic Press, London, pp. 419–33.

Silva, P.C. (1993) Continuity, an essential ingredient of modern taxonomy. *Korean Journal of Phycology*, **8**, 83–9.

Sleigh, M.A. (1989) *Protozoa and Other Protists,* 2nd edn, Arnold, London.

Sleigh, M.A. (1995) Progress in understanding the phylogeny of flagellates. *Cytology*, **37**, 985–1009.

Smith, R.McK. and Patterson, D.J. (1986) Analyses of heliozoan interrelationships: an example of the potentials and limitations of ultrastructural approaches to the study of protistan phylogeny. *Proceedings of the Royal Society of London, Series B,* **227**, 325–66.

Smothers, J.F., von Dohlen, C.D., Smith, Jr., L.H. and Spall, R.D. (1994) Molecular evidence that the myxozoan protists are metazoans. *Science*, **265**, 1719–21.

Sogin, M.L. (1991) Early evolution and the origin of eukaryotes. *Current Opinion in Genetics and Development*, **1**, 457–63.

Sogin, M.L. (1994) The origin of eukaryotes and evolution into major kingdoms, in *Early Life on Earth* (ed. S. Bengtson), Columbia University Press, New York.

Sogin, M.L., Morrison, H.G, Hinkle, G. and Silberman, J.D. (1996) Ancestral relationships of the major eukaryotic lineages. *Microbiologia SEM*, **12**, 17–28.

Taylor, F.J.R. (1987) An overview of the status of evolutionary cell symbiosis theories, in *Endocytobiology II, Annals of New York Academy of Sciences* (eds J.J. Lee and J.F. Frederick), **503**, 1–16.

Triemer, R.E. and Farmer, M.A. (1991) An ultrastructural comparison of the mitotic apparatus, feeding apparatus, flagellar apparatus and cytoskeleton in euglenoids and kinetoplastids. *Protoplasma*, **164**, 91–104.

Vickerman, K. (1992) The diversity and ecological significance of Protozoa. *Biodiversity and Conservation*, **1**, 334–41.

Vossbrinck, C.R., Maddox, J.V., Friedman, S. *et al.* (1987) Ribosomal RNA sequence suggests microsporidia are extremely ancient eukaryotes. *Nature*, **326**, 411–4.

Wainright, P.O., Hinkle, G., Sogin, M.L. and Stickel, S.K. (1993) Monophyletic origins of the metazoa: an evolutionary link with fungi. *Science*, **260**, 340–2.

Whittaker, R.H. (1959) On the broad classification of organisms. *Quarterly Review of Biology*, **34**, 210–26.

Whittaker, R.H. (1969) New concepts of kingdoms of organisms. *Science*, **163**, 150–60.

Whittaker, R.H. and Margulis, L. (1978) Protist classification and the kingdoms of organisms. *BioSystems*, **10**, 3–18.

Wolters, J. (1991) The troublesome parasites – molecular and morphological evidence that Apicomplexa belong to the dinoflagellate-ciliate clade. *BioSystems*, **25**, 75–83.

Wray, C.G., Langer, M.R., Desalle, R. *et al.* (1995) Origin of the foraminifera. *Proceedings of the National Academy of Sciences USA*, **92**, 141–5.

Index

Note: page numbers in *italics* refer to tables, those in **bold** refer to figures.

Systematics Association Publications

1. Bibliography of key works for the identification of the British fauna and flora, 3rd edition (1967)†
Edited by G.J. Kerrich, R.D. Meikle and N. Tebble
2. Function and taxonomic importance (1959)†
Edited by A.J. Cain
3. The species concept in palaeontology (1956)†
Edited by P.C. Sylvester-Bradley
4. Taxonomy and geography (1962)†
Edited by D. Nichols
5. Speciation in the sea (1963)†
Edited by J..P. Harding and N. Tebble
6. Phenetic and phylogenetic classification (1964)†
Edited by V.H. Heywood and J. McNeill
7. Aspects of Tethyan biogeography (1967)†
Edited by C.G. Adams and D.V. Ager
8. The soil ecosystem (1969)†
Edited by H. Sheals
9. Organisms and continents through time (1973)†
Edited by N.F. Hughes
10. Cladistics: a practical course in systematics (1992)
P.L. Forey, C.J. Humphries, I.J. Kitching, R.W. Scotland, D.J. Siebert and D.M. Williams

†Published by the Association (out of print)

Systematics Association Special Volumes

1. The new systematics (1940)
Edited by J.S. Huxley (reprinted 1971)
2. Chemotaxonomy and serotaxonomy (1968)*
Edited by J.G. Hawkes
3. Data processing biology and geology (1971)*
Edited by J.L. Cutbill
4. Scanning electron microscopy (1971)*
Edited by V.H. Heywood
Out of print
5. Taxonomy and ecology (1973)*
Edited by V.H. Heywood
6. The changing flora and fauna of Britain (1974)*
Edited by D.L. Hawksworth
Out of print

7. Biological identification with computers (1975)*
Edited by R.J. Pankhurst
8. Lichenology: progress and problems (1976)*
Edited by D.H. Brown, D.L. Hawksworth and R.H. Bailey
9. Key works to the fauna and flora of the British Isles and north-western Europe, 4th edition (1978)*
Edited by G.J. Kerrich, D.L. Hawksworth and R.W. Sims
10. Modern approaches to the taxonomy of red and brown algae (1978)
Edited by D.E.G. Irvine and J.H. Price
11. Biology and systematics of colonial organisms (1979)*
Edited by G. Larwood and B.R. Rosen
12. The origin of major invertebrate groups (1979)*
Edited by M.R. House
13. Advances in bryozoology (1979)*
Edited by G.P. Larwood and M.B. Abbott
14. Bryophyte systematics (1979)*
Edited by G.C.S. Clarke and J.G. Duckett
15. The terrestrial environment and the origin of land vertebrates (1980)
Edited by A.L. Pachen
16. Chemosystematics: principles and practice (1980)*
Edited by F.A. Bisby, J.G. Vaughan and C.A. Wright
17. The shore environment: methods and ecosystems (2 volumes) (1980)*
Edited by J.H. Price, D.E.G. Irvine and W.F. Farnham
18. The Ammonoidea (1981)*
Edited by M.R. House and J.R. Senior
19. Biosystematics of social insects (1981)*
Edited by P.E. House and J.-L. Clement
20. Genome evolution (1982)*
Edited by G.A. Dover and R.B. Flavell
21. Problems of phylogenetic reconstruction (1982)*
Edited by K.A. Joysey and A.E. Friday
22. Concepts in nematode systematics (1983)*
Edited by A.R. Stone, H.M. Platt and L.F. Khalil
23. Evolution, time and space: the emergence of the biosphere (1983)*
Edited by R.W. Sims, J.H. Price and P.E.S. Whalley
24. Protein polymorphism: adaptive and taxonomic significance (1983)*
Edited by G.S. Oxford and D. Rollinson
25. Current concepts in plant taxonomy (1983)*
Edited by V.H. Heywood and D.M. Moore
26. Databases in systematics (1984)*
Edited by R. Allkin and F.A. Bisby
27. Systematics of the green algae (1984)*
Edited by D.E.G. Irvine and D.M. John

28. The origins and relationships of lower invertebrates (1985)‡
Edited by S. Conway Morris, J.D. George, R. Gibson and H.M. Platt
29. Infraspecific classification of wild and cultivated plants (1986)‡
Edited by B.T. Styles
30. Biomineralization in lower plants and animals (1986)‡
Edited by B.S.C. Leadbeater and R. Riding
31. Systematic and taxonomic approaches in palaeobotany (1986)‡
Edited by R.A. Spicer and B.A. Thomas
32. Coevolution and systematics (1986)‡
Edited by A.R. Stone and D.L. Hawksworth
33. Key works to the fauna and flora of the British Isles and north-western Europe, 5th edition (1988)‡
Edited by R.W. Sims, P. Freeman and D.L. Hawksworth
34. Extinction and survival in the fossil record (1988)‡
Edited by G.P. Larwood
35. The phylogeny and classification of the tetrapods (2 volumes) (1988)‡
Edited by M.J. Benton
36. Prospects in systematics (1988)‡
Edited by D.L. Hawksworth
37. Biosystematics of haematophagous insects (1988)‡
Edited by M.W. Service
38. The chromophyte algae: problems and perspective (1989)‡
Edited J.C. Green, B.S.C. Leadbeater and W.L. Diver
39. Electrophoretic studies on agricultural pests (1989)‡
Edited by Hugh D. Loxdale and J. den Hollander
40. Evolution, systematics, and fossil history of the Hamamelidae (2 volumes) (1989)‡
Edited by Peter R. Crane and Stephen Blackmore
41. Scanning electron microscopy in taxonomy and functional morphology (1990)‡
Edited by D. Claugher
42. Major evolutionary radiations (1990)‡
Edited by P.D. Taylor and G.P. Larwood
43. Tropical lichens: their systematics, conservation and ecology (1991)‡
Edited by G.J. Galloway
44. Pollen and spores: patterns of diversification (1991)‡
Edited by S. Blackmore and S.H. Barnes
45. The biology of free-living heterotrophic flagellates (1991)‡
Edited by D.J. Patterson and J. Larsen
46. Plantanimal interactions in the marine benthos (1992)‡
Edited by D.M. Johns, S.J. Hawkins and J.H. Price
47. The Ammonoidea: environment, ecology and evolutionary change (1993)‡
Edited by M.R. House

48. Designs for a global plant species information system‡
Edited by F.A. Bisby, G.F. Russell and R.J. Pankhurst
49. Plant galls: organisms, interactions, populations‡
Edited by Michèle A.J. Williams
50. Systematics and conservation evaluation‡
Edited by P.L. Forey, C.J. Humphries and R.I. Vane-Wright
51. The Haptophyte algae‡
Edited by J.C. Green and B.S.C. Leadbeater
52. Models in phylogeny reconstruction‡
Edited by R. Scotland, D.J. Siebert and D.M. Williams
53. The Ecology of Agricultural Pests: biochemical approaches**
Edited by W.O.C. Symondson and J.E. Liddell
54. Species: the units of diversity**
Edited by M.F. Claridge, H.A. Dawah and M.R. Wilson
55. Arthropod Relationships**
Edited by R.A. Fortey and R.H. Thomas
56. Evolutionary Relationships Among Protozoa**
Edited by G.H. Coombs, K. Vickerman, M.A. Sleigh and A. Warren

*Published by Academic Press for the Systematics Association
†Published by the Palaeontological Association in conjunction with Systematics Association
‡Published by the Oxford University Press for the Systematics Association
**Published by Chapman & Hall for the Systematics Association